Nanostructured Catalysts for Environmental Applications

Marco Piumetti • Samir Bensaid
Editors

Nanostructured Catalysts for Environmental Applications

Springer

Editors
Marco Piumetti
Department of Applied Science and Technology
Politecnico di Torino
Torino, Italy

Samir Bensaid
Department of Applied Science and Technology
Politecnico di Torino
Torino, Italy

ISBN 978-3-030-58933-2 ISBN 978-3-030-58934-9 (eBook)
https://doi.org/10.1007/978-3-030-58934-9

This Springer imprint is published by the registered company Springer Nature Switzerland AG
The registered company address is: Gewerbestrasse 11, 6330 Cham, Switzerland

Preface

Over or during the last few decades, environmental catalysis has received much attention in scientific community and industry not only in terms of the worldwide catalyst market but also as a driver for developing new technologies that can be effective toward environmental protection, in agreement with the 2030 Agenda for Sustainable Development and the Paris Agreement on climate change. In this scenario, environmental catalysis emerges as a combination of theory and practice in which a synergy of fundamental principles, synthesis procedures, characterization techniques, and modeling stimulates the development of new technologies in biocatalysis, green chemistry, and environmental remediation. Indeed, catalytic solutions concerning the environmental protection and sustainable development can be achieved via combined fields of heterogeneous and homogeneous catalysis, photocatalysis, and biocatalysis. In other words, the field of environmental catalysis is highly interdisciplinary and requires the cooperation of scientists and engineers with expertise in different research domains, including automotive exhaust catalysis, catalysis for clean air, water, and soil, catalysis for sustainable energy production, environmentally benign routes to industrial chemicals, and the use of renewable sources. On the other hand, advances in heterogeneous catalysis have been recently made possible thanks to the synthesis of new multicomponent and nanostructured materials particularly effective because of their small size and their unique electronic features.

The present volume has contributions in different areas of environmental catalysis using nanostructured materials, namely the catalytic and photocatalytic abatement of environmentally hazardous effluents from stationary or mobile sources, the valorization of waste, the production of sustainable energy, and so on. Therefore, this book provides an overview of modern environmental and energy-related applications with a particular emphasis on nano-sized catalytic materials. Recent concepts, experimental data, and advanced theories are reported in this book to give evidence of the multitude of environmental and sustainable applications that can be found in the highly interdisciplinary field of catalysis. Consequently, this volume provides a brief introduction to modern environmental catalysis for researchers, scientists, and engineers operating in the field of catalysis for environment and energy. In fact, this book aims at being as a tool for the study of recent advances in environmental catalysis through nanomaterials.

The editors would like to express their gratitude to all the authors who have contributed to the realization of the book, for the time and dedication they devoted to provide interesting and high-quality chapters. They are also thankful to the reviewers of the chapters and to the Springer staff.

Torino, Italy

Marco Piumetti
Samir Bensaid

Contents

Chapter 1
Nanocrystalline Spinel Catalysts for Volatile Organic Compounds Abatement

Adriana Urdă, Ionel Popescu, and Ioan-Cezar Marcu

1.1 Introduction

Volatile organic compounds (VOC) are chemicals with relatively low molecular weight and high vapor pressure at normal temperatures, which are emitted into atmosphere from many sources [1]. Since they are linked to global environmental problems (e.g., photochemical smog, stratospheric ozone depletion, formation of tropospheric ozone), and also to health problems such as toxicity and carcinogenicity [2], their presence in atmosphere is monitored and regulated and their emission is controlled by different conventional or emerging methods [1, 3, 4], such as adsorption, scrubbing, and oxidation processes (thermal, catalytic, etc.).

Due to their net advantages, the catalytic processes are widely used for the abatement of VOC [1, 5–8]. In order to be able to convert VOC from gases with dilute levels of such compounds, the catalysts need to be both very active and selective [2]. Most industrial scale applications use noble metals (Pt, Pd) with high activity for the catalytic combustion of VOC [1], but because of their price, high sintering rate, and sensitivity to poisoning, many efforts are dedicated to the research of metal oxide catalysts that could be able to replace them [5, 8]. Indeed, a high number of research papers have been published on this subject, leading, in the last decade, to several review papers and book chapters focusing either on catalytic combustion of VOC [2,

A. Urdă · I.-C. Marcu (✉)
Laboratory of Chemical Technology and Catalysis, Department of Organic Chemistry, Biochemistry and Catalysis, Faculty of Chemistry, University of Bucharest, Bucharest, Romania

Research Center for Catalysts and Catalytic Processes, Faculty of Chemistry, University of Bucharest, Bucharest, Romania
e-mail: ioancezar.marcu@chimie.unibuc.ro

I. Popescu
Research Center for Catalysts and Catalytic Processes, Faculty of Chemistry, University of Bucharest, Bucharest, Romania

M. Piumetti, S. Bensaid (eds.), *Nanostructured Catalysts for Environmental Applications*, https://doi.org/10.1007/978-3-030-58934-9_1

9–11] and methane [12], or on different types of oxide catalysts, such as mesoporous silica-supported catalysts [13], ordered porous transition metal oxides [14], cobalt oxides [15], pillared clays [16, 17], and layered double hydroxide-derived mixed oxides [18]. Although the number of papers reporting spinel-based catalysts for total oxidation constantly increased in the last decade, only two review papers contain sections dedicated to them [10, 12].

Oxide spinels are magnesium aluminate ($MgAl_2O_4$)-related mixed oxides with the general formula AB_2O_4, with cations A and B located in tetrahedral and octahedral interstices of the cubic close-packed lattice of oxide anions [19]. The A cations can be all located in tetrahedral sites (normal spinels), in octahedral sites (inverse spinels), or they can be distributed between these two types of sites (random spinels) [19] depending on the cation size, covalent bonding effects, and crystal field stabilization energies. Spinels are used in numerous applications, among others, as magnetic materials [20–23], electronic equipment [24, 25], materials for solid oxide fuel cells [26], gas sensors [27], photocatalysts [28–31], or adsorbents [32, 33] for the removal of contaminants from water, catalysts [6, 23, 34–39], and cores for magnetically separable catalytic materials [23, 40, 41]. Catalytic combustion [5, 42, 43], water gas-shift reaction (WGS) [23], methanol or ethanol reforming [23, 44, 45], oxidative dehydrogenation of alkenes [46, 47], hydrocarbon selective oxidation [23, 34, 48], oxidative coupling of methane [43], esterification [49], CO oxidation [23, 50–53], selective reduction [54], N_2O decomposition [55], and water photoreduction [56, 57] represent reactions catalyzed by spinel compounds. Among this type of solids, ferrospinels (AFe_2O_4), manganites (AMn_2O_4), and cobaltites (ACo_2O_4) attracted most interest in the catalytic combustion of VOC. The recent achievements in the design and preparation of these spinel-based materials in correlation with their main physicochemical characteristics and the consequences on their catalytic performance in complete oxidation of VOC are discussed below through the most relevant examples from the literature published in the last decade.

1.2 Iron-Based Spinel Catalysts

According to their crystalline structure, ferrites can be classified into hexagonal ($MFe_{12}O_{19}$), garnet ($M_3Fe_5O_{12}$), spinel (MFe_2O_4, where M is a divalent, usually transition metal, e.g., Fe, Co, Ni, Cu, Mn, Zn), and orthoferrite ($REFeO_3$, where RE is a rare earth cation) compounds [23, 33].

Most spinel ferrites have a structure between normal and inverse spinel lattice, the structural formula being usually written as $\left(M^{2+}_{1-c}Fe^{3+}_{c}\right)\left[M^{2+}_{c}Fe^{3+}_{2-c}\right]O^{2-}_{4}$, where c represents the degree of inversion (the fraction of tetrahedral sites occupied by Fe^{3+} cations), the round brackets denote tetrahedral (A) sites, while the square brackets show the octahedral (B) sites [58].

Ferrites, both doped and undoped, are studied for the total oxidation of VOC due to their redox ability and stability in oxidative conditions [38, 59, 60]. A mini-review discussing their preparation and applications, mainly for organic reactions, has been recently published by Kharisov et al. [23].

Many organic compounds were tested in catalytic combustion reactions catalyzed by spinel ferrites: methane [42, 43, 61, 62], ethane [63], propane [5, 7, 8, 64], propene [65], butane [63], 1-butene [65], acetone [7, 66], ethanol [60], and aromatics such as benzene [7] and toluene [4, 35, 67].

Magnetite, Fe_3O_4, has an inverse spinel structure, with the 32 oxygen anions forming a close-packed cubic lattice, and the iron cations placed both in 8 of the 64 tetrahedral (A) and 16 of the 32 octahedral (B) sites [23, 25]. It can be represented by the formula: $Fe^{3+}\left[Fe^{2+}_{1-y}Fe^{3+}_{1-y}Fe^{3+}_{1.67y}V_{0.33y}\right]O_4$, where V represents cationic vacancies in the lattice, and $y = 0$ is for pure magnetite, while $y = 1$ is for maghemite (γ-Fe_2O_3, fully oxidized magnetite). At temperatures lower than 860 K, the tetrahedral sites in magnetite are occupied by Fe^{3+} ions, while the octahedral sites are populated with Fe^{2+} and Fe^{3+} ions in equal proportions, ensuring a ratio of 2 between Fe^{3+} and Fe^{2+} cations [23]. Substitution with Mg^{2+}, Zn^{2+}, or Al^{3+} cations improves the stability of the ferrites, while transition metals such as Ni^{2+}, Cu^{2+}, Mn^{2+}, or Co^{2+} strongly modify their redox properties [29]. Some ferrites where Fe^{3+} was partly substituted by M^{3+} cations (e.g., Gd^{3+}) [68] showed enhanced electrical properties.

Cobalt ferrite has a cubic spinel structure, with the Co^{2+} ions preferably placed in the octahedral sites, but not exclusively, the distribution reportedly depending on the thermal treatment of the sample [69]. When manganese or chromium ions substitute some of the Fe^{3+} ions, they enter in the octahedral sites and displace some of the Co^{2+} intro tetrahedral sites [69, 70], and chromium is more effective than manganese for this displacement [69]. $NiFe_2O_4$ ferrite has an inverse spinel cubic structure, with the tetrahedral sites completely occupied by Fe^{3+} ions, while the octahedral sites are occupied by Ni^{2+} and Fe^{3+} cations [20, 36, 71]. When zinc replaces some of the Ni cations, a mixed spinel structure, $Ni_{1-x}Zn_xFe_2O_4$, is obtained [72], with Zn^{2+} ions strongly preferring the tetrahedral sites and Ni^{2+} the octahedral ones. The normal spinel structure of $ZnFe_2O_4$ results when all the nickel ions are replaced by Zn. Manganese ferrite has a cubic structure [30] with a partial inverse cation distribution: ca. 20% of the Mn^{2+} cations are situated in octahedral sites, while the rest occupy tetrahedral positions [63]. $CuFe_2O_4$ has a stable, low temperature phase with tetragonal distortion due to the Jahn-Teller effect of the Cu(II) ions, but above 760 °C the structure becomes cubic [64]. When the solid is cooled down slowly to room temperature, the tetragonally distorted spinel structure is observed, while rapid quenching from high temperature maintains the cubic lattice [64]. The tetragonal Cu ferrite is an inverse spinel, with the Cu^{2+} ions sitting mainly in the octahedral sites and the Fe^{3+} found in almost equal proportions in both octahedral and tetrahedral sites [64].

The catalytic properties of ferrites together with the nonstoichiometry and number of vacancies depend significantly on the cation distribution in the tetrahedral or octahedral sites in the spinel structure [7, 38, 73] and on the preparation method [35, 44, 64]. Since the octahedral sites are almost exclusively exposed at the surface, cation species that occupy them are determinant for the catalytic activity [7, 8, 54]. It is believed that the cations occupying tetrahedral sites are inactive or less active from catalytic point of view due to stronger metal-oxygen bonds and lack of accessibility to reactants [54]. The redox and electronic properties that are of interest in catalysis can be tuned by varying the chemical composition of ferrites, by substituting part of the M^{2+} or Fe^{3+} cations with different cations to obtain mixed ferrites [38, 39, 43, 74].

1.2.1 Preparation Methods

Several methods are mentioned in the literature for the preparation of ferrite nanoparticles, as follows:

- Coprecipitation [5, 6, 28, 35, 36, 44, 45, 47, 60, 64, 74–76], by far the most common route, usually starting from the metal nitrates, but also from chloride [46, 60, 77], acetate [75], sulphate [50, 51], or industrial Cu-rich sludge [78]. NaOH [36], Na_2CO_3 [44], NH_4OH [21, 60, 77], or $H_2C_2O_4$ [60] were used as precipitating agents. Different sizes were reported for the nanoparticles (as small as 2 nm [23], but usually above 20 nm [21, 44, 75]) and with medium to high surface areas. However, impurity phases (e.g., α-Fe_2O_3 [21, 28, 46, 60]) were sometimes reported besides the spinel ferrite phase, possibly due to loss of M^{2+} cation during washing [75]. This method can be used in combination with other techniques, such as sonochemical, high-energy ball milling [23, 44], or combustion [27]. In order to avoid the formation of secondary phases, it was proposed [40, 58, 79] to prepare first a layered double hydroxide precursor with a $M^{2+}/(Fe^{2+} + Fe^{3+})$ molar ratio of 1/2, which, upon calcination, would lead to a pure ferrite phase;
- Citrate precursor [20, 21, 29, 64, 72, 80], with citric acid added usually in excess [49, 70, 81] for the complete coordination of metallic ions, and sometimes followed by self-combustion [8, 49, 70, 72] or freeze drying [81]; from this method, pure ferrite phases were reported [8, 21, 49, 70, 72, 81];
- Solvothermal, mainly using water (hydrothermal) [23, 29, 31, 36, 47, 48, 67, 76, 82], but also organic solvents (ethylene glycol [30], triethylene glycol [22]). The synthesis is conducted in autoclave reactors at temperatures above 100 °C (usually above 200 °C), in some cases in the presence of surfactants (the same as the solvent, but sometimes different, e.g., cetyl trimethyl ammonium bromide, CTAB [31]), and yields high surface area solids [31, 48]. Some authors [82] reported the formation of a Fe_2O_3 side phase besides the ferrite product;
- Sol-gel [7, 29, 36, 37, 59, 66], sometimes followed by self-combustion [34, 39, 83, 84] or supercritical drying [58]. Common fuels in self-combustion are dextrose ($C_6H_8O_7$), glucose ($C_6H_{12}O_6$), sucrose ($C_{12}H_{22}O_{11}$), urea (CH_4N_2O), and glycine ($C_2H_5NO_2$) [84]. Alternatively, self-combustion can be ignited by an electrically heated wire [83]. These fuels form complexes with the metal cations in solution, enabling homogeneous distribution, and generate heat in exothermic process. The solids need to be calcined in a final step to eliminate all traces of organic compounds;

- Solid-state reaction (ceramic technique) [25, 29, 52, 64, 67, 69, 71, 85], starting from the metal oxides that are calcined at high temperatures (usually above 1000 °C). Because of the high temperatures and long thermal treatment times required for the preparation, in some cases phase separation and nonstoichiometry occur [58]. If temperature is not high enough, or time is not sufficiently long, incomplete reactions in the solid state lead to impurity phases (e.g., α-Fe_2O_3), [85];
- Thermal decomposition of polynuclear coordination compounds [6, 32, 43, 61, 86–88], using complexing agents such as tartaric acid [6, 43]. Pure ferrite phases were obtained in many cases [32, 61, 87, 88], but some reports mention the presence of side oxide phases [6];
- Mechanochemical [5], starting from the metal salts (nitrates [5] or chlorides [89]) in stoichiometric ratio. The salts are grounded in a mechanical mortar until they dissolve in their crystallization water, then crystalline ammonium carbonate [5] or sodium borohydride [89] is added. In the case of $(NH_4)_2CO_3$, a paste is formed, then calcined to form the ferrite phase [5], while the use of $NaBH_4$ leads directly to ferrite [89];
- Combustion [23, 24, 27, 43, 55, 63, 90] is used because the heat generated in the exothermic reaction accelerates the process, and it produces nanosized oxide particles [42, 91]. Typically, a concentrated solution of the precursors, such as metal nitrates and urea [42], glycine [63, 91], or glycerol [63], is heated at high temperatures for a short time, so the excess water is evaporated and then ignition occurs to produce the very fast synthesis reactions. Pure ferrite catalysts were obtained using this procedure [42, 63, 91]. The nanoscale crystalline-sized solid obtained by this method was explained [91], taking into consideration two factors: (1) an atomic or molecular dispersion of the reactants before combustion, so when ignition occurs only a short-distance diffusion of atoms is necessary for the nucleation to take place; and (2) the rate of the combustion reaction is so high that there is not enough time for the long-distance diffusion of atoms that would lead to a growth of the crystallites. The characteristics of the obtained solids can be tuned by modifying the intensity of the combustion reaction through the variation of the oxidizer-to-fuel ratio or by mixing different fuels, which determine the combustion reaction temperature [63];
- Water-in-oil microemulsion [21, 92, 93], starting from metal salts (e.g., nitrates) and using an organic solvent (e.g., *n*-octane [92], xylene [93] or heptane [21]), surfactants such as cetyl trimethyl ammonium bromide (CTAB) [92], sodium dodecylbenzenesulfonate (NaDBS) [93], or poly(oxyethylene)nonylphenyl ether [21] for the preparation of the microemulsion, and NH_4OH [21, 92] or hydrazine [93] as precipitating agent. The metal hydroxides precipitate is formed, being confined in the microemulsion droplets. These droplets act as nanosized reactors for the precipitation reaction, since the surfactant layer surrounding them prevents the coagulation of the particles [92]. From this preparation route, pure ferrite phases were reported, with particle sizes below 20 nm [21, 92, 93], the size being controlled by varying the relative concentration of the metal salts, surfactant, and solvent [93].

Ferrites prepared by different methods lead to different physical (e.g., specific surface area) [20, 36, 76] and catalytic properties [36, 76]. For hydrothermal [36, 67, 76], sol-gel [7, 36], and citrate precursor self-combustion [20] methods, single-phase pure ferrite was reported, while solid-state reaction and coprecipitation produced in some cases [36, 67, 76] mixtures of ferrites with the corresponding oxides. For $NiFe_2O_4$ ferrites obtained by coprecipitation, the surface Fe/Ni ratio observed by XPS analysis was very high as compared to theoretical values, confirming the presence of impurity oxides besides the ferrite phase [36, 76]. Also, as the particle size decreases, a change in the ferrite structure is observed, from inverse spinel to mixed spinel [20].

Nanosized spinel ferrites proved to be a viable alternative for combustion catalysts, even for stable and nonreactive molecules [5, 8, 59].

One drawback of the ferrite catalysts is their rather small specific surface area [7, 8, 49, 66, 77, 84]. Some studies used ferrites supported on high surface area supports (e.g., silica [59]) in order to improve dispersion and, therefore, the catalytic activity. Other studies used core-shell multicomponent structures that prevent sintering and aggregates forming at high temperatures [77]: e.g., a $NiFe_2O_4$ core, with a SiO_2 shell porous enough to allow reactants and products transport to and from the catalytic core surface.

1.2.2 Ni–Fe Spinels

Benrabaa et al. [36] observed that in $NiFe_2O_4$ prepared by sol-gel method some of the Ni^{2+} ions were placed in tetrahedral sites, where they are less stable than in octahedral sites. This influences the H_2-TPR behavior of the ferrite, with Ni^{2+} being reduced at lower temperatures, and in turn, influences the catalytic activity of the ferrite [36].

H_2-TPR experiments of $NiFe_2O_4$ prepared by coprecipitation route [45] showed two peaks, one at low temperature (350 °C) attributed to reduction to Ni and Fe_3O_4, the second one at higher temperatures (ca. 520 °C), belonging to the reduction of Fe_3O_4. Magnetite prepared by the same route as the Ni ferrite showed three main reduction events: the low temperature one (400 °C) corresponded to FeO(OH) impurity species being reduced to Fe_3O_4, the second (550 °C) and the third one (650 °C) significantly overlapping were attributed to the reduction of magnetite to FeO and Fe. By comparing the reduction profiles of the two ferritic materials, it has been concluded that the presence of Ni^{2+} ions increases the reducibility of magnetite. This feature has a positive effect on the catalytic activity of ferrites when lattice oxygen is involved in oxidation reactions.

Nickel, cobalt, and magnesium ferrites were prepared from layered double hydroxides as precursors, and then coated with SiO_2 and Al_2O_3 to be used in applications as catalysts or catalyst supports [79]. The layered double hydroxides precursors were prepared by coprecipitation from nitrates, using NaOH and Na_2CO_3 as precipitating agents. By calcination at 900 °C, pure spinel ferrites were obtained, which were covered with a SiO_2 coating (1–3 μm) by using a Na_2SiO_3 solution. The final porous Al_2O_3 coating was deposited from an alumina sol using hexamethylenetetramine solution as coagulating agent, the final solid having a ferrite/SiO_2/Al_2O_3 mass ratio of 7.6/7.3/85.1. Since the silica and alumina shells are porous (average pore size ca. 20 nm) and have high specific surface areas (ca. 190 m^2 g^{-1}), the final solid can be used as catalyst or catalyst support and can also be magnetically recovered from the reaction due to the ferrite core.

A $NiFe_2O_4$@SiO_2 core-shell structure was prepared [77] in a one-pot synthesis, starting from $FeCl_2$, $Ni(NO_3)_2$ and NH_4OH, in the presence of a surfactant (polyethylene glycol hexadecyl ether dissolved in cyclohexane). After the ferrite was precipitated, a tetraethyl orthosilicate solution in methanol was added to provide the shell. It has been observed that the concentration of surfactant determined the shape of the final nanoparticles: nanorods were obtained instead of nanospheres when the concentration was doubled. Nickel ferrite nanoparticles with inverse spinel structure were prepared at 500 °C, with a secondary impurity phase of amorphous Ni oxide, but at 900 °C a Fe_2O_3 phase was also observed. The H_2-TPR profile for the uncoated $NiFe_2O_4$ shows that the reduction at low temperatures leads to the formation of metallic Ni + Fe_2O_3. The SiO_2 coating stabilized the ferrite and delayed the reduction until higher temperatures. It was proposed that the Fe_2O_3 phase formed by aggressive reduction of the core-shell structure may segregate to the surface and delay reduction of the underlying species. The porous shell allows transport to and from the core surface, but also prevents or retards sintering of the core particles, which can be used as catalysts over a large range of temperatures.

1.2.3 Co–Fe Spinels

The structural and chemical evolution of a $CoFe_2O_4$ spinel, prepared by solid-state reaction, was investigated by in situ neutron diffraction in isothermal conditions (900 °C) when the partial pressure of oxygen was lowered from $10^{-0.9}$ to 10^{-19} atm [85]. The composition of the solid was determined to be CoO $n$$Fe_2O_3$ (n = 1 and 2), in two phases: a $Co_{0.27}Fe_{0.73}(Co_{0.35}Fe_{0.65})O_4$ spinel, and a second phase containing a mixture of the spinel with unreacted α-Fe_2O_3. The spinel (content of the octahedral sites shown in brackets) had a composition with a slightly higher inversion degree, but close to the stoichiometric ferrite (Co:Fe = 1:2), and the second phase contained 64 wt. % spinel (with identical composition as the first phase) and 36 wt. % hematite. As the partial pressure of oxygen decreased (corresponding to a reduction reaction), the evolution observed for the second phase was: α-Fe_2O_3 → (Fe, Co)$_3O_4$ spinel → (Fe, Co)$_{1-x}$O wustite → (Fe, Co)O rock salt → γ-(Fe, Co) alloy → α-(Fe, Co) alloy. In the first step of the reduction, hematite is converted to magnetite with a spinel structure:

$$6Fe_2O_3\,(\text{hematite}) \rightarrow 4Fe_3O_4\,(\text{magnetite}) + O_2 \tag{1.1}$$

From diffraction patterns during reduction, it was proposed that the transformation involved a homogeneous internal process, in which the Fe spinel spontaneously precipitated at the interface between hematite and the Co spinel from adjacent grains.

$CoFe_2O_4$ was prepared by Tong et al. [34] by a self-combustion method starting from nitrates and citric acid, with different molar ratios between Co^{2+} and Fe^{3+}, i.e., 1:4.0; 1:2.8; 1:2.0; 1:1.5; and 1:1.1. The purpose of the study was to prepare nonstoichiometric spinel ferrites. Pure spinel cobalt ferrite was obtained only for the stoichiometric ratio (1:2.0). The other solids did not contain nonstoichiometric ferrites, but just either Fe_2O_3 (first two samples) or Co_2O_3 (the last two samples) additional phases. The mean particle sizes, determined from the XRD pattern, decreased from 34 to 21 nm as the Co:Fe molar ratio decreased.

The same self-combustion route was used by Bhagwat et al. [84] to prepare $CoFe_2O_4$, starting from the metal nitrates and using different fuels for the combustion step: ethylene glycol, glycine, and urea. All obtained solids were pure ferrites, without impurity phases and had similar sponge-like morphologies. The crystallite size (calculated from the XRD patterns) and specific surface areas were influenced by the organic compound used as fuel: the smallest particles (~15 nm) were obtained with urea, while glycine yielded the largest crystallites (~22 nm). The differences were attributed to the different ignition temperature and heat of combustion generated by the fuels.

A different study investigated the Ni and Co ferrites prepared by the citrate method [81] using four different procedures: (1) the metal citrate precursor solutions (0.3 M metal salt solution, metal-to-citric acid ratio = 1:1, 10% excess citrate, pH = 7) were freezed at −80 °C, then sublimated at −48 °C and 150 mTorr for 48 h; (2) the same procedure, but with metal-to-citric acid ratio = 1:2; (3) the same procedure as in (1), but with pH = 10; (4) the same procedure as in (1), but with diluted metal salt solutions (0.15 M). The FT-IR study of the precursors showed that, as the ratio between metal and citric acid increased, the bands attributed to citric acid are progressively replaced with those belonging to deprotonated citrate groups, while those related to hydrogen bonding in citric acid molecule are substituted by a diffuse band corresponding to a protonated hydroxyl group. These findings led the authors to conclude that a triionized citrate ligand was chelating the transition metal ions. The XRD patterns showed the formation of highly crystalline ferrites, while in the case of the sample obtained with metal-to-citric acid ratio = 2, a small diffraction line corresponding to hematite impurity was observed. The preparation procedure had a marked effect on the morphology of the solids: bulky particles resulting by the aggregation of primary particles were observed when the amount of citric acid was doubled, while at higher pH a corrugated web of microfibers was obtained. As the calcination temperature was increased from 600 to 1000 °C, increased crystallinity and crystallite sizes were observed due to sintering.

The synthesis of $CoFe_2O_4$ was investigated by Yan [91] using three routes: combustion, sol-gel, and solid-state reaction, the resulted solids having similar characteristics. In the combustion method, glycine, serving as a fuel for the combustion reaction, was oxidized by the nitrate ions. In order to completely decompose the residual organic compounds, the solids were calcined afterwards. By modifying the glycine-to-nitrate (*G/N*) ratio, the reaction temperature can be controlled and, consequently, the crystallite size of the obtained ferrite. The authors observed that, for *G/N* values in the investigated range (0.2–2.5), the lower the *G/N* ratio, the smaller the sizes for the ferrite crystallites.

$CoFe_2O_4$ and $MgFe_2O_4$ were studied as catalysts for the total oxidation of methane, together with several Cr-based spinels prepared by solution combustion route [42]. The synthesis method was adopted for the pure spinels that it produces, with relatively high specific surface areas compared to other methods. T_{50} (temperature corresponding to 50% CH_4 conversion) was used as an index for the catalytic activity. The two ferrite catalysts showed poor results (T_{50} of 549 °C and 580 °C, respectively) and total conversion at ca. 620 °C, while the Cr-based spinels had higher activity (T_{50} of 369 °C and total conversion below 500 °C, similar to Pd/γ-Al_2O_3), mainly due to their higher specific surface areas. The authors observed a strong correlation between the catalytic activity and the oxygen desorption capacity, namely the amount of α oxygen desorbed at low temperature (below 450 °C), where the most active spinels displayed their combustion activity [42].

Methane combustion was also investigated on Co ferrites obtained by thermal decomposition of ferrioxalate coordination compounds [61]. The oxalate precursor was prepared in situ from 1,2-ethanediol and nitric acid (1 M or 2 M), with Fe and Co nitrates, in aqueous medium. When HNO_3 1 M was used, a $[Fe_2Co(C_2O_4)_3(OH)_2(OH_2)_4]\cdot 2H_2O$ (I) coordination compound was obtained, while $[Fe_2Co(C_2O_4)_4(OH_2)_6]$ (II) compound was synthesized when HNO_3 2 M was used. The corresponding pure ferrites, $CoFe_2O_4$ (I) and $CoFe_2O_4$ (II), were obtained by thermal decomposition of the coordination compounds precursors. Both solids showed mesoporous structures, with nanosized particles (below 12 nm). When tested in the combustion of methane, they displayed good catalytic activities, with $CoFe_2O_4$ (I) showing conversion values only slightly lower than a Pt/Al_2O_3 reference catalyst (Engelhard), but better than $CoFe_2O_4$ (II). However, in stability tests the $CoFe_2O_4$ (I) solid lost about 40% of its activity in the first 3 h on stream. The authors attributed this decline to a collapse of the pore structure and proposed to disperse the high surface area ferrites on thermally stable supports such as barium hexaaluminate [61].

The influence of the preparation method for $NiFe_2O_4$ and $CoFe_2O_4$ was investigated by Urda et al. [5], the prepared ferrites being tested in the combustion of propane. Nickel ferrite was prepared by coprecipitation (method M1) from nitrates using NaOH, while $CoFe_2O_4$ was synthesized, besides the M1 method, by the mechanochemical method (M2) using Co and Fe nitrates and crystalline ammonium carbonate. All solids had cubic spinel structures with, for $CoFe_2O_4$ samples, a small amount of α-Fe_2O_3 side phase. The sample prepared by the M2 method showed higher crystallinity and lower specific surface area. $CoFe_2O_4$-M1 was the most active, leading to complete conversion of propane into CO_2 and H_2O at 400 °C, the order of activity being as follows: $CoFe_2O_4$-M1 > $NiFe_2O_4$-M1 > $CoFe_2O_4$-M2 (Fig. 1.1). $CoFe_2O_4$ prepared by both M1 and M2 methods were stable for 24 h on stream, and their physico-structural characteristics remained unchanged after the catalytic tests [5].

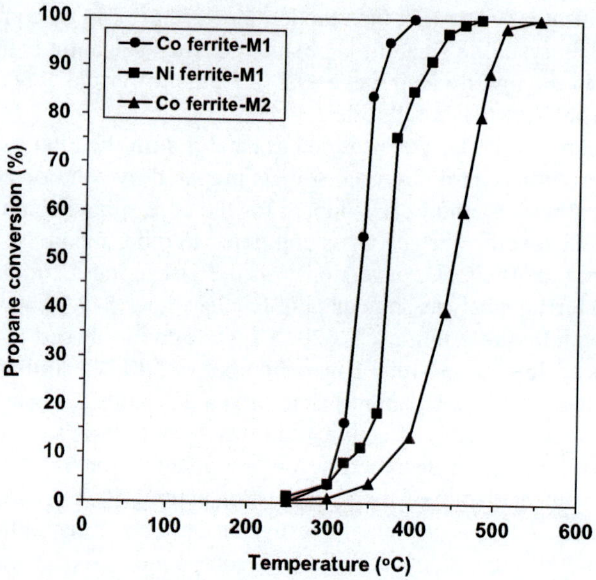

Fig. 1.1 Propane conversion as a function of the reaction temperature for the Co and Ni ferrospinel samples (Reaction conditions: 2% vol. propane in air, total VHSV = 6000 h^{-1}). (Reproduced from ref. [5] with permission)

Pure cobalt ferrite, prepared by pulsed spray evaporation chemical vapor deposition (PSE-CVD) from metal acetylacetonates, was used as catalyst for the low temperature combustion of olefins and dimethyl ether [65]. $CoFe_2O_4$ proved to have high activity for the combustion of propene, 1-butene, and dimethyl ether, without producing CO or hydrocarbon secondary products. This was attributed to the high mobility of oxygen species in the ferrite lattice.

To find correlations between the synthesis procedure and the catalytic activity, $CoFe_2O_4$ was prepared by different coprecipitation procedures and studied in the total oxidation of ethanol [60]. Metal chlorides or nitrates were used as starting materials, and precipitation was achieved with NaOH, NH_4OH, or $H_2C_2O_4$. When NaOH was used as precipitating agent, the formation of spinel structure was observed even before the calcination step, as also observed by Cheng et al. [82], and a pure ferrite phase was obtained after the thermal treatment. A pure ferrite phase was also observed for the samples precipitated with oxalic acid, the spinel phase being formed at temperatures as low as 500 °C. When ammonia was used as precipitating agent, a small amount of Fe_2O_3 side phase was obtained, the sample had smaller specific surface area and, consequently, lower catalytic activity in ethanol oxidation. Solids prepared from chloride salts also showed poor performances, attributed to the poisoning effect of chloride ions. The best results (total conversion at 310 °C) were obtained starting from nitrate salts, using NaOH as precipitating agent and a calcination temperature of 500 °C.

1.2.4 Cu–Fe Spinels

A similar, very detailed study for the influence of the preparation method on the catalytic activity in propane combustion was performed for $CuFe_2O_4$ [64]. The copper ferrite was prepared by three different ways: solid-phase reaction (SP) starting from CuO and Fe_2O_3, coprecipitation from nitrate precursors (NIT), and decomposition of the citric acid precursor (CIT). As expected, the obtained materials had different activities in the catalytic combustion of propane, strongly influenced by the crystal size, cation distribution in the spinel structure, and physicochemical properties (specific surface area, porosity, reducibility) derived from the different preparation procedures. The wet chemical methods (NIT, CIT) produce a spinel structure at lower temperatures than the solid-state reaction, with higher homogeneity and smaller particle sizes. The authors inferred that the presence of Cu ions, both in the octahedral and tetrahedral sites in the structure, keeps the active sites separated, maintaining the stability of the ferrite catalysts. The XRD pattern for the catalyst prepared by SP route showed the formation of ferrite lattice only above 800 °C, and after calcination at 900 °C, displayed the single-phase cubic structure. The samples obtained by the wet chemical methods, which were calcined at 700 °C, showed the tetragonal structure. For the CIT route, a very thorough study of the transformations occurring with increasing calcination temperature was performed using XRD, IR, and TG-DTA techniques. The XRD pattern of a poorly crystallized tetragonal $CuFe_2O_4$ lattice was visible even at 150 °C, and the diffraction lines became more intense with increasing calcination temperature. IR spectra also showed the disappearance of most of the lines attributed to the citrate precursor and the presence of bands belonging to copper ferrite at 150 °C. The TG curve showed a smooth weight loss below 150 °C, attributed to water vaporization, followed by a gradual mass loss between 150 and 300 °C, corresponding to the thermal decomposition of the citrate precursor. Above 300 °C, only a slight mass loss and no thermal effects are observed in the formed ferrite phase. The particle size of the ferrite obtained by CIT route was smaller compared to the ferrites obtained by other methods. The Mössbauer spectra confirmed the presence of Fe^{3+} cation both in the octahedral and tetrahedral sites with almost equal proportions, but in the ferrite from the CIT route the octahedral presence has a slight predominance. EDX scanning of the ferrite obtained from citrate precursor confirmed that Cu and Fe are well-dispersed on the surface of the solid, with Fe/Cu ratio of 2. Consequently, the authors concluded that the ferrites have inverse spinel structures, the degree of inversion depending on the preparation method and calcination temperature. The preparation method had a significant influence on the catalytic activity in propane combustion, the conversion values decreasing in the order: NIT > CIT > SP. It was observed that the presence of Fe^{3+} ions in equal proportions in the typically inverted $CuFe_2O_4$ spinel obtained by the NIT method favored the catalytic combustion of propane compared to the partially inverted spinel obtained by CIT route. Total conversion was obtained for the NIT sample at ca. 650 °C.

Tu et al. [78] investigated the preparation of $CuFe_2O_4$ ferrite from industrial Cu-rich waste sludge and tested it in the catalytic combustion of isopropyl alcohol. The authors studied the influence of reaction temperature (30–500 °C), oxygen content in the reaction mixture (10%, 15%, and 21%), space velocity (GHSV = 8000–30,000 h^{-1}), and alcohol inlet concentration (500–2000 ppm). The best results were obtained at 180 °C, GHSV = 30,000 h^{-1}, oxygen concentration 21%, and 2000 ppm alcohol in the reaction mixture. The catalyst could be recycled five times without any treatment, while stability tests for 100 h showed only a minor decline in the catalytic conversion during the first 24 h on stream, then reaching stable values.

The catalytic combustion of toluene was also investigated on copper ferrites obtained from the respective nitrates by coprecipitation with Na_2CO_3 solution at pH = 9 [35]. The influence of the annealing temperature was investigated between 300 and 800 °C. Below 500 °C, only amorphous materials were obtained, while at higher temperatures the product was a mixture of cubic and tetragonal ferrites, the latter increasing in importance with the annealing temperature. The distortion of the cubic structure with increasing temperature to form the tetragonal phase was attributed to the higher proportion of Cu in octahedral positions. From Mössbauer spectra of the solids annealed below 500 °C, the presence of finely dispersed iron containing particles with sizes below 10–12 nm was noticed. At higher temperatures, the spectra showed the typical octahedral and tetrahedral coordination of Fe^{3+} ions in ferrites and also the presence of Fe^{3+} in a hematite side phase. In the catalytic tests, the higher the annealing temperature, the lower the catalytic activity, an effect that was attributed to the transformation of cubic spinel to the tetragonal one and also to the increase in particle sizes with the annealing temperature. Under reaction conditions, no significant phase change was observed, only further crystallization. The authors observed a good correlation between the activity and the reduction ability determined from H_2-TPR measurements.

1.2.5 Zn–Fe Spinels

The influence of the pH during the synthesis of Zn ferrite by coprecipitation was studied by Lee et al. [46]. The authors used aqueous buffer solutions as coprecipitation medium in order to control the pH and prepared several $ZnFe_2O_4$ samples at pH values between 6 and 12. All preparations produced the desired pure ferrite phase, except for that performed at pH = 6, which also gave α-Fe_2O_3 side phase. The experimental Fe/Zn ratio was in good agreement with the theoretical value, except for the sample prepared at pH = 6 for which it is 3.24, likely due to the formation of Fe_2O_3 phase.

$MeFe_2O_4$ spinels (Me = Zn, Mn, Ni, Co, Cu) were investigated as potential oxygen carriers or reactive support in Chemical Looping Combustion (CLC) of methane [62]. In CLC, direct contact between fuel and air is avoided by using oxygen carriers (such as metal oxides) to transport oxygen from air to the fuel. Combustion takes place in one reactor, where fuel is oxidized by the oxygen carrier, and the reduced carrier is reoxidized with air in a different reactor. The advantage is that the effluent of the combustion reactor consists only of CO_2 and H_2O; therefore, pure CO_2 can be obtained after steam condensation, without being diluted with N_2. The oxygen carrier must have high oxygen transfer capacity, ferrites being considered good candidates for this process. The pure spinel ferrites were obtained by solid-state reaction at different temperatures (1000 °C for Me = Ni, Zn, and Co; 1165 °C for Mn and 750 °C for Cu), based on the phase diagrams of the corresponding ferrites. Since the CLC process requires redox cycles, the solids were tested to determine their oxygen transfer capacity and the stability during repeated cycles. In this respect, $ZnFe_2O_4$ and $NiFe_2O_4$ showed that they can deliver larger amounts of lattice oxygen compared to $MnFe_2O_4$, but they do not regain all of it during reoxidation; e.g., Ni ferrite is very active during the first cycle (80% CH_4 conversion at 1000 °C), but deactivates very fast (38% conversion in the second cycle) due to sintering and loss of active surface area. $ZnFe_2O_4$ loses activity due to an almost complete decomposition to hematite and metallic Zn. Co and Cu ferrites regain most of the transferred oxygen and are more stable, $CuFe_2O_4$ showing the best performance, with high methane conversions (>99%) and low CO selectivity (max. 6%), although some decomposition to $CuFeO_2$ was also observed in this case. Therefore, the authors concluded that these ferrites can be used either as oxygen carriers (Cu ferrite) or reactive supports for other oxygen carriers (the rest of the ferrites) in the CLC of methane [62].

1.2.6 *Mn–Fe Spinels*

C_1-C_4 light alkanes were used as test molecules to investigate the catalytic activity of $MnFe_2O_4$ obtained by solution combustion method [63]. It was suggested that the existence of manganese ions in various oxidation states (Mn^{2+} and Mn^{3+} were detected by XPS) could be useful in redox catalytic reactions. The fuel for the combustion consisted of a mixture of glycerol and glycine in various proportions, i.e., 1:0, 0.75:0.25, 0.5:0.5, and 0.25:0.75, and after ignition, the obtained solid was thermally treated in argon flow at 400 °C. Single-phase ferrites were obtained in all cases, with crystallite sizes increasing with glycine proportion in the fuel (from ca. 8 to 73 nm), due to the fact that combustion proceeds very rapidly and a higher amount of heat is released by glycine. The solids obtained with fuel mixtures containing up to 50% glycine showed similar textural and catalytic properties, while the ferrite obtained with 75% glycine had different properties, such as lower specific surface area, larger pores, different morphology (polyhedron shape, compared to spherical for the other three solids), different surface Mn/Fe atomic ratio, different reduction behavior, and lower catalytic activity. The temperature corresponding to 50% conversion decreased from C_1 to C_4 (450 °C for methane, 300 °C for ethane, 240 °C for propane, and 225 °C for butane), but, for each tested compound, the observed values were identical or very close for the ferrites prepared with up to 50% glycine.

1.2.7 Ternary and Quaternary Ferrites

$Ni_{0.5}Co_{0.5}Fe_2O_4$, $CuFe_2O_4$, and $MgFe_2O_4$ were tested by Rezlescu et al. [66] in the catalytic combustion of acetone. The ferrites, prepared by sol-gel self-combustion route starting from nitrate salts, had spinel structures without impurity phases: Mg- and Ni/Co samples showed cubic spinel structures, while Cu-ferrite had a tetragonal distorted lattice due to Cu^{2+} ions. They showed a rather poor activity in the catalytic combustion of acetone, none of them exhibiting complete conversion up to 550 °C. Cu ferrite, with ca. 95% acetone conversion at 550 °C, showed the best activity. The authors concluded that the higher activity observed for the Cu- and Ni/Co samples is due to the variable oxidation state of the Cu, Ni, and Co ions distributed in the octahedral sites of the spinels.

$Ni_{0.5}Co_{0.5}Sc_xFe_{2-x}O_4$ ($x = 0.05$; 0.1 and 0.2) [7] and $MgFe_2O_4$ [83] ferrites, prepared by a sol-gel self-combustion method starting from metal nitrates, polyvinyl alcohol, and NH_4OH, were tested in the catalytic combustion of acetone, propane, and benzene. In $Ni_{0.5}Co_{0.5}Sc_xFe_{2-x}O_4$, only low amounts of iron can be replaced by scandium, due to the large difference between their ionic radii (0.069 nm for Fe vs. 0.088 nm for Sc). An inverse spinel structure was observed for all samples, in which the Sc^{3+} ions preferred the octahedral coordination. With scandium incorporation, the crystallite size decreased while the specific surface area increased, attributed to the structural disorder produced by the voluminous Sc ions leading to a delay in crystallite growth. The catalytic tests showed a temperature shift towards lower values upon scandium incorporation, the activity increasing substantially with Sc loading. The best results were obtained for the solid with $x = 0.2$, and for propane and acetone, without deactivation for 40 h on stream. These results were attributed to the smaller crystallite size, larger surface area, and the presence of oxygen vacancies in the ferrite structure due to scandium ions present in the octahedral sites [7]. Similar results were obtained for $MgFe_2O_4$ in the combustion of propane, acetone, and benzene [83] and of Pb-free gasoline vapors [90].

A similar study was conducted for $Ni_{0.5}Co_{0.5}Sc_xFe_{2-x}O_4$ ferrites ($x = 0.1$ and 0.2), prepared by the same citrate route followed by self-combustion [8]. Their activity was compared with that of some perovskites with related compositions, in the catalytic combustion of propane. The authors observed the same preference of Sc^{3+} ions for the octahedral site in the ferrite structure, but they observed a decrease in propane conversion as scandium loading increased, contrary to the findings of Rezlescu et al. [7].

Total oxidation of toluene was investigated on several nickel and manganese ferrite catalysts obtained by two different methods: (1) hydrothermal synthesis and (2) ceramic method involving oxides calcination [67]. Cubic spinel $Ni_{0.5}Fe_{2.5}O_4$ and $Mn_{0.65}Fe_{2.35}O_4$ ferrites were obtained by hydrothermal synthesis procedure. $NiFe_2O_4$,

$Ni_{0.5}Zn_{0.5}Fe_2O_4$, and $MnFe_2O_4$ were intended to be prepared by the ceramic method, but incomplete ferritization was observed, leading to a complex mixture of phases. For the first two samples, NiO and ZnO, respectively, were observed in different amounts, besides the ferritic phase, while the last sample consisted of a mixture of α-Fe_2O_3 and Mn_2O_3. The well-structured, pure ferrites obtained hydrothermally showed a very low catalytic activity in the total oxidation of toluene, while the mixture of α-Fe_2O_3 and Mn_2O_3 was the most active catalyst, with conversion values over 80% at 500 °C. The authors presumed an intrinsic, intimate interaction between the individual oxidic phases, since a test performed in similar conditions on a physical mixture of the two oxides, or of the ferrites with the corresponding oxides, showed poor catalytic activity.

Carp et al. [86] explained the differences resulted in the structure and properties of the Ni-substituted Zn ferrites obtained by several preparation methods. There are two problems that have to be overcome when preparing modified zinc ferrites: (1) zinc is volatile at high temperatures, changing the stoichiometry of the final product and also leading to Fe^{2+} formation, and (2) the presence of three different cations, possibly resulting in different intermediate and final phases. When solid-state reaction is used for the preparation, a mixture of simple ferrites is formed instead of $Zn_{1-x}Ni_xFe_2O_4$, while a homogeneous ferrite is obtained only above 750 °C. The hydrothermal route produces either individual ferrites or defective, nonstoichiometric oxides, and the real stoichiometric ferrites are formed at 850 °C. The combustion method leads to a pure Ni-Zn ferrite at 1100 °C, but the decomposition of a polynuclear coordination compound (with the malate anion as ligand) produced the mixed $Zn_{1-x}Ni_xFe_2O_4$ ferrite (prepared with $x = 0.25$, 0.50, and 0.75) at relatively low temperature, i.e., 500 °C, while evolving nontoxic compounds (CO_2 and H_2O). IR spectra suggested that the malic acid ($C_4H_6O_5$) is coordinated to the metal ions both through its two carboxyl groups and its OH group. The final ferrites were obtained with crystallite sizes in the range of 65–85 nm.

$ZnFe_{2-x}Nd_xO_4$ ($x = 0$; 0.25; 0.5; 0.75; and 1) were prepared by Papa et al. [43] using the combustion route from tartrate precursors and then tested in the catalytic combustion of methane. As the Fe^{3+} ions are progressively replaced by Nd^{3+}, the cubic structure of the zinc ferrite coexists with the newly formed orthorhombic phase of $NdFeO_3$. The authors explained the formation of the latter phase by a limited solubility of Nd^{3+} ions with larger ionic radius (1.16 Å compared to 0.87 Å for Fe^{3+}) in the spinel lattice. The $ZnNd_2O_4$ oxide showed a distorted tetragonal structure, with $(ZnO_2)^{2-}$ and $(Nd_2O_2)^{2+}$ alternating units, in which Nd^{3+} are surrounded by $8O^{2-}$ ions, while Zn^{2+} only by $4O^{2-}$ ions. The catalytic activity in methane combustion showed a maximum (in the kinetic regime) at $x = 0.5$, where the authors postulated the highest concentration of lattice defects. Since Fe^{3+} and Nd^{3+} have the same oxidation state, these defects are not electrically charged, but due to significant differences in their ionic radii, lattice strains appear and a new crystalline phase is formed. In this way, the reduction behavior changes and the oxygen mobility within the lattice is affected. The catalytic combustion activity is therefore related to the degree of Fe^{3+} substitution with Nd^{3+} in the octahedral positions of the spinel.

$Cu_xCe_{1-x}Fe_2O_4$ ferrites ($x = 0$; 0.2; 0.5; 08; 1) prepared by a sol-gel method followed by self-combustion were also studied as catalysts in toluene combustion [39]. The XRD patterns showed the inverse spinel structure gradually disappearing as Ce loading increased, while CeO_2 diffraction lines were observed starting from $x = 0.8$. A decrease in the pore sizes was observed with increasing Ce content, while the BET surface areas remained constant and low (10 $m^2\ g^{-1}$). The catalytic activity decreased in the order: $Cu_{0.8}Ce_{0.2}Fe_2O_4 > Cu_{0.5}Ce_{0.5}Fe_2O_4 > CuFe_2O_4 > Cu_{0.2}Ce_{0.8}Fe_2O_4 > CeFe_2O_4$. The authors found a correlation between the order of activity and the amount of surface electrophilic oxygen and oxygen vacancies, with a maximum concentration at $x = 0.8$. At the same x value, the XPS measurements showed a maximum for the Ce^{3+}/Ce^{4+} ratio that decreased at higher Ce content. This suggested that the presence of Ce^{3+} facilitates the formation of oxygen vacancies and promotes the catalytic activity in toluene combustion.

The different ferrospinel catalysts hereby described are summarized in Table 1.1 together with their preparation methods and catalytic applications in VOC combustion.

1.2.8 The Redox Properties of Ferrites

The reducibility of ferrites can be modified by incorporation of different M^{2+} cations in the spinel structure and, in turn, it significantly influences the catalytic activity due to altered redox properties [73]. Khan and Smirniotis studied the H_2-TPR profiles of some MFe_2O_4 spinels (M = Cr, Mn, Co, Ni, Cu, Zn, and Ce) obtained by coprecipitation and compared them with the individual oxides. The authors observed that the positions of the temperature maxima were strongly influenced by the nature of the cation: Cu^{2+} showed the strongest shift of the $Fe_2O_3 \rightarrow Fe_3O_4$ reduction peak with ca. 160 °C to lower temperatures, followed by Mn^{2+} (50 °C) > Ce (ca. 40 °C) > Ni ≈ Co (30 °C) > Cr ≈ Zn (no shift). However, the maximum for the reduction of Fe_3O_4 to FeO was less influenced, while the FeO → Fe reduction peak was not shifted. These features can be correlated with the catalytic activity when redox reactions are involved [73].

Electrical conductivity measurements of spinel ferrites allowed to find correlations between their semiconductive and redox properties and their catalytic activity in oxidation reactions. Mathe and Kamble [75] studied $Ni_{0.4}Co_{0.6}Fe_2O_4$ ferrite and observed five domains with different slopes in the plot of $\ln(\rho)$ versus $1/T$, in the temperature range 300–900 K (where ρ represents the electrical resistivity). Besides the usual domains attributed to conduction by impurities, ferromagnetic and paramagnetic conduction, the authors identified a temperature interval (500–833 K) where some of the Fe^{3+} ions in the spinel lattice slowly migrate from tetrahedral to octahedral sites. This alteration of the cation distribution between the two types of sites will not lead to a modified crystal structure, but will affect the electrical conductivity of the solid.

Table 1.1 Preparation and catalytic performance of the ferrospinel catalysts in VOC total oxidation

Catalyst	Preparation method	Tested chemical compound (reaction conditions)	Best results	Ref.
Co ferrite and Mg ferrite	SolC[a] synthesis	Methane (200–750 °C, 2.5 or 0.4% CH_4/7.5% O_2/He balance; W/F = 0.12 g s cm^{-3})	$T_{100} \approx 620$ °C on Co ferrite	[42]
Co ferrite	TD[b] of ferrioxalate coordination compounds	Methane (300–650 °C, 100 cm^3 min^{-1} (60,000 mL g^{-1} h^{-1}), 1 vol. % CH_4, 10% O_2, Ar balance)	$T_{100} = 600$ °C	[61]
$ZnFe_{2-x}Nd_xO_4$ ($x = 0$; 0.25; 0.5; 0.75; 1)	TD[b] of polynuclear coordination compounds	Methane (400–700 °C, GHSV = 60,000 h^{-1}, 2% CH_4, 10% O_2, Ar balance)	$T_{100} \approx 675$ °C on $ZnFe_{2-x}Nd_xO_4$ with $x = 0.25$ and 0.5	[43]
$CuFe_{2-x}Mn_xO_4$ ($x = 0$; 0.4; 0.8; 1.6; 2)	SSR[c]	Methane (500–770 °C, GHSV = 15,000 h^{-1}, 1% CH_4 in air)	Conversion of ca. 65% at 770 °C on Cu ferrite ($x = 0$)	[38]
Mn ferrite	SolC[a]	Methane (GHSV = 60,000 h^{-1}, 0.01% CH_4, 20% O_2, N_2 balance)	$T_{50} = 450$ °C	[63]
Me ferrite (Me = Mn, Ni, Zn, Co, Cu)	SSR[c]	Methane (chemical looping combustion at 1000 °C, in a pulse reactor)	Conversion > 99% for Cu ferrite	[60]
Mn ferrite	SolC[a]	Ethane (GHSV = 60,000 h^{-1}, 0.01% C_2H_6, 20% O_2, N_2 balance)	$T_{50} = 300$ °C	[63]
Ni ferrite, Co ferrite	CoPP[d] and mechanochemical methods	Propane (250–600 °C, VHSV = 6000–11,000 h^{-1}, 2% propane in air)	$T_{100} = 400$ °C for CoPP Co ferrite at 6000 h^{-1}	[5]
$Ni_{0.5}Co_{0.5}Sc_x Fe_{2-x}O_4$ ($x = 0.05$; 0.1 and 0.2)	SG[e] + SelfC[f]	Propane (50–550 °C, GHSV = 5100 h^{-1}, 1–2‰ C_3H_8 in air)	Conversion of ca. 92% at 500 °C	[7]
$Ni_{0.5}Co_{0.5}Sc_x Fe_{2-x}O_4$ ($x = 0.05$; 0.1 and 0.2)	SG[e] + SelfC[f]	Propane (50–500 °C, GHSV = 5100 h^{-1}, 1–2‰ C_3H_8 in air)	$T_{90} \approx 370$ °C on $Ni_{0.5}Co_{0.5}Fe_2O_4$	[8]

(continued)

Table 1.1 (continued)

Catalyst	Preparation method	Tested chemical compound (reaction conditions)	Best results	Ref.
Cu ferrite	SSR[c], CoPP[d], and CitP[g]	Propane (200–750 °C)	T_{100} = 650 °C for the ferrites prepared by CoPP and CitP	[64]
Cu ferrite	CoPP[d]; TD[b] of polynuclear coordination compounds	Propane (25–450 °C, C_3H_8:O_2 = 1:0.9–1:6 (mol/mol), contact time = 1 and 2 s)	T_{100} = 390 °C for the sample obtained by TD	[6]
Mg ferrite	SG[e] + SelfC[f]	Propane (20–550 °C, GHSV = 5100 h^{-1})	T_{50} = 312 °C	[83]
Mn ferrite	SolC[a]	Propane (GHSV = 60,000 h^{-1}, 0.01% C_3H_8, 20% O_2, N_2 balance)	T_{50} = 240 °C	[63]
Mn ferrite	SolC[a]	Butane (GHSV = 60,000 h^{-1}, 0.01% C_4H_{10}, 20% O_2, N_2 balance)	T_{50} = 225 °C	[63]
Co ferrite	PSE-CVD[h]	Propene (220–460 °C, WHSV = 45,000 mL g_{cat}^{-1} h^{-1}, 1% propene, 10% O_2, Ar balance)	T_{50} = 348 °C T_{90} = 382 °C	[65]
Co ferrite	PSE-CVD[h]	1-butene (220–460 °C, WHSV = 45,000 mL g_{cat}^{-1} h^{-1}, 1% butene, 10% O_2, Ar balance)	T_{50} = 358 °C T_{90} = 402 °C	[65]
$Ni_{0.5}Fe_{2.5}O_4$ $Mn_{0.65}Fe_{2.35}O_4$ Ni ferrite $Ni_{0.5}Zn_{0.5}Fe_2O_4$	HT[i], SSR[c]	Toluene (150–500 °C, GHSV = 60,000 mL g_{cat}^{-1} h^{-1}, 1700 ppm toluene)	Conversion of ca. 80% at 500 °C on $MnFe_2O_4$	[67]
Cu ferrite	CoPP[d] and calcination at different temperatures	Toluene ($p_{toluene}$ = 0.9 kPa, WHSV = 1.2 h^{-1})	T_{100} = 325 °C on the sample annealed at 300 °C	[35]
$Cu_xCe_{1-x}Fe_2O_4$ (x = 0; 0.2; 0.5; 0.8; 1)	SG[e] + SelfC[f]	Toluene (100–300 °C, 300 mg m^{-3} toluene in air, $p_{toluene}$ = 0.8 MPa, GHSV = 6000 h^{-1})	T_{90} = 215 °C for $Cu_{0.8}Ce_{0.2}Fe_2O_4$	[39]

(continued)

Table 1.1 (continued)

Catalyst	Preparation method	Tested chemical compound (reaction conditions)	Best results	Ref.
Co ferrite	CoPP[d]	Ethanol (200–400 °C, GHSV = 30 m^3 h^{-1} kg^{-1})	T_{100} = 310 °C on $CoFe_2O_4$ prepared from nitrates, precipitated with NaOH and calcined at 500 °C	[60]
Cu ferrite	CoPP[d]	Isopropyl alcohol (30–500 °C, GHSV = 8000–30,000 h^{-1}, 500–2000 ppm alcohol, 10–21% O_2)	T_{100} = 180 °C, GHSV = 30,000 h^{-1}, 21% O_2, 2000 ppm alcohol	[78]
Cu ferrite, Mg ferrite, $Ni_{0.5}Co_{0.5}Fe_2O_4$	SG[e] + SelfC[f]	Acetone (50–550 °C, GHSV = 5100 h^{-1}, 1–2‰ in air)	Conversion of ca. 95% at 550 °C on Cu ferrite	[66]
$Ni_{0.5}Co_{0.5}Sc_x Fe_{2-x}O_4$ ($x = 0.05$; 0.1 and 0.2)	SG[e] + SelfC[f]	Acetone (50–550 °C, GHSV = 5100 h^{-1}, 1–2‰ acetone in air)	Conversion of ca. 92% at 550 °C	[7]
Mg ferrite	SG[e] + SelfC[f]	Acetone (20–550 °C, GHSV = 5100 h^{-1})	T_{50} = 250 °C	[83]
Co ferrite	PSE-CVD[h]	Dimethyl ether; olefins (220–460 °C, WHSV = 45,000 mL g_{cat}^{-1} h^{-1}, 1% propene, 10% O_2, Ar balance)	T_{50} = 356 °C T_{90} = 409 °C	[65]
$Ni_{0.5}Co_{0.5}Sc_x Fe_{2-x}O_4$ ($x = 0.05$; 0.1 and 0.2)	SG[e] + SelfC[f]	Benzene (50–550 °C GHSV = 5100 h^{-1}, 1–2‰ benzene in air)	Conversion of ca. 67% at 550 °C	[7]
Mg ferrite	SG[e] + SelfC[f]	Benzene (20–550 °C, GHSV = 5100 h^{-1})	T_{50} = 425 °C	[83]
$Ni_{0.5}Co_{0.5}Sc_xFe_{2-x}O_4$ ($x = 0$, 0.05, 0.10, and 0.20)	SG[e] + SelfC[f]	Pb-free gasoline vapors (50–600 °C, 1–2‰ gasoline in air, gas flow = 3 cm^3 s^{-1})	T_{50} = 360 °C for $Ni_{0.5}Co_{0.5}Sc_{0.2}Fe_{1.8}O_4$	[90]

[a]*SolC* solution combustion
[b]*TD* thermal decomposition
[c]*SSR* solid-state reaction
[d]*CoPP* coprecipitation
[e]*SG* sol-gel synthesis
[f]*SelfC* self-combustion
[g]*CitP* citrate precursor
[h]*PSE-CVD* pulsed spray evaporation chemical vapor deposition
[i]*HT* hydrothermal synthesis

The study of electrical conductivity could provide evidence in support of their catalytic properties in VOC combustion reactions [38]. When a *p*-type semiconductor is exposed to oxygen, its electrical conductivity increases compared to that in an inert atmosphere, due to the creation of holes (charge carriers) simultaneously with new oxygen anions in the lattice, according to the equation [38]:

$$O_{2(gas)} \Leftrightarrow O^{-}_{2(ads)} + h^{+} \Leftrightarrow 2O^{-}_{(ads)} + h^{+} \Leftrightarrow 2O^{2-}_{(ads)} + 2h^{+} \Leftrightarrow 2O^{2-}_{lattice} \quad (1.2)$$

where h^+ denotes a positive hole. However, if the same *p*-type semiconductor is exposed to a reducing atmosphere, e.g., methane, lattice oxygen is consumed and oxygen vacancies are formed, with electrons released in the valence band and the number of positive holes diminishing, therefore decreasing the electrical conductivity. For the *n*-type semiconductors, the inverse mechanism is observed: exposure to oxygen leads to electrons from the conduction band (charge carriers in this case) being consumed, therefore decreasing the electrical conductivity, while exposure to methane consumes some of the lattice oxygen, with the electrons being released into the conduction band and leading to higher electrical conductivity [38].

Popescu et al. [38] used in situ electrical measurements of some manganese-modified copper ferrites to study the correlation between the catalytic activity in methane total oxidation and their redox properties. The $CuFe_{2-x}Mn_xO_4$ spinels ($x = 0$; 0.4; 0.8; 1.2; 1.6; and 2) were prepared by solid-state reactions between CuO, Fe_2O_3, and Mn_2O_3 oxides, at 800 °C. $CuFe_2O_4$ and $CuMn_2O_4$ nanocrystalline spinels were identified in the XRD patterns, together with smaller amounts of individual oxides as side phases. The catalytic activity expressed as both the temperature corresponding to 10% methane conversion and the specific reaction rates at 710 °C decreased with increasing x, indicating a detrimental effect of manganese incorporation. The authors determined that as Mn is introduced in the structure the activation energy increases up to $x = 0.8$, then decreases with increasing x, while the pre-exponential factor varies in the opposite way accounting for the decreasing density of the active sites. This compensation effect explains the decrease of the specific catalytic activity when manganese concentration is increased [38]. From electrical conductivity measurements, $CuFe_2O_4$ was determined to be a *p*-type semiconductor, while $CuMn_2O_4$ an *n*-type semiconductor. The ternary spinel catalysts show both types of conduction mechanisms, with one type predominant as a function of temperature: at temperatures below ca. 500 °C they behave as *n*-type semiconductors, while at higher temperatures as *p*-type [38]. Since the transition from *n*- to *p*-type took place at temperatures lower than those needed for 50% methane conversion, all the ternary spinels were of *p*-type in the reaction temperature domain. For $CuFe_2O_4$, the presence of the two competing conduction mechanisms was explained [24] by the simultaneous existence of electrons (e^-) and holes (h^+) as charge carriers:

$$Fe^{2+} \leftrightarrows Fe^{3+} + e^{-} \quad (1.3)$$

$$Cu^{2+} \leftrightarrows Cu^{+} + h^{+} \quad (1.4)$$

The kinetics of these two reactions determine the type and concentrations of the charge carriers, and hence, the conduction mechanism [6, 38].

Electrical conductivity measurements under conditions as close as possible to those of catalysis (640 °C, sequential periods under air, under methane-air reaction mixtures, and under methane-nitrogen mixtures) were performed for the $CuFe_{2-x}Mn_xO_4$ with $x = 0$; 0.8 and 2 (Fig. 1.2) [38]. Under air, for $x = 0$ and 0.8, an n-type character was observed, while for $x = 2$ the behavior corresponded to the p-type. Notably, when methane was contacted with the samples, the nature of the semiconductivity changed for all three samples: those with $x = 0$ and 0.8 switched from p to n-type, while $CuMn_2O_4$ passed from n to p-type.

For the solids with $x = 0$ and 0.8, these variations were reversible when the atmosphere was changed again to air. This "breathing" behavior was explained by the oxidation of methane with lattice oxygen anions from the solid surface, leading to a partially reduced surface on which anion vacancies were created, in agreement with a Mars-van Krevelen mechanism. The free electrons thus produced are released into

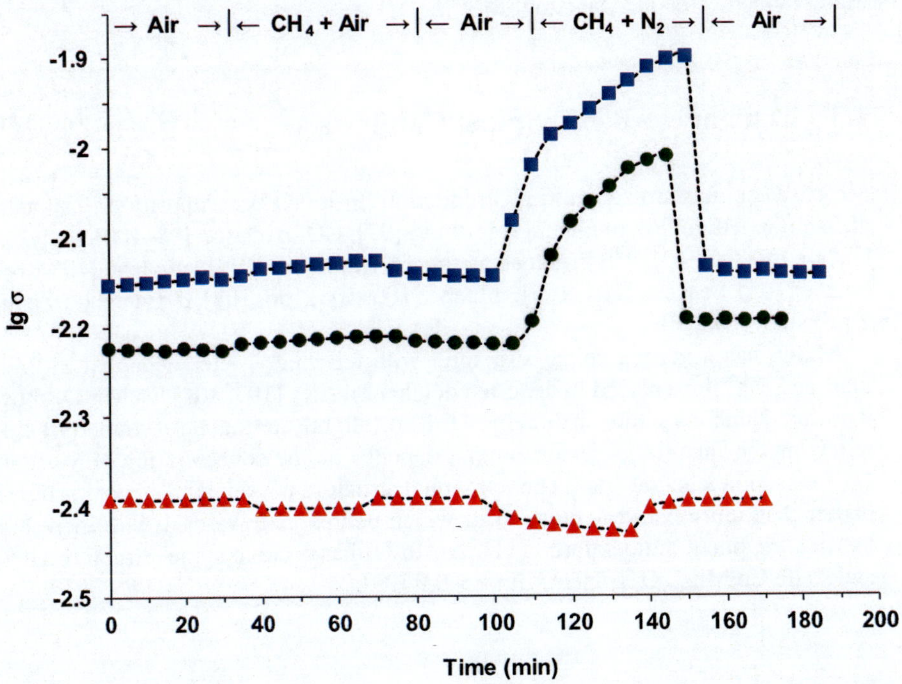

Fig. 1.2 Variation of the electrical conductivity under sequential exposures to air, methane–air mixture (reaction mixture), and methane–nitrogen mixture for $CuFe_2O_4$ (■), $CuFe_{1.2}Mn_{0.8}O_4$ (●), and $CuMn_2O_4$ (▲) at 640 °C (σ in ohm^{-1} cm^{-1}). (Reproduced from ref. [38] with permission)

the conduction band, thus increasing the electrical conductivity for the *n*-type semiconductors. When contacted with air again, the electrical conductivity values decrease, indicating that the free electrons are used for the formation of lattice oxygen anions by dioxygen dissociation, thus refilling the anion vacancies (Fig. 1.2). Similar conclusions were obtained by Munteanu et al. [6] in a study of propane combustion on $CuFe_2O_4$.

In the case of $CuMn_2O_4$ ($x = 2$), which is a *p*-type semiconductor with holes as charge carriers, each positive hole can be considered an electron vacancy in the valence band of the lattice oxygen anions or, from a chemical point of view, an O^- lattice species. This species can also activate methane molecules, and oxygen vacancies are generated again when water and CO_2 are produced. These vacancies are consistent with the observed reduction of the solid in methane-air and methane-nitrogen atmospheres, and they are filled when gaseous oxygen reoxidizes the solid. This is similar with a Mars-van Krevelen mechanism, but involving O^- lattice species in line with the *p*-type character of the solid. However, due to the smaller variations of the electrical conductivity values observed for this catalyst, the authors suggested a low concentration of O^- species for this solid, therefore explaining its low catalytic activity. As a conclusion, the in situ electrical conductivity measurements provided clear and direct evidence for a redox mechanism in which surface lattice oxygen from the spinel solid is consumed, which can be assimilated to a Mars-van Krevelen mechanism [38].

1.3 Manganese-Based Spinel Catalysts

Mn-based spinels are active in NO reduction [94], NO decomposition [95], water gas-shift reaction [96], methanol reforming [97], CO oxidation [98–100], and combustion of VOC [101, 102] such as methane [4], propane [99], ethylene [103], propylene [103], benzene [104, 105], toluene [100, 104, 106, 107], *o*-xylene [105], and 2-propanol [108, 109].

Mn_3O_4 has a normal spinel structure, with tetrahedral sites occupied by Mn^{2+} ions, and Mn^{3+} ions placed in distorted octahedral sites [103, 104]. $CuMn_2O_4$ shows a normal spinel structure with cubic [110] or slightly tetragonal symmetry [111], reflecting the Jahn-Teller distortion that depends on the concentration of Mn^{3+} and Cu^{2+} ions in octahedral sites. The tetragonal structure has an axial *c*/*a* ratio slightly higher than unity but very close to it, which decreases towards 1 with increasing thermal treatment temperature [111]. $NiMn_2O_4$ has a cubic spinel structure [108], and so do $Co_xMn_{3-x}O_4$ spinels with $x > 0.9$ [99].

1.3.1 Preparation Methods

The preparation methods of manganites are similar to those already discussed for ferrites:

- Coprecipitation [4, 100, 105, 109, 112, 113] is a commonly used method that usually starts from nitrates [4, 100, 109, 112, 113], but also from acetates [113] or chlorides [105] and uses NaOH [4, 100, 112], alkaline carbonates [105, 109], or tetramethylammonium hydroxide (TMAH) [113] as precipitating agents. An alternative of this method, the redox-precipitation procedure, uses $KMnO_4$ as oxidant [105, 113]. The redox-precipitation method leads to metal cation loss (e.g., Cu^{2+}) due to the fact that these cations are not involved in the redox reaction; therefore, a high Mn/M ratio is obtained in the final solid, far exceeding the value of 2 obtained by coprecipitation [113]. Also, minor side phases such as CuO [113] were reported besides the expected spinel. In some cases [105], the redox-precipitation does not lead to spinel mixed oxides, but to a mixture of oxides with different structures. To obtain solids with high specific surface areas, silica aquagel confined coprecipitation (SACOP) procedure uses a sodium silicate solution to obtain a high dispersion of the metallic cations in the silica aquagel. During calcination that takes place confined inside the pores of the silica matrix, the growth of the solid nanoparticles is restricted and, hence, high surface areas are obtained. The silica matrix is removed in a final step by dissolution in NaOH solution [112];
- Solid-state reaction [104, 111] starts from the corresponding oxides that are sintered at high temperatures in order for the spinel to be formed;
- Chemical complexation method uses metal salts (e.g., nitrates [99] or acetates [114]) and an organic acid such as $H_2C_2O_4$ [99, 114, 115]. The specific surface area and the particle size of the spinel depend on the conditions used for the decomposition of the precursor (oxalate) [99, 114];
- Sol-gel method uses metal nitrates and citric acid [108] or glycine [109] to obtain a gel. The procedure may include a self-combustion step that increases the temperature above 1000 °C for a short time when the gel is heated [108, 109]. Although the method is used for obtaining pure-phase mixed oxides [109], minor impurity phases containing single oxides such as CuO and NiO were sometimes reported [108];
- Solution combustion synthesis uses salts (e.g., nitrates) as oxidants and a fuel (e.g., $C_2H_5NO_2$) as reducing agent. The homogeneous solution of the two reagents is heated in an oven at constant temperature for the reaction to take place, then the resultant powder is washed and dried. Pure spinel phase was reported for this method [103];
- Dispersion-precipitation method is a modified precipitation procedure that uses glacial acetic acid added after precipitation to the obtained slurry to form a stable homogeneous dispersion of nanoparticles; in the second step, the suspension is diluted with water and precipitation gradually occurs by agglomeration of the nanoparticles [100];

– Spray deposition followed by thermal decomposition [110] starts with a solution of salts (e.g., nitrates) that is sprayed at temperatures above 400 °C on a surface (e.g., quartz plate) and heated afterwards at high temperatures for the decomposition to take place completely.

1.3.2 Mn_3O_4

Toluene combustion was studied on Mn_3O_4 obtained by the dispersion-precipitation (DP) method [100]. When compared with Mn_3O_4 obtained by direct precipitation (P) or by decomposition of manganese acetate (D), the solid obtained by DP method showed smaller particle size (5–25 nm compared to 30–55 nm and more than 200 nm, respectively) and larger specific surface areas (110 $m^2\ g^{-1}$, compared to 37 and 5 $m^2\ g^{-1}$, respectively), better reducibility and, therefore, higher catalytic activity in toluene total oxidation. T_{90} for the three catalysts were 245 °C, 260 °C, and 280 °C, respectively. The smaller particle size and better reducibility of the catalyst synthesized by the DP method accounted for its higher activity.

Combustion of benzene and toluene was investigated on 0.5 wt. % K, Ca, or Mg-modified Mn_3O_4 catalysts obtained by impregnation followed by calcination [104]. All modified samples showed better activity for the combustion of toluene than the parent Mn_3O_4: T_{50}/T_{90} values were 230/250 °C for both 0.5% K/Mn_3O_4 and 0.5% Ca/Mn_3O_4, and 232/255 and 245/270 °C for 0.5% Mg/Mn_3O_4 and Mn_3O_4, respectively. When benzene was used, T_{50} and T_{90} values were shifted by 10–20 °C to lower temperatures, indicating that benzene is more easily combusted than toluene. When benzene-toluene mixtures were tested, a mutual inhibitory effect was observed and the conversion curves were shifted to higher temperatures. A strong correlation between the conversion values and the presence of a hydroxyl-like group defect with higher mobility than the lattice oxygen in the manganese oxide structure was observed, detected by XPS and TPR. This defect was present in higher concentrations in the promoted catalysts; therefore, it has been concluded that K, Ca, and Mg have an important role in the formation of the mentioned defect.

Mn_3O_4 spinel prepared by solution combustion synthesis, together with other manganese oxides (Mn_2O_3 and Mn_xO_y—a mixture of Mn_2O_3 and MnO_2 phases), were tested in the total oxidation of some VOC test molecules: ethylene, propylene, toluene, and their mixture [103]. O_2-TPD and XPS experiments showed the presence of both chemisorbed oxygen (O^- and O_2^-, named α species, with high mobility) and structural oxygen (named β-species) in Mn_3O_4, which exhibited higher O_α/O_β ratio compared to the other manganese oxides. FTIR spectroscopy by means of NH_3 adsorption at room temperature indicated the presence of different types of hydroxyl groups on the surface of Mn_3O_4, and also Mn sites acting as Lewis acid centers. At higher temperatures, the surface hydroxyls can be converted into Lewis acid sites by dehydroxylation, while water interaction with such centers reconverts them into hydroxyls. These nanodefects on the surface can favor oxidation reactions, especially at low temperatures (below 400 °C). The order of activity for all individually tested VOC and for their mixture was $Mn_3O_4 > Mn_2O_3 > Mn_xO_y$. A Mn_3O_4-based monolith was prepared by solution combustion synthesis and showed comparable performances to the Mn_3O_4 powder, which were stable for 10 h on stream without deactivation. The authors highlighted the importance of the surface defects leading to a high amount of adsorbed oxygen species on the surface (α species) that promote total oxidation.

1.3.3 Cu–Mn Spinels

The valence state of copper and manganese ions in $CuMn_2O_4$ was investigated by Ghare et al. [110], Gillot et al. [111], and by Valdes-Solis et al. [112] in order to clarify if copper is present as Cu^+ or Cu^{2+} in the structure. Ghare et al. [110] studied the spinel structure by electrical conductivity and differential thermal analysis methods, while Gillot et al. [111] and Valdes-Solis et al. [112] used XPS and FTIR spectroscopy measurements. While literature proposed either the formula $Cu^+[Mn^{3+}Mn^{4+}]O_4$ or $Cu^{2+}[Mn^{3+}{}_2]O_4$, Ghare et al. [110] were able to observe that at ca. 600 K the following transformation occurs:

$$Cu_A{}^{+} + Mn_B{}^{4+} \leftrightarrows Cu_B{}^{2+} + Mn_A{}^{3+} \tag{1.5}$$

that was identified as a source of high activity in oxidation reactions [107, 112]. This modification in valence states accounted for a change of the slope in $\log R = f(1/T)$ plots (R = resistivity, the inverse of conductivity) and a small endothermic change in the DTA curve. It has been concluded that both formulas are correct, but at different temperatures. In $Cu^+[Mn^{3+}Mn^{4+}]O_4$, the numbers of Mn^{3+} and Mn^{4+} ions are equal, so the number of charge carriers is large and, hence, the activation energy of conduction is low; this feature was observed at low temperatures. For $Cu^{2+}[Mn^{3+}{}_2]O_4$, all octahedral positions are occupied by Mn^{3+}, so the number of charge carriers is small and, consequently, the activation energy of conduction is high; this behavior was observed above 600 K. Therefore, Ghare et al. [110] concluded that the solid behaves as $Cu^+[Mn^{3+}Mn^{4+}]O_4$ at temperatures up to 600 K and changes to $Cu^{2+}[Mn^{3+}{}_2]O_4$ above this value. XPS and FTIR measurements performed by Gillot et al. [111] on samples treated at 850–930 °C and then rapidly quenched showed the presence of Cu^+ in tetrahedral sites and Cu^{2+} (in lower amount) in octahedral sites, which led to the conclusion that, regardless of the preparation temperature, most copper ions are placed in tetrahedral positions and are monovalent, with the Cu^+/Cu^{2+} ratio close to 1.5. Another electronic exchange that influences the electronic configuration has been identified:

$$Mn_A{}^{3+} + Mn_B{}^{3+} \leftrightarrows Mn_A{}^{2+} + Mn_B{}^{4+} \tag{1.6}$$

the ionic configuration being best represented by the formula:

$$\left(Cu_{\alpha}^{+}Mn_{\beta}^{2+}Mn_{1-\alpha-\beta}^{3+}\right)_{A}\left(Cu_{1-\alpha}^{2+}Mn_{1-\beta}^{3+}Mn_{\alpha+\beta}^{4+}\right)_{B}O_{4}^{2-}$$

The XPS spectra of the $CuMn_2O_4$ prepared by Valdes-Solis et al. [112] with the silica aquagel confined coprecipitation method also showed the presence of Cu^+ and Mn^{2+} in tetrahedral positions, while Cu^{2+}, Mn^{3+}, and Mn^{4+} were identified in octahedral sites, but no Mn^{3+} was observed in tetrahedral positions.

A study about the influence of the preparation variables on the properties of $CuMn_2O_4$ was published by Hutchings et al. [98]. The investigated variables were: precipitate aging time (30 min–24 h), pH (7.5–10.0), and temperature (25–80 °C) of the precipitation, [Cu]/[Mn] ratio in the initial solution (from 100% Cu to 100% Mn), and calcination temperature (300–800 °C). The unaged precursor obtained during precipitation was identified as a mixture of copper hydroxyl nitrate and manganese carbonate, hence Cu and Mn are found in separate phases. During aging, the $CuMn_2O_4$ was formed together with CuO. The BET surface areas were not influenced by the aging time or aging temperature, but by pH and the [Cu]/[Mn] ratio. The Cu/Mn ratio in the bulk is significantly influenced by the pH of the precipitation step, since the onset for Cu^{2+} and Mn^{2+} precipitation from solution is found at 7 and 8, respectively, and higher pH is needed for a complete precipitation of Mn^{2+}. At pH = 7.5, a Cu-rich amorphous solid was produced, while at higher pH values Cu-Mn spinels were obtained together with Mn_2O_3 and CuO secondary phases. The composition of the initial solution also has a marked influence, with the stoichiometric ratio leading to the desired spinel. Calcination at low temperatures (300 °C) leads to poorly crystalline $MnCO_3$ and CuO, and at 500 °C, $CuMn_2O_4$ was formed. At higher temperatures, $Cu_{1.4}Mn_{1.6}O_4$ and, then, $Cu_{1.2}Mn_{1.8}O_4$ phases were identified, together with a Mn_2O_3 minority phase. The optimum parameters for the synthesis were found to be 12 h aging time, pH = 9, precipitation temperature of 80 °C, [Cu]/[Mn] = 1/2, and calcination temperature of 500 °C.

While studying catalytic methane combustion over Cu-(Mn)-Zn-Mg-Al mixed oxides obtained from layered double hydroxides (LDH) precursors, Răciulete et al. [4] observed the formation of some spinel phases. Four Cu(Mn)ZnMgAl-LDH precursors were prepared by coprecipitation, with (Cu + Mn + Zn + Mg)/Al atomic ratio equal to 3, Cu/Zn = 1 and Mn/Cu = 0, 0.5, 1, and 2. During the thermal decomposition of the LDH precursors, several spinel phases were identified, such as $Cu_{1.5}Mn_{1.5}O_4$ and $CuMn_2O_4$. When solids with Mn/Cu = 1 were further studied for the influence of the calcination temperature, the formation of the spinel phases was already observed at 550 °C, besides the periclase-like mixed oxide phase, and increasing the thermal treatment temperature up to 800 °C resulted in a higher crystallinity (Fig. 1.3). TPR studies showed that the reduction of copper in these solids is promoted by the presence of manganese, while XPS spectra confirmed the presence of Cu and Mn species with different valence states and the formation of $Cu_{1.5}Mn_{1.5}O_4$ spinel [4]. In catalytic methane combustion tests, the incorporation of manganese strongly increased the activity up to Mn/Cu = 1, but only this latter sample was able to lead to high methane conversion values (94% at 640 °C) (Fig. 1.4.). It has been suggested that at temperatures below 500 °C, the active phases are CuO

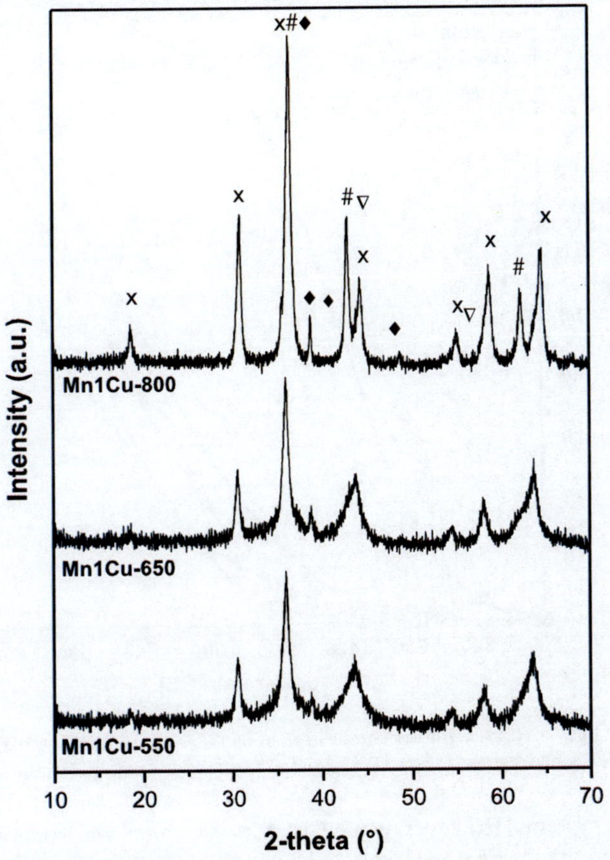

Fig. 1.3 XRD patterns of Mn1Cu-*T* mixed oxide catalysts calcined at different temperatures. Symbols: ×—$Cu_{1.5}Mn_{1.5}O_4$, $MnAl_2O_4$, and $CuMn_2O_4$, ◆—CuO, #—periclase-like phase. (Reproduced from ref. [4] with permission)

and $Cu_{1.5}Mn_{1.5}O_4$, while the $CuMn_2O_4$ spinel becomes active only at higher temperatures. The high intrinsic activities observed were associated with a high reducibility both in terms of hydrogen consumption and easiness of reduction, related to the presence of $Cu_{1.5}Mn_{1.5}O_4$ spinel [4].

$Cu_xMn_{2-x}O_4$ ($0 < x < 1$) spinels prepared using an alginate precursor were used by Behar et al. [107] for the complete oxidation of toluene as model VOC. Alginate is a natural polysaccharide with carboxylic groups, extracted from brown algae, which has a high affinity for divalent cations and provides nanoscale oxide dispersion and relatively high specific surface areas. $Cu_{1.5}Mn_{1.5}O_4$ spinel was obtained when the [Cu]/([Cu] + [Mn]) ratio in the initial solution varied from 0.1 to 0.57, together with Mn_3O_4 (for $x = 0$) and variable amounts of CuO (for $x > 0.7$). The high activity of the $Cu_{1.5}Mn_{1.5}O_4$ spinel in toluene combustion was correlated with its low reduction temperature, starting at 130 °C, and the presence of Cu^+/Cu^{2+} and $Mn^{3+}/$

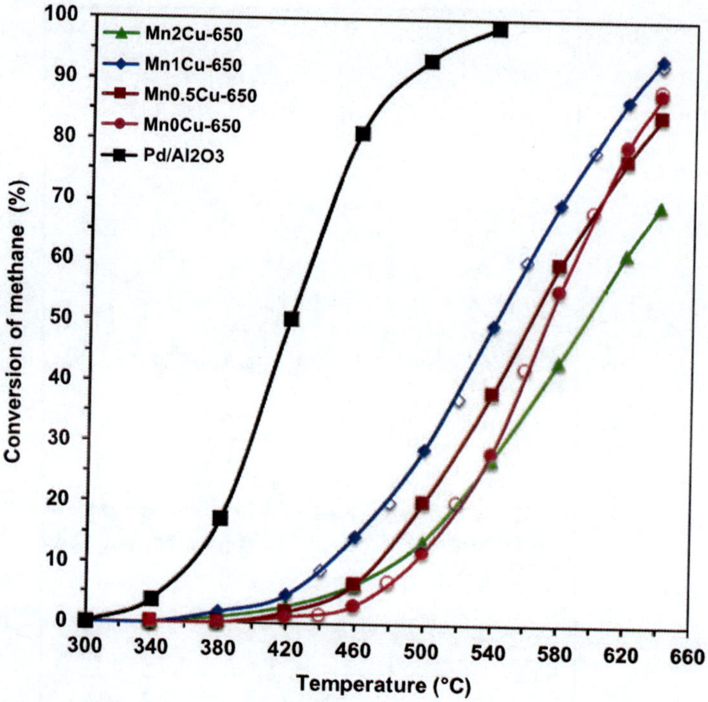

Fig. 1.4 The light-off curves for the combustion of methane over Mn*X*Cu-650 (open symbols correspond to the second light-off test). (Reproduced from ref. [4] with permission)

Mn^{4+} redox systems. However, when $Cu_{1.5}Mn_{1.5}O_4$ spinel was supported on TiO_2, its catalytic activity decreased, likely due to an inhibition effect of the support. A kinetic study [116] showed that the Mars-van Krevelen model describes the total oxidation of toluene on $Cu_{1.5}Mn_{1.5}O_4$ better than other kinetic models, such as power law and Langmuir-Hinshelwood.

The same $Cu_{1.5}Mn_{1.5}O_4$ spinel was prepared by coprecipitation [105] and tested, together with other Cu-Mn mixed oxides with different structures, in the complete oxidation of benzene, toluene, and *o*-xylene. Although the performances of the $Cu_{1.5}Mn_{1.5}O_4$ spinel were lower compared with the other tested catalysts, it showed reasonably good activity: T_{90} for benzene, toluene, and *o*-xylene were 295 °C, 280 °C, and 286 °C, respectively. However, CO was produced besides CO_2, which was related to a poor activity for CO oxidation. Slow deactivation was observed for long time on stream, while the presence of water, SO_2, and CO_2 had an important inhibiting effect, which was reversible in the case of water. It has been concluded that the presence of surface oxygen species and oxygen vacancies, observed by XPS and H_2-TPR measurements, plays a major role in the catalytic combustion of VOC.

Vu et al. [106] studied toluene combustion on $Cu_yMn_{1-y}O_x$ mixed oxides ($y = 0$, 0.25, 0.33, 0.50, 0.66, 0.75, 1.0) obtained by incipient wetness impregnation of Cu and Mn nitrates on a TiO_2 support, followed by calcination. Formation of

$Cu_{1.4}Mn_{1.6}O_4$ mixed spinel was identified by XRD in all samples except for Mn_2O_3 ($y = 0$) and CuO ($y = 1$). A decrease of specific surface area with increasing the Cu content (y values) was noticed. On all the catalysts, toluene was completely and selectively converted into CO_2 at temperatures above 250 °C, with conversion values maintained at 100% for 14 h on stream. However, at lower temperatures, the conversion decreased due to deactivation: at 200 °C, it reached 48% after 1 h and 21% after 4 h, but the activity could be completely restored by treatment in air at 300 °C. A direct correlation between the presence of the $Cu_{1.4}Mn_{1.6}O_4$ spinel in higher proportion and the high activity in toluene conversion was observed.

The influence of the preparation method on the performances of $CuMnO_x$ mixed oxides in the total oxidation of toluene was studied by Ye et al. [113]. The Cu-Mn oxides were prepared either by coprecipitation using tetramethylammonium hydroxide ($CuMn_2O_4$), or by redox-precipitation with $KMnO_4$ ($CuMnO_x$). The second method ensured a better dispersion of copper in close interaction with manganese, providing more active sites at the surface and, hence, a higher activity was obtained compared with $CuMn_2O_4$ prepared by coprecipitation or with the single Cu and Mn oxides. Copper was detected by XPS to be present both as Cu^+ and Cu^{2+}, with a Cu^+/Cu^{2+} ratio close to 0.5 in $CuMn_2O_4$, while manganese was present as Mn^{3+} and Mn^{4+}, the latter being predominant. Surface oxygen species with low coordination (surface oxygen vacancies) were also observed by XPS in high concentration in solids obtained by both methods. The order of activity for total oxidation of toluene was $CuMnO_x > CuMn_2O_4 > MnO_x > CuO_x$. The good performance of the catalysts obtained by the redox route was ascribed to improved textural properties such as high specific surface areas, amorphous state, and presence of oxygen vacancies. The oxidation of toluene was observed to occur also in the absence of gaseous oxygen and was attributed to the participation of surface and lattice oxygen species. XRD patterns of the catalysts after the reaction without gaseous oxygen revealed the presence of MnO and Cu^0 for $CuMn_2O_4$, while $Cu_{0.1}Mn_{2.9}O_4$ and $Cu_{1.5}Mn_{1.5}O_4$ were observed for $CuMnO_x$. It is worth noting that the $Cu_{1.5}Mn_{1.5}O_4$ active phase was also observed in other studies [4, 105, 116] (see above). Some organic reaction intermediates were observed by ToF-SIMS spectroscopy, such as benzene, phenol, benzaldehyde, and benzoic acid [113]. All the catalysts deactivated with time on stream, but modifying them with 0.5% Pd or Pt by impregnation allowed a better resistance to deactivation, without improving activity.

1.3.4 Co-Mn Spinels

Castaño et al. [109] studied the influence of the synthesis method on the performances of some Co-Mn spinel mixed oxides in the total oxidation of 2-propanol and toluene. The solids were prepared by coprecipitation (CP) or self-combustion (SC) of hydrotalcite-type precursors in order to obtain high specific surface areas and good inter-dispersion of the cations. The prepared catalysts contained simple spinel structures of Mn and Co and mixed $Co_xMn_{3-x}O_4$ ($x = 1$ or 2) spinel oxides. Those prepared by SC method showed smaller particle sizes, but also smaller specific surface areas and larger pores compared to those prepared by CP. The latter method seemed to favor oxygen mobility due to a cooperative effect between Mn and Co cations. The results demonstrate the fundamental role of the synthesis method on the oxygen mobility and, therefore, on the catalytic activity. In the total oxidation of 2-propanol, the catalysts derived from CP procedure had a lower T_{90} (242 °C, compared to 262 °C for SC), while in toluene complete oxidation they behaved similarly (T_{90} = 280 °C). In combustion tests of both compounds, the co-precipitated solids showed better performance than the single Mn and Co oxides, while those prepared by SC were intermediary between the manganese oxide and the cobalt oxide. Three complementary factors determined the catalytic behavior of the studied catalysts: oxygen mobility, redox properties, and oxygen species, both adsorbed on the surface (O_2^- and O^-), and lattice oxygen ions (O^{2-}), whose presence was demonstrated by O_2-TPD, ^{18}O isotope exchange, and H_2-TPR.

$Co_xMn_{3-x}O_4$ ($0 \leq x \leq 3$) spinels were prepared by controlled thermal decomposition of mixed oxalates and tested in the total oxidation of propane [99]. The synthesis method ensured large specific surface areas due to restrained shrinkage of the crystallites during thermal treatment, the highest value being obtained for $x = 1.6$ (270 m^2 g^{-1}). At x values above 0.9, the activity in propane combustion starts to increase, up to $x = 2.3$; $Co_{2.3}Mn_{0.7}O_4$ showed the best performance, with complete conversion of C_3H_8 at 300 °C. Stability tests for this composition indicated the conversion was stable for 14 h on stream at 160 °C. The obtained results were explained by the presence of catalytically active Co^{3+} ions in octahedral sites: at low x values, the activity is not influenced much because the substitution of Mn with Co takes place in the inactive tetrahedral sites, but when these sites are fully occupied by Co^{2+} (at $x > 1$) the substitution continues in the octahedral sites, where Mn^{3+} is replaced by Co^{2+} and Co^{3+}. It is worth noting that, for each Co^{2+} in an octahedral site, the oxidation of one Mn^{3+} to Mn^{4+} takes place in order to preserve the lattice neutrality. A similar study by Tang et al. [115] concerning total oxidation of ethyl acetate and hexane over Co-Mn spinel obtained from oxalate decomposition reached the same conclusions: the high specific surface area and small particle size led to a better performance than that of the single Mn or Co oxides, and complete oxidation for the two model compounds was achieved under 230 °C.

$Co_xMn_{3-x}O_4$ spinel catalysts with different Mn/Co atomic ratios (1:1, 1:2, 1:4, 2:1 and 4:1) were also studied by Huang et al. [117] for the total oxidation of formaldehyde at low temperatures. The solids were prepared by anodic electrodeposition on a flexible carbon textile substrate from a solution of manganese nitrate, cobalt chloride, and dimethyl sulfoxide at 70 °C, followed by the annealing of the Co-Mn precursor nanosheets in air at 550 °C. By adjusting the Mn/Co ratio in the initial electrolyte, x was varied between 0.43 and 0.78 in the final spinel structures. The $Co_{0.65}Mn_{2.35}O_4$ sample had the highest amounts of Co^{3+} content and adsorbed oxygen on the surface, a significant increase in reducibility compared to the other spinels, and more oxygen vacancies on the surface. It was concluded that all these properties account for its highest performance among the investigated spinels and single Mn and Co oxides, i.e., 100% HCOH conversion at 100 °C. The order of activity and the temperatures for complete oxidation of formaldehyde followed the order: $Co_{0.65}Mn_{2.35}O_4$ (100 °C) > $Co_{0.70}Mn_{2.30}O_4$ (110 °C) > $Co_{0.58}Mn_{2.42}O_4$ (130 °C) > $Co_{0.78}Mn_{2.22}O_4$ (140 °C) = $Co_{0.43}Mn_{2.57}O_4$ (140 °C) > MnO_x (170 °C) > CoO_x (180 °C). The $Co_{0.65}Mn_{2.35}O_4$ catalyst also showed excellent stability for 300 h on stream, even in humid atmosphere. A mechanism was proposed for the process that includes oxidation of formaldehyde to formate and hydrocarbonate species, which decompose to CO_2 and water.

Copper, nickel, and cobalt manganites prepared by the sol-gel self-combustion method were used in the total oxidation of two model VOC compounds, i.e., toluene and 2-propanol [108]. For both molecules, the order of activity was $NiMn_2O_4$ > $CuMn_2O_4$ > $CoMn_2O_4$, the better performance of nickel manganite being attributed to a higher content of Mn^{3+} and a synergetic effect between manganese and nickel in the spinel structure. Toluene was completely oxidized at 350 °C on $NiMn_2O_4$, while for 2-propanol a temperature of 250 °C was necessary on the same catalyst.

Cobalt manganite prepared by chemical complexation with oxalic acid was studied in the toluene combustion process and proved to be more active than the metal oxides, either single (Co_3O_4, MnO_x) or the mixed oxide Co_3O_4/MnO_x obtained by impregnation [114]. The authors claimed that the high performance was due to a larger specific surface area and abundant defects in the spinel structure, leading to a higher number of active sites for toluene adsorption and activation. Co^{2+}, Mn^{3+}, Mn^{4+}, chemisorbed oxygen, and a high number of oxygen vacancies were identified in the spinel by XPS. $CoMn_2O_4$ showed a better stability compared with the other oxides, with conversion values above 98% at 220 °C for 700 min in the presence or absence of water vapor (2%). The surface lattice oxygen was proposed to be the main active oxygen species in the catalytic reaction, an observation similar with that of Piumetti et al. [103] and Ye et al. [113], while migration of bulk lattice oxygen promotes the formation of the surface oxygen species. This claim was supported by the fact that the combustion reaction continued for the spinel for more than 45 min after the oxygen supply was stopped. When oxygen was reintroduced in the feed, the oxygen vacancies were replenished and the activity recovered. From the reaction intermediates identified by DRIFT spectroscopy, the authors concluded that toluene combustion takes place on $CoMn_2O_4$ through a benzyl alcohol—benzoic acid—maleic anhydride—acetic acid pathway, finally leading to CO_2 and H_2O, with anhydride conversion identified as the rate-controlling step. It was shown that the Co_3O_4/MnO_x mixed oxide catalyzes the combustion reaction on a different pathway due to low migration of bulk oxygen to the surface and benzoate dissociation as the rate-controlling step. The differences in structure and reducibility between $CoMn_2O_4$ and the other Co-Mn oxides result in different oxygen mobility from bulk to surface and number of active oxygen species and, hence, different catalytic performances [114].

Table 1.2 summarizes the different manganite spinel catalysts hereby described together with their preparation methods and catalytic applications in VOC combustion.

Table 1.2 Preparation and catalytic performance of the manganite spinel catalysts in VOC total oxidation

Catalyst	Preparation method	Tested chemical compound (reaction conditions)	Best results	Ref.
$Cu_{1.5}Mn_{1.5}O_4$ in a periclase-like matrix	TD[a] of LDH precursors	Methane (300–660 °C, 1% CH_4 in air, GHSV = 20,000 h^{-1})	94% conversion at 640 °C	[4]
Mn_3O_4	SolC[b]	Ethene (100–350 °C, 1000 ppm ethene in air, GHSV = 19,100 h^{-1}, W/F = 0.03 g h L^{-1})	T_{100} = 265 °C	[103]
$Co_xMn_{3-x}O_4$ ($0 \leq x \leq 3$)	TD[a] of oxalate precursors	Propane (30–300 °C, 0.4% propane, 20% O_2 in Ar, 1.63 mL s^{-1})	T_{100} = 300 °C on $Co_{2.3}Mn_{0.7}O_4$	[99]
Mn_3O_4	SolC[b] synthesis	Propene (100–350 °C, 1000 ppm propene in air, GHSV = 19,100 h^{-1}, W/F = 0.03 g h L^{-1})	T_{100} = 250 °C	[103]
Co manganite	TD[a] of oxalate precursors	Hexane (100–250 °C 1000 ppm hexane in air, 100 mL min^{-1}, 120,000 mL g^{-1} h^{-1})	T_{100} = 230 °C	[115]
Mn_3O_4 doped with 0.5 wt.% K, Ca, and Mg	IMP[c] of Mn_3O_4 with K, Ca, or Mg nitrates, followed by calcination	Benzene (200–400 °C, 2000 ppm C_6H_6 in air, 100 mL min^{-1})	T_{100} = 260 °C on 0.5% Ca/Mn_3O_4	[104]
$Cu_{1.5}Mn_{1.5}O_4$	CoPP[d]	Benzene (200–350 °C, 500 ppm benzene in humid air, 200 mL min^{-1}, GHSV = 100,000 h^{-1})	T_{90} = 295 °C	[105]
Mn_3O_4	SolC[b] synthesis	Toluene (100–350 °C, 1000 ppm toluene in air, GHSV = 19,100 h^{-1}, W/F = 0.03 g h L^{-1})	T_{100} = 265 °C	[103]
Mn_3O_4	DPP[e] method	Toluene (180–300 °C, 1000 ppmv toluene in air, GHSV = 180,000 h^{-1})	T_{90} = 245 °C	[100]

(continued)

Table 1.2 (continued)

Catalyst	Preparation method	Tested chemical compound (reaction conditions)	Best results	Ref.
Mn_3O_4 doped with 0.5 wt. % K, Ca, and Mg	IMP[c] of Mn_3O_4 with K, Ca, or Mg nitrates, followed by calcination	Toluene (200–400 °C, 2000 ppm toluene in air, 100 mL min^{-1})	T_{100} = 280 °C on 0.5% Ca/Mn_3O_4	[104]
$Cu_yMn_{1-y}O_x$/TiO_2 (y = 0, 0.25, 0.33, 0.50, 0.66, 0.75, 1)	IWI[f] of Mn and Cu nitrates on TiO_2, followed by calcination	Toluene (150–300 °C, 500 ppm toluene in air, GHSV = 5000 h^{-1})	T_{100} = 240 °C on the sample with y = 0.5	[106]
$Cu_xMn_{3-x}O_4$ ($0 < x < 1$)	Xerogels obtained via alginate hydrogel, calcined at 450 °C	Toluene (25–450 °C, 1000 ppm toluene in air, 100 mL min^{-1})	T_{50} = 239 °C, T_{100} = 315 °C on $Cu_{1.5}Mn_{1.5}O_4$	[107]
$CuMnO_x$	CoPP[d] and RedoxPP[g], calcined at different temperatures	Toluene (100–300 °C, 800 ppmv toluene in air, 100 mL min^{-1}, GHSV = 30,000 mL g^{-1} h^{-1})	T_{90} = 190 °C for $CuMnO_x$ prepared by RedoxPP and calcined at 200 °C	[113]
$Cu_{1.5}Mn_{1.5}O_4$	CoPP[d]	Toluene (200–350 °C, 500 ppm toluene in humid air, 100 mL min^{-1}, GHSV = 50,000 h^{-1})	T_{90} = 280 °C	[105]
Cu, Co, and Ni manganite	SG[h] + SelfC[i]	Toluene (150–400 °C, 0.2% toluene in air)	T_{100} = 350 °C on Ni manganite	[108]
$Co_xMn_{3-x}O_4$ (x = 1 or 2)	CoPP[d] and SelfC[i] of hydrotalcite-like precursors	Toluene (100–400 °C, 600 ppm toluene in air, 280 mL min^{-1})	T_{90} = 280 °C for $CoMn_2O_4$	[109]
Co manganite	Chemical complexation with $H_2C_2O_4$	Toluene (160–250 °C, 500 ppm toluene, 20% O_2/Ar, 75 mL min^{-1}, GHSV = 22,500, and 45,000 mL g^{-1} h^{-1})	T_{50} = 202 °C, T_{90} = 210 °C, T_{100} = 220 °C, at GHSV = 22,500 mL g^{-1} h^{-1}	[113]
$Cu_{1.5}Mn_{1.5}O_4$	CoPP[d]	*o*-xylene (200–350 °C, 200 ppm *o*-xylene in humid air, 100 mL min^{-1}, GHSV = 50,000 h^{-1})	T_{90} = 286 °C	[105]

(continued)

Table 1.2 (continued)

Catalyst	Preparation method	Tested chemical compound (reaction conditions)	Best results	Ref.
$Co_xMn_{3-x}O_4$ (x = 0.43; 0.58; 0.65; 0.70; and 0.78)	ED[h] on a carbon textile substrate, followed by annealing in air at 550 °C	Formic aldehyde (4100 mL min^{-1}, GHSV = 120,000 mL g^{-1} h^{-1})	T_{100} = 100 °C	[117]
Cu, Co, and Ni manganite	SG[i] + SelfC[j]	2-propanol (150–400 °C, 0.2% 2-propanol in air)	T_{100} = 250 °C on Ni manganite	[108]
$Co_xMn_{3-x}O_4$ (x = 1 or 2)	CoPP[d] and SelfC[j] of hydrotalcite-like precursors	2-propanol (50–400 °C, 1000 ppm 2-propanol in air, 280 mL min^{-1})	T_{90} = 242 °C for $CoMn_2O_4$ obtained by CoPP	[109]
Co manganite	TD[a] of oxalate precursors	Ethyl acetate (100–250 °C, 1000 ppm ethyl acetate in air, 100 mL min^{-1}, 120,000 mL g^{-1} h^{-1})	T_{100} = 220 °C	[115]
Mn_3O_4	SolC[b] synthesis	Ethylene + propylene + toluene (100–350 °C, 1000 ppm mixture in air, GHSV = 19,100 h^{-1}, W/F = 0.03 g h L^{-1})	T_{100} = 310 °C	[103]

[a]*TD* thermal decomposition
[b]*SolC* solution synthesis
[c]*IMP* impregnation
[d]*CoPP* coprecipitation
[e]*DPP* dispersion-precipitation
[f]*IWI* incipient wetness impregnation
[g]*RedoxPP* redox-precipitation
[h]*ED* electrodeposition
[i]*SG* sol-gel synthesis
[j]*SelfC* self-combustion

1.4 Cobalt-Based Spinel Catalysts

Supported and unsupported Co_3O_4 spinel is one of the most active mixed-valence metal oxides catalysts for the total oxidation reaction of VOC mainly due to its excellent reduction ability and high concentration of surface electrophilic oxygen species [2, 11, 15]. Introducing another metallic component into the Co_3O_4 spinel to form a binary cobaltite is an effective way to improve its combustion catalytic properties due to the synergy between the two metallic cations [118]. The most relevant papers published in the last decade focusing on Co-based spinel catalysts for complete oxidation of VOC are reviewed below.

1.4.1 Unsupported and Supported Co_3O_4

In an attempt to enhance the catalytic performance by tuning the catalyst morphology, spinel-type cobalt oxide (Co_3O_4) nanocrystals with different morphologies, i.e., cubic (C), hexagonal plate-like (P), hexagonal sheet-like (S) and flower-like (F), denoted as Co_3O_4-C, Co_3O_4-P, Co_3O_4-S, and Co_3O_4-F, respectively, were controllably prepared via a hydrothermal method [119]. Pure and highly crystalline spinel phases were obtained in all cases. In spite of their different morphologies, Co_3O_4-S, Co_3O_4-P, and Co_3O_4-F catalysts predominantly expose {111} planes, while Co_3O_4-C catalyst exposes {001} planes. The ratio between the surface-adsorbed oxygen (O_{ads}) and surface lattice oxygen (O_{latt}) species in the Co_3O_4 catalysts, determined by XPS, increases following the order: Co_3O_4-C < Co_3O_4-S < Co_3O_4-F ≈ Co_3O_4-P. Their activity in methane combustion follows the same order (Fig. 1.5), which is consistent with the fact that, in general, the lattice oxygen is involved in selective oxidation, while adsorbed electrophilic species such as O_2^{2-}, O_2^-, and O^- participate preferentially in the complete oxidation. On the other hand, the Co_3O_4 catalysts exposing dominantly {111} plane (flower-like, hexagonal plate-like, and hexagonal sheet-like) were more active than cubic Co_3O_4, which exposes mainly {001} planes.

Co_3O_4 with different morphologies was also studied in the total oxidation of formaldehyde. Thus, rod-like, sheet-like, and cube-like pure Co_3O_4 spinel catalysts were prepared by a modified precipitation for the former or a hydrothermal method for the two later [120]. As expected, the oxides have different physicochemical characteristics. Indeed, the Co_3O_4-rod sample had a significantly larger specific surface area (83.5 $m^2\ g^{-1}$) compared to that of Co_3O_4-sheet (8.5 $m^2\ g^{-1}$) and Co_3O_4-cube (3.7 $m^2\ g^{-1}$). Also, Co_3O_4-rod showed $(OH)_s/O_{latt}$ and Co^{3+}/Co^{2+} surface ratios larger and reducibility and basicity greater than its sheet-like and cube-like counterparts. Accordingly, the Co_3O_4-rod showed the best performance for the formaldehyde total oxidation, the T_{50} temperatures being ranked in the order Co_3O_4-rod < Co_3O_4-sheet < Co_3O_4-cube. Moreover, the Co_3O_4-rod catalyst completely converted formaldehyde at 120 °C, with excellent time on stream stability.

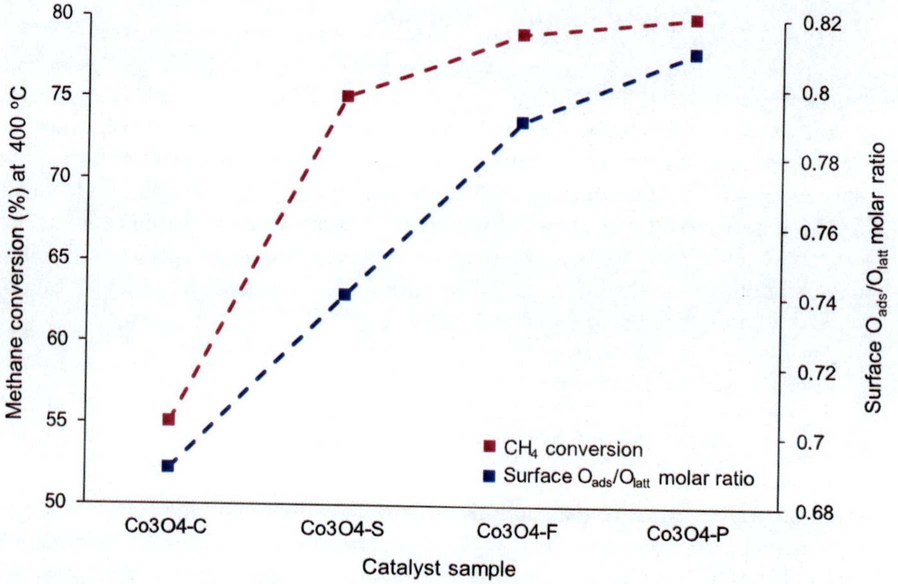

Fig. 1.5 Correlation between the catalytic activity in methane combustion and surface O_{ads}/O_{latt} molar ratio for Co_3O_4 nanocrystals with different morphologies: cubic (C), hexagonal sheet-like (S), flower-like (F), and hexagonal plate-like (P) (Reaction conditions: 0.2 vol. % CH_4 in air; GHSV = 110,000 h^{-1}). (Data from ref. [119])

High surface area ordered Co_3O_4 catalysts have been successfully prepared by a nanocasting route using different mesoporous KIT-6 silica as a hard template and tested in the complete oxidation of propane, as a model of short chain alkane, and toluene, as a model of monoaromatic hydrocarbon [121]. It has been shown that the synthesis parameters, i.e., aging and calcination temperatures, of the silica template strongly influence the physicochemical properties and, hence, catalytic performances of the final catalysts. Thus, siliceous KIT-6 hard templates were prepared at three different aging temperatures (40, 70, and 100 °C) and calcined at 550 °C (denoted KIT6-40–550, KIT6-70–550, and KIT6-100–550, respectively). Additionally, the silica sample aged at 100 °C was calcined at 700, 800, and 900 °C (samples KIT6-100–700, KIT6-100–800, and KIT6-100–900, respectively). They were impregnated with the Co-precursor (Co nitrate), then calcined at 550 °C for 6 h, and finally the silica template was eliminated using a concentrated aqueous solution of NaOH to obtain the corresponding mesoporous cobalt oxides C40–550, C70–550, C100–550, C100–700, C100–800, and C100–900, respectively. These catalytic materials were shown to be more active in the total oxidation reaction compared with a conventional Co_3O_4 prepared by simple evaporation of cobalt nitrate followed by calcination. The best C100–550 catalyst showed T_{50} temperatures (corresponding to 50% conversion) of 196 and 164 °C in propane and toluene combustion, respectively, which are significantly lower compared to the conventional Co_3O_4 (235 °C and 260 °C, respectively) (Fig. 1.6). Interestingly, an inverse

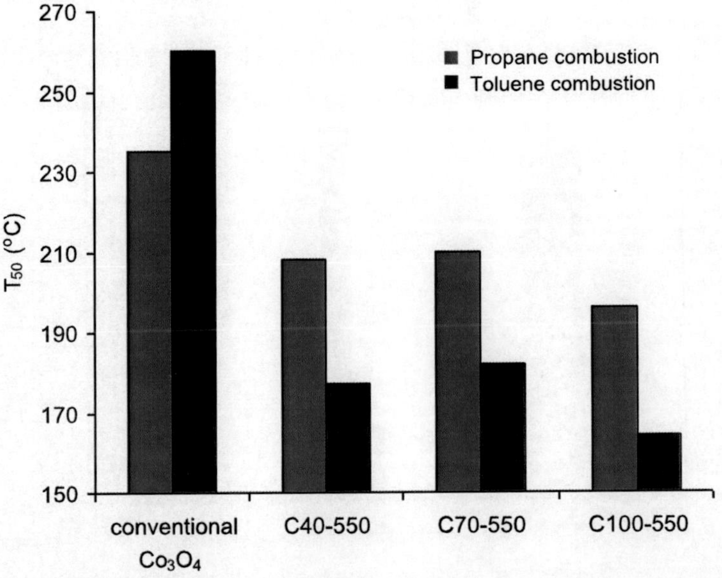

Fig. 1.6 Influence of the aging temperature during the synthesis of the silica template (40, 70, and 100 °C) on the catalytic activity of the corresponding mesoporous Co_3O_4 catalysts (expressed as T_{50}) in complete oxidation of propane and toluene and comparison with Co_3O_4 prepared by simple evaporation of Co nitrate (Reaction conditions: 8000 ppm C_3H_8 in air; GHSV = 12,000 mL g_{cat}^{-1} h^{-1} and 100 ppm C_7H_8 in air; GHSV = 60,000 mL g_{cat}^{-1} h^{-1}, respectively). (Data from ref. [121])

relationship between the ordered structure of these catalysts and their activity was found. For example, the C100–550 catalyst, which possesses a partly ordered structure, leads to a higher conversion per surface area in propane combustion than the C40–550 sample which has a highly ordered, bimodal structure. Since the reducibility was not very different among ordered and non-ordered catalysts, it was concluded that the enhanced activity of the non- or partly ordered catalysts is due to a higher concentration of highly reactive oxygen defects (a higher concentration of surface Co^{2+} species) in these materials. Notably, it has been shown that these catalysts exhibit high stability at moderate temperatures, regardless of the test molecule.

Ordered mesoporous Co_3O_4 spinel prepared by the nano-replication method using either SBA-15 (Co_3O_4-S) or KIT-6 (Co_3O_4-K) templates was studied in the catalytic oxidation of dibromomethane (CH_2Br_2) as a model pollutant for Br-VOC [122]. The ordered mesoporous spinel samples were shown to be the replica structures of their templates with unimodal pore size distribution centered at 3.9 and 3.6 nm for Co_3O_4-S and Co_3O_4-K, respectively, and have lower crystallinity and smaller crystallite size compared with a precipitated counterpart (Co_3O_4-B). Also, the ordered mesoporous spinel samples have significantly larger surface areas and better redox properties compared to that of Co_3O_4-B. Both the Co^{3+}/Co^{2+} surface ratio and the ratio of the surface-adsorbed oxygen to lattice oxygen (O_{ads}/O_{latt}) increased following the order: Co_3O_4-B < Co_3O_4-S < Co_3O_4-K. In the same order,

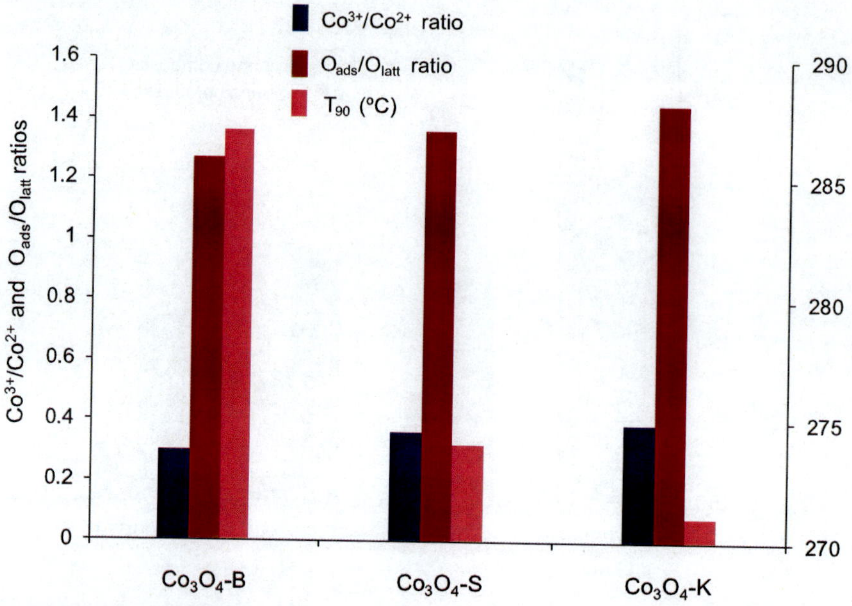

Fig. 1.7 Catalytic activity (in terms of T_{90} values) in the oxidation of CH_2Br_2 and Co^{3+}/Co^{2+} and O_{ads}/O_{latt} ratios for the different Co_3O_4 materials (Reaction conditions: 500 ppm CH_2Br_2, 10% O_2 and balancing N_2; GHSV = 112,500 $cm^3\ g_{cat}^{-1}\ h^{-1}$). (Data from ref. [122])

their catalytic performance oxidation reaction of CH_2Br_2 increased. The superior catalytic activity of ordered mesoporous spinel catalysts was attributed to their high specific surface areas and strong redox property together with their higher content of superficial Co^{3+} species and surface active oxygen (Fig. 1.7). Notably, the most active catalyst in this series, i.e., Co_3O_4-K, also showed the highest selectivity to CO_2 at low temperature and a good time-on-stream stability for at least 30 h.

A series of Co_3O_4 catalysts were synthesized at different pH values by a facile precipitation method, the influence of preparation pH on the catalytic performance in methane combustion being investigated [123]. The final catalysts were denoted as Co_3O_4-X, where X (8.0; 8.5; 9.0; 9.5; 10) represents the pH values at the end point of titration, which strongly influenced their physicochemical characteristics and, hence, their catalytic performance. Indeed, their surface area passes through a minimum (25.2 $m^2\ g^{-1}$) and the crystallite size through a maximum (33.4 nm) for Co_3O_4-9.0 sample, which also shows maxima of surface Co^{2+}/Co^{3+} (0.46) and O_{ads}/O_{latt} (0.51) ratios. Notably, the catalytic activity of the Co_3O_4-X samples in terms of intrinsic reaction rate passes through a maximum for Co_3O_4-9.0 and correlates well with the surface Co^{2+}/Co^{3+} and O_{ads}/O_{latt} ratios (Fig. 1.8). Co_3O_4-9.0 catalyst also shows the lowest T_{50} and T_{90} temperatures (corresponding to 50 and 90% methane conversion, respectively), i.e., 344 °C and 413 °C, respectively, and possesses good long-term stability, no loss of catalytic activity being recorded after 60 h on stream.

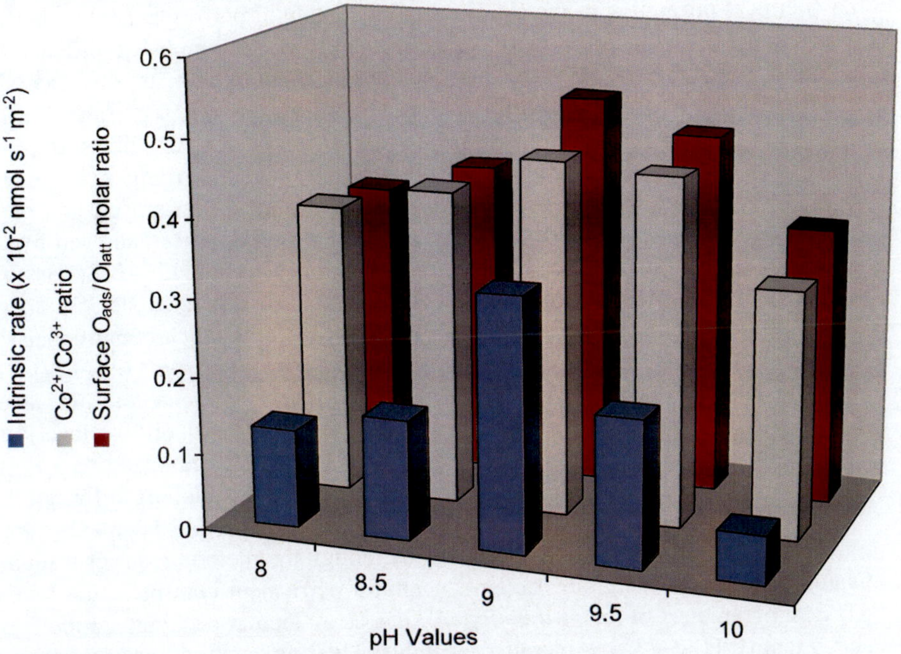

Fig. 1.8 Influence of preparation pH on the surface Co^{2+}/Co^{3+} and O_{ads}/O_{latt} ratios of Co_3O_4-*X* catalysts and on their activity in methane combustion (Reaction conditions: 0.5 vol. % CH_4, 8 vol. % O_2 and balancing N_2; GHSV = 18,000 mL g_{cat}^{-1} h^{-1}). (Data from ref. [123])

A nanosized Co_3O_4 catalyst for the complete oxidation of propane and CO was prepared by dispersion-precipitation (DP) [124]. Compared with the conventional alkali-induced precipitation (AP) method, this method involves the reaction between wet cobalt hydroxide and acetic acid, with the formation of a colloidal dispersion, followed by a subsequent dilution which destabilizes and precipitates the colloidal particles. The oxide thus obtained (CoDP) had smaller particle size, lower crystallinity, higher surface area, and larger pore volume compared to its conventionally precipitated counterpart (CoAP). Moreover, CoDP was shown to be more reducible and to possess larger amount of reactive surface-adsorbed oxygen species. These superior properties of CoDP make it a significantly better catalyst than CoAP for the complete oxidation of both C_3H_8 and CO. Thus, C_3H_8 was completely oxidized on CoDP catalyst at 240 °C, whereas only ca. 21% conversion was achieved on CoAP, even at 300 °C. Regarding the catalytic oxidation of CO, complete conversion was achieved at ca. 110 °C on CoDP catalyst and at ca. 170 °C on CoAP. Notably, CoDP catalyst showed a good long-term stability in the total oxidation of propane.

Three different methods, i.e., nonequilibrium plasma deposition (NEP), Langmuir-Blodgett (LB), and impregnation, were used to prepare a series of cobalt oxide catalysts deposited on precalcined metallic carriers with two different geometries, i.e., wire gauze and sheets, in order to compare the efficiency of deposition, the chemical properties of the catalyst, and its catalytic performance in n-hexane combustion [125]. First, it is worth describing shortly the principles of NEP and LB methods, which are less usual for the preparation of catalytic materials. The NEP technique is used to obtain thin-film materials deposited on various carriers, starting from metal-organic precursors in a plasma reactor operated at low pressure [126–128]. Thus, the metal-organic precursor is transported with a carrier gas (argon and/or oxygen) through the region of low temperature nonequilibrium plasma of an electric discharge where its decomposition/oxidation takes place followed by the deposition of the metal or metal oxide on the support material [126–128]. Usually, precalcined Cr-Al stainless steel leaves, steel wire gauzes, kanthal steel, and steel sheets are used as support materials, while cyclopentadienyldicarbonyl-cobalt (I), Cu(II)-acetylacetate, etc. are examples of metal-organic precursors [127–129]. The LB deposition is a technique used to prepare highly ordered, defectless film layers of controlled thickness and architecture of a compound on a solid surface [130, 131]. In this method, a single layer of molecules is initially organized on a liquid surface, usually water, before being transferred onto a solid support to form a thin film having the thickness of a constituent molecule. If the process is repeated, multilayered films can be obtained [130, 132]. Although the method requires a sophisticated equipment (Langmuir trough), it allows permanent control of the several physical parameters of layers transferred, such as molecular packing, composition, and quantity [130–132]. Among the catalyst layering methods used to prepare deposited cobalt oxide catalysts, the NEP technique was found to be the most adequate for the deposition of materials on the structural reactors for catalytic applications, since it allows the control of the dispersion of the deposited material, maintaining at the same time the geometry of the carriers [125]. Contrarily, in the case of LB method, the deposition efficiency was too low to obtain an active catalyst within reasonable time, while the impregnation method did not allow the control of the thickness of the deposited layers [125]. The NEP layering method led to a well-dispersed cobalt oxide catalyst with spinel structure, which turned out to be only slightly less active than the commercial Pt catalyst in *n*-hexane combustion [125].

A series of structured cobalt spinel catalysts prepared by NEP deposition of thin films on precalcined metallic supports (kanthal sheets) was studied in *n*-nonane catalytic combustion reaction [133]. The different catalysts were prepared under different atmospheres (oxygen-free and oxygen/argon) and residence times of the precursor (Co(I)-cyclopentadienyldicarbonyl) in the plasma reactor. This induced significant changes in the catalysts' structure and morphology, including the amount and the shape of the particles deposited on the metallic support. The catalytic tests showed that the most active systems in the *n*-nonane combustion reaction are the samples synthesized under oxygenless conditions. Contrarily, the activity is significantly lower for the catalysts prepared under synthetic air conditions for which the dominant form of the cobalt oxide is its amorphous phase highly dispersed on alumina.

Incipient wetness impregnation (IWI) and a combination of IWI and subsequent combustion synthesis (CS) were used to prepare a series of $Co_3O_4/\gamma\text{-}Al_2O_3$ catalysts with different Co_3O_4 loadings, tested in the catalytic combustion of CH_4 [134]. The

IWI method consisted of dripping an aqueous solution of cobalt nitrate onto γ-Al_2O_3 followed by water evaporation, drying, and calcination at 700 °C in air, while for the IWI/CS method, the impregnation was realized with aqueous solutions of cobalt nitrate and urea (molar ratio = 2:1), then the dried sample was heated to initiate the exothermic reactions at 230–250 °C for 5 min. The self-propagating flameless combustion led to the complete IWI/CS catalyst formation within 1–15 min. It has been shown that the IWI/CS-X catalysts (*X* = 5, 10, 30, and 50, representing the Co_3O_4 loading as wt %) have lower degree of crystallization, smaller average crystallite size, greater surface area, and higher degree of dispersion of Co_3O_4 on the γ-Al_2O_3 support, in comparison with the IWI-X catalysts with similar Co_3O_4 content. The large quantities of vapors (e.g., CO_2, H_2O, and NO_2) rapidly released in the combustion process, which is completed in a very short time, limit the growth of spinel crystals. Moreover, for the IWI/CS-X catalysts, the synthesis process determined the formation of abundant easily reducible surface Co^{3+} species. Accordingly, the IWI/CS-X catalysts have excellent catalytic performance in CH_4 combustion, being significantly more active than their IWI-X counterparts (Fig. 1.9). In particular, IWI/CS-30 catalyst, which possesses the smallest average particle size (6.5 nm) and the highest surface $Co^{3+}/(Co^{2+}+Co^{3+})$ ratio, exhibited the best catalytic activity with total conversion of CH_4 achieved at 550 °C. On the other hand, the excessive loading of Co_3O_4 in IWI/CS-50 sample led to a diminished catalytic activity.

A series of nanoflower-like nickel foam monolith-supported Co_3O_4 catalysts, prepared via a green and facile electrodeposition method, was studied in the propane total oxidation [135]. The amount and interaction of Co_3O_4 loaded on the

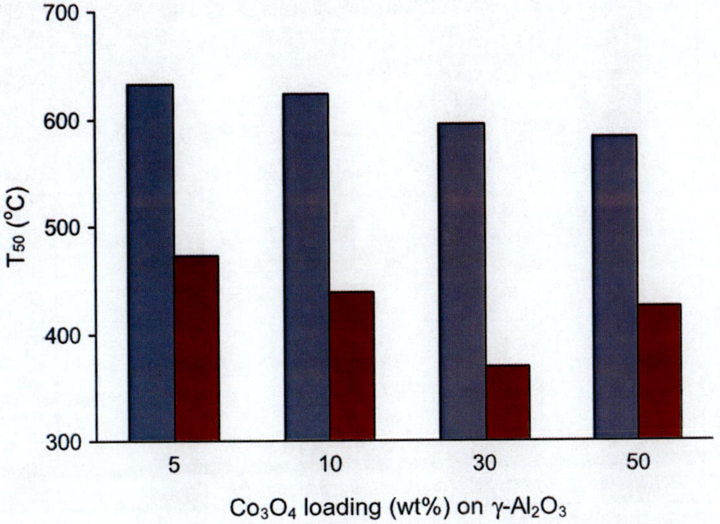

Fig. 1.9 Comparison between T_{50} temperatures in CH_4 combustion on catalysts with similar Co_3O_4 loadings prepared via two methods: IWI (■) and IWI/CS (■) (Reaction conditions: 0.2 vol % CH_4, 10 vol % O_2, and N_2 as the balance gas; GHSV = 36,000 mL g_{cat}^{-1} h^{-1}). (Data from ref. [134])

Ni-foam monolith were controlled by the time duration of potential deposition. The resulted catalysts were denoted as Co-NF-*t*, where *t* represents the corresponding electrodeposition time duration in seconds, i.e., Co-NF-300, Co-NF-600, Co-NF-900, and Co-NF-3600. For comparison, a pure Co_3O_4 catalyst was also synthesized by thermal decomposition of cobalt nitrate, using the same calcination conditions. For example, it has been shown that Co-NF-300 nanosheet (diameter 1–2 μm) stacked by Co_3O_4 nanoparticles (10–30 nm) presents an abundance of pores. Such 3D hierarchical microstructures with interconnected pore channels expose more active sites (larger cobalt oxide surface), without any reactant/product diffusion limitations. All of the Co-NF-*t* are flower-like sheets with the same morphology and microstructure, the electrodeposition time duration having no obvious effect on these characteristics. On the other hand, the morphology of pure Co_3O_4 is different from that of Co-NF-*t*, its shape being elongated and particle size significantly larger (10–50 nm). Co_3O_4 loading and the intensity of the diffraction lines corresponding to Co_3O_4 with spinel structure increased with the deposition time duration. The spinel particle size for the Co-NF-*t* is small and varies between 18 and 21 nm. Strong interactions between Co_3O_4 and Ni-foams were shown to be formed during the electrodeposition process, enhancing the reducibility of Co_3O_4. The Co^{2+}/(Co^{2+} + Co^{3+}) ratio calculated from the areas of reduction peaks appears to depend strongly on the deposition time duration, Co_3O_4 supported on Ni-foam having

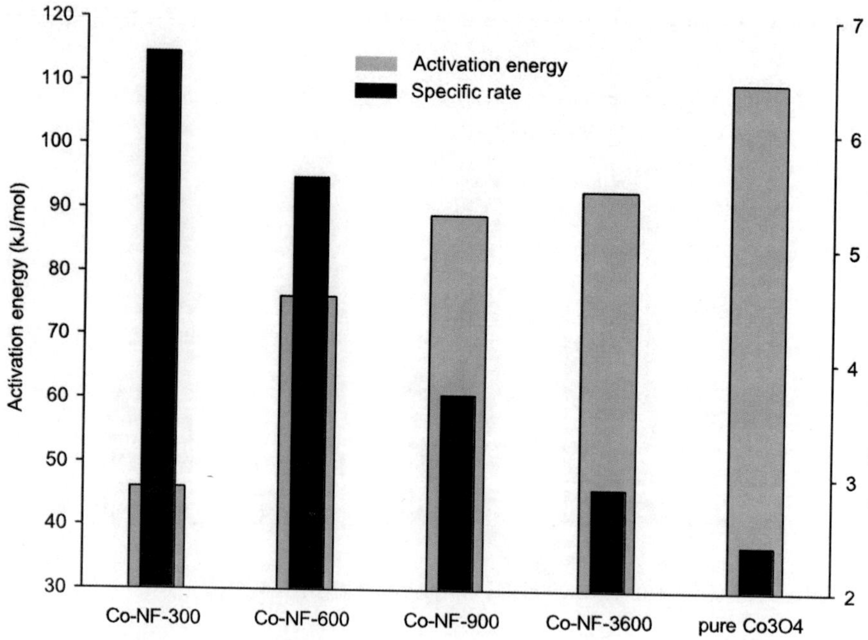

Fig. 1.10 Variation of the specific reaction rates (per gram of Co_3O_4) at 220 °C and of the activation energies for the Co-NF-*t* catalysts in comparison with pure Co_3O_4 (Reaction conditions: 1000 ppm C_3H_8 in air; GHSV = 30,000 mL g_{cat}^{-1} h^{-1}). (Data from ref. [135])

higher Co^{2+} content compared with pure Co_3O_4 sample. Both the Co^{2+} content and catalytic activity increase in the order: pure Co_3O_4 < Co-NF-3600 < Co-NF-900 < Co-NF-600 < Co-NF-300. With the highest specific activity and the lowest activation energy, Co-NF-300 sample was found to be the most active in total oxidation of propane reaction (Fig. 1.10). Moreover, Co-NF-300 sample also possesses good time on stream stability for 40 h of reaction.

To separately study the geometrical-site-dependent catalytic activity of tetrahedrally coordinated Co^{2+} sites ($Co^{2+}{}_{Td}$), octahedrally coordinated Co^{2+} sites ($Co^{2+}{}_{Oh}$), and octahedrally coordinated Co^{3+} sites ($Co^{3+}{}_{Oh}$) in VOC oxidation, a series of Co-based catalysts, i.e., Co_3O_4, CoO, $CoAl_2O_4$, $ZnCo_2O_4$, and $CoFe_2O_4$, was prepared by the nanocasting method using mesoporous silica KIT-6 as a hard template [3]. Thus, starting from three-dimensionally ordered mesoporous Co_3O_4, Co^{2+} or Co^{3+} sites were replaced with catalytically inactive or less active Zn^{2+} (d^0), Al^{3+} (d^0), and Fe^{3+} (d^5), respectively. It has been shown that not only $Co^{3+}{}_{Oh}$ species act as active sites, but also $Co^{2+}{}_{Oh}$ sites exhibit high catalytic activity due to the fact that they are easily oxidized to active Co^{3+} species which were shown to be responsible for the formation of carboxylate species, the main intermediate in the oxidation of benzene. Therefore, $ZnCo_2O_4$ with $Co^{3+}{}_{Oh}$ and CoO with $Co^{2+}{}_{Oh}$ species exhibit good catalytic activity in benzene oxidation and high TOF_{Co} values at low temperature. Moreover, $ZnCo_2O_4$ exhibits good durability at 500 °C and strong water resistance ability. Contrarily, $CoAl_2O_4$ with $Co^{2+}{}_{Td}$ sites show poor catalytic activity and a low TOF_{Co} value.

1.4.2 M-Co Spinels (M = Transition Metal)

A series of ordered mesoporous MCo_2O_4 (M = Cu, Zn, and Ni) spinel catalysts, synthesized via nano-replication method using mesoporous silica KIT-6 as a hard template, was tested in the complete oxidation of methane reaction and compared with bulk MCo_2O_4 spinel catalysts prepared by coprecipitation [136]. The ordered structure of mesoporous MCo_2O_4 (m-MCo_2O_4) spinel catalysts was clearly confirmed by the combined results of N_2 adsorption-desorption and TEM and XRD analysis. Indeed, the N_2 adsorption-desorption measurements revealed that all the m-MCo_2O_4 samples show a type IV isotherm with H1 hysteresis loop and a narrow pore size distribution, confirming the formation of mesoporous structure responsible for their higher surface area and pore volume than their bulk counterpart (b-MCo_2O_4). Also, TEM analysis clearly suggests that m-MCo_2O_4 spinels are well-replicated from the KIT-6 template, their wall thickness corresponding well with the pore size diameter of KIT-6 template. Moreover, the low angle XRD patterns prove the formation of ordered mesoporous structures of the m-MCo_2O_4 spinels, in line with the results of N_2 adsorption-desorption and TEM analysis. At the same time, wide angle XRD analysis shows that m-MCo_2O_4 spinels possess smaller size of crystallites than b-MCo_2O_4 spinels, which leads to greater resistance to decomposition and higher resistance to sintering during calcination of the spinel phase. XPS analysis revealed that the normalized amount of Co^{3+} cations on the surface decreases for m-MCo_2O_4 spinels as follows: m-$CuCo_2O_4$ > m-$ZnCo_2O_4$ > m-$NiCo_2O_4$. The catalytic activity of the meso catalysts in methane combustion, expressed as T_{50} and T_{90} temperatures (corresponding to 50% and 90% conversion, respectively), is considerably lower than that of their bulk counterpart, obviously due to the unique ordered mesostructure and the excellent thermal stability of the former. Notably, the activity of the meso catalysts increases with increasing the normalized amount of surface Co^{3+} cations (Fig. 1.11), clearly suggesting that they play a key role in improving the catalytic activity.

A series of spinel MCo_2O_4 (M = Co, Ni, Cu) hollow mesoporous spheres (HMS) was prepared by solvothermal alcoholysis, in order to induce a cation-substituting effect on Co_3O_4 spinel with the aim of improving the activity in total oxidation of VOC [137]. The cubic spinel was the main phase present in all the catalysts, with tiny amounts of NiO and CuO side phases in NiHMS and CuHMS, respectively. It has been shown that the number of the defective sites determined by Raman spectroscopy, the surface O_{ads}/O_{latt} and Co^{3+}/Co^{2+} molar ratios, and the catalyst reducibility, both in terms of hydrogen consumption and easiness of reduction, increased in the following order: CoHMS < NiHMS < CuHMS. Accordingly, the activity in

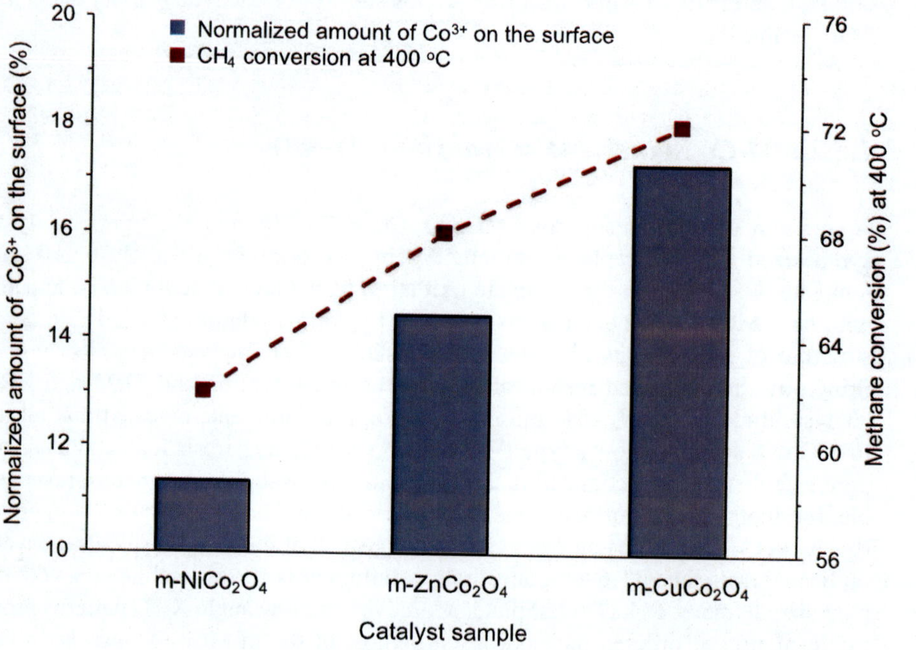

Fig. 1.11 Correlation between the catalytic activity in methane combustion and the normalized amount of Co^{3+} cations on the surface of m-$NiCo_2O_4$, m-$ZnCo_2O_4$, and m-$CuCo_2O_4$ spinel catalysts (Reaction conditions: 1500 ppm CH_4, 12.12 vol. % O_2 and balancing N_2, total flow rate of 200 mL min^{-1}, GHSV = 60,000 mL g_{cat}^{-1} h^{-1}). (Data from ref. [136])

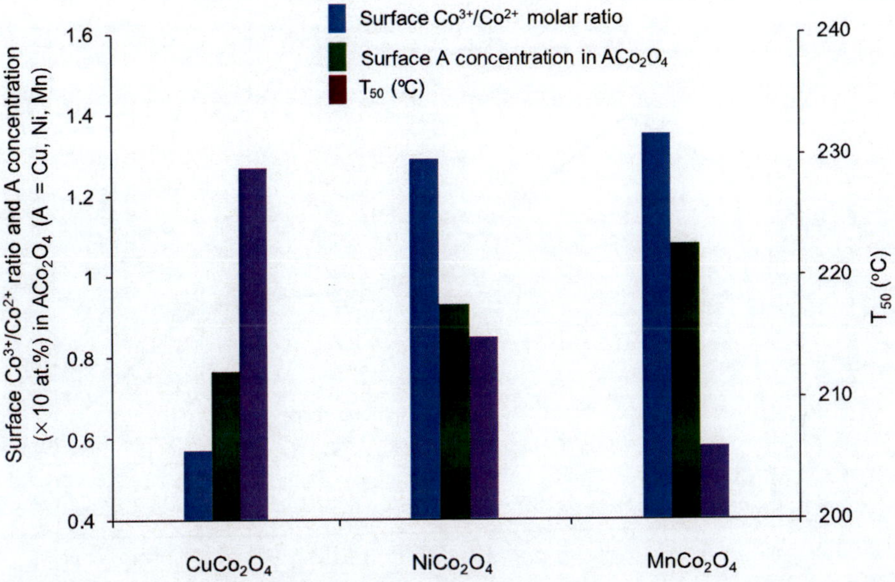

Fig. 1.12 Co^{3+}/Co^{2+} surface ratio and the surface concentration of A cation in ACo_2O_4 (A = Cu, Ni, Mn) catalysts and their activity (in terms of T_{50} values) in the complete oxidation of benzene (Reaction conditions: 1000 ppm C_6H_6 in air; GHSV = 90,000 mL g_{cat}^{-1} h^{-1}). (Data from ref. [118])

the total oxidation of acetone as model VOC followed the same order. The most active catalyst, CuHMS, also showed long-term stability and good water tolerance.

Nanocrystalline ordered mesoporous ACo_2O_4 (A = Cu, Ni and Mn) catalysts prepared by a simple co-nanocasting method using SBA-15 as a hard template were studied in benzene total oxidation reaction [118]. It has been shown that both the surface A content and Co^{3+}/Co^{2+} molar ratio in ACo_2O_4 decrease as follows: $MnCo_2O_4$ > $NiCo_2O_4$ > $CuCo_2O_4$. These characteristics were correlated with the catalytic performance in benzene total oxidation (Fig. 1.12), $MnCo_2O_4$ being the best catalyst in this series. Moreover, it showed high thermal stability and good tolerance against water vapor. It is worth noting that the oxidation of benzene over these ACo_2O_4 catalysts involves both the Mars-van Krevelen and Langmuir-Hinshelwood mechanisms [118].

A series of $ZrO_2(x)$-Co_3O_4 mixed oxide catalysts with $x = 0$; 0.5; 2; 5; 8; and 10 wt. % and forming a solid solution with the spinel structure was prepared by coprecipitation and tested in the catalytic combustion of lean methane at low temperature [138]. Their BET surface area increases monotonous with increasing ZrO_2 content, while the crystallite size decreases. The surface O_{ads}/O_{latt} molar ratio increased with increasing ZrO_2 content from 0 to 2% and then decreased with further increasing ZrO_2. This is consistent with the variation of the catalytic activity in methane combustion which passes through a maximum for 2 wt % ZrO_2 (Fig. 1.13). Similarly, the surface ratio between the Co^{2+} in tetrahedral coordination and Co^{3+} in octahedral coordination passes through a maximum for the $ZrO_2(x)$-Co_3O_4 system

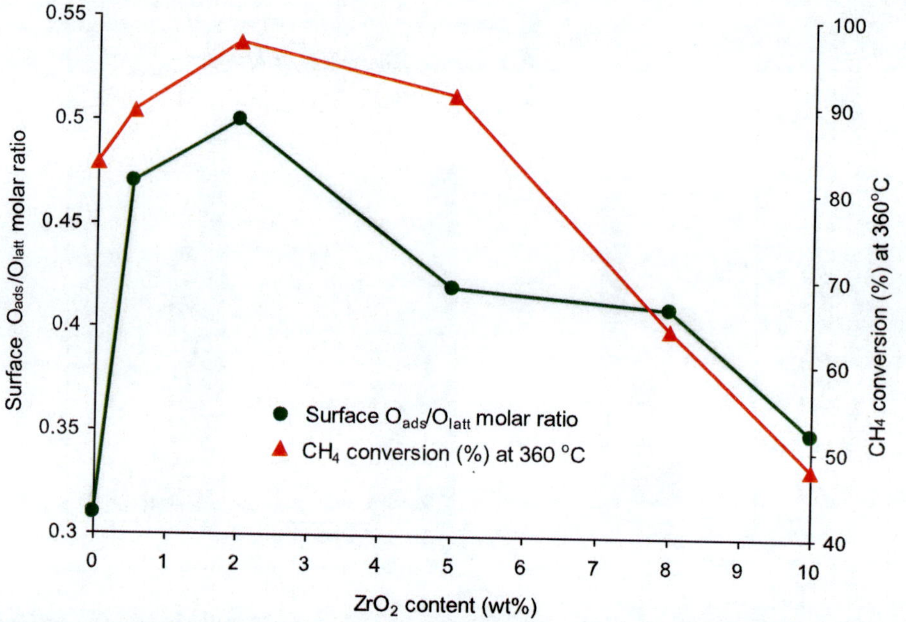

Fig. 1.13 Variation of surface O_{ads}/O_{latt} molar ratio and of the catalytic activity in methane combustion as a function of ZrO_2 content in $ZrO_2(x)$-Co_3O_4 catalysts (Reaction conditions: 0.5% CH_4, 8.0% O_2, 91.5% N_2; GHSV = 12,000 mL g_{cat}^{-1} h^{-1}). (Data from ref. [138])

with $x = 2\%$, suggesting that the Co^{2+} tetrahedral sites also play an important role in methane combustion (Fig. 1.14). Notably, no deactivation phenomenon was observed during the catalytic tests, proving the superior stability of these catalysts.

A series of Cu-containing mixed oxides with spinel structure, i.e., $CuCo_2O_4$, $CuMn_2O_4$, and $CuCr_2O_4$, prepared by a sol-gel combustion method, was studied in the catalytic oxidation of 2-propanol as a model molecule of oxygenated VOC [139]. Cu-Cr mixed oxide was pure $CuCr_2O_4$ spinel, while Cu-Mn and Cu-Co mixed oxides consisted of mixtures of spinel and copper oxide phases. The catalysts containing both spinel and CuO phases, i.e., Cu-Mn and Cu-Co mixed oxides, were shown to be more active than the pure $CuCr_2O_4$ spinel. This behavior was attributed to a synergistic cooperation between CuO and spinel phases which enhances the catalytic performance.

The different Co-based spinel catalysts hereby described, their preparation methods, and their catalytic applications together with the best performance are summarized in Table 1.3.

The main conclusion of the analysis presented above is that the most important factors controlling the catalytic performance of Co-based spinels in the complete oxidation reaction of VOC are the Co^{3+}/Co^{2+} surface mol ratio and the ratio of the surface-adsorbed oxygen to lattice oxygen. They are influenced by the method of preparation used, including the procedure and conditions adopted for a given

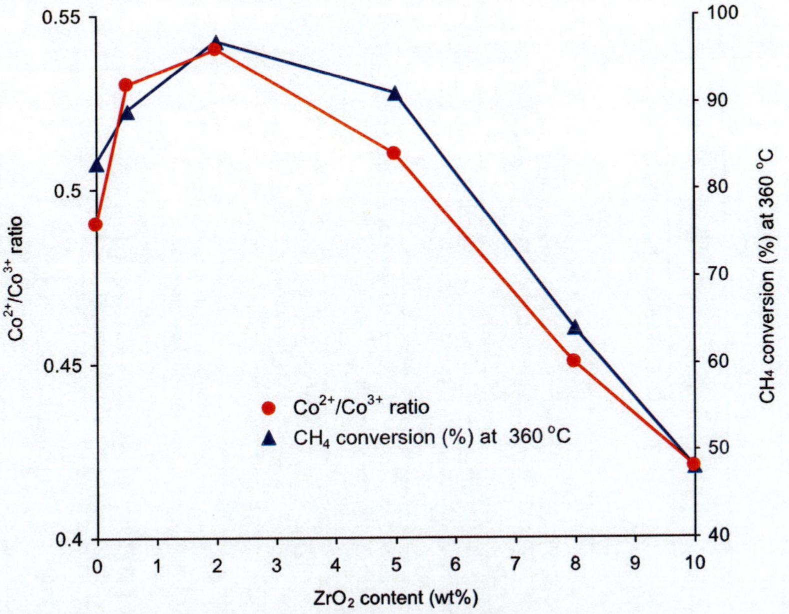

Fig. 1.14 Variation of Co^{2+}/Co^{3+} surface ratio and of the catalytic activity in methane combustion as a function of ZrO_2 content in ZrO_2(x)-Co_3O_4 catalysts (Reaction conditions: 0.5% CH_4, 8.0% O_2, 91.5% N_2; GHSV = 12,000 mL g_{cat}^{-1} h^{-1}). (Data from ref. [138])

method, and by the nature and surface concentration of cation M in MCo_2O_4 cobaltites.

1.5 General Conclusion

Fe-, Mn-, and Co-based spinel oxides catalysts show high catalytic performances, including stability on stream, for the total oxidation of a wide range of VOC, and some of them were shown to have activities close to that of the noble metal catalysts. Their catalytic properties are complex functions of their bulk and surface characteristics which are determined by their composition and preparation method. Indeed, the method of preparation used to synthesize the spinel oxides and, for a given method, the nature of the starting materials and reagents, the pH, the aging conditions as well as the thermal treatment of the catalyst precursors and the post-synthesis modification strongly influence their phase purity and homogeneity, textural properties, particle size and morphology, cation distribution in the spinel structure, density of surface defects and reduction ability and, hence, their catalytic performance. Also, the redox and electronic properties of the spinels, which are key factors controlling their activity in total oxidation reaction, can be tuned by varying their chemical composition. Indeed, the nature of both cations A and M in AM_2O_4

Table 1.3 Preparation and catalytic performance of the cobaltite spinel catalysts in VOC total oxidation

Catalysts	Preparation Method	Tested chemical compound (reaction conditions)	Best results	Ref.
Co_3O_4 nanocrystals with different morphologies	HT[a] method	Methane (0.2 vol. % CH_4 in air; GHSV = 110,000 h^{-1})	T_{50} = 346 °C on Co_3O_4 with flower-like morphology	[119]
Co_3O_4 with different morphologies	Modified PP[b]/HT[a]	Formic aldehyde (100 ppm HCHO, 21 vol. % O_2 and balancing N_2; GHSV = 69,000 h^{-1})	T_{50} = 72 °C on Co_3O_4 with rod-like morphology	[120]
Nano-replicated ordered Co_3O_4 on KIT-6 hard templates synthesized at different aging and calcination temperatures	NR[c] using mesoporous silica KIT-6	Propane (8000 ppm C_3H_8 in air; GHSV = 12,000 mL g_{cat}^{-1} h^{-1}) Toluene (100 ppm C_7H_8 in air; GHSV = 60,000 mL g_{cat}^{-1} h^{-1})	T_{50} = 196 and 164 °C for propane and toluene, respectively, on Co_3O_4 replicated on KIT-6 aged and calcined at 100 °C and 550 °C, respectively	[121]
Ordered mesoporous Co_3O_4/bulk Co_3O_4	NR[c] using SBA-15 or KIT-6/CoPP[d]	Dibromomethane (500 ppm CH_2Br_2, 10% O_2 and balancing N_2; GHSV = 112,500 cm^3 g_{cat}^{-1} h^{-1})	T_{50} = 235 °C on Co_3O_4-K (synthesized using KIT-6 template)	[122]
Co_3O_4	PP[b] at different pH values	Methane (0.5 vol. % CH_4, 8 vol. % O_2 and balancing N_2; GHSV = 18,000 mL g_{cat}^{-1} h^{-1})	T_{50} = 344 °C on Co_3O_4-9.0 (synthesized at pH = 9.0)	[123]
Nanosized Co_3O_4 catalysts	DP[e]/conventional PP[b]	Propane (1000 ppm C_3H_8 in air; GHSV = 24,000 h^{-1}) CO (10,000 ppm CO in air; GHSV = 12,000 h^{-1})	T_{50} = 194 and 83 °C for propane and CO, respectively, on nanosized Co_3O_4 prepared by DP	[124]
Co_3O_4 deposited on wire gauze and metallic sheets carriers	NEP[f], LB[g], and IMP[h]	*n*-hexane (0.75 mol. % C_6H_{14} in air; Total flow = 100 mL min^{-1})	Best results obtained on Co_3O_4 deposited by NEP on wire gauze carrier	[125]
Structured cobalt catalysts, deposited on metallic supports	NEP[f] under different conditions	*n*-nonane (2000 ppm C_9H_{20}, 20 vol. % O_2 and balancing He; Total flow = 25 mL min^{-1})	T_{50} = 320 °C on the catalyst prepared under oxygen-free conditions, with the highest value of the monomer flow	[133]
Co_3O_4/γ-Al_2O_3 with different Co_3O_4 loadings (5, 10, 30, and 50 wt%)	IWI[i] and a combination of IWI and CS[j]	Methane (0.2 vol. % CH_4, 10 vol. % O_2, and balancing N_2; GHSV = 36,000 mL g_{cat}^{-1} h^{-1})	T_{50} = 371 °C on Co_3O_4/γ-Al_2O_3 with 30 wt% Co_3O_4 prepared by IWI/CS	[134]
Nanoflower-like nickel foam-supported Co_3O_4 monolith	In situ ED[k] using different deposition times	Propane (1000 ppm C_3H_8 in air; GHSV = 30,000 mL g_{cat}^{-1} h^{-1})	Best results obtained on Co-NF-300 s (deposition time = 300 s)	[135]

(continued)

Table 1.3 (continued)

Catalysts	Preparation Method	Tested chemical compound (reaction conditions)	Best results	Ref.
Co_3O_4, CoO, $CoAl_2O_4$, $ZnCo_2O_4$, and $CoFe_2O_4$	NC[l] using mesoporous silica KIT-6	Benzene (498 ppm C_6H_6, 20 vol. % O_2 and balancing N_2; GHSV = 90,000 mL g_{cat}^{-1} h^{-1})	T_{50} = 196 °C on CoO	[3]
Mesoporous M cobaltite (M = Cu, Zn, and Ni)	NR[c] using mesoporous silica KIT-6	Methane (1500 ppm CH_4, 12.12 vol. % O_2 and balancing N_2; GHSV = 60,000 mL g_{cat}^{-1} h^{-1})	T_{50} = 369 °C on mesoporous $CuCo_2O_4$	[136]
M cobaltite (M = Co, Ni, Cu) hollow mesoporous spheres (HMS)	Solvothermal alcoholysis	Acetone (1000 ppm acetone, 20% O_2 and balancing N_2; WHSV = 93,000 mL g_{cat}^{-1} h^{-1})	T_{50} = 157 °C on $CuCo_2O_4$ (CuHSM)	[137]
Mesoporous ACo_2O_4 (A = Cu, Ni, and Mn)	NR[c] using SBA-15	Benzene (1000 ppm C_6H_6 in air; GHSV = 90,000 mL g_{cat}^{-1} h^{-1})	T_{50} = 206 °C on $MnCo_2O_4$	[118]
$ZrO_2(x)$-Co_3O_4 (x = 0; 0.5; 2; 5; 8; and 10 wt.%)	CoPP[d]	Methane (0.5% CH_4, 8.0% O_2, 91.5% N_2; GHSV = 12,000 mL g_{cat}^{-1} h^{-1})	T_{50} = 279 °C on $ZrO_2(2)$-Co_3O_4	[138]
Cu cobaltite, Cu manganite, Cu chromite	SG[m] + SelfC[n]	2-propanol (0.085 mol mol^{-1} C_3H_8O in air; GHSV = 2400 h^{-1})	T_{50} = 150 °C on $CuCo_2O_4$	[139]

[a]*HT* hydrothermal method
[b]*PP* precipitation
[c]*NR* nano-replication
[d]*CoPP* coprecipitation
[e]*DP* dispersion-precipitation
[f]*NEP* nonequilibrium plasma deposition
[g]*LB* Langmuir–Blogett method
[h]*IMP* impregnation
[i]*IWI* incipient wetness impregnation
[j]*CS* combustion synthesis
[k]*ED* electrodeposition
[l]*NC* nanocasting method
[m]*SG* sol-gel method
[n]*SelfC* self-combustion

(M = Fe, Mn and Co) binary spinels and the addition of a third cation, which can lead to a ternary spinel or to phase separation as a function of its ionic size, strongly influence the catalytic behavior due to synergistic effects between the different cations or phases rationalized on the basis of enhanced electron transfer and oxygen nonstoichiometry.

Finally, through the presented examples, this chapter clearly demonstrates the high potential of Fe-, Mn-, and Co-based spinels as total oxidation catalysts for VOC.

References

1. C. He, J. Cheng, X. Zhang, M. Douthwaite, S. Pattisson, Z. Hao, Recent advances in the catalytic oxidation of volatile organic compounds: a review based on pollutant sorts and sources. Chem. Rev. **119**, 4471–4568 (2019). https://doi.org/10.1021/acs.chemrev.8b00408
2. P. Kuśtrowski, A. Rokicińska, T. Kondratowicz, Abatement of volatile organic compounds emission as a target for various human activities including energy production. Adv. Inorg. Chem. **72**, 385–419 (2018). https://doi.org/10.1016/bs.adioch.2018.05.004
3. X. Wang, Y. Liu, T. Zhang, Y. Luo, Z. Lan, K. Zhang, J. Zuo, L. Jiang, R. Wang, Geometrical-site-dependent catalytic activity of ordered mesoporous Co-based spinel for benzene oxidation: in situ DRIFTS study coupled with Raman and XAFS spectroscopy. ACS Catal. **7**, 1626–1636 (2017). https://doi.org/10.1021/acscatal.6b03547
4. M. Răciulete, G. Layrac, F. Papa, C. Negrilă, D. Tichit, I.C. Marcu, Influence of Mn content on the catalytic properties of Cu-(Mn)-Zn-Mg-Al mixed oxides derived from LDH precursors in the total oxidation of methane. Catal. Today **306**, 276–286 (2018). https://doi.org/10.1016/j.cattod.2017.01.013
5. A. Urdă, A. Herraïz, Á. Rédey, I.C. Marcu, Co and Ni ferrospinels as catalysts for propane total oxidation. Catal. Commun. **10**, 1651–1655 (2009). https://doi.org/10.1016/j.catcom.2009.05.002
6. C. Munteanu, M. Caldararu, D. Gingasu, M. Feder, L. Diamandescu, N.I. Ionescu, In situ electrical conductivity and catalytic properties of copper ferrite synthesized by non-conventional methods. Reac. Kinet. Mech. Catal. **104**, 357–368 (2011). https://doi.org/10.1007/s11144-011-0357-5
7. N. Rezlescu, E. Rezlescu, P.D. Popa, C. Doroftei, M. Ignat, Scandium substituted nickel–cobalt ferrite nanoparticles for catalyst applications. Appl. Catal. B Environ. **158-159**, 70–75 (2014). https://doi.org/10.1016/j.apcatb.2014.03.052
8. C. Doroftei, L. Leontie, Synthesis and characterization of some nanostructured composite oxides for low temperature catalytic combustion of dilute propane. RSC Adv. **7**, 27863–27871 (2017). https://doi.org/10.1039/c7ra03916f
9. W.B. Li, J.X. Wang, H. Gong, Catalytic combustion of VOCs on non-noble metal catalysts. Catal. Today **148**, 81–87 (2009). https://doi.org/10.1016/j.cattod.2009.03.007
10. M. Tomatis, H.H. Xu, J. He, X.D. Zhang, Recent development of catalysts for removal of volatile organic compounds in flue gas by combustion: a review. J. Chem. **2016**, 8324826 (2016). https://doi.org/10.1155/2016/8324826
11. M.S. Kamal, S.A. Razzak, M.M. Hossain, Catalytic oxidation of volatile organic compounds (VOCs)—a review. Atmos. Environ. **140**, 117–134 (2016). https://doi.org/10.1016/j.atmosenv.2016.05.031
12. J. Chen, H. Arandiyan, X. Gao, J. Li, Recent advances in catalysts for methane combustion. Catal. Surv. Jpn. **19**, 140–171 (2015). https://doi.org/10.1007/s10563-015-9191-5

13. X.D. Zhang, Y. Wang, Y.Q. Yang, D. Chen, Recent progress in the removal of volatile organic compounds by mesoporous silica materials and supported catalysts. Acta Phys. Chim. Sin. **31**, 1633–1646 (2015). https://doi.org/10.3866/PKU.WHXB201507281
14. Y. Liu, J. Deng, S. Xie, Z. Wang, H. Dai, Catalytic removal of volatile organic compounds using ordered porous transition metal oxide and supported noble metal catalysts. Chin. J. Catal. **37**, 1193–1205 (2016). https://doi.org/10.1016/S1872-2067(16)62457-9
15. Z. Ma, Cobalt oxide catalysts for environmental remediation. Curr. Catal. **3**, 15–26 (2014). https://doi.org/10.2174/2211544711302999001 7
16. P. Mohapatra, T. Mishra, K.M. Parida, Pillared clay as an effective catalyst for low temperature VOCs decomposition. Key Eng. Mater. **571**, 71–91 (2013). https://doi.org/10.4028/www.scientific.net/KEM.571.71
17. J. Li, M. Hu, S. Zuo, X. Wang, Catalytic combustion of volatile organic compounds on pillared interlayered clay (PILC)-based catalysts. Curr. Opin. Chem. Eng. **20**, 93–98 (2018). https://doi.org/10.1016/j.coche.2018.02.001
18. I.C. Marcu, A. Urdă, I. Popescu, V. Hulea, Layered double hydroxides-based materials as oxidation catalysts (Ch. 3), in *Sustainable Nanosystems Development, Properties and Applications*, ed. by M. V. Putz, M. C. Mirica, (IGI Global, Hershey, PA, 2017), pp. 59–121. https://doi.org/10.4018/978-1-5225-0492-4.ch003
19. A.R. West, *Solid State Chemistry and Its Application*, 2nd edn. (Wiley, 2014), pp. 66–70
20. A. Ahlawat, V.G. Sathe, V.R. Reddy, A. Gupta, Mossbauer, Raman and X-ray diffraction studies of superparamagnetic $NiFe_2O_4$ nanoparticles prepared by sol–gel auto-combustion method. J. Magn. Magn. Mater. **323**, 2049–2054 (2011). https://doi.org/10.1016/j.jmmm.2011.03.017
21. J. Chandradass, A.H. Jadhav, K.H. Kim, H. Kim, Influence of processing methodology on the structural and magnetic behavior of $MgFe_2O_4$ nanopowders. J. Alloys Compd. **517**, 164–169 (2012). https://doi.org/10.1016/j.jallcom.2011.12.071
22. H. Sozeri, Z. Durmus, A. Baykal, Structural and magnetic properties of triethylene glycol stabilized $Zn_xCo_{1-x}Fe_2O_4$ nanoparticles. Mater. Res. Bull. **47**, 2442–2448 (2012). https://doi.org/10.1016/j.materresbull.2012.05.036
23. B.I. Kharisov, H.V. Rasika Dias, O.V. Kharissova, Mini-review: ferrite nanoparticles in the catalysis. Arab. J. Chem. **12**, 1234–1246 (2019). https://doi.org/10.1016/j.arabjc.2014.10.049
24. R.K. Selvan, C.O. Augustin, L.J. Berchmans, R. Saraswathi, Combustion synthesis of $CuFe_2O_4$. Mater. Res. Bull. **38**, 41–54 (2003). https://doi.org/10.1016/S0025-5408(02)01004-8
25. K. Kamala Bharathi, C.V. Ramana, Improved electrical and dielectric properties of La-doped Co ferrite. J. Mater. Res. **26**, 584–591 (2011). https://doi.org/10.1557/jmr.2010.37
26. T. Brylewski, A. Kruk, A. Adamczyk, W. Kucza, M. Stygar, K. Przybylski, Synthesis and characterization of the manganese cobaltite spinel prepared using two "soft chemical" methods. Mater. Chem. Phys. **137**, 310–316 (2012). https://doi.org/10.1016/j.matchemphys.2012.09.026
27. N. Iftimie, E. Rezlescu, P.D. Popa, N. Rezlescu, On the possibility of the use of a nickel ferrite as semiconducting gas sensor. J. Optoelectron. Adv. Mater. **7**, 911–914 (2005)
28. M.A. Valenzuela, P. Bosch, J. Jiménez-Becerrill, O. Quiroz, A.I. Páez, Preparation, characterization and photocatalytic activity of ZnO, Fe_2O_3 and $ZnFe_2O_4$. J. Photochem. Photobiol. A Chem. **148**, 177–182 (2002). https://doi.org/10.1016/S1010-6030(02)00040-0
29. E. Casbeer, V.K. Sharma, X.Z. Li, Synthesis and photocatalytic activity of ferrites under visible light: a review. Sep. Purif. Technol. **87**, 1–14 (2012). https://doi.org/10.1016/j.seppur.2011.11.034
30. P. Guo, G. Zhang, J. Yu, H. Li, X.S. Zhao, Controlled synthesis, magnetic and photocatalytic properties of hollow spheres and colloidal nanocrystal clusters of manganese ferrite. Colloid Surf. A Physicochem. Eng. Asp. **395**, 168–174 (2012). https://doi.org/10.1016/j.colsurfa.2011.12.027

31. M. Su, C. He, V.K. Sharma, M.A. Asi, D. Xia, X. Li, H. Deng, Y. Xiong, Mesoporous zinc ferrite: Synthesis, characterization, and photocatalytic activity with H_2O_2/visible light. J. Hazard. Mater. **211–212**, 95–103 (2012). https://doi.org/10.1016/j.jhazmat.2011.10.006
32. R. Voda, A. Negrea, L. Lupa, M. Ciopec, P. Negrea, C.M. Davidescu, M. Butnariu, Nanocrystalline ferrites used as adsorbent in the treatment process of waste waters resulted from ink jet cartridges manufacturing. Open Chem. **13**, 743–747 (2015). https://doi.org/10.1515/chem-2015-0092
33. K.K. Kefeni, B.B. Mamba, T.A.M. Msagati, Application of spinel ferrite nanoparticles in water and wastewater treatment: a review. Sep. Purif. Technol. **188**, 399–422 (2017). https://doi.org/10.1016/j.seppur.2017.07.015
34. J. Tong, L. Bo, Z. Li, Z. Lei, C. Xia, Magnetic $CoFe_2O_4$ nanocrystal: a novel and efficient heterogeneous catalyst for aerobic oxidation of cyclohexane. J. Mol. Catal. A Chem. **307**, 58–63 (2009). https://doi.org/10.1016/j.molcata.2009.03.010
35. T. Tsoncheva, E. Manova, N. Velinov, D. Paneva, M. Popova, B. Kunev, K. Tenchev, I. Mitov, Thermally synthesized nanosized copper ferrites as catalysts for environment protection. Catal. Commun. **12**, 105–109 (2010). https://doi.org/10.1016/j.catcom.2010.08.007
36. R. Benrabaa, A. Löfberg, A. Rubbens, E. Bordes-Richard, R.N. Vannier, A. Barama, Structure, reactivity and catalytic properties of nanoparticles of nickel ferrite in the dry reforming of methane. Catal. Today **203**, 188–195 (2013). https://doi.org/10.1016/j.cattod.2012.06.002
37. N. Rezlescu, E. Rezlescu, P.D. Popa, E. Popovici, C. Doroftei, M. Ignat, Preparation and characterization of spinel-type $MeFe_2O_4$ (Me = Cu, Cd, Ni and Zn) for catalyst applications. Mater. Chem. Phys. **137**, 922–927 (2013). https://doi.org/10.1016/j.matchemphys.2012.11.005
38. I. Popescu, A. Boudjemaa, N. Helaili, Y. Bessekhouad, M. Tudorache, K. Bachari, I.C. Marcu, Study of the electrical and catalytic properties of spinels with $CuFe_{2-x}Mn_xO_4$ composition (x = 0, 0.4, 0.8, 1.6 and 2). Appl. Catal. A Gen. **504**, 29–36 (2015). https://doi.org/10.1016/j.apcata.2014.09.048
39. Y. Wang, R. Xue, C. Zhao, F. Liu, C. Liu, F. Han, Effects of Ce in the catalytic combustion of toluene on $Cu_xCe_{1-x}Fe_2O_4$. Colloid Surf. A Physicochem. Eng. Asp. **540**, 90–97 (2018). https://doi.org/10.1016/j.colsurfa.2017.12.067
40. H. Zhang, R. Qi, D.G. Evans, X. Duan, Synthesis and characterization of a novel nano-scale magnetic solid base catalyst involving a layered double hydroxide supported on a ferrite core. J. Solid State Chem. **177**, 772–780 (2004). https://doi.org/10.1016/j.jssc.2003.09.009
41. T. Tatarchuk, B. Al-Najar, M. Bououdina, M.A. Aal Ahmed, Catalytic and photocatalytic properties of oxide spinels, in *Handbook of Ecomaterials*, ed. by L. M. Torres Martinez, O. V. Kharissova, B. I. Kharisov, vol. 3, (Springer, Cham, 2019), pp. 1701–1750. https://doi.org/10.1007/978-3-319-48281-1_158-1
42. D. Fino, S. Solaro, N. Russo, G. Saracco, V. Specchia, Catalytic removal of methane over thermal-proof nanostructured catalysts for CNG engines. Top. Catal. **42–43**, 449–454 (2007). https://doi.org/10.1007/s11244-007-0223-x
43. F. Papa, L. Patron, O. Carp, C. Paraschiv, I. Balint, Catalytic activity of neodymium substituted zinc ferrites for oxidative conversion of methane. J. Mol. Catal. A Chem. **299**, 93–97 (2009). https://doi.org/10.1016/j.molcata.2008.10.036
44. N. Velinov, K. Koleva, T. Tsoncheva, E. Manova, D. Paneva, K. Tenchev, B. Kunev, I. Mitov, Nanosized $Cu_{0.5}Co_{0.5}Fe_2O_4$ ferrite as catalyst for methanol decomposition: effect of preparation procedure. Catal. Commun. **32**, 41–46 (2013). https://doi.org/10.1016/j.catcom.2012.12.002
45. C. Trevisanut, M. Mari, J.M.M. Millet, F. Cavani, Chemical-loop reforming of ethanol over metal ferrites: an analysis of structural features affecting reactivity. Int. J. Hydrog. Energy **40**, 5264–5271 (2015). https://doi.org/10.1016/j.ijhydene.2015.01.054
46. H. Lee, J.C. Jung, H. Kim, Y.M. Chung, T.J. Kim, S.J. Lee, S.H. Oh, Y.S. Kim, I.K. Song, Preparation of $ZnFe_2O_4$ catalysts by co-precipitation method using aqueous buffer solution and their catalytic activity for oxidative dehydrogenation of n-butene to 1,3-butadiene. Catal. Lett. **122**, 281–286 (2008). https://doi.org/10.1007/s10562-007-9371-7

47. J.A. Toledo-Antonio, N. Nava, M. Martínez, X. Bokhimi, Correlation between the magnetism of non-stoichiometric zinc ferrites and their catalytic activity for oxidative dehydrogenation of 1-butene. Appl. Catal. A Gen. **234**, 137–144 (2002). https://doi.org/10.1016/S0926-860X(02)00212-0
48. D. Guin, B. Baruwati, S.V. Manorama, A simple chemical synthesis of nanocrystalline AFe_2O_4 (A = Fe, Ni, Zn): an efficient catalyst for selective oxidation of styrene. J. Mol. Catal. A Chem. **242**, 26–31 (2005). https://doi.org/10.1016/j.molcata.2005.07.021
49. P.P. Hankare, U.B. Sankpal, R.P. Patil, P.D. Lokhande, R. Sasikal, Synthesis, characterization and catalytic activity of chromium substituted cobalt ferrospinels. Mater. Sci. Eng. B Solid State Mater. Adv. Technol. **176**, 103–109 (2011). https://doi.org/10.1016/j.mseb.2010.10.005
50. M.M. Rashad, O.A. Fouad, Synthesis and characterization of nano-sized nickel ferrites from fly ash for catalytic oxidation of CO. Mater. Chem. Phys. **94**, 365–370 (2005). https://doi.org/10.1016/j.matchemphys.2005.05.028
51. O.A. Fouad, K.S. Abdel Halim, M.M. Rashad, Catalytic oxidation of CO over synthesized nickel ferrite nanoparticles from fly ash. Top. Catal. **47**, 61–65 (2008). https://doi.org/10.1007/s11244-007-9034-3
52. K.S. Abdel Halim, A.M. Ismail, M.H. Khedr, M.F. Abadir, Catalytic oxidation of CO gas over nanocrystallite $Cu_xMn_{1-x}Fe_2O_4$. Top. Catal. **47**, 66–72 (2008). https://doi.org/10.1007/s11244-007-9031-6
53. C.U. Aniz, T.D.R. Nair, A study on catalysis by ferrospinels for preventing atmospheric pollution from carbon monoxide. Open J. Phys. Chem. **1**, 124–130 (2011). https://doi.org/10.4236/ojpc.2011.13017
54. J.P. Jacobs, A. Maltha, J.G.H. Reintjes, J. Drimal, V. Ponec, H.H. Brongersma, The surface of catalytically active spinels. J. Catal. **147**, 294–300 (1994). https://doi.org/10.1006/jcat.1994.1140
55. R. Amrousse, T. Katsumi, Substituted ferrite $M_xFe_{1-x}Fe_2O_4$ (M = Mn, Zn) catalysts for N_2O catalytic decomposition processes. Catal. Commun. **26**, 194–198 (2012). https://doi.org/10.1016/j.catcom.2012.05.024
56. N. Hellaili, G. Mitran, I. Popescu, K. Bachari, I.C. Marcu, A. Boudjemaa, Photoelectrochemical properties of AFe_2O_4 (A = Co, Cu, Zn) ferrospinels for water photo-reduction. J. Electroanal. Chem. **742**, 47–53 (2015). https://doi.org/10.1016/j.jelechem.2015.01.018
57. A. Boudjemaa, I. Popescu, T. Juzsakova, M. Kebir, N. Helaili, K. Bachari, I.C. Marcu, M-substituted (M = Co, Ni and Cu) zinc ferrite photo-catalysts for hydrogen production by water photo-reduction. Int. J. Hydrog. Energy **41**, 11108–11118 (2016). https://doi.org/10.1016/j.ijhydene.2016.04.088
58. F. Li, J. Liu, D.G. Evans, X. Duan, Stoichiometric synthesis of pure MFe_2O_4 (M = Mg, Co and Ni) spinel ferrites from tailored layered double hydroxide (hydrotalcite-like) precursors. Chem. Mater. **16**, 1597–1602 (2004). https://doi.org/10.1021/cm035248c
59. J.B. Silva, C.F. Diniz, R.M. Lago, N.D.S. Mohallem, Catalytic properties of nanocomposites based on cobalt ferrites dispersed in sol-gel silica. J. Non-Cryst. Solids **348**, 201–204 (2004). https://doi.org/10.1016/j.jnoncrysol.2004.08.169
60. Y. Hammiche-Bellal, A. Benadda, L. Meddour-Boukhobza, S. Barama, A. Djadoun, A. Barama, Preparation and catalytic activity in ethanol combustion reaction of cobalt–iron spinel catalysts. Catal. Commun. **42**, 62–67 (2013). https://doi.org/10.1016/j.catcom.2013.07.042
61. R. Dumitru, F. Papa, I. Balint, D.C. Culita, C. Munteanu, N. Stanica, A. Ianculescu, L. Diamandescu, O. Carp, Mesoporous cobalt ferrite: a rival of platinum catalyst in methane combustion reaction. Appl. Catal. A Gen. **467**, 178–186 (2013). https://doi.org/10.1016/j.apcata.2013.07.013
62. A. Evdou, V. Zaspalis, L. Nalbandian, Ferrites as redox catalysts for chemical looping processes. Fuel **165**, 367–378 (2016). https://doi.org/10.1016/j.fuel.2015.10.049

63. T. Lazarova, D. Kovacheva, M. Georgieva, D. Tzankov, G. Tyuliev, I. Spassova, A. Naydenov, Tunable nanosized spinel manganese ferrites synthesized by solution combustion method. Appl. Surf. Sci. **496**, 143571 (2019). https://doi.org/10.1016/j.apsusc.2019.143571
64. J.E. Tasca, C.E. Quincoces, A. Lavat, A.M. Alvarez, M.G. González, Preparation and characterization of $CuFe_2O_4$ bulk catalysts. Ceram. Int. **37**, 803–812 (2011). https://doi.org/10.1016/j.ceramint.2010.10.023
65. Z.Y. Tian, P.M. Kouotou, A. El Kasmi, P.H.T. Ngamou, K. Kohse-Höinghaus, H. Vieker, A. Beyer, A. Gölzhäuser, Low-temperature deep oxidation of olefins and DME over cobalt ferrite. Proc. Combus. Inst. **35**, 2207–2214 (2015). https://doi.org/10.1016/j.proci.2014.06.111
66. N. Rezlescu, E. Rezlescu, P.D. Popa, C. Doroftei, M. Ignat, Some nanograined ferrites and perovskites for catalytic combustion of acetone at low temperature. Ceram. Int. **41**, 4430–4437 (2015). https://doi.org/10.1016/j.ceramint.2014.11.134
67. M. Florea, M. Alifanti, V.I. Parvulescu, D. Mihaila-Tarabasanu, L. Diamandescu, M. Feder, C. Negrila, L. Frunza, Total oxidation of toluene on ferrite-type catalysts. Catal. Today **141**, 361–366 (2009). https://doi.org/10.1016/j.cattod.2008.05.005
68. M.T. Rahman, C.V. Ramana, Impedance spectroscopic characterization of gadolinium substituted cobalt ferrite ceramics. J. Appl. Phys. **116**, 164108 (2014). https://doi.org/10.1063/1.4896945
69. S.J. Lee, C.C.H. Lo, P.N. Matlage, S.H. Song, Y. Melikhov, J.E. Snyder, D.C. Jiles, Magnetic and magnetoelastic properties of Cr-substituted cobalt ferrite. J. Appl. Phys. **102**, 073910 (2007). https://doi.org/10.1063/1.2794711
70. M.V. Chaudhari, S.E. Shirsath, A.B. Kadam, R.H. Kadam, S.B. Shelke, D.R. Mane, Site occupancies of Co–Mg–Cr–Fe ions and their impact on the properties of $Co_{0.5}Mg_{0.5}Cr_xFe_{2-x}O_4$. J. Alloy. Compd. **552**, 443–450 (2013). https://doi.org/10.1016/j.jallcom.2012.11.070
71. S.S. Ata-Allah, M.K. Fayek, H.S. Refai, M.F. Mostafa, Mössbauer effect study of copper containing nickel-aluminate ferrite. J. Solid State Chem. **149**, 434–442 (2000). https://doi.org/10.1006/jssc.1999.8577
72. A. Sutka, A. Borisova, J. Kleperis, G. Mezinskis, D. Jakovlev, I. Juhnevica, Effect of nickel addition on colour of nanometer spinel zinc ferrite pigments. J. Aust. Ceram. Soc. **48**, 150–155 (2012).
73. A. Khan, P.G. Smirniotis, Relationship between temperature-programmed reduction profile and activity of modified ferrite-based catalysts for WGS reaction. J. Mol. Catal. A Chem. **280**, 43–51 (2008). https://doi.org/10.1016/j.molcata.2007.10.022
74. A.S. Albuquerque, M.V.C. Tolentino, J.D. Ardisson, F.C.C. Moura, R. de Mendonça, W.A.A. Macedo, Nanostructured ferrites: structural analysis and catalytic activity. Ceram. Int. **38**, 2225–2231 (2012). https://doi.org/10.1016/j.ceramint.2011.10.071
75. V.L. Mathe, R.B. Kamble, Anomalies in electrical and dielectric properties of nanocrystalline Ni-Co spinel ferrite. Mater. Res. Bull. **43**, 2160–2165 (2008). https://doi.org/10.1016/j.materresbull.2007.09.001
76. R. Benrabaa, H. Boukhlouf, S. Barama, E. Bordes-Richard, R.N. Vannier, A. Barama, Structural, textural and acid-base properties of nano-sized $NiFe_2O_4$ spinel catalysts. Catal. Lett. **142**, 42–49 (2012). https://doi.org/10.1007/s10562-011-0726-8
77. N. Shukla, A. Ondeck, J.C. Lee, J.B. Miller, $NiFe_2O_4$@SiO_2 nanoparticles stabilized by porous silica shells. Catal. Lett. **142**, 582–587 (2012). https://doi.org/10.1007/s10562-012-0795-3
78. Y.J. Tu, C.K. Chang, C.F. You, Combustion of isopropyl alcohol using a green manufactured $CuFe_2O_4$. J. Hazard. Mater. **229-230**, 258–264 (2012). https://doi.org/10.1016/j.jhazmat.2012.05.100
79. J.T. Feng, Y.J. Lin, F. Li, D.G. Evans, D.Q. Li, Preparation, structure and properties of microspherical alumina with magnetic spinel ferrite cores. Appl. Catal. A Gen. **329**, 112–119 (2007). https://doi.org/10.1016/j.apcata.2007.06.032
80. B. Delmon, M. Devillers, Solid state reactions, in *Handbook of Heterogeneous Catalysis*, ed. by G. Ertl, H. Knozinger, F. Schuth, J. Weitkamp, 2nd edn., (Wiley-VCH, Weinheim, 2008), pp. 295–318

81. P. Lavela, J.L. Tirado, $CoFe_2O_4$ and $NiFe_2O_4$ synthesized by sol-gel procedures for their use as anode materials for Li ion batteries. J. Power Sources **172**, 379–387 (2007). https://doi.org/10.1016/j.jpowsour.2007.07.055
82. Y. Cheng, Y. Zheng, Y. Wang, F. Bao, Y. Qin, Synthesis and magnetic properties of nickel ferrite nano-octahedra. J. Solid State Chem. **178**, 2394–2397 (2005). https://doi.org/10.1016/j.jssc.2005.05.006
83. N. Rezlescu, E. Rezlescu, L. Sachelarie, P.D. Popa, C. Doroftei, Structural and catalytic properties of mesoporous nanocrystalline mixed oxides containing magnesium. Catal. Commun. **46**, 51–56 (2014). https://doi.org/10.1016/j.catcom.2013.11.021
84. V.R. Bhagwat, A.V. Humbe, S.D. More, K.M. Jadhav, Sol-gel auto combustion synthesis and characterizations of cobalt ferrite nanoparticles: different fuels approach. Mater. Sci. Eng. B Solid State Mater. Adv. Technol. **248**, 114388 (2019). https://doi.org/10.1016/j.mseb.2019.114388
85. Y. Li, E.R. Maxey, J.W. Richardson Jr., B. Ma, Structural and chemical evolution of Fe-Co-O based ceramics under reduction/oxidation—an in situ neutron diffraction study. Mater. Sci. Eng. B Solid State Mater. Adv. Technol. **106**, 6–26 (2004). https://doi.org/10.1016/j.mseb.2003.07.004
86. O. Carp, L. Patron, G. Pascu, I. Mindru, N. Stanica, Thermal investigations of nickel-zinc ferrites formation from malate coordination compounds. J. Therm. Anal. Calorim. **84**, 391–394 (2006). https://doi.org/10.1007/s10973-005-6920-4
87. J. Zhao, L. Mi, H. Hou, X. Shi, Y. Fan, The preparation of zinc ferrite nanorods by using single ferrocenyl complex as precursor. Mater. Lett. **61**, 4196–4198 (2007). https://doi.org/10.1016/j.matlet.2007.01.053
88. I.V. Vasylenko, K.S. Gavrylenko, V.G. Il'yin, V. Golub, G. Goloverda, V. Kolesnichenko, A.W. Addison, V.V. Pavlishchuk, The metamorphosis of heterometallic trinuclear antiferromagnetic complexes into nano-sized superparamagnetic spinels. Mater. Chem. Phys. **121**, 47–52 (2010). https://doi.org/10.1016/j.matchemphys.2009.12.040
89. B. Medina, M.G. Verdério Fressati, J.M. Gonçalves, F.M. Bezerra, F.A.P. Scacchetti, M.P. Moisés, A. Bail, R.B. Samulewski, Solventless preparation of Fe_3O_4 and Co_3O_4 nanoparticles: a mechanochemical approach. Mater. Chem. Phys. **226**, 318–322 (2019). https://doi.org/10.1016/j.matchemphys.2019.01.043
90. L. Leontie, C. Doroftei, Nanostructured spinel ferrites for catalytic combustion of gasoline vapors. Catal. Lett. **147**, 2542–2548 (2017). https://doi.org/10.1007/s10562-017-2164-8
91. C.H. Yan, Z.G. Xu, F.X. Cheng, Z.M. Wang, L.D. Sun, C.S. Liao, J.T. Jia, Nanophased $CoFe_2O_4$ prepared by combustion method. Solid State Commun. **111**, 287–291 (1999). https://doi.org/10.1016/S0038-1098(99)00119-2
92. V. Pillai, D.O. Shah, Synthesis of high-coercivity cobalt ferrite particles using water-in-oil microemulsions. J. Magn. Magn. Mater. **163**, 243–248 (1996). https://doi.org/10.1016/S0304-8853(96)00280-6
93. Y. Lee, J. Lee, C.J. Bae, J.G. Park, H.J. Noh, J.H. Park, T. Hyeon, Large-scale synthesis of uniform and crystalline magnetite nanoparticles using reverse micelles as nanoreactors under reflux conditions. Adv. Funct. Mater. **15**, 503–509 (2005). https://doi.org/10.1002/adfm.200400187
94. I. Spassova, M. Khristova, D. Panayotov, D. Mehandjiev, Coprecipitated CuO-MnO_x catalysts for low-temperature CO-NO and CO-NO-O_2 reactions. J. Catal. **185**, 43–57 (1999). https://doi.org/10.1006/jcat.1998.2347
95. G. Fierro, R. Dragone, G. Ferraris, NO and N_2O decomposition and their reduction by hydrocarbons over Fe-Zn manganite spinels. Appl. Catal. B Environ. **78**, 183–191 (2008). https://doi.org/10.1016/j.apcatb.2007.09.021
96. K. Zhi, Q. Liu, Y. Zhang, S. He, R. He, Effect of precipitator on the texture and activity of copper-manganese mixed oxide catalysts for the water gas shift reaction. J. Fuel Chem. Technol. **38**, 445–451 (2010). https://doi.org/10.1016/S1872-5813(10)60038-2

97. Y.H. Huang, S.F. Wang, A.P. Tsai, S. Kameoka, Reduction behaviors and catalytic properties for methanol steam reforming of Cu-based spinel compounds CuX_2O_4 (X= Fe, Mn, Al, La). Ceram. Int. **40**, 4541–4551 (2014). https://doi.org/10.1016/j.ceramint.2013.08.130
98. G.J. Hutchings, A.A. Mirzaei, R.W. Joyner, M.R.H. Siddiqui, S.H. Taylor, Effect of preparation conditions on the catalytic performance of copper manganese oxide catalysts for CO oxidation. Appl. Catal. A Gen. **166**, 143–152 (1998). https://doi.org/10.1016/S0926-860X(97)00248-2
99. B. Faure, P. Alphonse, Co-Mn-oxide spinel catalysts for CO and propane oxidation at mild temperature. Appl. Catal. B Environ. **180**, 715–725 (2016). https://doi.org/10.1016/j.apcatb.2015.07.019
100. J. Li, L. Li, F. Wu, L. Zhang, X. Liu, Dispersion-precipitation synthesis of nanorod Mn_3O_4 with high reducibility and the catalytic complete oxidation of air pollutants. Catal. Commun. **31**, 52–56 (2013). https://doi.org/10.1016/j.catcom.2012.11.013
101. C. Lahousse, A. Bernier, P. Grange, B. Delmon, P. Papaefthimiou, T. Ionnides, X. Verykios, Evaluation of γ-MnO_2 as a VOC removal catalyst: comparison with a noble metal catalyst. J. Catal. **178**, 214–225 (1998). https://doi.org/10.1006/jcat.1998.2148
102. M. Baldi, V.S. Escribano, J.M. Gallardo Amores, F. Milella, G. Busca, Characterization of manganese and iron oxides as combustion catalysts for propane and propene. Appl. Catal. B Environ. **17**, L175–L182 (1998). https://doi.org/10.1016/S0926-3373(98)00013-7
103. M. Piumetti, D. Fino, N. Russo, Mesoporous manganese oxides prepared by solution combustion synthesis as catalysts for the total oxidation of VOCs. Appl. Catal. B Environ. **163**, 277–287 (2015). https://doi.org/10.1016/j.apcatb.2014.08.012
104. S.C. Kim, W.G. Shim, Catalytic combustion of VOCs over a series of manganese oxide catalysts. Appl. Catal. B Environ. **98**, 180–185 (2010). https://doi.org/10.1016/j.apcatb.2010.05.027
105. Y. Wang, D. Yang, S. Li, L. Zhang, G. Zheng, L. Guo, Layered copper manganese oxide for the efficient catalytic CO and VOCs oxidation. Chem. Eng. J. **357**, 258–268 (2019). https://doi.org/10.1016/j.cej.2018.09.156
106. V.H. Vu, J. Belkouch, A. Ould-Dris, B. Taouk, Catalytic oxidation of volatile organic compounds on manganese and copper oxides supported on titania. AICHE J. **54**, 1585–1591 (2008). https://doi.org/10.1002/aic.11482
107. S. Behar, P. Gonzalez, P. Agulhon, F. Quignard, D. Świerczyński, New synthesis of nanosized Cu–Mn spinels as efficient oxidation catalysts. Catal. Today **189**, 35–41 (2012). https://doi.org/10.1016/j.cattod.2012.04.004
108. S.A. Hosseini, A. Niaei, D. Salari, S.R. Nabavi, Nanocrystalline AMn_2O_4 (A = Co, Ni, Cu) spinels for remediation of volatile organic compounds—synthesis, characterization and catalytic performance. Ceram. Int. **38**, 1655–1661 (2012). https://doi.org/10.1016/j.ceramint.2011.09.057
109. M.H. Castaño, R. Molina, S. Moreno, Cooperative effect of the Co-Mn mixed oxides for the catalytic oxidation of VOCs: influence of the synthesis method. Appl. Catal. A Gen. **492**, 48–59 (2015). https://doi.org/10.1016/j.apcata.2014.12.009
110. D.B. Ghare, A.P.B. Sinha, L. Singh, Change in the valency state of ions in $CuMn_2O_4$ at high temperatures. J. Mater. Sci. **2**, 389–394 (1968). https://doi.org/10.1007/BF00550982
111. B. Gillot, S. Buguet, E. Kester, Oxidation mechanism and valence states of copper and manganese in tetragonal $CuMn_2O_4$. J. Mater. Chem. **7**, 2513–2517 (1997). https://doi.org/10.1039/a703731g
112. T. Valdés-Solis, I. López, G. Marban, Copper manganite as a catalyst for the PROX reaction. Deactivation studies. Int. J. Hydrog. Energy **35**, 1879–1887 (2010). https://doi.org/10.1016/j.ijhydene.2009.12.117
113. Z. Ye, J.M. Giraudon, N. Nuns, P. Simon, N. De Geyter, R. Morent, J.F. Lamonier, Influence of the preparation method on the activity of copper-manganese oxides for toluene total oxidation. Appl. Catal. B Environ. **223**, 154–166 (2018). https://doi.org/10.1016/j.apcatb.2017.06.072

114. C. Dong, Z. Qu, Y. Qin, Q. Fu, H. Sun, X. Duan, Revealing the highly catalytic performance of spinel $CoMn_2O_4$ for toluene oxidation: involvement and replenishment of oxygen species using in situ designed-TP techniques. ACS Catal. **9**, 6698–6710 (2019). https://doi.org/10.1021/acscatal.9b01324
115. W. Tang, X. Wu, S. Li, W. Li, Y. Chen, Porous Mn-Co mixed oxide nanorod as a novel catalyst with enhanced catalytic activity for removal of VOCs. Catal. Commun. **56**, 134–138 (2014). https://doi.org/10.1016/j.catcom.2014.07.023
116. S. Behar, N.A. Gómez-Mendoza, M.Á. Gómez-García, D. Świerczyński, F. Quignard, N. Tanchoux, Study and modelling of kinetics of the oxidation of VOC catalyzed by nanosized Cu-Mn spinels prepared via an alginate route. Appl. Catal. A Gen. **504**, 203–210 (2015). https://doi.org/10.1016/j.apcata.2014.12.021
117. Y. Huang, K. Ye, H. Li, W. Fan, F. Zhao, Y. Zhang, H. Ji, A highly durable catalyst based on $Co_xMn_{3-x}O_4$ nanosheets for low-temperature formaldehyde oxidation. Nano Res. **9**, 3881–3892 (2016). https://doi.org/10.1007/s12274-016-1257-9
118. X. Wang, W. Zhao, X. Wu, T. Zhang, Y. Liu, K. Zhang, Y. Xiao, L. Jiang, Total oxidation of benzene over ACo_2O_4 (A = Cu, Ni and Mn) catalysts: In situ DRIFTS account for understanding the reaction mechanism. Appl. Surf. Sci. **426**, 1198–1205 (2017). https://doi.org/10.1016/j.apsusc.2017.07.269
119. Z. Chen, S. Wang, W. Liu, X. Gao, D. Gao, M. Wang, S. Wang, Morphology-dependent performance of Co_3O_4 via facile and controllable synthesis for methane combustion. Appl. Catal. A Gen. **525**, 94–102 (2016). https://doi.org/10.1016/j.apcata.2016.07.009
120. Z. Fan, W. Fang, Z. Zhang, M. Chen, W. Shangguan, Highly active rod-like Co_3O_4 catalyst for the formaldehyde oxidation reaction. Catal. Commun. **103**, 10–14 (2018). https://doi.org/10.1016/j.catcom.2017.09.003
121. T. Garcia, S. Agouram, J.F. Sánchez-Royo, R. Murillo, A.M. Mastral, A. Aranda, I. Vázquez, A. Dejoz, B. Solsona, Deep oxidation of volatile organic compounds using ordered cobalt oxides prepared by a nanocasting route. Appl. Catal. A Gen. **386**, 16–27 (2010). https://doi.org/10.1016/j.apcata.2010.07.018
122. J. Mei, J. Xie, Z. Qu, Y. Ke, X. Hu, N. Yan, Ordered mesoporous spinel Co_3O_4 as a promising catalyst for the catalytic oxidation of dibromomethane. Mol. Catal. **461**, 60–66 (2018). https://doi.org/10.1016/j.mcat.2018.10.001
123. Y. Zheng, Y. Liu, H. Zhou, W. Huang, Z. Pu, Complete combustion of methane over Co_3O_4 catalysts: influence of pH values. J. Alloy. Compd. **734**, 112–120 (2018). https://doi.org/10.1016/j.jallcom.2017.11.008
124. W. Zhang, F. Wu, J. Li, Z. You, Dispersion–precipitation synthesis of highly active nanosized Co_3O_4 for catalytic oxidation of carbon monoxide and propane. Appl. Surf. Sci. **411**, 136–143 (2017). https://doi.org/10.1016/j.apsusc.2017.03.162
125. J. Łojewska, A. Kołodziej, T. Łojewski, R. Kapica, J. Tyczkowski, Structured cobalt oxide catalyst for VOC combustion. Part I: catalytic and engineering correlations. Appl. Catal. A Gen. **366**, 206–211 (2009). https://doi.org/10.1016/j.apcata.2009.07.006
126. H. Biederman, *Plasma Polymer Films* (Imperial College Press, London, 2004). https://doi.org/10.1142/p336
127. J. Tyczkowski, R. Kapica, J. Łojewska, Thin cobalt oxide films for catalysis deposited by plasma-enhanced metal-organic chemical vapor deposition. Thin Solid Films **515**, 6590–6595 (2007). https://doi.org/10.1016/j.tsf.2006.11.056
128. J. Łojewska, A. Kolodziej, R. Kapica, A. Knapik, J. Tyczkowski, In search for active non-precious metal catalyst for VOC combustion. Evaluation of plasma deposited Co and Co/Cu oxide catalysts on metallic structured carriers. Catal. Today **147S**, S94–S98 (2009). https://doi.org/10.1016/j.cattod.2009.07.021
129. D.K. Chlebda, P.J. Jodłowski, R.J. Jędrzejczyk, J. Łojewska, 2D-COS of in situ μ-Raman and in situ IR spectra for structure evolution characterisation of NEP-deposited cobalt oxide catalyst during n-nonane combustion. Spectrochim. Acta A Mol. Biomol. Spectrosc. **186**, 44–51 (2017). https://doi.org/10.1016/j.saa.2017.06.009

130. M.C. Petty, *Langmuir-Blodgett Films: An Introduction* (Cambridge University Press, Cambridge, 1996). https://doi.org/10.1002/adma.19970091019
131. J. Łojewska, P. Dynarowicz-Łatka, A. Kołodziej, Preparation, characterization and deposition of Langmuir-Blodgett Co, Al organic films for the catalytic applications. Thin Solid Films **495**, 299–307 (2006). https://doi.org/10.1016/j.tsf.2005.08.202
132. S. Deb, S. Biswas, S.A. Hussain, D. Bhattacharjee, Spectroscopic characterizations of the mixed Langmuir-Blodgett (LB) films of 2,20-biquinoline molecules: Evidence of dimer formation. Chem. Phys. Lett. **405**, 323–329 (2005). https://doi.org/10.1016/j.cplett.2005.02.060
133. P.J. Jodłowski, R.J. Jędrzejczyk, D. Chlebda, J. Tyczkowsk, J. Kryca, A. Kołodziej, J. Łojewska, Structure effects on activity of plasma deposited cobalt oxide catalysts for VOC combustion. Top. Catal. **60**, 318–325 (2017). https://doi.org/10.1007/s11244-016-0618-7
134. Q. Wang, Y. Peng, J. Fu, G.Z. Kyzas, S.M.R. Billah, S. An, Synthesis, characterization, and catalytic evaluation of $Co_3O_4/\gamma\text{-}Al_2O_3$ as methane combustion catalysts: significance of Co species and the redox cycle. Appl. Catal. B Environ. **168-169**, 42–50 (2015). https://doi.org/10.1016/j.apcatb.2014.12.016
135. B. Xiao, K. Zhao, L. Zhang, T. Cai, X. Zhang, Z. Wang, J. Yuan, L. Yang, P. Gao, D. He, A green and facile synthesis of Co_3O_4 monolithic catalyst with enhanced total oxidation of propane performance. Catal. Commun. **116**, 1–4 (2018). https://doi.org/10.1016/j.catcom.2018.07.013
136. T.H. Lim, S.B. Park, J.M. Kim, D.H. Kim, Ordered mesoporous MCo_2O_4 (M = Cu, Zn and Ni) spinel catalysts with high catalytic performance for methane combustion. J. Mol. Catal. A Chem. **426**, 68–74 (2017). https://doi.org/10.1016/j.molcata.2016.11.002
137. C. Zhang, J. Wang, S. Yang, H. Liang, Y. Men, Boosting total oxidation of acetone over spinel MCo_2O_4 (M = Co, Ni, Cu) hollow mesoporous spheres by cation-substituting effect. J. Colloid Interf. Sci. **539**, 65–75 (2019). https://doi.org/10.1016/j.jcis.2018.12.061
138. Z. Pu, Y. Liu, H. Zhou, W. Huang, Y. Zheng, X. Li, Catalytic combustion of lean methane at low temperature over ZrO_2-modified Co_3O_4 catalysts. Appl. Surf. Sci. **422**, 85–93 (2017). https://doi.org/10.1016/j.apsusc.2017.05.231
139. S.A. Hosseini, A. Niaei, D. Salari, M.C. Alvarez-Galvan, J.L.G. Fierro, Study of correlation between activity and structural properties of Cu-(Cr, Mn and Co)$_2$ nano mixed oxides in VOC combustion. Ceram. Int. **40**, 6157–6163 (2014). https://doi.org/10.1016/j.ceramint.2013.11.068

Chapter 2
Catalytic Oxidation of Volatile Organic Compounds over Porous Manganese Oxides Prepared via Sol-Gel Method

Miguel Jose Marin Figueredo, Marco Piumetti, Samir Bensaid, Debora Fino, and Russo Nunzio

2.1 Introduction

Volatile organic compounds (VOCs) are a set of substances that are emitted from anthropogenic or biogenic sources [1]. These chemical compounds may participate in several atmospheric reactions initiated by the action of solar radiation and thus enhance the formation of compounds (e.g., ozone) that may be harmful for human health [1–3]. Thus, the pollutants formed in the atmosphere and their precursors (i.e., the VOCs) may represent a health risk for the people if present at the ground level or in indoor environments [4].

In order to minimize the several impacts and harm caused by the eventual gaseous emissions of organic compounds, different techniques have been investigated over the years. The abatement procedures exploited chemical and/or physical phenomena in order to reach acceptable levels of VOCs in the output stream. As an example, techniques like physical or chemical adsorption, condensation, absorption, thermal oxidation, and catalytic oxidation have been studied [5]. In accordance, the catalytic oxidation has shown excellent performances under different operating conditions [6, 7]. Materials like noble metals (e.g., Pt, Pd, Rh, etc.,) and metal oxides (e.g., CeO_2, CuO, MnO_x, etc.,) have been object of research in the heterogeneous catalysis field due to their excellent catalytic performance at low temperatures [8–10]. On the other hand, various mixed oxides have shown a good performance in the abatement of different VOCs [10–16].

In our previous investigations, manganese oxides evidenced a remarkable catalytic activity in the oxidation of different VOCs (i.e., ethylene, propylene, and toluene) at low temperatures. In the best catalyst, a high amount of electrophilic oxygens adsorbed on the surface, as well as the enhanced presence of Brønsted acidic sites were fundamental for the elevated catalytic activity [17]. As well, other studies

M. J. Marin Figueredo · M. Piumetti (✉) · S. Bensaid · D. Fino · R. Nunzio
Department of Applied Science and Technology DISAT, Politecnico di Torino, Torino, Italy
e-mail: marco.piumetti@polito.it

M. Piumetti, S. Bensaid (eds.), *Nanostructured Catalysts for Environmental Applications*, https://doi.org/10.1007/978-3-030-58934-9_2

concerning cerium-copper-mixed oxides showed that the synergistic interactions taking place between CuO_x-dispersed clusters and ceria enhanced the catalytic activity. This was associated with the improved reducibility and the enhanced presence of structural defects (i.e., oxygen vacancies) [13]. Accordingly, these synergistic interactions were further verified during the study of binary and ternary cerium-copper-manganese-mixed-oxide catalysts. The improvement of the catalytic activity of ceria was correlated with the interactions between cerium and the foreign metals (Cu and/or Mn) in solid solutions, as well as with dispersed Mn/Cu oxide phases interacting with the bulk ceria [18].

Elements like iron, manganese, and copper are among the ten most abundant transition metals in the earth crust [19], and therefore, could be used as a sustainable resource for the preparation of catalytic materials. Consistently, the synthesis of oxide materials composed of these abundant elements has been investigated using several preparation techniques, like sol-gel [20–24]. This evidences that these are materials of potential interest for their application in the catalytic oxidation of VOCs.

In this work, a set of binary and ternary manganese-copper-iron oxides was prepared via the citrate sol-gel synthesis. This facile technique allowed the preparation of mesoporous materials with promising performances for VOC oxidation in relatively short times and at mild temperatures. The physico-chemical properties of the samples were investigated by means of X-ray diffraction analysis, N_2 physisorption at −196 °C, field emission scanning electron microscopy, X-ray photoelectron spectroscopy, and H_2 temperature-programmed reduction. The catalytic activity of the prepared samples was evaluated in the oxidation reaction of two VOC probe molecules (i.e., ethylene and propylene). Additionally, the reproducibility of the catalytic performance over the samples was studied through the realization of consecutive VOC catalytic oxidation runs.

2.2 Experimental

2.2.1 Catalyst Preparation

Pure manganese oxide (Mn_2O_3) and three manganese-mixed oxides (namely $MnCu_{15}$, $MnFe_{15}$, and $MnCu_{7.5}Fe_{7.5}$) were prepared by means of the sol-gel synthesis. As a whole, the manganese-mixed oxides were prepared with a nominal foreign metal content of 15 at. % and the balance in manganese.

2.2.1.1 Pure and Mixed Manganese Oxides

During a typical synthesis, the corresponding metal precursor(s), namely $Mn(NO_3)_2{\cdot}4H_2O$; $Cu(NO_3)_2{\cdot}3H_2O$; and $Fe(NO_3)_3{\cdot}9H_2O$ (Sigma-Aldrich), was/were dissolved in 50 mL of Milli-Q water in order to prepare a solution with a total molar concentration of 0.2 M. The amounts of nitrates were consistent with the defined metal-to-manganese ratio. On the other hand, citric acid (CA) ($C_6H_8O_7{\cdot}H_2O$ Sigma-Aldrich) was added to the solution assuring a 1:1 molar ratio with the nitrates. The mixture was continuously stirred in a water bath at r.t. and the initial pH of the solution was modified adding ammonium hydroxide solution (NH_4OH, Sigma-Aldrich) dropwise.

Until pH = 5 was reached. Then, the temperature of the water bath was increased slowly (ca. 1 °C min^{-1}) until 60 °C and isothermal condition was maintained for 2 h. During this period, the formation and aggregation of colloidal particles was observed in the suspension. Afterward, the particles were separated from the suspension using vacuum filtration and washed with water several times. Then, it was dried at 60 °C overnight. Finally, it was heated in an oven (5 °C min^{-1}) at 550 °C for 2 h.

2.2.1.2 Pure CuO and Fe_2O_3 Were Synthesized for Comparison Purposes

The corresponding metal precursor was dissolved in 50 mL of Milli-Q water to prepare a 0.2 M solution. Molar ratios of 1:1 between CA:$Cu(NO_3)_2$ and 1.5:1 between CA:$Fe(NO_3)_3$ were assured. The solution was continuously stirred in a water bath at r.t. and the initial pH was increased until pH = 5 (for the solution containing copper ions) and to pH = 7 (for the solution containing iron ions). Afterward, the temperature of the water bath was increased until 80 °C and maintained constant for 4 h. Subsequently, the viscous liquid was dried at 60 °C overnight. The dry gel was heated in an oven at 550 °C for 2 h.

2.2.2 Catalyst Characterization

X-ray diffraction patterns were obtained by means of a X'Pert Philips PW3040 diffractometer using Cu K_α radiation (2θ range 20–80°, step = 0.05° 2θ, and time per step = 0.02 s). The patterns were indexed using the Powder Data File Database (PDF 2000, International Centre of Diffraction Data, Pennsylvania). The average crystallite size (D) was calculated using the Scherrer formula, $D = 0.9\ (\lambda/b\ \cos\theta)$, where λ: wavelength of the Cu K_α radiation, b: full width and half maximum (in radians), 0.9: shape factor for spherical particles, and θ: angle of the diffraction peaks.

The specific surface area (S_{BET}), the total pore volume (V_P) and the average pore width (D_P) were evaluated by means of N_2 physisorption analysis at −196 °C (Micromeritics Tristar II 3020) on powders pretreated for 2 h at 200 °C in order to remove atmospheric pollutants. S_{BET} was calculated according to the Brunauer-Emmett-Teller (BET) method, while the total pore volume and the average pore diameter using the Barrett-Joyner-Halenda (BJH) method.

The morphology of the catalysts was studied by means of a field emission scanning electronic microscope (FESEM) Zeiss Merlin, Gemini-II column under the following conditions: an extra high tension (EHT) of 3 kV, a working distance (WD) of 2.5 mm, and a probe intensity of 100 pA.

The temperature-programmed reduction/desorption analyses were performed using a ThermoQuest TPD/R/O 1100 analyzer, equipped with a thermal conductivity detector (TCD). Before the H_2-TPR analysis, 20 mg of the powder was set inside a quartz tubular reactor and was pretreated with He (40 mL min^{-1}) at 500 °C for 1 h. During the H_2-TPR analysis, a reducing gas mixture containing 5 vol. % H_2 in Ar (20 mL min^{-1}) was fed in the reactor and the programmed heating raised the temperature from r.t. to 800 °C (5 °C min^{-1}).

XPS (X-ray photoelectron spectroscopy) measurements were carried out on an XPS PHI 5000 Versa probe apparatus under the following analysis conditions: a band-pass energy of 187.85 eV, a 45° take off angle, and a 100 μm diameter X-ray spot size. The curve-fitting was realized using the Multipack 9.0 software.

2.2.3 *Catalytic Activity Tests*

The catalytic activity testing of the powder samples was realized in a classical temperature-programmed oxidation setup. The setup consisted in a fixed-bed quartz U-tube reactor (internal diameter = 4 mm) hosting the catalytic powder. The reactor was heated by means of a PID-controlled furnace and the temperature of the fixed-bed was measured using a K-type thermocouple.

Prior to the catalytic test, 0.1 g of the pelletized sample (particles diameter: 212–300 μm) was pretreated under a N_2 flow (30 mL min^{-1}) at 150 °C for 1 h in order to desorb atmospheric pollutants. In the catalytic tests, a VOC containing flow (propylene/ethylene: 500 ppm; O_2: 10 vol. %; N_2: balance) was fed to the reactor. The tests were performed maintaining a gas hourly space velocity (GHSV) of 20,000 h^{-1} and a catalyst weight-to-volumetric flow rate ratio (W/F) of 0.050 g h L^{-1}. The temperature was varied in the 100–400 °C range, performing isothermal steps every 30 °C. The composition of the reactor outlet gas was measured by means of a nondispersive infrared (NDIR) analyzer (ABB Uras 14).

The reproducibility of the catalytic performance was evaluated realizing three consecutive catalytic tests. During this screening, the temperature was raised slowly (ca. 2 °C min^{-1}). The other operative conditions were the same utilized during the catalytic screening.

2.3 Results and Discussion

2.3.1 *Material Textural Properties*

The incorporation of foreign metals (i.e., Cu or/and Fe) into the structure of Mn oxide and its consequent influence on the physicochemical properties is assessed in this section. Figure 2.1 shows the diffraction patterns of the manganese oxide powders (namely Mn_2O_3, $MnCu_{15}$, $MnFe_{15}$, and $MnCu_{7.5}Fe_{7.5}$) measured using the XRD technique.

As observed in Fig. 2.1a, the diffraction patterns collected for the samples containing manganese correspond to Mn_2O_3 with a cubic crystal system (reference code 00-041-1442). This suggests that Fe and/or Cu were effectively inserted inside the Mn_2O_3 lattice, and thus, the formation of solid solutions.

On the other hand, Fig. 2.1b shows a magnification of the diffraction patterns. It can be observed that the peak centered at 33.09° translates to lower 2θ values in the case of the powder samples containing iron (i.e., the $MnFe_{15}$ and $MnCu_{7.5}Fe_{7.5}$). This behavior suggests that an expansion of the Mn_2O_3 lattice took place, probably due to strain caused by the presence of Fe ions with bigger ionic radii [r_{ion} of Fe^{2+}/Fe^{3+} (coordination number or CN = 6) is 0.061/0.055 nm] with respect to the radii of Mn [r_{ion} of Mn^{3+}/Mn^{4+} (CN = 6) is 0.058/0.053 nm] and Cu [r_{ion} of Cu^{+}/Cu^{2+} (CN = 6) is 0.077/0.073 nm] species [25].

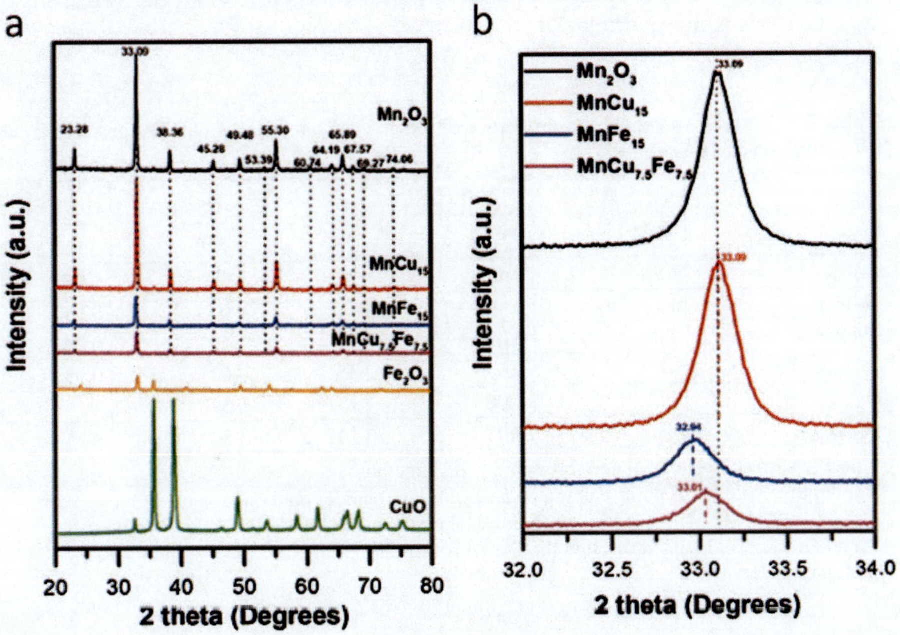

Fig. 2.1 X-ray diffraction patterns of the prepared powder catalysts (**a**) and their magnification (**b**)

Consistently, as the content of Fe ions increases, the lattice parameter (*a*) of the cubic cell increases (see Table 2.1). Moreover, the insertion of Fe ions into the lattice of Mn_2O_3 caused a slight reduction (see Table 2.1) of the crystallite size, as calculated from the Scherrer formula. While the contemporary presence of Fe and Cu ions promoted the formation of relatively smaller particles (ca. 62 nm). On the other hand, the diffraction patterns of the prepared copper (green line) and iron oxide (orange line) matched with the crystalline system of the CuO (ref. code 01-072-0629) and the α-Fe_2O_3 (ref. code 00-024-0072).

As a whole, the BET surface areas of the manganese oxides are in the range of 8–15 $m^2\ g^{-1}$. Particularly, the insertion of Fe and/or Cu into the lattice of manganese oxide seems to promote a decrease of the BET surface area. Generally, the cumulative volume of pores ($V_P = 0.07$–$0.12\ cm^3\ g^{-1}$) and the pore width ($D_P = 29$–35 nm) shows a decreasing behavior when the foreign metals were inserted inside the Mn_2O_3 crystalline structure. Instead, the BET analysis of the Fe_2O_3 and the CuO powders showed S_{BET}: 11 and 3 $m^2\ g^{-1}$, respectively, as well as lower values of V_P and D_P with respect to the catalysts containing manganese.

The FESEM micrographs (see Fig. 2.2) of the manganese oxides evidenced, in general, a sheet-like morphology. The sheets presented a relatively thick shell and a hollowed (honeycomb type) interior, thus confirming the formation of mesoporous structures after the synthesis procedure. The pores of these samples have similar lengths and are comparable to the values obtained through BET calculations. On the other hand, the micrograph of the Fe_2O_3 shows a more compact sheet structure with respect to one of the manganese oxides. Instead, the image of the CuO evidenced the aggregation of nanoparticles, suggesting that the pore width derived from the S_{BET} analysis is interparticle.

Table 2.1 Textural and structural properties of the powder catalysts, as obtained from XRD and N_2 adsorption analysis at −196 °C

Catalyst	S_{BET} [a] m^2g^{-1}	V_p [b] cm^3g^{-1}	D_p [c] nm	CS [d] nm	*a* [e] nm
Mn_2O_3	15	0.12	32	67	0.937
$MnCu_{15}$	14	0.10	30	67	0.937
$MnFe_{15}$	11	0.09	29	65	0.941
$MnCu_{7.5}Fe_{7.5}$	8	0.07	35	62	0.939
Fe_2O_3	11	0.06	23	57	–
CuO	3	0.01	14	34	–

[a]Specific surface area calculated through the BET method
[b]Cumulative volume of pores calculated according to the BJH method
[c]Pore diameter calculated according to the BJH method
[d]Crystallite size calculated through the Scherrer formula
[e]Cell parameter

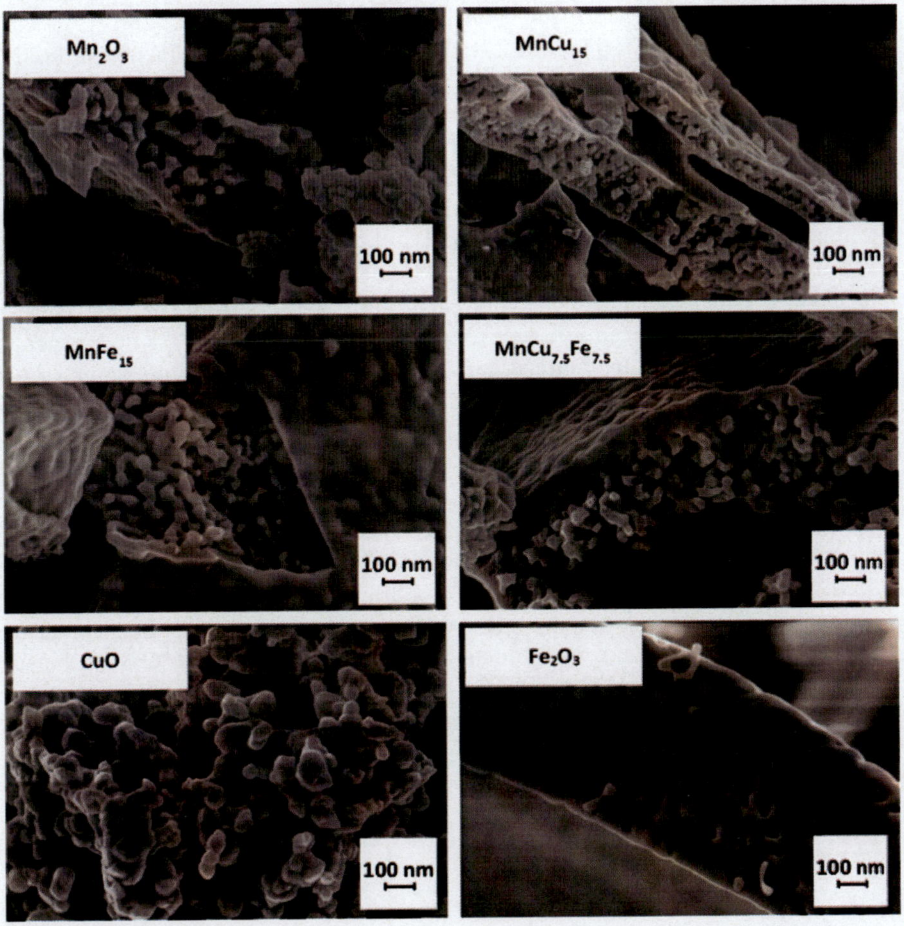

Fig. 2.2 FESEM micrographs of the fresh powder catalysts

2.3.2 H_2-TPR Analysis

The H_2-TPR profiles of the powder catalysts are summarized in Fig. 2.3. The reduction of Mn_2O_3 started at about 200 °C and shows two intense peaks: the first at 351 °C and other with higher intensity at 487 °C. The former was attributed to a first reduction step: $Mn_2O_3 \rightarrow Mn_3O_4$, while the latter was assigned to the reduction step: $Mn_3O_4 \rightarrow MnO$ [12, 14, 17, 26]. The deconvolution of the whole profile revealed a total H_2 consumption of 7.16 mmol g^{-1}, which is higher than the theoretical uptake of the whole reduction process (6.33 mmol g^{-1}). However, this consumption is lower than the one required for the reduction of MnO_2 (11.50 mmol g^{-1}), thus suggesting the presence of low amounts of Mn^{4+} species coupled with the Mn^{3+} in the Mn_2O_3 catalyst.

The reduction profile of CuO shows a sharp reduction peak at 296 °C in agreement with the literature [13, 15, 27]. On the other hand, the reduction of Fe_2O_3 is characterized by a complex signal. According to the literature, the side-peak located at 301 °C seems related to the reduction of the hydroxylated oxide. On the other hand, the deconvolution of the peak at 392 °C evidenced a H_2 consumption of ca. 2.35 mmol g^{-1} in fair agreement with the theoretical uptake (2.09 mmol g^{-1}) for the reduction step $Fe_2O_3 \rightarrow Fe_3O_4$. While the complex reduction taking place at higher temperatures was attributed to the reduction steps $Fe_3O_4 \rightarrow FeO \rightarrow Fe$ [28, 29].

The reduction profiles of the binary oxides follow, in general, the two-peak behavior of the parent Mn_2O_3. Consistently, in $MnCu_{15}$ (red line, Fig. 2.3), the reduction peaks occurred at lower temperatures (ca. 308 and 487 °C). Since the peak at the lowest temperature occurs close to the reduction temperature of Cu^{2+} species, it was attributed to the contemporary reduction of Cu^{2+} and Mn^{3+} species. Instead, manganese species reduce further at higher temperatures. These evidences suggests that the Mn^{3+} was effectively substituted by copper in the lattice [12, 30].

The wide blue profile (see Fig. 2.3) evidences the reduction of the $MnFe_{15}$ catalyst, starting at $T > 300$ °C and with a maximum at 522 °C. As a whole, the profile shape is identical to the one of the parent oxide (Mn_2O_3) but with a signal extended to higher temperatures. This suggests that the insertion of Fe^{x+} species led to the formation of an oxide structure more stable and thus more resistant to the reduction. According to the literature, the reduction occurring below ca. 530 °C is attributed to

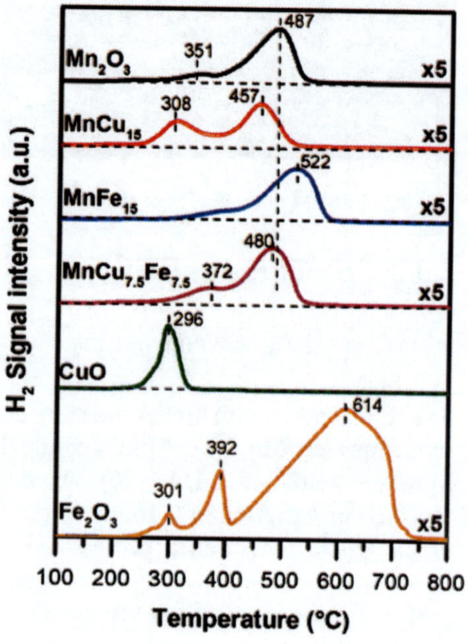

Fig. 2.3 H_2-TPR profiles of the fresh powder catalysts

the contemporary reduction of $Mn^{3+} \rightarrow Mn^{2+}$ and $Fe^{3+} \rightarrow Fe^{2+}$ species, while the reduction at higher temperature corresponds to the final reduction of $Fe^{2+} \rightarrow Fe$ species [31, 32].

$MnCu_{7.5}Fe_{7.5}$ showed a reduction profile (magenta line, Fig. 2.3) equivalent to the one of Mn_2O_3, but with a slight translation of the peak maximum signal and an increased signal intensity (higher H_2 consumption) at ca. 372 °C. This seems to be an intermediate behavior between $MnCu_{15}$ and the $MnFe_{15}$. Accordingly, the peak at the lower temperature was assigned to the reduction of copper species to Cu^0 and the initial reduction of Fe^{3+} and Mn^{3+} species to the oxidation state 2+. While the reduction taking place at 480 °C was assigned to the continuing reduction of manganese and iron species.

On the other hand, the reducibility variation of the catalysts (the most performing ones) was assessed after their use for the oxidation of the probe VOC molecules (i.e., ethylene and propylene). As observed in Fig. 2.4a, the reduction profiles of the fresh (black line) and the used (red line) Mn_2O_3 catalyst are analogous. However, the reduction of the latter occurs at a temperature slightly higher. This may be ascribed to the reduction of bigger manganese oxide particles [33] formed due to the relatively elevated temperature (ca. 300 °C) used during the catalytic screenings.

Similarly, the reduction peaks of the used $MnCu_{7.5}Fe_{7.5}$ catalyst (see Fig. 2.4b), and the whole profile occurs at temperatures slightly higher than those of the fresh catalyst. Suggesting that the formation of larger particles also took place in the case of the ternary catalyst. Interestingly, the peak located at ca. 384 °C indicates an increased amount of species that are reduced in that temperature range, thus Cu^{2+}, Fe^{3+}, or Mn^{3+}. Since the oxidation state of Cu retrieved in XPS analysis is 2+ (vide infra), this suggests that either Mn^{2+} or Fe^{2+} species oxidized to Mn^{3+} or Fe^{3+}, respectively, during the catalytic screening.

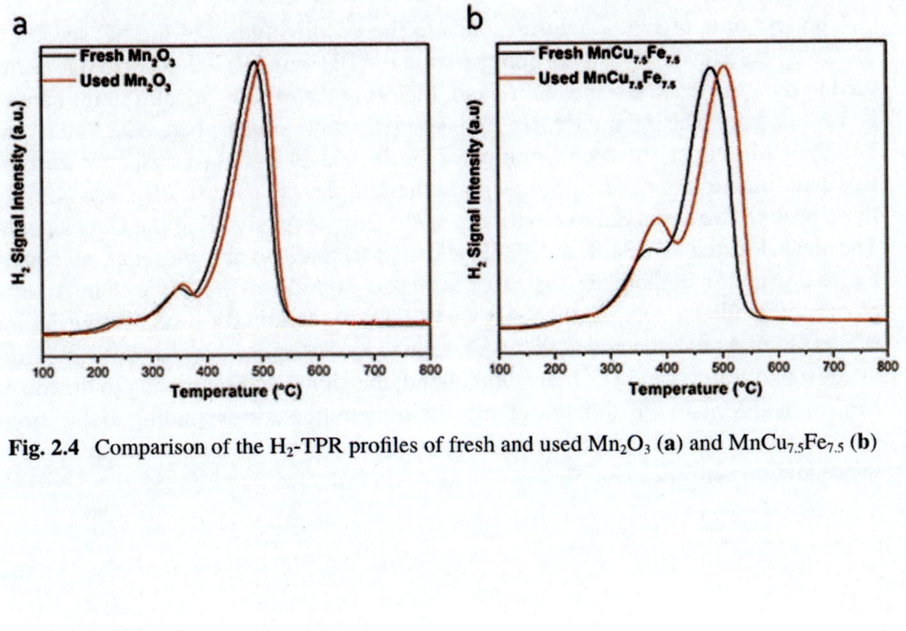

Fig. 2.4 Comparison of the H_2-TPR profiles of fresh and used Mn_2O_3 (**a**) and $MnCu_{7.5}Fe_{7.5}$ (**b**)

2.3.3 X-Ray Photoelectron Spectroscopy

The X-ray photoelectron spectra of the various elements present in the prepared catalysts are summarized in Fig. 2.5. Particularly, Fig. 2.5a includes the deconvolution of the XP spectra of the prepared catalysts in the O 1 *s* core level. The spectra were mainly characterized by two peaks, each one associated with a specific oxygen specie. First, the peak located in the binding energy (BE) range of 529.5–529.8 eV was attributed to oxygen species (O_β) located in the lattice (O^{2-}), and thus bonded to Mn, Cu, or Fe [34]. On the other hand, the peaks occurring in the range 531.1–531.4 eV were assigned to chemisorbed oxygen species (O_α), e.g., O^-, O_2^-, and OH groups [35, 36].

The data observed in Table 2.2 show that the Mn_2O_3 catalyst has the highest abundance of adsorbed species, and thus the highest O_α/O_β ratio of the catalysts set. Then, it is followed by the mixed oxides containing copper (i.e., $MnCu_{15}$ and $MnCu_{7.5}Fe_{7.5}$) and CuO with intermediate values (0.62 and 0.55, respectively). Finally, it is the binary $MnFe_{15}$ and the Fe_2O_3 samples with the lowest relative presence of chemisorbed species.

Figure 2.5b shows the deconvolution of the spectra in the Mn 2*p* core level. The spin peaks comprised in the BE range of 638.0–647.5 correspond to the $2p_{3/2}$ level, while the signal appearing in the 647.5–655.5 range is attributed to the $2p_{1/2}$ level. According to the literature, the abundances of the Mn^{2+}, Mn^{3+}, and Mn^{4+} can be estimated through the deconvolution of the $2p_{3/2}$ level, [17, 35, 37]. Table 2.3 includes the estimated relative abundances of the Mn^{x+} species in the catalysts. With respect to the parent oxide (Mn_2O_3), the relative amount of the Mn^{4+} specie increases when foreign metals are introduced in the sample. This indicates that the presence of Cu or/and Fe enhances the formation of oxidized Mn^{4+} species. On the other hand, the deconvolution evidenced a marked amount of reduced Mn^{2+} in the $MnCu_{15}$ sample.

The section c of Fig. 2.5 corresponds to the deconvolution of the XP spectra in the Cu 2*p* core level. The signal comprised in the BE range 930.0–939.0 eV is attributed to the $2p_{3/2}$ level. On the other hand, the presence of Cu^{2+} species in the sample is revealed by the characteristic "shake-ups" that appear between 940.0 and 946.0 eV. Moreover, the deconvolution of the $2p_{3/2}$ level signal was used for the relative quantification of the copper species in the samples [13, 33, 38, 39]. Accordingly, three peaks were observed to contribute to the $2p_{3/2}$ level signal in the CuO sample. The peak located at 934.8 and 933.6 eV is attributed to the presence of copper hydroxide and/or carbonates. Instead, the peak centered at 933.6 eV is characteristic of copper bonded to oxygen in the lattice (Cu-O). Interestingly, a small contribution (6.3 at.%) of a peak located at 932.6 eV related to Cu^+ species was observed during the deconvolution [33, 38]. On the other hand, the signal corresponding to the mixed $MnCu_{15}$ and $MnCu_{7.5}Fe_{7.5}$ increased mostly in the range corresponding to the strong Cu^{2+} satellite. This suggests that the copper species present in those samples are in the oxidation state +2.

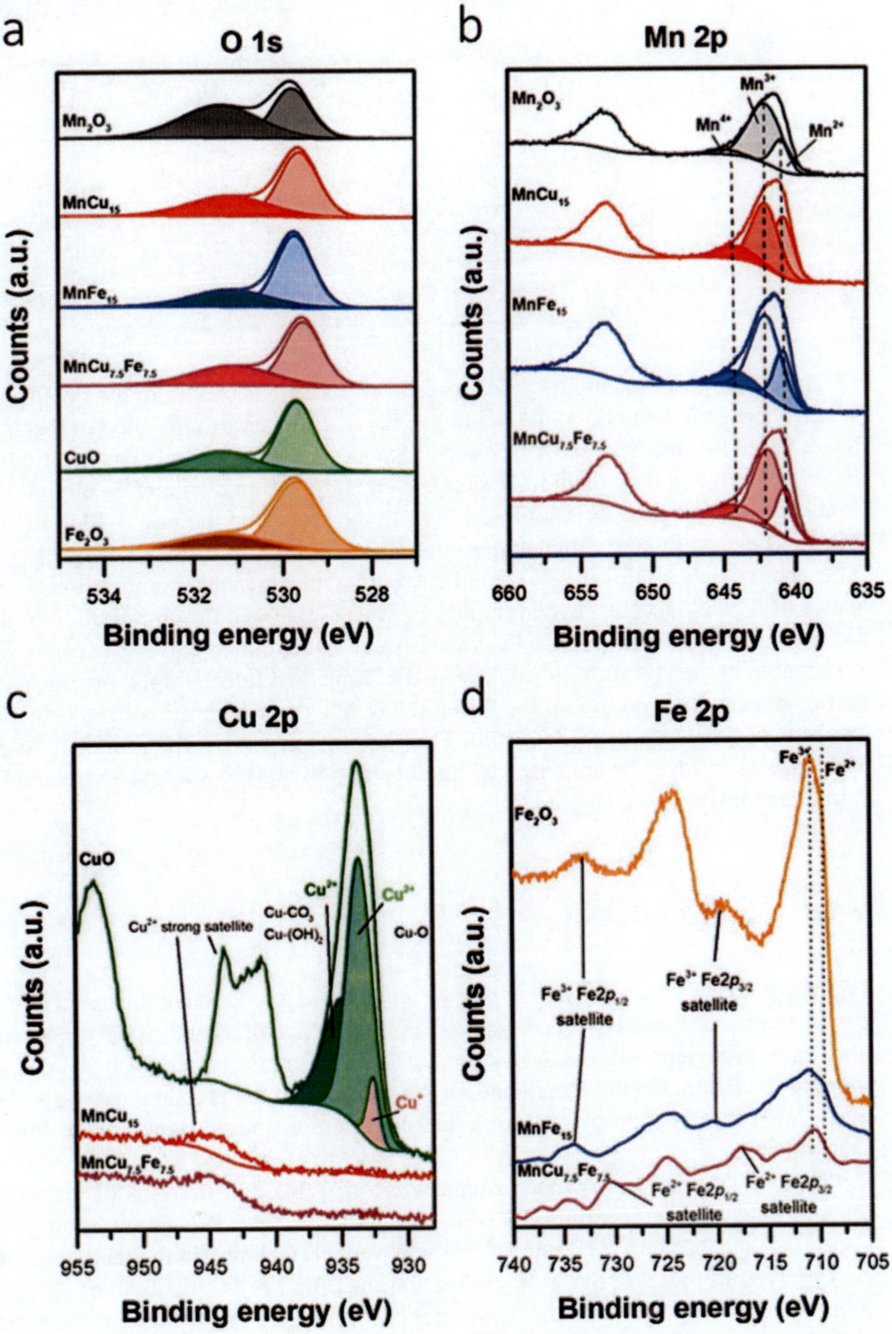

Fig. 2.5 XPS spectra of the prepared samples in the O (1*s*), Mn (2*p*), Cu (2*p*), and Fe (2*p*) core level

Table 2.2 Abundance of oxygen species (at.%) and their ratio obtained from the deconvolution of the XPS spectra: O 1 *s* core level

Catalyst	O_α,OH BE eV	O_α %at.	O_β BE eV	O_β %at.	O_α/O_β
Mn_2O_3	531.4	57.7	529.8	42.3	1.36
$MnCu_{15}$	531.2	38.3	529.6	61.7	0.62
$MnFe_{15}$	531.1	31.8	529.7	68.2	0.47
$MnCu_{7.5}Fe_{7.5}$	531.1	38.3	529.5	61.7	0.62
CuO	531.4	35.3	529.7	64.7	0.55
Fe_2O_3	531.4	28.4	529.8	71.6	0.40

The signal observed in Fig. 2.5d corresponds to the XP spectra in the Fe 2*p* core level. The signal obtained in the BE range 707.5–714.8 eV is attributed to the $2p_{3/2}$ level, while the one between 721.7 and 728.5 eV to the $2p_{1/2}$ level. The position of these peaks, along with their corresponding satellites aid in the identification of the iron species present in the sample [37, 40]. In the spectra of the Fe_2O_3, $MnFe_{15}$ and $MnCu_{7.5}Fe_{7.5}$, an intense signal appeared at 710.6, 711.0 and 710.8 eV, respectively. This and the Fe^{3+} satellites observed in the spectra unequivocally confirm the dominance of the Fe^{3+} species in the samples. However, although the contribution of the Fe^{2+} species to the signal of the $2p_{3/2}$ level is not evident, their satellites seem to be observable in the spectrum of the $MnCu_{7.5}Fe_{7.5}$ catalyst. Interestingly, the presence of Fe^{2+} species coincides with the formation of a high amount of oxidized Mn^{4+} in the $MnCu_{7.5}Fe_{7.5}$ sample (vide supra). This suggests that the formation of reduced Fe^{2+} species seems to be enhanced by the substitution of Mn^{3+} species by Cu and Fe (simultaneously) in the lattice.

2.3.4 *Catalytic Activity*

Figure 2.6 shows the conversion of propylene VOCs as a function of the temperature on the synthesized catalysts. As observed, in the absence of a catalyst only 9% conversion of propylene was reached at the highest temperature tested (400 °C) due to thermal oxidation. On the other hand, in the presence of the prepared catalysts, the total conversion of propylene took place at temperatures significantly lower than 400 °C.

The catalytic performances are summarized in Table 2.4, in terms of the VOC reaction rate. The temperatures at which a 10%, 50%, and 90% conversion (i.e., $T_{10\%}$, $T_{50\%}$, and $T_{90\%}$) of VOC into CO_2 is achieved are included in the table as well. As observed in Fig. 2.6, the VOC conversion over the Mn_2O_3 (black line) catalyst took place at lower temperature with respect to those of Fe_2O_3 (orange line) and CuO (green line). Accordingly, the catalytic activity for these catalysts decreases as follows: $Mn_2O_3 > CuO > Fe_2O_3$. On the other hand, the conversion of propylene in the mixed manganese-oxides occurs at temperatures slightly higher than those

Table 2.3 Relative abundances of manganese species in the prepared catalysts and BE values obtained from the deconvolution of the Mn 2*p* core level spectra

Catalyst	Mn^{4+} BE eV	Mn^{4+} %at.	Mn^{3+} BE eV	Mn^{3+} %at.	Mn^{2+} BE eV	Mn^{2+} %at.	$\frac{Mn^{4+}}{Mn^{3+}}$	$\frac{Mn^{3+}}{Mn^{2+}}$
Mn_2O_3	644.8	5.5	642.0	75.3	640.9	19.2	0.07	3.92
$MnCu_{15}$	644.1	13.7	642.1	51.3	640.8	35.1	0.27	1.46
$MnFe_{15}$	644.2	13.9	642.0	60.6	640.8	25.5	0.23	2.38
$MnCu_{7.5}Fe_{7.5}$	644.2	14.9	641.8	62.2	640.7	22.9	0.24	2.71

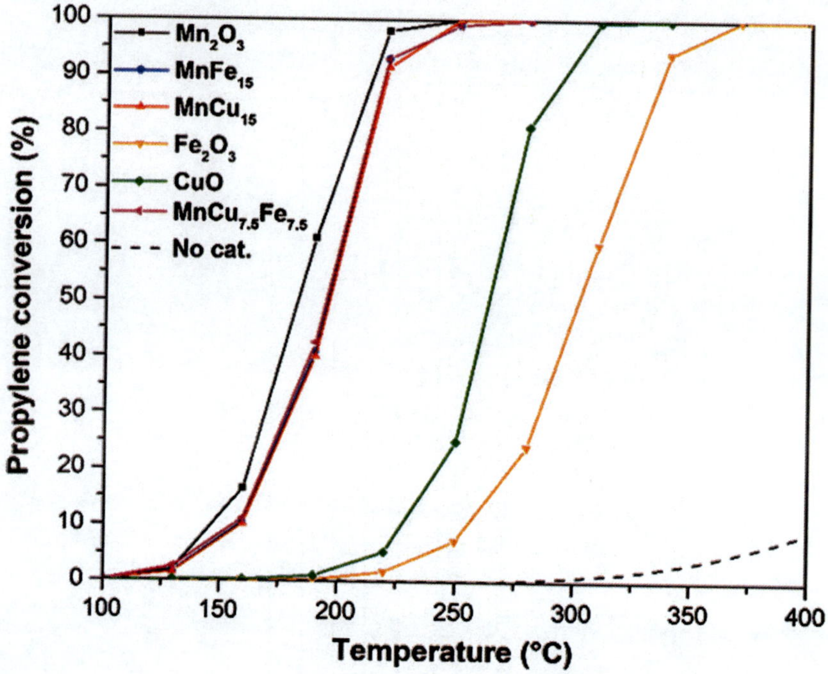

Fig. 2.6 Catalytic conversions (%) of propylene achieved over the prepared samples

Table 2.4 Catalytic performances of the catalysts for the oxidation of propylene and ethylene

Sample	Propylene oxidation				Ethylene oxidation			
	$T_{10\%}$/°C	$T_{50\%}$/°C	$T_{90\%}$/°C	r_p/μmol h^{-1}m^{-2} [a]	$T_{10\%}$/°C	$T_{50\%}$/°C	$T_{90\%}$/°C	r_e/μmol h^{-1}m^{-2} [b]
Mn_2O_3	152	183	212	0.94	168	204	237	1.67
$MnCu_{15}$	160	196	219	0.51	176	222	250	1.69
$MnFe_{15}$	159	196	217	0.60	175	213	243	1.53
$MnCu_{7.5}Fe_{7.5}$	158	195	217	1.47	165	208	244	4.41
CuO	231	263	289	0.25	235	271	303	0.85
Fe_2O_3	257	302	335	0.05	281	342	386	0.22

[a]Calculated at 130 °C
[b]Calculated at 160 °C

required for the parent oxide Mn_2O_3 (i.e., the oxidation occurs at higher $T_{10\%}$, $T_{50\%}$, and $T_{90\%}$). Their catalytic performance seems to be practically identical in the whole temperature range studied. However, the catalytic kinetic analysis is realized at low conversions (i.e., at the light-off temperatures), while at higher temperatures other phenomena are involved (namely, mass-transfer phenomena) [41]. The data in Table 2.4 evidence that only the binary oxides presented a lower reaction rate (r_p) at 130 °C, with respect to the parent oxide, and thus, a worsening of the catalytic activity. In this sense, both $MnCu_{15}$ and $MnFe_{15}$ show almost equivalent catalytic activities. Remarkably, the highest r_p was observed in the $MnCu_{7.5}Fe_{7.5}$ (1.47 μmol h^{-1} m^{-2}) catalyst despite the fact of having a surface area lower than that of Mn_2O_3. According to these results, the following catalytic trend can be drawn $MnCu_{7.5}Fe_{7.5} > Mn_2O_3 > MnFe_{15} \approx MnCu_{15}$.

The performances of the synthesized powders in the catalytic oxidation of ethylene are shown in Fig. 2.7. The noncatalytic abatement of ethylene (black-dashed line) reached a 1.2% conversion at the highest temperature tested (ca. 400 °C). The pure oxides demonstrated the same catalytic performance trend observed in the oxidation of propylene, thus $Mn_2O_3 > CuO > Fe_2O_3$. The temperatures required for the catalytic oxidation of ethylene were higher than those needed for propylene, thus evidencing the higher stability of the former.

On the other hand, the high stability of ethylene toward oxidation allows better comparing the catalytic performance of the catalysts. Above 190 °C, the parent oxide, thus Mn_2O_3 showed a better performance with respect to the mixed oxides. However, the catalytic reaction rates of ethylene calculated at 160 °C (see Table 2.4) clearly evidence that the catalytic activity of Mn_2O_3 is equivalent to that of the binary $MnCu_{15}$. While the presence of Fe species seems to diminish the catalytic activity of the Mn species. This suggests that better interactions take place between Cu and Mn species, while those occurring between Mn and Fe alone seem to compromise the catalytic activity. The possibility of beneficial interactions occurring

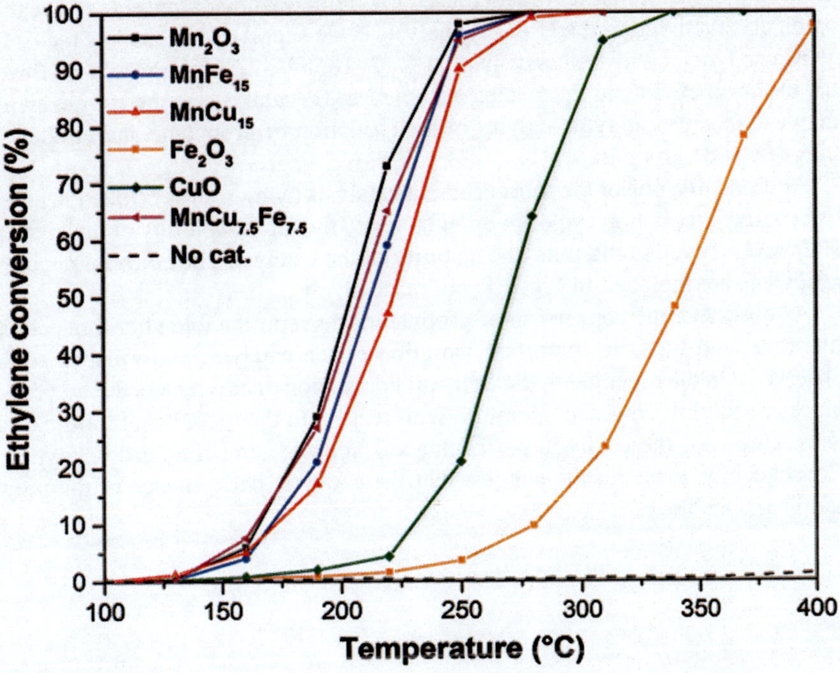

Fig. 2.7 Catalytic conversions (%) of ethylene achieved over the prepared samples

between Mn and Cu species in the oxidation of ethylene has been reported/observed in our previous work as well [18]. Interestingly, in the oxidation of ethylene, the ternary $MnCu_{7.5}Fe_{7.5}$ showed as well the highest catalytic reaction rate (4.41 μmol h^{-1} m^{-2}) among the prepared catalysts. This interesting result highlights the probability that beneficial interactions may be taking place between all three Mn, Cu, and Fe species, in manganese oxides prepared using the sol-gel technique. On the other hand, the catalytic activity of the ternary oxide in ethylene oxidation seems to be strongly related to the presence of high amounts of chemisorbed oxygen and enhanced reducibility at low temperatures like in propylene oxidation. Consistently, the following trend of catalytic activity for the oxidation of ethylene can be confirmed: $MnCu_{7.5}Fe_{7.5} > Mn_2O_3 > MnCu_{15} > MnFe_{15}$.

Several authors have shown that an enhanced amount of chemisorbed oxygen species is beneficial in the catalytic oxidation of VOCs, due to their higher mobility with respect to the oxygen in the lattice [17, 35, 36]. In accordance, the enhanced catalytic performance of the ternary catalyst may be attributed to increased amount of weakly-bonded oxygen species (i.e., chemisorbed oxygen), as observed during XPS analysis (see Table 2.2).

As well, the enhanced presence of coupled manganese species with different oxidation states (i.e., $Mn^{4+}|Mn^{3+}|Mn^{2+}$), has been correlated with enhancements of the catalytic activity of manganese oxides [17, 35]. However, the catalytic improvement of our samples in the oxidation of propylene seems to be more compromised by the increased amounts of the reduced Mn^{2+} species (see Table 2.2). Additionally, according to the literature, the increased low-temperature reducibility of the ternary oxide observed during the H_2-TPR analysis (vide supra) can be the key factor for its VOC catalytic activity enhancement [10–12, 18, 35, 42]. These results confirm that the insertion of Cu and Fe species into the Mn_2O_3 catalyst via the sol-gel synthesis may enhance its catalytic activity in the oxidation of propylene and ethylene, and thus of a wide group of VOCs.

Additionally, one of the most efficient catalysts (Mn_2O_3) was utilized in consecutive catalytic oxidation cycles in order to verify the reproducibility of the conversion after several cycles, and thus, the stability of the catalyst. The conversion curves of propylene are included in Fig. 2.8a, b, respectively.

As observed, the conversion of propylene was reproducible after three consecutive runs, and thus, no important variation of the catalytic performance could be observed. On the other hand, the light-off conversion of ethylene in the first run took place at a slightly lower temperature with respect to the second and third oxidation tests. However, the catalytic performance stabilized after the third catalytic runs. Consequently, these results suggest that the catalytic performance of the prepared samples is stable.

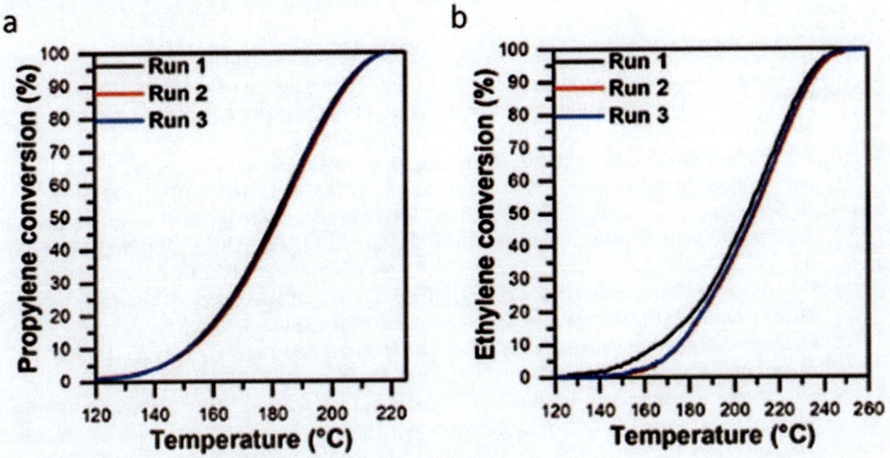

Fig. 2.8 Stability screening of the Mn_2O_3 catalyst in the oxidation of propylene (**a**) and ethylene (**b**)

2.4 Conclusions

In this chapter, several pure and mixed transition metal oxides were prepared via a simple sol-gel synthesis: one set of pure oxides (Mn_2O_3, CuO, and Fe_2O_3) and one set of mixed oxides ($MnCu_{15}$, $MnFe_{15}$, and $MnCu_{7.5}Fe_{7.5}$). The syntheses were realized using typical reagents like metal nitrates and citric acid. The various physicochemical properties of the catalytic materials were investigated by means of the different characterization techniques: XRD, N_2-physisorption at −196 °C, H_2-TPR, FESEM, and XPS.

The first analyses evidenced that mesoporous structures resulted from the sol-gel preparations, characterized by the crystalline structure of the parent oxide Mn_2O_3. The effective insertion of the foreign metals (Cu and/or Fe) in the crystalline structure was confirmed by the variation of the lattice parameter and the shifting of the characteristic XRD pattern. The results of the catalytic screening evidenced the excellent performance and catalytic activity of the pure Mn_2O_3 and the ternary $MnCu_{7.5}Fe_{7.5}$ in the oxidation of either ethylene or propylene. It was observed that the enhanced catalytic activity in the oxidation of these VOC molecules is strongly related to two key aspects: (1) the elevated amount of surface-chemisorbed oxygen species and (2) the enhanced reducibility of the oxide at lower temperatures. Accordingly, the Mn_2O_3 catalyst showed an excellent catalytic performance. On the other hand, the synergistic interaction between the Mn, Cu, and Fe species in $MnCu_{7.5}Fe_{7.5}$ assured one of the best catalytic activities among the materials synthesized. As well, the catalytic performance of Mn_2O_3 showed a good stability in the oxidation of the probe VOCs after consecutive runs.

References

1. R. Koppmann, *Volatile Organic Compounds in the Atmosphere* (Blackwell, Hoboken, 2007)
2. R. Atkinson, J. Arey, Atmospheric degradation of volatile organic compounds. Chem. Rev. **103**, 4605–4638 (2003). https://doi.org/10.1021/cr0206420
3. B.J. Finlayson-Pitts, J.N. Pitts Jr., Tropospheric air pollution: ozone, airborne toxics, polycyclic aromatic hydrocarbons, and particles. Science **276**(5315), 1045–1051 (1979)
4. N. Ramírez, A. Cuadras, E. Rovira, et al., Chronic risk assessment of exposure to volatile organic compounds in the atmosphere near the largest Mediterranean industrial site. Environ. Int. **39**, 200–209 (2012). https://doi.org/10.1016/J.ENVINT.2011.11.002
5. U.S. Environmental Protection Agency. Office of Air and Radiation. *Control Techniques for Volatile Organic Compound Emissions from Stationary Sources.* (1992)
6. D. Duprez, F. Cavani, *Handbook of Advanced Methods and Processes in Oxidation Catalysis: From Laboratory to Industry* (Imperial College Press, London, 2014)
7. E.C. Moretti, American Institute of Chemical Engineers. Center for Waste Reduction Technologies, *Practical Solutions for Reducing Volatile Organic Compounds and Hazardous Air Pollutants* (Center for Waste Reduction Technologies, American Institute of Chemical Engineers, New York, NY, 2001)
8. L.F. Liotta, Catalytic oxidation of volatile organic compounds on supported noble metals. Appl. Catal. B Environ. **100**, 403–412 (2010)
9. H. Huang, Y. Xu, Q. Feng, D.Y.C. Leung, Low temperature catalytic oxidation of volatile organic compounds: a review. Cat. Sci. Technol. **5**, 2649–2669 (2015)
10. M.S. Kamal, S.A. Razzak, M.M. Hossain, Catalytic oxidation of volatile organic compounds (VOCs)—a review. Atmos. Environ. **140**, 117–134 (2016) https://doi.org/10.1016/J.ATMOSENV.2016.05.031
11. H.C. Genuino, S. Dharmarathna, E.C. Njagi, et al., Gas-phase total oxidation of benzene, toluene, ethylbenzene, and xylenes using shape-selective manganese oxide and copper manganese oxide catalysts. (2012). https://doi.org/10.1021/jp301342f
12. D.A. Aguilera, A. Perez, R. Molina, S. Moreno, Cu-Mn and Co-Mn catalysts synthesized from hydrotalcites and their use in the oxidation of VOCs. Appl. Catal. B Environ. **104**, 144–150 (2011). https://doi.org/10.1016/j.apcatb.2011.02.019
13. M. Piumetti, S. Bensaid, T. Andana, et al., Cerium-copper oxides prepared by solution combustion synthesis for total oxidation reactions: from powder catalysts to structured reactors. Appl. Catal. B Environ. **205**, 455–468 (2017). https://doi.org/10.1016/j.apcatb.2016.12.054
14. D. Delimaris, T. Ioannides, VOC oxidation over MnOx-CeO2 catalysts prepared by a combustion method. Appl. Catal. B Environ. **84**, 303–312 (2008). https://doi.org/10.1016/j.apcatb.2009.02.003
15. D. Delimaris, T. Ioannides, VOC oxidation over CuO-CeO2 catalysts prepared by a combustion method. Appl. Catal. B Environ. **89**, 295–302 (2009). https://doi.org/10.1016/j.apcatb.2009.02.003
16. H. Lu, X. Kong, H. Huang, et al., Cu–Mn–Ce ternary mixed-oxide catalysts for catalytic combustion of toluene. J. Environ. Sci. **32**, 102–107 (2015). https://doi.org/10.1016/J.JES.2014.11.015
17. M. Piumetti, D. Fino, N. Russo, Mesoporous manganese oxides prepared by solution combustion synthesis as catalysts for the total oxidation of VOCs. Appl. Catal. B Environ. **163**, 277–287 (2015). https://doi.org/10.1016/j.apcatb.2014.08.012
18. M.J. Marin Figueredo, T. Andana, S. Bensaid, et al., Cerium–copper–manganese oxides synthesized via solution combustion synthesis (SCS) for total oxidation of VOCs. Catal. Letters. **150**, 1821–1840 (2020). https://doi.org/10.1007/s10562-019-03094-x
19. W.M. Haynes, *CRC Handbook of Chemistry and Physics: A Ready-Reference Book of Chemical and Physical Data*, 95th edn. (CRC Press, Boca Raton, 2014)
20. C. Julien, M. Massot, R. Baddour-Hadjean, et al., Raman spectra of birnessite manganese dioxides. Solid State Ionics **159**, 345–356 (2003). https://doi.org/10.1016/S0167-2738(03)00035-3

21. S. Bach, M. Henry, N. Baffier, J. Livage, Sol-gel synthesis of manganese oxides. J. Solid State Chem. **88**, 325–333 (1990). https://doi.org/10.1016/0022-4596(90)90228-P
22. X. Tang, Y. Li, X. Huang, et al., MnOx-CeO2 mixed oxide catalysts for complete oxidation of formaldehyde: effect of preparation method and calcination temperature. Appl. Catal. B Environ. **62**, 265–273 (2006). https://doi.org/10.1016/j.apcatb.2005.08.004
23. A. Pintar, J. Batista, S. Hočevar, TPR, TPO, and TPD examinations of Cu0.15Ce0.85O 2-y mixed oxides prepared by co-precipitation, by the sol-gel peroxide route, and by citric acid-assisted synthesis. J. Colloid Interface Sci. **285**, 218–231 (2005). https://doi.org/10.1016/j.jcis.2004.11.049
24. S.E. Shirsath, D. Wang, S.S. Jadhav, et al., Ferrites Obtained by Sol–Gel Method, in *Handbook of Sol-Gel Science and Technology*, (Springer International Publishing, New York, 2018), pp. 1–41
25. R.D. Shannon, Revised effective ionic radii and systematic studies of interatomic distances in halides and chalcogenides. Acta. Crystallogr. Sect. A. **32**, 751–767 (1976). https://doi.org/10.1107/S0567739476001551
26. F. Kapteijn, L. Singoredjo, A. Andreini, J.A. Moulijn, Activity and selectivity of pure manganese oxides in the selective catalytic reduction of nitric oxide with ammonia. Appl. Catal. B Environ. **3**, 173–189 (1994). https://doi.org/10.1016/0926-3373(93)E0034-9
27. G. Avgouropoulos, T. Ioannides, H. Matralis, Influence of the preparation method on the performance of CuO-CeO 2 catalysts for the selective oxidation of CO. Appl. Catal. B Environ. **56**, 87–93 (2005). https://doi.org/10.1016/j.apcatb.2004.07.017
28. G. Munteanu, L. Ilieva, D. Andreeva, Kinetic parameters obtained from TPR data for α-Fe2O3 and Au/α-Fe2O3 systems. Thermochim. Acta **291**, 171–177 (1997). https://doi.org/10.1016/S0040-6031(96)03097-3
29. J. Zieliński, I. Zglinicka, L. Znak, Z. Kaszkur, Reduction of Fe2O3 with hydrogen. Appl. Catal. A Gen. **381**, 191–196 (2010). https://doi.org/10.1016/j.apcata.2010.04.003
30. A. Wöllner, F. Lange, H. Schmelz, H. Knözinger, Characterization of mixed copper-manganese oxides supported on titania catalysts for selective oxidation of ammonia. Appl. Catal. A Gen. **94**, 181–203 (1993). https://doi.org/10.1016/0926-860X(93)85007-C
31. T. Li, Y. Yang, C. Zhang, et al., Effect of manganese incorporation manner on an iron-based catalyst for Fischer-Tropsch synthesis. J. Nat. Gas Chem. **16**, 244–251 (2007). https://doi.org/10.1016/S1003-9953(07)60055-3
32. I.R. Leith, M.G. Howden, Temperature-programmed reduction of mixed iron-manganese oxide catalysts in hydrogen and carbon monoxide. Appl. Catal. **37**, 75–92 (1988). https://doi.org/10.1016/S0166-9834(00)80752-6
33. L. Kundakovic, M. Flytzani-Stephanopoulos, Reduction characteristics of copper oxide in cerium and zirconium oxide systems. Appl. Catal. A Gen. **171**, 13–29 (1998). https://doi.org/10.1016/S0926-860X(98)00056-8
34. J.F. Moulder, W.F. Stickle, P.E. Sobol, K.D. Bomben, *Handbook of X-Ray Photoelectron Spectroscopy* (Perkin-Elmer Corporation, Physical Electronics Division, Eden Prairie, 1992)
35. V.P. Santos, M.F.R. Pereira, J.J.M. Órfão, J.L. Figueiredo, The role of lattice oxygen on the activity of manganese oxides towards the oxidation of volatile organic compounds. Appl. Catal. B Environ. **99**, 353–363 (2010). https://doi.org/10.1016/j.apcatb.2010.07.007
36. S.C. Kim, W.G. Shim, Catalytic combustion of VOCs over a series of manganese oxide catalysts. Appl. Catal. B Environ. **98**, 180–185 (2010). https://doi.org/10.1016/j.apcatb.2010.05.027
37. M.C. Biesinger, B.P. Payne, A.P. Grosvenor, et al., Resolving surface chemical states in XPS analysis of first row transition metals, oxides and hydroxides: Cr, Mn, Fe, Co and Ni. Appl. Surf. Sci. **257**, 2717–2730 (2011). https://doi.org/10.1016/j.apsusc.2010.10.051
38. M.C. Biesinger, L.W.M. Lau, A.R. Gerson, R.S.C. Smart, Resolving surface chemical states in XPS analysis of first row transition metals, oxides and hydroxides: Sc, Ti, V, Cu and Zn. Appl. Surf. Sci. **257**, 887–898 (2010). https://doi.org/10.1016/j.apsusc.2010.10.051

39. W. Liu, M. Flytzani-Stephanopoulos, Total oxidation of carbon monoxide and methane over transition metal-fluorite oxide composite catalysts.pdf. J. Catal. **153**, 317–332 (1995)
40. T. Yamashita, P. Hayes, Analysis of XPS spectra of Fe 2+ and Fe 3+ ions in oxide materials. Appl. Surf. Sci. **254**, 2441–2449 (2008). https://doi.org/10.1016/j.apsusc.2007.09.063
41. M. Piumetti, N. Russo, *Notes on Catalysis for Environment and Energy* (CLUT-Politecnico di Torino, Torino, 2017)
42. E.C. Njagi, H.C. Genuino, C.K. King, et al., Applied catalysis A: general catalytic oxidation of ethylene at low temperatures using porous copper manganese oxides. Appl. Catal. A Gen. **421-422**, 154–160 (2012). https://doi.org/10.1016/j.apcata.2012.02.011

Chapter 3
Preferential Oxidation of Carbon Monoxide in Hydrogen-Rich Streams over CuO/CeO_2 Catalysts: How Nano (and Subnano) Structure Affects Catalytic Activity and Selectivity

Almerinda Di Benedetto, Gianluca Landi, and Luciana Lisi

Abbreviations

(HR-)TEM	(High-resolution) transmission electron microscopy
(MW)CNT	(Multiwall) carbon nanotubes
(R)GO	(Reduced) Graphene oxide
[Bmim][BF_4]	1-Butyl-3-methylimidazolium tetrafluoroborate
BET	Brunauer-Emmet-Teller
BTC	1,3,5-Benzene-tricarboxylate
CO-PROX	Carbon monoxide preferential oxidation
CO-SMET	Carbon monoxide selective methanation
EDTA	Ethylenediaminetetraacetic acid
H_2-TPR	Hydrogen temperature programmed reduction
IL	Ionic liquid
MW	Microwave
NC	Nanocubes
NF	Nanofibers
NP	Nanoparticles
NR	Nanorods
NS	Nanospheres
PEM-FC	Proton exchange membrane fuel cell
SACOP	Silica aquagel coprecipitation
SFRD	Supercritical fluid reactive deposition

A. Di Benedetto
Dipartimento di Ingegneria Chimica, dei Materiali e della Produzione Industriale, University of Naples Federico II, Naples, Italy

G. Landi (✉) · L. Lisi
Institute of Sciences and Technologies for Sustainable Energy and Mobility CNR, Naples, Italy
e-mail: gianluca.landi@cnr.it

M. Piumetti, S. Bensaid (eds.), *Nanostructured Catalysts for Environmental Applications*, https://doi.org/10.1007/978-3-030-58934-9_3

TOF	Turnover frequency
TWC	Three way catalyst
WGS	Water gas shift
XPS	X-ray photoelectron spectroscopy
XRD	X-ray diffraction

3.1 Introduction

The need of green, renewable, transportable, and storable fuel is nowadays commonly accepted. In this context, hydrogen represents an ideal energy vector; it can be produced from several sources, such as solar, wind, biomass, and biogas, by different processes, such as electrolysis [1–3], reforming [4–7], thermochemical splitting [8, 9], photoelectrochemical splitting [10], etc. Moreover, hydrogen can be reacted with captured carbon dioxide to produce both fuels and fine chemicals.

The interest toward hydrogen as a fuel has been increased together with the development of proton exchange membrane fuel cells (PEM-FCs). PEM-FCs directly convert hydrogen and oxygen into water vapor and electricity; they show high efficiency, low operating temperature, and no moving parts, and are a mature technology [11]. Hydrogen streams produced by renewable and/or fossil fuel conversion (i.e., reforming, gasification, pyrolysis, etc.,) also contain CO, CO_2, and H_2O in different ratios. The hydrogen content is generally increased by one or two water gas shift (WGS) stages, according to Eq. (3.1).

$$CO + H_2O \leftrightarrow CO_2 + H_2 \tag{3.1}$$

Due to thermodynamic constraints, the CO concentration cannot be lowered below 0.5–2 vol.%. However, PEM-FCs suffer from poisoning by carbon monoxide even at very low concentration (≤100 ppm); as a matter of fact, CO strongly adsorbs onto Pt active sites of the PEM-FC anode [11]. This means that a purification stage is necessary.

Two main processes have been proposed in order to convert CO: selective methanation [CO-SMET; (Eq. 3.2)] and preferential oxidation [CO-PROX; (Eq. 3.3)] [12–14].

$$CO + 3H_2 \leftrightarrow CH_4 + 2H_2O \tag{3.2}$$

$$CO + 0.5O_2 \rightarrow CO_2 \tag{3.3}$$

Both of them present pros and cons. Both reactions are exothermic ($\Delta H_{CO\text{-}SMET} \approx -206$ kJ/mol$_{CO}$; $\Delta H_{CO\text{-}PROX} \approx -283$ kJ/mol$_{CO}$), and thus, no external heating is required (but heat management is needed). Both reactions compete with undesired reactions; CO_2 methanation and reverse WGS can occur in parallel with CO-SMET, while H_2 combustion and reverse WGS can occur in parallel with CO-PROX. H_2 consumption, even if undesired, is unavoidable in both processes. During CO-SMET, 3 moles of hydrogen per mole of carbon monoxide are consumed even at 100% selectivity (i.e., no CO_2 consumption by methanation or reverse WGS occurs). CO-PROX cannot be carried out under stoichiometric conditions, generally O_2 excess being necessary. However, oxygen cannot be fed to the Pt anode, because hydrogen should be directly burned without the production of protons to be transferred to the cathode side. Therefore, during CO-PROX, H_2 consumption is directly related to O_2 excess [15]. The typical equivalence ratio (λ; it corresponds to the ratio between the actual O_2/CO ratio and the stoichiometric one) is 2; this means that 1 mole of hydrogen per mole of carbon monoxide is consumed at complete CO and O_2 conversions. Accordingly, CO-PROX is characterized by a lower H_2 loss than CO-SMET.

On the other hand, O_2 supply for CO-PROX occurs via air addition, related to safety issues in the mixer; moreover, the low but not negligible hydrogen dilution with nitrogen is to be considered as a con absent for CO-SMET.

A definitive preference between CO-PROX and CO-SMET has not yet been defined; however, some prototype systems have been developed with one or more CO-PROX stages to purify the hydrogen stream [16–19].

The first paper on CO-PROX was published in 1922 [20], followed by other two papers in 1930 [21, 22]. In these works, CO-PROX was suggested as the purification stage of hydrogen streams for the Haber process. As is known, CO-PROX is not used in the Haber process and, thus, the interest toward this reaction dramatically dropped.

Fundamental studies on the platinum activity toward CO and H_2 oxidation [23–25] led to renewed interest toward CO-PROX [26]. As is known, platinum can be considered as an intrinsic preferential CO oxidation catalyst due to the strong CO adsorption at low temperature, which is also the reason for the high sensitivity of PEM Pt anodes to low CO partial pressures in the H_2 streams.

Starting from these works, the number of papers related to CO-PROX significantly increased, as shown in Fig. 3.1. Obviously, the increase was strictly related to the development of low-temperature PEM-FC.

Catalysts for preferential oxidation of carbon monoxide should show (1) good activity toward CO oxidation, (2) high selectivity (i.e., hydrogen oxidation rate should be lower than CO oxidation rate), (3) resistance to the inhibiting effect of carbon dioxide and water vapor (unavoidable in reformate streams), and (4) long life-time. Noble metal-based catalysts were first studied and are still of interest [27–33]. Their performance toward CO-PROX was also reviewed [12, 34]. In addition to platinum, gold nanoparticles were also proposed for CO-PROX showing interesting activity, selectivity, and resistance to the inhibiting effect of CO_2 and H_2O [27, 35–40]. A review of the most interesting results was published by Lakshmanan et al. [41].

Transition metal oxides, mainly supported on ceria, represent the last class of catalysts proposed for CO-PROX. Despite several attempts, copper oxide supported on ceria is undoubtedly the most interesting catalytic system. For instance, Cwele

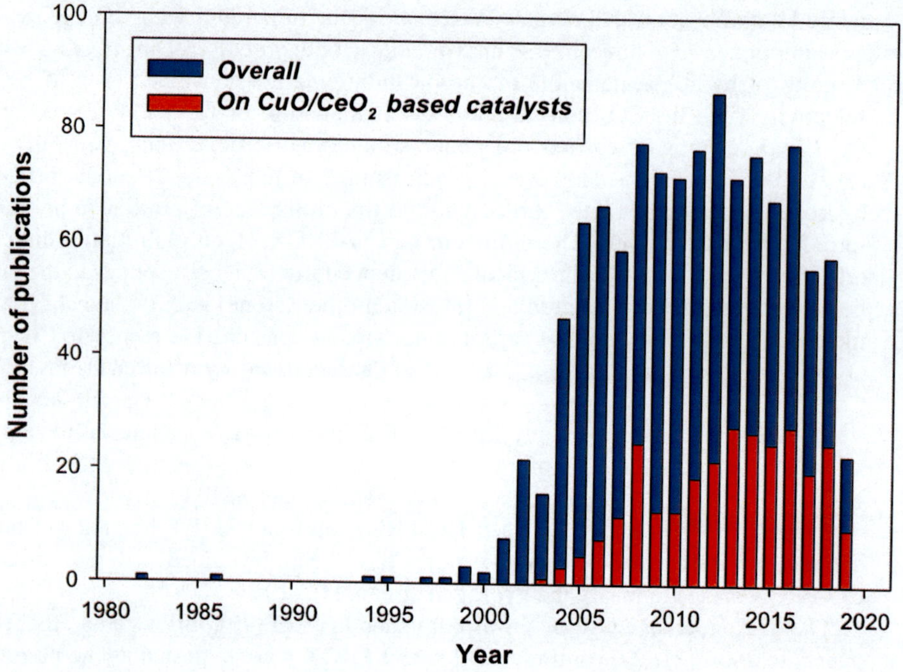

Fig. 3.1 Number of publications related to CO-PROX (blue bars) and focused on CuO/CeO_2 catalysts for CO-PROX (red bars) during the last 40 years (Source: Scopus. Access: May 2019)

et al. [42] and Chagas et al. [43] studied the promoting effect of copper on cobalt substituted ceria and Ni-CeO_2, respectively, and concluded that copper/ceria catalysts, i.e., the system without Co or Ni, showed the best catalytic properties.

As shown in Fig. 3.1, the number of papers devoted to CuO/CeO_2 catalysts for CO-PROX increased rapidly in the mid of 2000s and represents a significant fraction of the overall scientific production related to CO-PROX. It is widely accepted that a key feature, for the good activity of CuO/CeO_2 catalysts, is the strong interactions between copper and ceria resulting in an enhanced reducibility of both oxides [44–46]. It has been evidenced that copper reducibility depends on its dispersion on CeO_2, highly dispersed copper showing the best CO oxidation activity and selectivity [47]. Indeed, copper oxides clusters, not interacting with ceria, are recognized as active centers for the (undesired) H_2 oxidation [48, 49]. According to the above considerations, the nano and subnano structure of CuO/CeO_2 catalysts appears of fundamental interest in order to understand and tune the features affecting catalytic activity and selectivity.

In this chapter, the most interesting results on the effect of nano and subnano structures of CuO/CeO_2 catalysts on the CO-PROX performance are reviewed and critically discussed.

3.2 CuO/CeO_2 Catalysts

As reported above, the interest toward CuO/CeO_2 catalysts has grown during the last few years. Contemporarily, the research was devoted to the comprehension of the nature of the active sites and of the origin of their selectivity. A brief discussion on this topic is necessary before introducing how nanostructure affects the catalytic activity.

To date, a general consensus on the exact nature of the active sites has not been attained. Discussion is open on the oxidation state of active copper [50–55], results supporting both Cu^{2+} and Cu^{+} as active sites for CO adsorption and oxidation. In particular, it is not well established if the active sites for the CO adsorption step are isolated small Cu^{2+} clusters in close contact with CeO_2 [50] or Cu^{+} sites [51]. Moreover, copper active sites are assumed as well as dispersed CuO_x clusters [52], while surface $Cu_yCe_{1-y}O_{2-x}$ has been suggested as active centers by Elias et al. [53] and Arango-Diaz et al. [56].

Nevertheless, some points have been considered accepted and/or more reasonable. First, the key factor for good CO-PROX catalysts is a strong copper-ceria interaction, generally related to a high copper dispersion [57–60]. Moreover, good CO-PROX catalysts require a high (mainly surface and/or subsurface) oxygen mobility related to ceria sites, generally due to the formation of oxygen vacancies [59, 61–63]. It is worth noting that the development of novel in-situ and extra-situ characterization techniques and of more reliable theoretical models allowed a deep understanding of the role of copper-ceria interaction and oxygen vacancies in catalysis and, specifically, in the CO-PROX reaction [64].

In order to improve copper dispersion and to strengthen copper-ceria interaction, several preparation and post-treatment technique were proposed. Preparations included deposition–precipitation [65], incipient wetness impregnation [47, 48, 66, 67], coprecipitation [65, 68–70], sol–gel precipitation [70, 71], solution combustion synthesis [72], urea gelation/coprecipitation [71, 73], "solvent free" synthesis [71]. Examples of post-treatments are reported by Avgouropoulos and coworkers; they showed redispersion of copper species and improved oxygen vacancies under both acid [74] and basic [75] conditions. Du et al. [76] treated the as-calcined CuO/CeO_2 samples with copper loadings as high as 40 at.% with ammonium carbonate aqueous solutions showing that the weakly bonded copper species were removed by the leaching process and replaced by strongly bonded species (active toward CO-PROX) migrating during reduction or reaction condition.

Another key feature related to the strong copper-ceria interaction is the selectivity of the CuO/CeO_2 systems toward CO rather than H_2 oxidation. Several studies reported that the formation of metallic copper enhances H_2 reactivity [47, 69, 77–79], as sketched in Fig. 3.2; thus, over-reduction of copper under reaction conditions leads to the formation of unselective sites. Accordingly, the shift of copper reduction under H_2 at temperature significantly higher than the operation temperature of CO-PROX is beneficial to the catalyst selectivity [80]. Lopez Camara et al. [81] reported that the reduction of ceria particle dimension in an inverse CeO_2/CuO

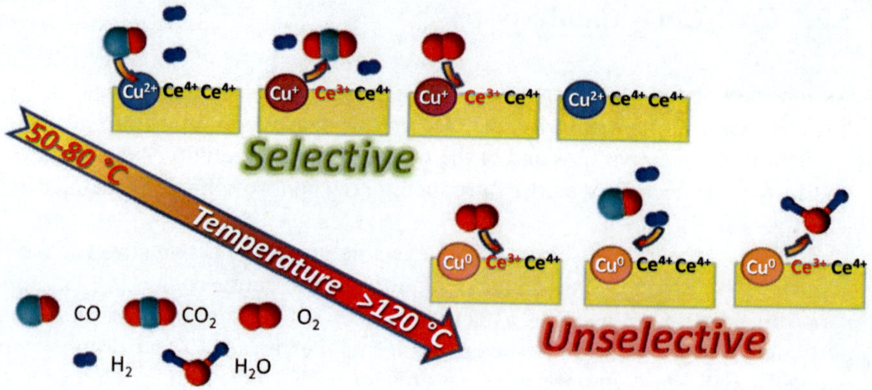

Fig. 3.2 Sketch of the effect of temperature on the copper oxidation state and its influence on the selectivity

catalyst (i.e., a system where ceria is dispersed on copper oxide) delayed copper reduction under H_2 due to a more favored interfacial electron transfer from copper to ceria.

The above considerations suggest that the use of nanoparticles with a high number of copper-ceria contact points can lead to more active and selective CO-PROX catalysts [82]. Indeed, more active structures are formed in the transition to the nanometer range than with the bulk analogs [83]. This is due to change in the unit cell parameters, the mobility of the atoms, the electronic structure, the surface morphology, the reactivity of active centers at the faces, edges, and points of the crystal lattice, etc. The systems formed during the incorporation into nanostructures or inclusion in porous matrices acquire new properties in addition to the individual characteristics of isolated nanosized particles. Such interaction can result in the appearance of additional active centers at the contact points of the nanoparticles. In these systems, a synergic effect seems to be due to the decrease in the energy of electronic transitions in the zones of contact between nanoparticles of different chemical nature [84].

As reported above, this chapter is focused on the relationship between nanostructure and catalytic activity. Results of catalysts characterization, carried out using both conventional and advanced techniques (H_2-TPR, XRD, N_2 physisorption, XPS, and (HR-)TEM), reported in the literature are generally easily comparable. In contrast, catalytic activity is tested under a wide range of experimental conditions differing for CO and/or O_2 concentrations (and thus for CO/O_2 ratio), contact time, and CO_2 and/or H_2O addition (and their concentrations), the only common point generally being the investigated temperature window (i.e., 50–200 °C). The catalytic results are, then, difficult to compare out of a single paper. An attempt was done by López et al. more than 10 years ago [85]; in subsequent years, few works reported a comparison with literature results and generally in a very limited way (for instance, see Di Benedetto et al. [66]). Therefore, in the next sections, catalysts will be not ranked with respect to their activity and selectivity; the "fil rouge" will be the link among nanostructure, physicochemical feature, and catalytic activity.

3.2.1 *How Nanostructure Affects Catalytic Activity*

The role of particle dimension has been widely explored for copper-ceria catalysts. Sciré et al. [65] highlighted the effect of ceria nanoparticles in determining the catalytic performance. For instance, Di Benedetto et al. [55] reported the occurrence of CO oxidation at temperature as low as room temperature over a structured catalyst washcoated with CuO/CeO_2 having ceria nanoparticles as the support. The associated transient production of carbon dioxide was related to the high redox properties of the catalyst and, particularly, to the large availability of surface and labile oxygen. The same Authors studied the effect of copper loading on a commercial ceria nanopowder [66]. Outstanding catalytic performances were achieved due to the support nanodimension. In order to observe the effect of copper loading on CO conversion, a contact time as low as 0.027 g s cm^{-3} was necessary. The Authors suggested that on these very active catalysts under certain reaction conditions, the limiting step is not dependent on the copper sites, but more likely on the CO_2 desorption from ceria sites in the neighborhood of copper centers.

A large variety of methods have been used to synthesize ceria-based catalysts with a size as small as possible and, sometimes, with a special morphology, which will be briefly described below.

3.2.1.1 Preparation Techniques

Sciré et al. [65] found that CuCe catalysts prepared by coprecipitation technique have a high surface area and lower ceria particle dimension, which generate higher amounts of defective sites (corners and edges), where oxygen release occurs with a lower energy barrier.

As a consequence, different preparation techniques were proposed in order to produce nanometric catalyst particles.

Maciel et al. [44] obtained CuCe catalysts with particles as small as 5 nm by hydrothermal synthesis with copper content of about 1%, which showed the highest activity with respect to catalysts with higher copper load and larger particles. The same authors [86] prepared both CeO_2 and Cu/CeO_2 samples by hydrothermal synthesis comparing their properties to samples obtained by precipitation with the same composition. The high dispersion of copper in the nanometric support particles prepared by the hydrothermal method led to very good catalytic performance. Moreover, the small crystallites size of ceria (<10 nm) enhanced redox capability and ionic conductivity of CeO_2 due to the higher mobility, primarily of the oxygen ions, important for the CO-PROX mechanism.

Zhang et al. [87] used the urea gelation method to obtain Cu-Ce mixed catalyst with a high surface area and particle dimension of about 10 nm. Furthermore, they obtained mixed CuO and CeO_2 with the same particle size also by mechanical mixing of the two oxides prepared by urea gelation.

Araújo et al. [88] used a microwave assisted hydrothermal synthesis to obtain $Ce_{0.97}Cu_{0.03}O_2$ nanospheres with an average diameter of 20 nm and nanorods with an average diameter of 8 nm and 40 nm in length. Moreover, they observed size reduction of the nanoparticles from 20 to approximately 15 nm with increasing synthesis temperature. Catalysts synthesized at the maximum temperature (160 °C) were those with the best copper dispersion and best CO-PROX performance.

Arango-Diaz et al. [89] used the freeze-drying method to obtain CeO_2 particles with a regular size of 9.5 nm. Copper was successively dispersed. The freeze drying method consisted in a flask freezing of cerium nitrate and Ethylenediaminetetraacetic acid (EDTA) solution using liquid nitrogen to obtain an amorphous precursor which was then calcined.

The same authors used the freeze-drying method to synthesize ceria-zirconia and ceria-alumina supports.

Kosmambetova [84] reported that CeO_2 synthesized from ammonium nitrate precursor was characterized by greater defect structure, lesser occupation of the oxygen positions (increased amount of vacancies), and smaller average size of the crystallites (11 nm) compared with the sample from simple nitrate precursor (13 nm).

One of the effective methods to prepare nanosized catalysts is nanocasting generally using mesoporous silica as a sacrificial template [90]. Advantages of nanocasting method are: (a) uniform and small crystallite size, usually smaller than 10 nm; (b) high surface area and uniform pore size; (c) adjustable structure; and (d) crystalline framework, with domain sizes that do not increase with increase in synthesis temperature [68]. Gu et al. [68] synthesized ordered mesoporous Cu–Ce–O catalysts with different copper contents (5–70 mol%) using mesoporous silica KIT-6 as a hard template in one-step nanocasting method with a surface area as high as 159 $m^2\ g^{-1}$. The catalysts nanocasted in one step show better performance than the ones prepared by a two-step impregnation method.

A more unusual technique was also proposed [91, 92] using the highly ordered HKUST-1 with a microporous structure and a long copper–copper dimer distance (well separated by the 1,3,5-benzene-tricarboxylate (BTC) ligand with a distance of 0.8 nm) as the hard template to disperse cerium ions, which can be uniformly adsorbed into the channel of $Cu_3(BTC)_2$ (Fig. 3.3). After removing the organic ligand by thermal treatment, copper and ceria remain well distributed. The use of a nanoporous host provided 10–11 nm CuCe crystals after thermal treatment at 500–600 °C with very high CO-PROX performance.

Li et al. [93] used a special technique to produce CeO_2 nanoparticles (5 nm average size) with highly dispersed copper: the melt infiltration method into silica hollow nanospheres starting from cerium and copper nitrate precursors. They reported that this method allows an effective dispersion of precursor salts in the pores of the silica hollow nanospheres providing the maximum activity for a Cu/Ce ratio 1/8.

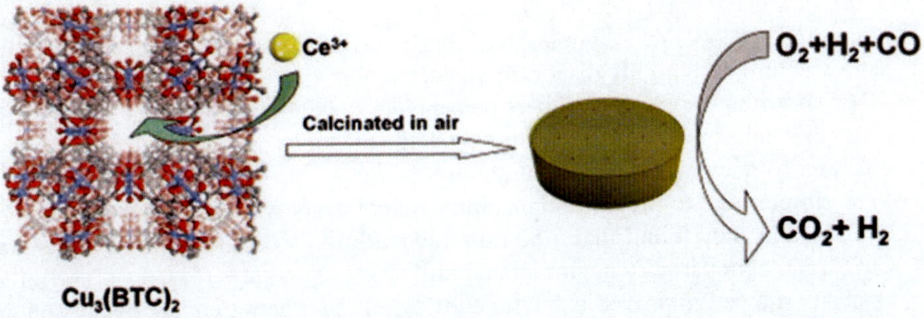

Fig. 3.3 CuO/CeO_2 catalyst prepared by MOF precursor for CO-PROX (From Zhang et el. [91] Copyright © 2012 Elsevier B.V. Reproduced with permission)

Tang et al. [94] proposed a facile solid state impregnation method to prepare $CuO\text{-}CeO_2/SBA\text{-}15$ catalysts. Both copper and ceria were well dispersed on SBA-15, and the CO-PROX activity was greatly enhanced, which could be attributed to the distinct reduction of ceria size.

Gong et al. [95] used the surfactant template method to produce both CeO_2, MnO_x and $CeO_2\text{-}MnO_x$ with nanometric size (<8 nm), which was preserved also after dispersion of CuO by wet impregnation. The same authors [95] obtained similar dimension of $CeO_2\text{-}MnO_x$ mixed oxides using both deposition-precipitation and surfactant template methods. The same effect was obtained by Zeng et al. [96] always on $MnO_x\text{-}CeO_2$ synthesized by the hydrothermal method but for the preparation of inverse catalysts. The $MnO_x\text{-}CeO_2$ particles on CuO have maximum 7 nm dimension.

Liu et al. [97] prepared the $CuO–CeO_2$ catalyst by coprecipitation and ethanol washing obtaining nanosized samples with a high surface area. They also found lattice defects over the surface of $CuO–CeO_2$, which are beneficial to enhance catalytic performance.

Chung and Yeh [98] prepared $CuO–CeO_2$ nanocomposites by coprecipitation of nitrates at changing pH values. They found that precipitation at high pH led to a high concentration of CeO_2 nucleation and significant number of Cu sites at interface between CuO and CeO_2. They also demonstrated that the samples obtained at high pH values exhibited very high CO conversion and selectivity.

Marban et al. [99] synthesized CuO_x/CeO_2 nanocatalysts using the silica aquagel coprecipitation (SACOP). They found very good performance in terms of both catalytic activity and selectivity. However, they also found catalyst deactivation. From their analysis it turns out that the reason for the catalyst deactivation is related to the formation of surface carbonates and to the aggregation of dispersed Cu^{2+} in CuO clusters.

Conversely, Sedmak et al. [100] showed that the nanostructured $Cu_{0.1}Ce_{0.9}O_{2-y}$ catalyst prepared by a sol–gel method exhibits very stable operation also under severe reducing conditions.

Xu et al. [101] prepared Cu-doped cerium oxide nanofiber catalyst via electrospinning. They obtained samples with high surface areas and a special fiber-like nanostructure with highly dispersed and active Cu^{2+} ions.

As shown in Fig. 3.4, nanofibers present length of about 50 μm and a diameter of about 200 nm. The TEM images also show the presence of a large quantity of nanofibers with an average diameter of about 8.5 nm.

A simple way to prepare nanocomposite catalysts was suggested by Borchers et al. [102]. They found that a 60 min ball milling procedure allowed to generate CuO/CeO_2 nanoparticles agglomerated into larger aggregates (Fig. 3.5). The milled catalysts showed improved catalytic activity due to a better copper dispersion and larger Cu^+ fraction.

In Table 3.1 the performances and the average nanometric size of most of the best catalysts discussed above are reported.

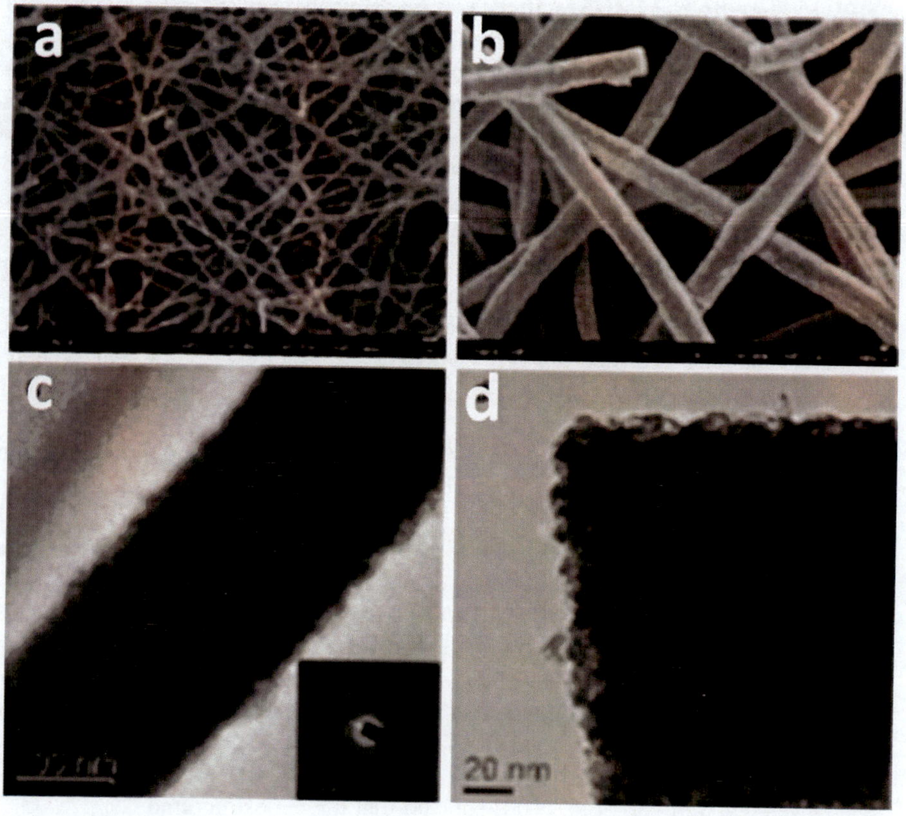

Fig. 3.4 SEM images of $Cu_{0.1}Ce_{0.9}O_{2-x}$ nanofibers (NFs) at low magnification (**a**) and high magnification (**b**); TEM images of a single $Cu_{0.1}Ce_{0.9}O_{2-x}$ nanofiber at low magnification (**c**) and high magnification (**d**) (From Xu et al. [101] Copyright © 2011 Elsevier B.V. Reproduced with permission)

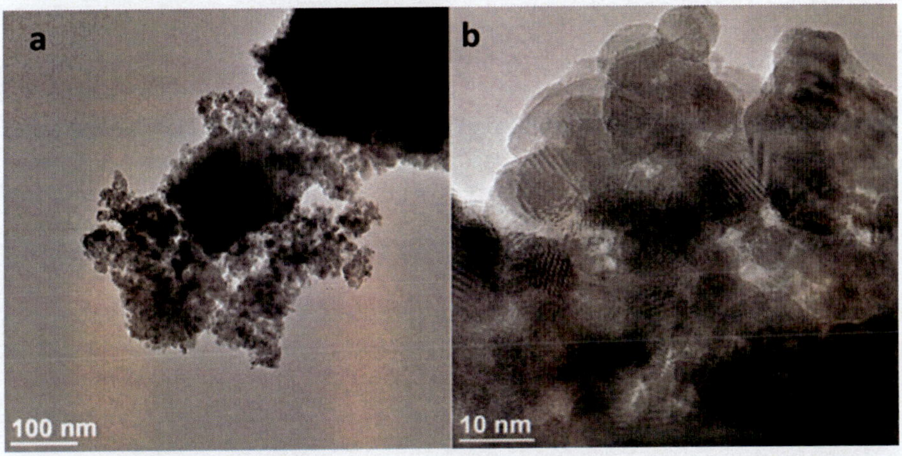

Fig. 3.5 (**a**) TEM micrograph of $CuO-CeO_2$ powder milled for 60 min; (**b**) enlargement showing dimension and shape of agglomerated nanoparticles (From Borchers et al. [102] Copyright © 2016 Elsevier B.V. Reproduced with permission)

Table 3.1 Performance of catalysts prepared by different preparation techniques: average particle size (*d*, nm), best CO conversion (*x*, %) and corresponding selectivity to CO_2 (*s*, %), temperature (*T*, °C), and contact time (*W*/*F*, g s cm^{-3})

Preparation techniques	*d*	x; s^a	T^b; W/F	References
Coprecipitation	6	95; 62	150; 0.038	[65]
Wet impregnation	<20	100; 100	110; 0.027	[55]
Hydrothermal synthesis	5–8	100; 100	110; n. a.	[44]
Urea gelation	11	n.a.; n.a.	60; n. a.	[87]
MW assisted hydrothermal synthesis	20	92; n. a.	150; n.a.	[88]
freeze-drying method	12	100; 97	90; 0.18	[89]
Nanocasting with mesoporous silica	8–9 (crystal)	100; 100	40; 0.097	[90]
Nanocasting with mesoporous silica	23 (crystal)	99; 50	170; 0.06	[68]
Thermal treat. of $Cu_3(BTC)_2$ precursor	11 (crystal)	100; 100	90; 0.2	[91]
Melt infiltration	5	90; 60	175; 0.0045	[93]
Surfactant template method	5 (crystal)	100; 50	140; 0.06	[95]
Ethanol washing	10 (crystal)	100; 95	120; 0.03	[97]
Coprecipitation at various pH	6	100; 100	120; 0.006	[98]
Silica aquagel coprecipitation (SACOP)	2–3 (crystal)	95; 68	208; 0.06	[99]
Electrospinning	8.5 (crystal)	100; 92	120; 0.3	[101]

n.a. not available (catalyst mass or flow rate not reported)

[a]Values of conversion and selectivity can be indicative in some cases because extracted from graphs. CO conversion reported is the best conversion obtained with the best performing catalyst. The value of selectivity is that corresponding to the best conversion

[b]Temperature corresponding to the best conversion

3.2.1.2 Particle Morphology

As reported by Trovarelli and Llorca [83], not only the dimension but also the shape of CeO_2 nanoparticles can significantly affect the metal-support interaction, thus providing enhancement of activity and selectivity. Thus, at the beginning of the last decade, in addition to the nanometric dimension the effect of ceria morphology was also investigated producing CeO_2 particles as stars, cubes, rods, etc.

Han et al. [103] synthesized ceria nanocrystals in different shapes: octahedra, rods and cubes with 10–20 nm dimension. They found that copper was deposited at higher concentration on octahedra but it better migrates from bulk crystal to the surface for rods. This determined a different CO-PROX activity at low and high temperature, rods working better at high temperature and cubes at lower temperature. On the other hand, Monte et al. [104] found that ceria nanocubes were the most selective due to the greater exposure of {001} facets, coordinatively more unsaturated, promoting a stronger interaction with copper.

Gamarra et al. [105] investigated the catalysts morphologies and obtained both CeO_2 support and Cu/CeO_2 catalysts with particles dimension as low as 7 nm as nanospheres (Fig. 3.6). These catalysts showed a better CO-PROX activity at low temperature with respect to nanocubes or nanorods with a larger dimension and lower surface area although they also found an effect of crystalline faces exposure on selectivity.

The effect of the phase exposure was also reported by Guo and Zhou [106]. They compared CuO/CeO_2 catalysts synthesized by the impregnation method on differently shaped ceria (rod, cube, plate and polyhedral), showing that CuO/CeO_2-rod and CuO/CeO_2-polyhedra exhibit a higher low-temperature catalytic oxidation activity coupled with a broader operating temperature window, while CuO/CeO_2-cube showed the worst performance. The reason was found in the improved dispersion of copper with a stronger interaction on the ceria rods and polyhedra.

Accordingly, Xie et al. [107] demonstrated that copper-ceria interaction is favored onto {111} and {002} ceria planes; so, nanospheres and nanospindles are

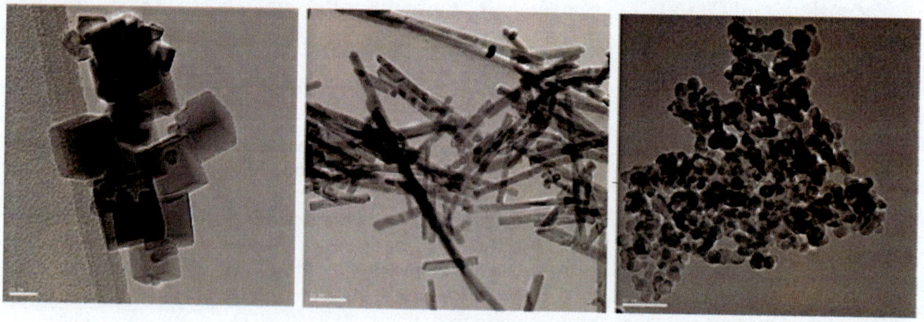

Fig. 3.6 TEM images of the different CeO_2 supports: CeO_2-nanocubes (left), CeO_2-nanorods (middle) and CeO_2-nanospheres (right). The horizontal scale bar shown corresponds to 50 nm except for CeO_2-NC for which it represents 20 nm (From Gamarra et al. [105] Copyright © 2013 Elsevier B.V. Reproduced with permission))

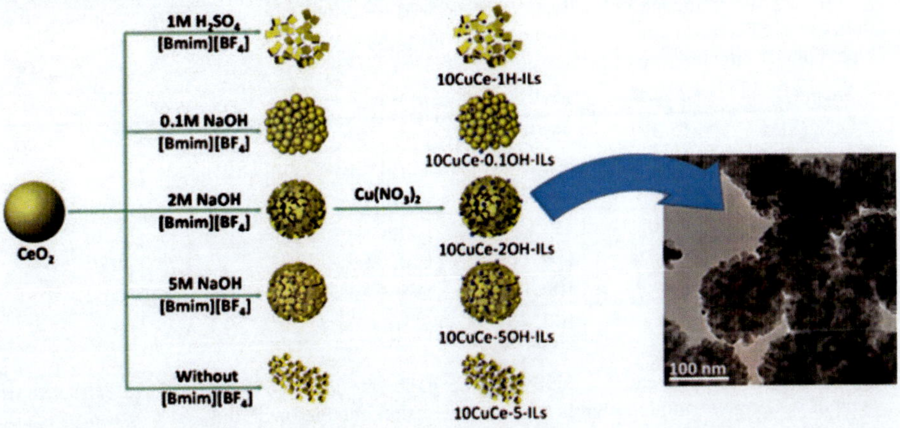

Fig. 3.7 Schematic illustration of the synthetic procedures of Cu_xO/CeO_2 nanocube catalysts and TEM image of the sample providing the best catalytic activity. Adapted from Gong et al. [109] (Copyright © 2017 Elsevier B.V. Reproduced with permission)

the preferred nanoshaped supports. Onto these supports in the presence of copper, the oxygen vacancies fraction is increased, thus improving the oxygen transfer from copper to ceria, according to the following equation.

$$Cu^{2+} + Ce^{3+} \rightarrow Cu^{+} + Ce^{4+} \tag{3.4}$$

Kou et al. [108] prepared CuO/CeO_2 nanowires starting from $Al_{90}Cu_{2.5}Ce_{7.5}$ (at%, nominal concentration) alloy ribbons by a de-alloying procedure in 5 M NaOH aqueous solution. The so-prepared catalysts preferentially expose the {110} ceria plane and showed good activity toward CO-PROX due to good copper dispersion and oxygen mobility.

Gong et al. [109] prepared ceria nanocubes by acid or basic treatment of nanospheres with an ionic liquid (IL) of 1-butyl-3-methylimidazolium tetrafluoroborate ([Bmim][BF$_4$]) under hydrothermal condition and used them as support for CuO/ CeO_2 catalysts (Fig. 3.7). The best results were obtained with a 2 M NaOH solution, providing the best copper-ceria interaction. Figure 3.7 shows a TEM image of this catalyst.

In Table 3.2 the performances and the average nanometric size of most of the best catalysts discussed above are reported.

3.2.1.3 Inverse CuO/CeO_2 Catalysts

In the same years also inverse catalysts were proposed. In these catalysts CeO_2, generally as nanoparticles, is deposited on CuO in order to enhance the number of contact points between the two oxides where active sites are supposed to be located. In 2010 Martinez-Arias and coworkers [110] first reported the inverse CuO/CeO_2

Table 3.2 Performance of catalysts characterized by different shapes: average particle size (d, nm), best CO conversion (x, %) and corresponding selectivity to CO_2 (s, %), temperature (T, °C), and contact time (W/F, g s cm^{-3})

Shape	d	x; s^a	T^b; W/F	References
Octahedra	15.3	95; 50	140; 0.1	[103]
Rods	n.a.	85; 70	170; 0.12 (in situ Drift)	[104]
Cubes	46	>99;70	170; 0.12	[105]
Polyhedra	5–10	>99; 100	85; 0.06	[106]
Spheres	180	100; 100	80; 0.09	[107]
Wires	20	100; 93	95; 0.45	[108]
Cubes	14.1	100; 70	150; 0.06	[109]

n.a. not available

[a]Values of conversion and selectivity can be indicative in some cases because extracted from graphs. CO conversion reported is the best conversion obtained with the best performing catalyst. The value of selectivity is that corresponding to the best conversion

[b]Temperature corresponding to the best conversion

catalyst for preferential CO oxidation. They showed that the amount and properties of copper-ceria interfacial sites in the inverse system keep a high level of CO oxidation activity due to the structure sensitivity of these catalysts.

The inverse CeO_2/CuO catalysts show a wider CO conversion window and higher CO_2 selectivity in comparison with the classical CuO/CeO_2 catalyst [111]. A synergistic effect occurs at the interface of the CeO_2-CuO catalyst and the particle sizes of CuO and CeO_2 directly determine the perimeter and area of this contact interface.

Different preparation techniques were proposed to prepare the inverse catalysts. Zeng et al. [111] prepared inverse catalysts using the solvothermal method to obtain the CuO precursor with cotton-ball-like morphology and CeO_2/CuO catalysts with a high BET surface area. The authors confirmed that the CO-PROX reaction takes place at the interface of the nanometric (<10 nm) CeO_2 particles and CuO.

Zeng et al. [112] compared direct and inverse copper cerium catalysts both prepared by the hydrothermal method. TEM analyses indicated that the inverse CeO_2/CuO catalyst has CeO_2 crystallites with a dimension of about 3–8 nm gathered into many small islands dispersed on the bulk CuO with 20–50 nm size. The traditional CuO/CeO_2 catalyst shows better activity at lower temperature and the inverse CeO_2/CuO catalyst present higher CO_2 selectivity when the CO conversion reaches 100% since it can still supply sufficient CuO for CO oxidation $T < 200$ °C. In the inverse sample the bulk CuO can chemisorb CO and H_2 at the suitable temperature and the small islands of CeO_2 can provide oxygen for CO and H_2 oxidation. This confirms that the reaction occurs at the contact interface of the CeO_2 islands and bulk CuO.

Zeng et al. [113] also proposed CeO_2/CuO catalysts prepared by hydrothermal method using Na_2CO_3 as precipitant with different Cu/Ce ratio. They obtained CeO_2 particles of about 10 nm and CuO particles with about 20 nm dimension. The CeO_2 particles self-assemble into the rod-like structures during the hydrothermal procedure and the rods of CeO_2 become shorter with the decrease of Ce/Cu molar ratio.

Furthermore, copper cations enter into the lattice of CeO_2 and the highly dispersed CuO are favorable for CO oxidation at lower temperature.

The same authors [114] used different precipitants (urea, NaOH, tetramethylammonium hydroxide) for a reverse microemulsion impregnation method to obtain inverse catalysts with an average particle size of 5 nm. They reported that the catalytic performance was closely associated to the extent of the contact interface between CeO_2 and CuO depending on the crystal sizes of CeO_2 and CuO. In a subsequent work [115] they found that CeO_2/CuO catalyst starting from urea precipitant was that with the smallest CuO crystallites. The particle-like structure consists of bulk CuO and filamentous and bowl-like structures of 5 nm CeO_2 crystallites.

Zeng et al. [116] also proposed CeO_2 nanoparticles supported on CuO with sphere-flower and petal-like morphologies. The particle size of CuO support was about 20 nm whilst that of CeO_2 dispersed particles was about 5 nm. They supposed a semi-spherical shape of CeO_2 particles because they observed an increase of TOF with increasing CeO_2 size. In fact, this maximizes the copper-cerium contact confirming that active sites are located at cerium-copper interface.

The same authors prepared rod-like CeO_2 particles with an average size of 10 nm dispersed on CuO by hydrothermal method with different Cu/Ce molar ratio. The smaller CeO_2 easily entered into the crystal lattice of CuO. Moreover, the rod-like CeO_2 exposed more {111} planes, representing those with the highest oxygen storage and transportation capacity. The contact interface of highly dispersed ceria and bulk CuO was more favorable to CO oxidation than to H_2 oxidation thus providing the best catalytic performance.

Zeng et al. [117] used the surfactant template method to synthesize both CuO/CeO_2 and inverse CeO_2/CuO catalysts to investigate the active sites that they supposed located at the contact interface between the two oxides. They obtained both CuO and CeO_2 with nanometric dimension.

The hydrothermal in combination with impregnation method was used to prepare the CeO_2/CuO catalysts with spherical structure [118]. The microspheres of CuO (20–25 nm) consist of the sheet-like CuO and the way of arrangement results in the formation of a shell structure. The core in the middle of shell structure was composed of the nanosized CuO particles. CeO_2 particles are supported on the surface of the CuO microspheres or embedded in the pores of sheet-like CuO and their size was 6–7 nm.

Catalysts in inverse configuration (CeO_2/CuO) were also proposed by López Cámara [119] with CeO_2 nanoparticles (4.9 nm) deposited on larger (12–20 nm) CuO nanoparticles. They also reported that the addition of ZnO nanoclusters further reduced both CeO_2 (3.9 nm) and CuO particle dimension and that resulted most favorable to the CO-PROX catalytic properties. On the basis of their results, they suggested that the crystal size of CeO_2 was a relevant parameter to tune CO-PROX characteristics of this type of inverse CeO_2/CuO catalysts.

Chen et al. [120] proposed a multistep process for synthesizing CeO_2 nanoparticles on Cu with a large concentration of oxygen vacancies using urea as the additive and NaOH as the precipitation agent. The HR-TEM analysis confirmed the presence on Cu surface of pillar-shaped CeO_2 nanoparticles with a diameter less than 10 nm

Table 3.3 Performance of different inverse CeO_2/CuO catalysts: best CO conversion (x, %) and corresponding selectivity to CO_2 (s, %), temperature (T, °C), and contact time (W/F, g s cm^{-3})

Catalyst	x; s^a	T^b; W/F	References
CeCu	>99; 100	90; 0.06	[110]
CeCu	100; 75	110; 0.09	[111]
CeCu	>99; 100	100; 0.09	[112]
CeCu	>99; 100	140; 0.09	[114]
CeCu	100; 80	160; 0.09	[115]
CeCu	100; 100	150; 0.09	[116]
CeCu	100; 100	100; 0.09	[117]
CeCu	100; 90	130; 0.09	[118]
CeCu-Zn	100; 100	150; 0.06	[81]
CeCu	100; 93	120; 0.2	[123]
CeCu	90; 60	160; 0.075	[121]
CuCe	100; 88	120; 0.09	[122]

[a]Values of conversion and selectivity can be indicative in some cases because extracted from graphs. CO conversion reported is the best conversion obtained with the best performing catalyst. The value of selectivity is that corresponding to the best conversion
[b]Temperature corresponding to the best conversion

with a fluorite-type structure. Oxygen vacancies were generated as a consequence of electron donation from metal copper atoms to the CeO_2 acceptor and the subsequent reverse spillover of oxygen induced by electron transfer in a well-controlled nanoheterojunction. The anchored oxygen vacancies play a bridging role in electron capture or transfer and drive oxygen molecules into active oxygen species to interact with the CO molecules adsorbed at interfaces, thus leading to an excellent preferential CO oxidation performance.

Gu et al. [121] prepared ceria nanorods and investigated the role of the copper content. The best catalyst was characterized by a Cu/Ce ratio equal to about 2.5. With respect to the other samples in the best catalyst, the Cu^+ and the Ce^{3+} fractions were enhanced.

Xie et al. [122] prepared inverse CeO_2/CuO catalysts by using star-shaped Cu_2O particles. Ceria was detected as nanoparticles dispersed onto the Cu_2O surface. Despite the starting copper oxidation state was one, the Authors related the catalytic activity to the coexistence of Cu^{2+} and Cu^+; in particular, they linked the redox copper cycle to the redox cerium cycle.

In Table 3.3, the performances and the average nanometric size of most of the best catalysts discussed above are reported.

3.2.1.4 CuO/CeO_2 on Carbon

In order to get strong copper-ceria interaction and a very high surface area, CuO/CeO_2 catalysts were supported onto carbon nanotubes and similar structures. Zeng et al. [124] supported CuO/CeO_2 onto multiwall carbon nanotubes (MWCNTs). The catalyst with 20 wt% CuO and equimolar amounts of copper and cerium showed the best catalytic activity; the sample was characterized by a large specific surface area (about 150 m^2 g^{-1}). The pore diameter of carbon tubes was about 20 nm before active phase deposition, decreasing down to 12 nm on the catalyzed systems, while ceria particles showed wire shape with almost 7 nm characteristic dimension. Results showed a relation between copper interaction with ceria, Ce^{3+} fraction (i.e. oxygen vacancies) and catalytic activity, thus confirming other results on this topic. On the other hand, the use of multiwall carbon nanotubes with a high surface area seemed to weaken the poisoning effect of H_2O and CO_2.

Similarly, Gao et al. [125] supported CuO/CeO_2 catalysts on MWCNTs with different Cu/Ce ratios and compared with Cu/CeO_2 supported onto activated carbon, γ-alumina and silica. Best performance was measured on the catalyst showing an equimolar ratio between copper and cerium supported onto MWCNTs. The Authors related the catalytic properties to a unique interaction between the active phase and the support, providing also the highest Ce^{3+} fraction.

Dongil et al. [126] supported copper-ceria catalysts onto carbon nanotubes (CNT) and graphene oxide (GO) and investigated the effect of K addition. Figure 3.8 shows TEM images of the investigated samples. Catalysts deposited onto CNTs were more active than those supported onto GO due to a better active phase dispersion. Moreover, K doping ($K \leq 1.0$ wt.%) resulted in an improved ceria dispersion and stronger copper-ceria interaction, thus improving catalytic performance toward CO-PROX.

Zeng and coworkers used reduced graphene oxide (RGO) as CuO/CeO_2 support in combination with MWCNTs [127] or alone [128]. In the absence of MWCNTs [128], the best catalyst was characterized by a Cu/Ce ratio equal to 1. Good copper and cerium dispersions were detected, suggesting good interaction; moreover, significant fractions of Cu^+ and Ce^{3+} were measured and related to the good catalytic activity. In addition, the sample showed improved resistance to the inhibiting effect of carbon dioxide and water vapor, probably due to the formation of hydroxyl groups by dissociative adsorption of H_2O on the surface of reduced graphene oxide. MWCNTs added to this catalyst [127] acted as spacer between RGO sheets both improving copper oxide and ceria dispersion and enhancing the concentration of oxygen vacancies. So, the ternary nanocomposite showed better performance toward CO-PROX reaction.

Zhang et al. [129] also prepared CuO/CeO_2 ($Cu/Ce = 1$) catalysts supported on reduced graphene oxide according to the procedure sketched in the Fig. 3.9. It is nothing that the active phase is dispersed as nanoparticles onto the support; additionally, the Authors suggested that CuO/CeO_2 particles were wrapped up in RGO layers with close interfacial interaction, thus providing optimal spatial condition for charge transport between RGO layers and CuO/CeO_2.

Shi et al. [130] supported CuO/CeO_2 with different Cu/Ce ratios onto commercial MWCNT by a special ultrasound-aided impregnation. The sample with a Cu/Ce ratio equal to 0.67 showed the best catalytic properties due to an enhanced copper reducibility and a good dispersion of CuO and CeO_2 particles. They also accounted for a higher fraction of oxygen vacancies.

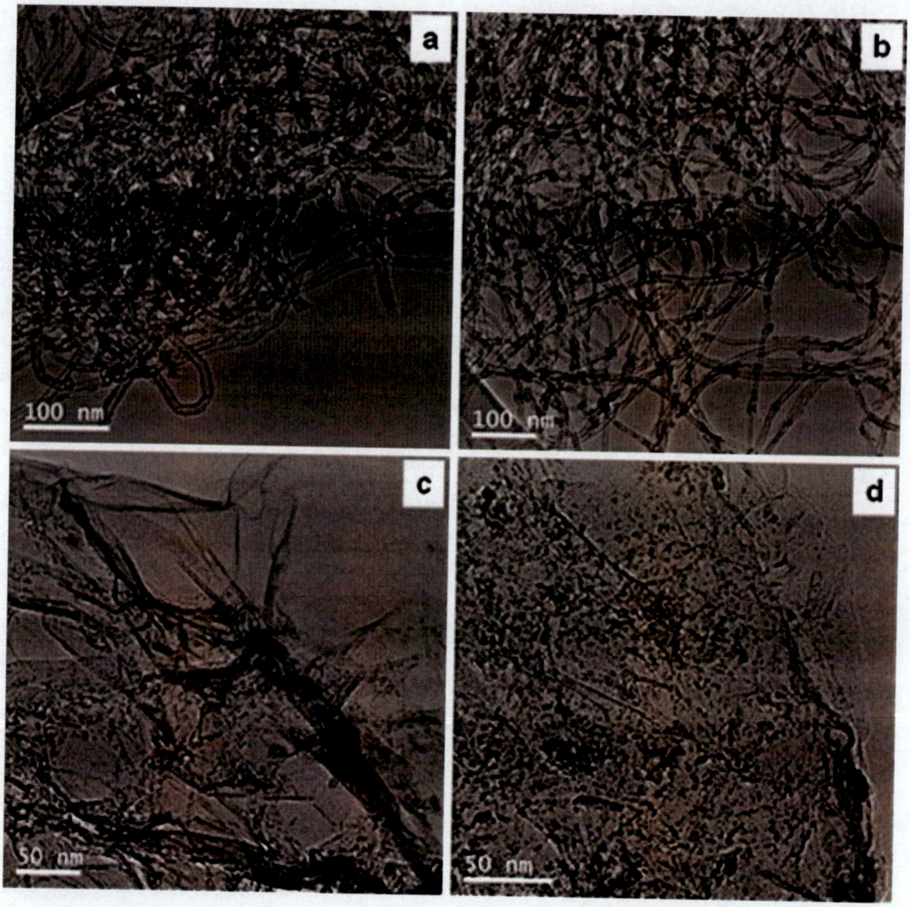

Fig. 3.8 TEM images of Cu/CeO_2 catalysts supported onto carbon nanotubes (**a**, **b**) and onto graphene oxide (**c**, **d**). Samples in figures **b** and **d** are doped with potassium (From Dongil et al. [126] Copyright © 2016 Royal Society of Chemistry Reproduced with permission)

In Table 3.4, the performances and the average nanometric size of most of the best catalysts discussed above are reported.

3.2.2 How Doping Affects Nano (and Subnano) Features

Improving catalytic activity and selectivity means improving copper dispersion and Cu-Ce interactions, thus reducing the typical dimension of the copper aggregates, i.e., the active sites. To this end, addition of other elements to Cu-Ce catalysts was used to improve catalytic properties.

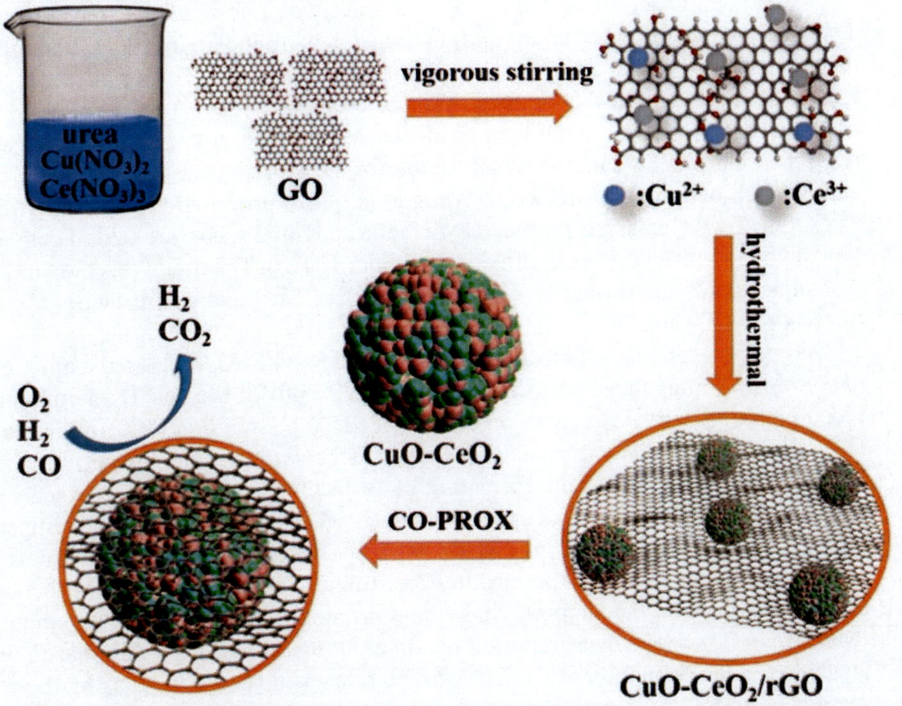

Fig. 3.9 Sketch of the preparation of RGO/CuO/CeO_2 catalysts (From Zhang et al. [129] Copyright © 2018 Springer Nature Reproduced with permission)

Table 3.4 Performance of CuCe catalysts supported onto different carbon-based supports: average particle size (*d*, nm), best CO conversion (*x*, %) and corresponding selectivity to CO_2 (*s*, %), temperature (*T*, °C), and contact time (*W*/*F*, g s cm^{-3})

Carbon support	*d*	*x*; *s*[a]	*T*[b]; *W*/*F*	References
MWCNT	7.0	100; 100	110; 0.09	[124]
MWCNT	4.3	100; 100	130; 0.03	[125]
CNT (+ K-doping)	4.2	100; 70	175; 0.03	[126]
RGO/MWCNT	2.8	100; 55	120; 0.09	[127]
RGO	3.2	95; 70	140; 0.09	[128]
RGO	100–220	100; 90	120; 0.06	[129]
CNT	n.a.	100; 100	120; 0.09	[130]

n.a. not available

[a]Values of conversion and selectivity can be indicative in some cases because extracted from graphs. CO conversion reported is the best conversion obtained with the best performing catalyst. The value of selectivity is that corresponding to the best conversion

[b]Temperature corresponding to the best conversion

Zr has been generally added to CeO_2 to improve performance, as occurring in TWCs. However, often different and, in some cases, opposite results are found in the literature. A positive effect was reported by Martinez-Arias et al. [131] and Reddy et al. [132]. On the contrary, Caputo et al. [133] and Ayastuy et al. [134] reported that adding Zr to the support decreases both activity and selectivity with respect to CuO/CeO_2 catalyst which outperforms with respect to $CuO/Ce_xZr_{1-x}O_2$ and is much better than CuO/ZrO_2. Wang et al. [135] proposed highly performing $CuO_x/Ce_{0.8}Zr_{0.2}O_2$ catalysts prepared by a facile and rapid route (urea grind combustion method) with a very small dimension (about 10 nm). The use of this preparation technique avoids the formation of bulk unselective CuO and also provides a good tolerance to CO_2 and H_2O.

Guo et al. [136] modified the redox properties of CuO supported onto CeO_2 nanorods by doping the ceria rods with Mn, Ti, Zr, and Ni. Mn and Ti addition provided higher activity, selectivity and resistance to the inhibition effect of CO_2 and H_2O, while Zr showed a marginal effect and Ni decreased the catalytic performance. Both Mn and Ti promoted the formation of surface defects and oxygen vacancies and incorporation of isolated copper. However, while Mn improved low temperature oxygen mobility due to its surface existence with multiple valence states, Ti promoted the formation of a perovskite-like structure enhancing copper dispersion and interaction with the support. An evident promoting effect of low Mn amounts ($Mn/(Mn + Ce) = 0.05$) was reported by Jin et al. too [137]. The main reason was attributed to the formation of Ce-Cu-Mn-O ternary solid solution, causing the formation of more oxygen vacancies, an improved reducibility and an increased amount of surface oxygen species. Further Mn addition decreased the catalytic performance and lowered the related physicochemical features.

Cecilia et al. [138] compared CuO/CeO_2 catalysts doped with Zr, La, Tb, and Pr to an undoped reference sample. Under “ideal” reaction conditions (i.e., in the absence of CO_2 and H_2O) the doped samples showed higher catalytic activity than the undoped one (except the La-doped sample), Zr-doped catalyst showing the best performance. These results suggested a relationship between the dopant charge and the catalytic activity, 4+ cations being thus preferable. However, under “real” reaction conditions (i.e., in the presence of CO_2 and H_2O) no significant difference among the sample was detected.

Also, the Nb addition was ineffective to improve the catalytic behavior of CuO/CeO_2 catalysts [139]. As a matter of fact, niobia did not modify the redox properties of the catalyst, i.e., it did not change the copper-ceria interaction, but it introduced acid sites useless for CO-PROX reaction.

As reported in the Sect. 3.1, Pt and Au based catalysts are very active toward CO-PROX. Accordingly, several efforts have been done in order to combine the good properties of CuO/CeO_2-based catalysts with those of Pt and Au.

Kugai et al. [140] reported that nanosized CuO particles or clusters interact with CeO_2 surface to exhibit synergistic effect in reducibility at the interfacial area. CO adsorbs on copper sites and oxygen is supplied from CeO_2 through the interface. Although CuO/CeO_2 shows high selectivity, it has drawback of slow kinetics at low CO concentration which makes noble metal still necessary for deep CO removal from H_2 stream. For this reason, they developed supported PtCu alloy nanoparticles (3–4 nm) synthesized by a unique radiation-induced synthesis process supported on CeO_2. This radiolytic process is described as a simple method to obtain nanoparticles with well-controlled structure and size through irradiating the aqueous ions of metal sources with no need for post-treatment by heat or other chemicals.

Cu-Pt alloy nanoparticles with a slightly higher dimension (4–5 nm) were deposited on CeO_2 or γ-Fe_2O_3 by the same authors by electron beam irradiation method [141, 142]. They also found that Pt-Cu with CeO_2 of small crystallite size had highest activity and selectivity in CO-PROX, preserved also when water was added in the mixture. The high oxygen transport capacity was attributed to small CeO_2 crystallite size (average size 24 nm).

In the work of Lang et al. [143] copper and platinum were loaded simultaneously on the ceria-coated alumina sponge by means of supercritical fluid reactive deposition (SFRD), which consists in adsorption and reduction of metal-organic complexes dissolved in supercritical CO_2. Nanosized copper inside and in close contact with the ceria layer provided a good reducibility and thus high activity of the catalyst.

About Au-doped CuO/CeO_2 catalysts, the formation of Au-Cu alloys has been reported [144–148]. In particular, alloys with nanometric dimension (5–10 nm) with preset stoichiometry dispersed on different support (SiO_2, CeO_2, γ-Al_2O_3) were prepared by Potemkin et al. [147] starting from $[Au(en)_2](NO_3)_3$ and $(NH_4)_2[Cu(C_2O_4)_2]\cdot 2H_2O$ complex compounds. These catalysts showed a higher selectivity in the realistic hydrogen-rich mixture containing CO_2 and H_2. Interestingly, Papavasiliou [148] reported decoration of CuO nanoparticles with gold (metallic and cationic) clusters; the presence of these species was related to the improved CO-PROX performance.

The promoting effect of iron addition to CeO_2 or CuO/CeO_2 mixed oxide has been studied in several papers [61, 80, 149–154]. Firsova et al. [149] reported a positive effect of iron addition due to the role of iron in the redox cycles, while Bao et al. [153] suggested that the improved catalytic activity of a Fe_2O_3–CeO_2 composite was related to the density of surface oxygen vacancies promoted by iron addition. Results reported in further papers confirmed that iron addition improved the catalytic properties by modifying the nanostructure of the catalysts. In particular, improved copper dispersion (and thus copper-ceria interaction) [80] and the generation of more oxygen vacancies, enhancing the Cu-Ce interaction, trapping the gas-phase oxygen, and promoting the mobility of subsurface lattice oxygen, [61] were claimed to explain the Fe-promoting effect. Recently, Dasireddy et al. [155] prepared bimetallic Cu-Fe/CeO_2 and Cu-Co samples supported onto carbon nanotubes (CNT) and compared their features and performance to those of monometallic ones. The Authors detected the formation of a $CuFe_2O_4$ phase enhancing the copper dispersion. From the catalytic point of view, the Fe-promoted sample showed a significant resistance to the inhibiting effect of CO_2 and H_2O. In particular, CO_2 acted as inhibitor, while H_2O as promoter. The latter effect was probably due to the formation of hydroxyl group on the surface, boosting the catalytic activity, in agreement with Bueno-Lopez and coworkers [54]. In contrast, on a Fe-promoted CuO/CeO_2

Table 3.5 Performance of CuCe catalysts doped with different elements: average particle size (*d*, nm), best CO conversion (*x*, %) and corresponding selectivity to CO_2 (*s*, %), temperature (*T*, °C), and contact time (*W*/*F*, g s cm^{-3})

Dopant	*d*	*x*; *s*[a]	*T*[b]; *W*/*F*	References
Zr (Ce/Zr = 4)	10	100; 100	120; 0.15	[135]
Mn (Mn/(Mn + Ce) = 0.05)	8	100; 100	100; 0.06	[136]
Pt (Cu/Pt = 9)	4 (PtCu particles)	100; 30	90; 0.12	[140]
Au (0.15 wt.%)	5–15	100; 90	120; 0.144	[148]
Fe (5 wt%)	n.a.	100; 75	140; 0.054	[80]
K (K/Cu = 0.68)	4.2	100; 75	175; 0.03	[158]

n.a. not available

[a]Values of conversion and selectivity can be indicative in some cases because extracted from graphs. CO conversion reported is the best conversion obtained with the best performing catalyst. The value of selectivity is that corresponding to the best conversion

[b]Temperature corresponding to the best conversion

catalyst, both CO_2 and H_2O lowered the catalytic performance [156]. The Authors reported that the CO_2 showed the most inhibiting effect, even higher than $CO_2 + H_2O$, while H_2O addition slightly decreased the catalytic activity. Therefore, iron-promoted catalysts seem to be less sensitive to H_2O addition. However, the promoting effect of H_2O detected by Dasireddy et al. [155] could also be due to the CNT support.

Doping with potassium was demonstrated to be effective to enhance the catalytic properties of CuO/CeO_2 catalysts, especially in terms of resistance to the inhibiting effect of CO_2 and H_2O [157]. More recently, Dongil et al. [158] investigated the whole group of alkalis (Li, Na, K, Cs), added to CuO/CeO_2 samples supported onto carbon nanotubes, thus expanding their previous study on K-doped catalysts [126]. The results suggested a significant role of the nature of the alkali element on the dispersion of ceria and the Cu-ceria interaction. This was also due to the formation of CeO_2 particles with small crystal size (4–6 nm) in the presence of alkalis (except Cs). These features were related to the improved catalytic performance of K-, Li-, and Na-doped samples, respectively. In contrast, Cs showed a detrimental effect on the CO-PROX activity.

In Table 3.5, the performances and the average nanometric size of most of the best catalysts discussed above are reported.

3.3 Conclusions and Outlooks

The preferential oxidation of carbon monoxide poses both fundamental and applicative issues. Research was, thus, devoted both to investigate the relationship between physicochemical features and catalytic properties and to engineer effective catalysts. From this point of view, copper-ceria catalysts are very intriguing. From the beginning of their exploration, it was clear that the copper-ceria interaction and, thus, the copper dispersion play a fundamental role, pushing the research at nanoscale levels.

Reduction of ceria size to nanoparticles showed a positive effect on the catalytic properties of CuO/CeO_2 systems and several preparation methods were proposed for this purpose. The nanodimensions improve both Cu dispersion and oxygen mobility and prevent, at the same time, the over-reduction of copper to its metallic state, generally identified as the H_2 oxidation site, thus affecting also the selectivity. Another interesting feature of nanosized copper-ceria catalysts is their easier desorption of carbon dioxide; this property is reflected into higher activity at low temperature, where CO_2 desorption is the rate-limiting step. An improved resistance toward the inhibiting effect of CO_2, typically present in the composition of the reformed stream, is a secondary but nonetheless beneficial effect.

Results reported above show that the ceria nanomorphology affects the catalytic properties as well as its dimension. It has been demonstrated that copper is preferentially dispersed onto defective planes, where copper-ceria interaction is boosted. Accordingly, catalysts prepared with specific morphologies are generally more active and selective.

The fundamental role of copper dispersion and its interaction with ceria is indirectly demonstrated by the development of "inverse" CeO_2/CuO catalysts. In these materials, ceria is deposited onto copper oxide; results clearly showed that the formation of nanostructures related to ceria and/or copper oxide, improving the contact points between the two oxides, is directly related to the catalytic activity and selectivity.

Deposition of both CeO_2 and CuO onto carbon-based supports intrinsically provides an intimate contact between copper and ceria. Moreover, the carbon-based supports offer large surface area, stable structure and they weaken the inhibiting effect of H_2O and CO_2 as they disadvantage their adsorption. It is worth noting that the optimal Ce/Cu ratio is around 1; evidently, the very small size of catalyst particles dispersed onto the support surface enhances the Cu-Ce distribution and interaction, thus also shifting the best composition to the equimolar ratio with respect to conventional CuO/CeO_2 catalysts.

Dopants also affect significantly the catalytic properties. In ceria-based catalysts, dopant addition is generally related to the modification of the redox properties and, in particular, to the improvement of bulk and/or surface oxygen vacancies. With regard to CuO/CeO_2 systems for CO-PROX, a larger number of oxygen vacancies is not straightforwardly linked to better catalytic performance. As a matter of fact, the improved redox properties should be coupled with a better copper dispersion in order to obtain more performing catalytic systems.

Figure 3.10 summarizes the above considerations.

Despite the absence of a general consensus on the exact nature of the active sites, the results reported in this chapter show that good catalysts based on CuO/CeO_2 for CO-PROX should have a copper-ceria interaction as wide as possible. Accordingly, the design of novel catalysts implies their engineering at the nanoscale level with the aim of a better utilization of the active components. Moreover, an opportune choice of additional components (dopants and/or supports as graphene and carbon

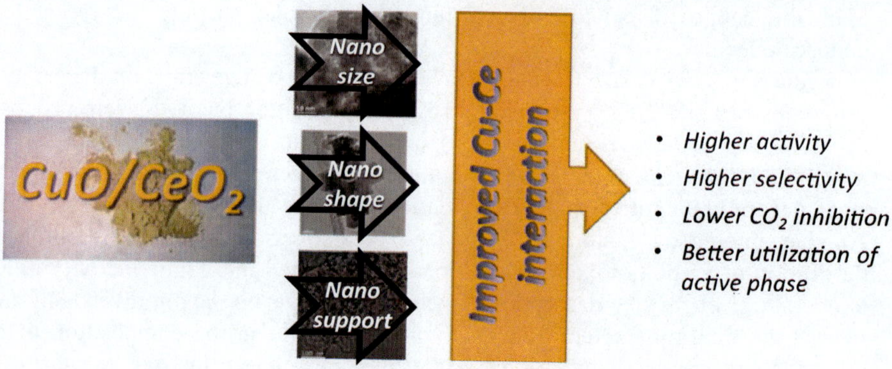

Fig. 3.10 How nanostructuring can affect the catalytic performance of CuO/CeO_2-based catalysts toward CO-PROX

nanotubes) can improve not only the intrinsic activity but also the resistance to the inhibiting effect of carbon dioxide and water vapor, unavoidable in reformate streams.

Finally, this chapter did not address the issue of structuring the active phase onto monolithic systems. However, structured catalysts are necessary under applicative conditions. Accordingly, another topic of interest is the development of active phases and preparation methods suitable for deposition onto substrates such as honeycombs, foams, and slabs.

References

1. D. Yadav, R. Banerjee, Economic assessment of hydrogen production from solar driven high-temperature steam electrolysis process. J. Clean. Prod. **183**, 1131–1155 (2018). https://doi.org/10.1016/j.jclepro.2018.01.074
2. K. Bareiß, C. de la Rua, M. Möckl, T. Hamacher, Life cycle assessment of hydrogen from proton exchange membrane water electrolysis in future energy systems. Appl. Energy **237**, 862–872 (2019). https://doi.org/10.1016/j.apenergy.2019.01.001
3. J. Moir, N. Soheilnia, P. O'Brien, A. Jelle, C.M. Grozea, D. Faulkner, M.G. Helander, G.A. Ozin, Enhanced hematite water electrolysis using a 3D antimony-doped tin oxide electrode. ACS Nano **7**, 4261–4274 (2013). https://doi.org/10.1021/nn400744d
4. E.C. Vagia, A.A. Lemonidou, Thermodynamic analysis of hydrogen production via autothermal steam reforming of selected components of aqueous bio-oil fraction. Int. J. Hydrog. Energy **33**, 2489–2500 (2008). https://doi.org/10.1016/j.ijhydene.2008.02.057
5. R. Rezaei, G. Moradi, Study of the performance of dry methane reforming in a microchannel reactor using sputtered Ni/Al2O3 coating on stainless steel. Int. J. Hydrog. Energy **43**, 21374–21385 (2018). https://doi.org/10.1016/j.ijhydene.2018.09.200
6. S. Kim, S.W. Yun, B. Lee, J. Heo, K. Kim, Y.T. Kim, H. Lim, Steam reforming of methanol for ultra-pure H2 production in a membrane reactor: techno-economic analysis. Int. J. Hydrog. Energy **44**, 2330–2339 (2018). https://doi.org/10.1016/j.ijhydene.2018.08.087

7. V. Fierro, O. Akdim, H. Provendier, C. Mirodatos, Ethanol oxidative steam reforming over Ni-based catalysts. J. Power Sources **145**, 659–666 (2005). https://doi.org/10.1016/j.jpowsour.2005.02.041
8. C.L. Muhich, B.D. Ehrhart, I. Al-Shankiti, B.J. Ward, C.B. Musgrave, A.W. Weimer, A review and perspective of efficient hydrogen generation via solar thermal water splitting. Wiley Interdiscip. Rev. Energ. Environ. **5**, 261–287 (2016). https://doi.org/10.1002/wene.174
9. G. Luciani, G. Landi, C. Imparato, G. Vitiello, F.A. Deorsola, A. Di Benedetto, A. Aronne, Improvement of splitting performance of Ce0.75Zr0.25O2 material: tuning bulk and surface properties by hydrothermal synthesis. Int. J. Hydrog. Energy **44**, 17565–17577 (2019). https://doi.org/10.1016/j.ijhydene.2019.05.021
10. P.M. Rao, I.S. Cho, X. Zheng, Flame synthesis of WO3 nanotubes and nanowires for efficient photoelectrochemical water-splitting. Proc. Combust. Inst. **34**, 2187–2195 (2013). https://doi.org/10.1016/j.proci.2012.06.122
11. A. Albarbar, M. Alrweq, *Proton Exchange Membrane Fuel Cells: Review* (Springer International Publishing, Cham, 2018)
12. E.D. Park, D. Lee, H.C. Lee, Recent progress in selective CO removal in a H2-rich stream. Catal. Today **139**, 280–290 (2009). https://doi.org/10.1016/j.cattod.2008.06.027
13. G. Ercolino, M.A. Ashraf, V. Specchia, S. Specchia, Performance evaluation and comparison of fuel processors integrated with PEM fuel cell based on steam or autothermal reforming and on CO preferential oxidation or selective methanation. Appl. Energy **143**, 138–153 (2015). https://doi.org/10.1016/j.apenergy.2014.12.088
14. M.A. Ashraf, G. Ercolino, S. Specchia, V. Specchia, Final step for CO syngas clean-up: comparison between CO-PROX and CO-SMET processes. Int. J. Hydrog. Energy **39**(31), 18109–18119 (2014)
15. Q. Zhang, L. Shore, R.J. Farrauto, Selective CO oxidation over a commercial PROX monolith catalyst for hydrogen fuel cell applications. Int. J. Hydrog. Energy **37**, 10874–10880 (2012). https://doi.org/10.1016/j.ijhydene.2012.04.032
16. F. Barrai, M.J. Castaldi, Experimental investigation of a JP8 fuel processor: autothermal reformer and CO-cleanup train. Ind. Eng. Chem. Res. **49**, 1577–1587 (2010). https://doi.org/10.1021/ie901735x
17. L. Du, J. Zhang, L. Sun, Z. Yuan, D. Li, A compact fuel processor integrated with 75kw PEM fuel cells, in *2011 International Conference on Electric Information and Control Engineering, ICEICE 2011—Proceedings. IEEE,* (2011), pp. 1906–1910
18. C. Fabiano, C. Italiano, A. Vita, L. Pino, M. Laganà, V. Recupero, Performance of 1.5 Nm3/h hydrogen generator by steam reforming of n-dodecane for naval applications. Int. J. Hydrog. Energy **41**, 19475–19483 (2016). https://doi.org/10.1016/j.ijhydene.2016.07.166
19. F. Cipitì, L. Pino, A. Vita, M. Laganà, V. Recupero, Performance of a 5 kWe fuel processor for polymer electrolyte fuel cells. Int. J. Hydrog. Energy **33**, 3197–3203 (2008). https://doi.org/10.1016/j.ijhydene.2008.02.042
20. A.B. Lamb, C.C. Scalione, G. Edgar, The preferential catalytic combustion of carbon monoxide in hydrogen. J. Am. Chem. Soc. **44**, 738–757 (1922). https://doi.org/10.1021/ja01425a007
21. W.E. Kuentzel, The preferential catalytic oxidation of carbon monoxide in the presence of hydrogen. I. The activity of two water gas conversion catalysts, of copper oxide, of manganese dioxide and of a mixture of these oxides. J. Am. Chem. Soc. **52**, 437–444 (1930). https://doi.org/10.1021/ja01365a001
22. W.E. Kuentzel, The preferential catalytic oxidation of carbon monoxide in the presence of hydrogen. II. The activity of 2-component hopcalites. J. Am. Chem. Soc. **52**, 445–455 (1930). https://doi.org/10.1021/ja01365a002
23. P.C. Liao, J.J. Carberry, T.H. Fleisch, E.E. Wolf, CO oxidation activity and XPS studies of PtCu γ-Al2O3 bimetallic catalysts. J. Catal. **74**, 307–316 (1982). https://doi.org/10.1016/0021-9517(82)90036-7
24. S. Akhter, J.M. White, The effect of oxygen islanding on Co and H2 oxidation on Pt(111). Surf. Sci. **171**, 527–542 (1986). https://doi.org/10.1016/0039-6028(86)91058-7

25. V. Gorodetskii, W. Drachsel, J.H. Block, The surface specificity of the oscillating CO oxidation on platinum investigated by field ion microscopy. Appl. Surf. Sci. **76–77**, 122–128 (1994). https://doi.org/10.1016/0169-4332(94)90332-8
26. A. Lesar, T. Yamanaka, Y. Ohno, T. Matsushima, Preferential occurrence of carbon monoxide oxidation on two reaction sites on platinum (113). Chem. Phys. Lett. **234**, 330–336 (1995). https://doi.org/10.1016/0009-2614(95)00061-8
27. G. Avgouropoulos, T. Ioannides, C. Papadopoulou, J. Batista, S. Hocevar, H.K. Matralis, A comparative study of Pt/γ-Al2O3, Au/α-Fe2O3 and CuO-CeO2 catalysts for the selective oxidation of carbon monoxide in excess hydrogen. Catal. Today **75**, 157–167 (2002). https://doi.org/10.1016/S0920-5861(02)00058-5
28. F. Mariño, C. Descorme, D. Duprez, Noble metal catalysts for the preferential oxidation of carbon monoxide in the presence of hydrogen (PROX). Appl. Catal. B Environ. **54**, 59–66 (2004). https://doi.org/10.1016/j.apcatb.2004.06.008
29. V. Recupero, L. Pino, A. Vita, F. Cipitì, M. Cordaro, M. Laganà, Development of a LPG fuel processor for PEFC systems: laboratory scale evaluation of autothermal reforming and preferential oxidation subunits. Int. J. Hydrog. Energy **30**, 963–971 (2005). https://doi.org/10.1016/j.ijhydene.2004.12.014
30. C.D. Dudfield, R. Chen, P.L. Adcock, A carbon monoxide PROX reactor for PEM fuel cell automotive application. Int. J. Hydrog. Energy **26**, 763–775 (2001). https://doi.org/10.1016/S0360-3199(00)00131-2
31. O. Görke, P. Pfeifer, Preferential CO oxidation over a platinum ceria alumina catalyst in a microchannel reactor. Int. J. Hydrog. Energy **36**, 4673–4681 (2011). https://doi.org/10.1016/j.ijhydene.2011.01.069
32. T.S. Nguyen, F. Morfin, M. Aouine, F. Bosselet, J.L. Rousset, L. Piccolo, Trends in the CO oxidation and PROX performances of the platinum-group metals supported on ceria. Catal. Today **253**, 106–114 (2015). https://doi.org/10.1016/j.cattod.2014.12.038
33. F. Morfin, T.S. Nguyen, J.L. Rousset, L. Piccolo, Synergy between hydrogen and ceria in Pt-catalyzed CO oxidation: an investigation on Pt–CeO2 catalysts synthesized by solution combustion. Appl. Catal. B Environ. **197**, 2–13 (2016). https://doi.org/10.1016/j.apcatb.2016.01.056
34. N. Bion, F. Epron, M. Moreno, F. Mariño, D. Duprez, Preferential oxidation of carbon monoxide in the presence of hydrogen (PROX) over noble metals and transition metal oxides: advantages and drawbacks. Top. Catal. **51**, 76–88 (2008). https://doi.org/10.1007/s11244-008-9116-x
35. S. Scirè, C. Crisafulli, S. Minicò, G.G. Condorelli, A. Di Mauro, Selective oxidation of CO in H2-rich stream over gold/iron oxide: an insight on the effect of catalyst pretreatment. J. Mol. Catal. A Chem. **284**, 24–32 (2008). https://doi.org/10.1016/j.molcata.2007.12.026
36. E. Quinet, F. Morfin, F. Diehl, P. Avenier, V. Caps, J.L. Rousset, Hydrogen effect on the preferential oxidation of carbon monoxide over alumina-supported gold nanoparticles. Appl. Catal. B Environ. **80**, 195–201 (2008). https://doi.org/10.1016/j.apcatb.2007.11.011
37. J.D.S.L. Fonseca, H.S. Ferreira, N. Bion, L. Pirault-Roy, M.D.C. Rangel, D. Duprez, F. Epron, Cooperative effect between copper and gold on ceria for CO-PROX reaction. Catal. Today **180**, 34–41 (2012). https://doi.org/10.1016/j.cattod.2011.06.008
38. A. Pitois, A. Pilenga, A. Pfrang, G. Tsotridis, Temperature-dependent CO desorption kinetics on supported gold nanoparticles: relevance to clean hydrogen production and fuel cell systems. Int. J. Hydrog. Energy **36**, 4375–4385 (2011). https://doi.org/10.1016/j.ijhydene.2010.12.123
39. P. Landon, J. Ferguson, B.E. Solsona, T. Garcia, A.F. Carley, A.A. Herzing, C.J. Kiely, S.E. Golunski, G.J. Hutchings, Selective oxidation of CO in the presence of H2, H2O and CO2 via gold for use in fuel cells. Chem. Commun., 3385–3387 (2005). https://doi.org/10.1039/b505295p
40. Y. Liu, B. Liu, Q. Wang, C. Li, W. Hu, Y. Liu, P. Jing, W. Zhao, J. Zhang, Three-dimensionally ordered macroporous Au/CeO2-Co 3O4 catalysts with mesoporous walls for enhanced CO

preferential oxidation in H2-rich gases. J. Catal. **296**, 65–76 (2012). https://doi.org/10.1016/j.jcat.2012.09.003

41. P. Lakshmanan, J.E. Park, E.D. Park, Recent advances in preferential oxidation of CO in H2 over gold catalysts. Catal. Surv. Asia **18**, 75–88 (2014)
42. T. Cwele, N. Mahadevaiah, S. Singh, H.B. Friedrich, Effect of Cu additives on the performance of a cobalt substituted ceria (Ce0.90Co0.10O2-δ) catalyst in total and preferential CO oxidation. Appl. Catal. B Environ. **182**, 1–14 (2016). https://doi.org/10.1016/j.apcatb.2015.08.043
43. C.A. Chagas, E.F. de Souza, R.L. Manfro, S.M. Landi, M.M.V.M. Souza, M. Schmal, Copper as promoter of the NiO-CeO2 catalyst in the preferential CO oxidation. Appl. Catal. B Environ. **182**, 257–265 (2016). https://doi.org/10.1016/j.apcatb.2015.09.033
44. C.G. Maciel, M.N. Belgacem, J.M. Assaf, Performance of CuO-CeO2 catalysts with low copper content in CO preferential oxidation reaction. Catal. Letters **141**, 316–321 (2011). https://doi.org/10.1007/s10562-010-0486-x
45. E. Moretti, L. Storaro, A. Talon, M. Lenarda, One-pot mesoporous Al-Ce-Cu oxide systems as catalysts for the preferential carbon monoxide oxidation (CO-PROX). Catal. Commun. **10**, 522–527 (2009). https://doi.org/10.1016/j.catcom.2008.10.023
46. A. Martínez-Arias, D. Gamarra, M. Fernández-García, A. Hornés, P. Bera, Z. Koppány, Z. Schay, Redox-catalytic correlations in oxidised copper-ceria CO-PROX catalysts. Catal. Today **143**, 211–217 (2009). https://doi.org/10.1016/j.cattod.2008.09.018
47. D. Gamarra, G. Munuera, A.B. Hungría, M. Fernández-García, J.C. Conesa, P.A. Midgley, X.Q. Wang, J.C. Hanson, J.A. Rodríguez, A. Martínez-Arias, Structure-activity relationship in nanostructured copper-ceria-based preferential CO oxidation catalysts. J. Phys. Chem. C **111**, 11026–11038 (2007). https://doi.org/10.1021/jp072243k
48. T. Caputo, L. Lisi, R. Pirone, G. Russo, On the role of redox properties of CuO/CeO2 catalysts in the preferential oxidation of CO in H2-rich gases. Appl. Catal. A Gen. **348**, 42–53 (2008). https://doi.org/10.1016/j.apcata.2008.06.025
49. A. Di Benedetto, G. Landi, L. Lisi, G. Russo, Role of CO2 on CO preferential oxidation over CuO/CeO2 catalyst. Appl. Catal. B Environ. **142–143**, 169–177 (2013). https://doi.org/10.1016/j.apcatb.2013.05.001
50. J.A. Cecilia, A. Arango-Díaz, F. Franco, J. Jiménez-Jiménez, L. Storaro, E. Moretti, E. Rodríguez-Castellón, CuO-CeO<inf>2</inf> supported on montmorillonite-derived porous clay heterostructures (PCH) for preferential CO oxidation in H<inf>2</inf>–rich stream. Catal. Today **253**, 126–136 (2015). https://doi.org/10.1016/j.cattod.2015.01.040
51. X. Guo, J. Li, R. Zhou, Catalytic performance of manganese doped CuO-CeO2 catalysts for selective oxidation of CO in hydrogen-rich gas. Fuel **163**, 56–64 (2016). https://doi.org/10.1016/j.fuel.2015.09.043
52. W.W. Wang, P.P. Du, S.H. Zou, H.Y. He, R.X. Wang, Z. Jin, S. Shi, Y.Y. Huang, R. Si, Q.S. Song, C.J. Jia, C.H. Yan, Highly dispersed copper oxide clusters as active species in copper-ceria catalyst for preferential oxidation of carbon monoxide. ACS Catal. **5**, 2088–2099 (2015). https://doi.org/10.1021/cs5014909
53. J.S. Elias, N. Artrith, M. Bugnet, L. Giordano, G.A. Botton, A.M. Kolpak, Y. Shao-Horn, Elucidating the nature of the active phase in copper/ceria catalysts for CO oxidation. ACS Catal. **6**, 1675–1679 (2016). https://doi.org/10.1021/acscatal.5b02666
54. A. Davó-Quiñonero, M. Navlani-García, D. Lozano-Castelló, A. Bueno-López, J.A. Anderson, Role of hydroxyl groups in the preferential oxidation of CO over copper oxide–cerium oxide catalysts. ACS Catal. **6**, 1723–1731 (2016). https://doi.org/10.1021/acscatal.5b02741
55. A. Di Benedetto, G. Landi, L. Lisi, CO reactive adsorption at low temperature over CuO/CeO2structured catalytic monolith. Int. J. Hydrog. Energy **42**, 12262–12275 (2017). https://doi.org/10.1016/j.ijhydene.2017.03.077
56. A. Arango-Díaz, J.A. Cecilia, L. Dos Santos-Gómez, D. Marrero-López, E.R. Losilla, J. Jiménez-Jiménez, E. Rodríguez-Castellón, Characterization and performance in preferential oxidation of CO of CuO-CeO2 catalysts synthesized using polymethyl metac-

rylate (PMMA) as template. Int. J. Hydrog. Energy **40**, 11254–11260 (2015). https://doi.org/10.1016/j.ijhydene.2015.04.094

57. C. Xu, S. Li, Y. Zhang, Y. Li, J. Zhou, G. Qin, Synthesis of CuO x –CeO 2 catalyst with high-density interfaces for selective oxidation of CO in H 2 -rich stream. Int. J. Hydrog. Energy **44**, 4156–4166 (2019). https://doi.org/10.1016/j.ijhydene.2018.12.152
58. J. Ding, L. Li, H. Li, S. Chen, S. Fang, T. Feng, G. Li, Optimum preferential oxidation performance of CeO 2 -CuO x -RGO composites through interfacial regulation. ACS Appl. Mater. Interfaces **10**, 7935–7945 (2018). https://doi.org/10.1021/acsami.7b15549
59. X. Liu, X. Li, H. Qian, J. Chi, B. Chen, S. Wang, C. Chen, N. Zhang, Preferential CO oxidation over CuO–CeO 2 catalyst synthesized from MOF with nitrogen-containing ligand as precursor. Int. J. Hydrog. Energy **43**, 23299–23309 (2018). https://doi.org/10.1016/j.ijhydene.2018.10.162
60. W. Hu, G. Li, J. Chen, F. Huang, M. Gong, L. Zhong, Y. Chen, Enhancement of activity and hydrothermal stability of Pd/ZrO2-Al2O3 doped by Mg for methane combustion under lean conditions. Fuel **194**, 368–374 (2017). https://doi.org/10.1016/j.fuel.2016.11.028
61. J. Lu, J. Wang, Q. Zou, D. He, L. Zhang, Z. Xu, S. He, Y. Luo, Unravelling the nature of the active species as well as the doping effect over Cu/Ce-based catalyst for carbon monoxide preferential oxidation. ACS Catal. **9**, 2177–2195 (2019). https://doi.org/10.1021/acscatal.8b04035
62. X. Zhang, X. Zhang, L. Song, F. Hou, Y. Yang, Y. Wang, N. Liu, Enhanced catalytic performance for CO oxidation and preferential CO oxidation over CuO/CeO2 catalysts synthesized from metal organic framework: effects of preparation methods. Int. J. Hydrog. Energy **43**, 18279–18288 (2018). https://doi.org/10.1016/j.ijhydene.2018.08.060
63. H. Choi, J. Kim, S.J. Choung, J. Kim, M.R. Othman, Complete removal of carbon monoxide by functional nanoparticles for hydrogen fuel cell application. Chem. Eng. Sci. **172**, 688–693 (2017). https://doi.org/10.1016/j.ces.2017.07.026
64. M. Konsolakis, The role of Copper–Ceria interactions in catalysis science: recent theoretical and experimental advances. Appl. Catal. B Environ. **198**, 49–66 (2016). https://doi.org/10.1016/j.apcatb.2016.05.037
65. S. Scirè, C. Crisafulli, P.M. Riccobene, G. Patanè, A. Pistone, Selective oxidation of CO in H 2-rich stream over Au/CeO 2 and Cu/CeO 2 catalysts: an insight on the effect of preparation method and catalyst pretreatment. Appl. Catal. A Gen. **417–418**, 66–75 (2012). https://doi.org/10.1016/j.apcata.2011.12.025
66. A. Di Benedetto, G. Landi, L. Lisi, Improved CO-PROX performance of CuO/CeO2 catalysts by using nanometric ceria as support. Catalysts **8**, 209 (2018). https://doi.org/10.3390/catal8050209
67. A.P. Jia, G.S. Hu, L. Meng, Y.L. Xie, J.Q. Lu, M.F. Luo, CO oxidation over CuO/Ce 1-xCu xO 2-δ and Ce 1-xCu xO 2-δ catalysts: synergetic effects and kinetic study. J. Catal. **289**, 199–209 (2012). https://doi.org/10.1016/j.jcat.2012.02.010
68. D. Gu, C.J. Jia, H. Bongard, B. Spliethoff, C. Weidenthaler, W. Schmidt, F. Schüth, Ordered mesoporous Cu-Ce-O catalysts for CO preferential oxidation in H2-rich gases: influence of copper content and pretreatment conditions. Appl. Catal. B Environ. **152–153**, 11–18 (2014). https://doi.org/10.1016/j.apcatb.2014.01.011
69. D. Gamarra, C. Belver, M. Fernández-García, A. Martínez-Arias, Selective CO oxidation in excess H2 over copper-ceria catalysts: identification of active entities/species. J. Am. Chem. Soc. **129**, 12064–12065 (2007). https://doi.org/10.1021/ja073926g
70. A. Gurbani, J.L. Ayastuy, M.P. González-Marcos, M.A. Gutiérrez-Ortiz, CuO-CeO 2 catalysts synthesized by various methods: comparative study of redox properties. Int. J. Hydrog. Energy **35**, 11582–11590 (2010). https://doi.org/10.1016/j.ijhydene.2010.04.045
71. J. Wang, H. Pu, G. Wan, K. Chen, J. Lu, Y. Lei, L. Zhong, S. He, C. Han, Y. Luo, Promoted the reduction of Cu2+ to enhance CuO–CeO2 catalysts for CO preferential oxidation in H2-rich streams: effects of preparation methods and copper precursors. Int. J. Hydrog. Energy **42**, 21955–21968 (2017). https://doi.org/10.1016/j.ijhydene.2017.07.122

72. P.S. Barbato, S. Colussi, A. Di Benedetto, G. Landi, L. Lisi, J. Llorca, A. Trovarelli, Origin of high activity and selectivity of CuO/CeO_2 catalysts prepared by solution combustion synthesis in CO-PROX reaction. J. Phys. Chem. C **120**, 13039–13048 (2016). https://doi.org/10.1021/acs.jpcc.6b02433
73. G. Avgouropoulos, T. Ioannides, Selective CO oxidation over CuO-CeO2 catalysts prepared via the urea-nitrate combustion method. Appl. Catal. A Gen. **244**, 155–167 (2003). https://doi.org/10.1016/S0926-860X(02)00558-6
74. J. Papavasiliou, J. Vakros, G. Avgouropoulos, Impact of acid treatment of CuO-CeO2 catalysts on the preferential oxidation of CO reaction. Catal. Commun. **115**, 68–72 (2018). https://doi.org/10.1016/j.catcom.2018.07.014
75. J. Papavasiliou, M. Rawski, J. Vakros, G. Avgouropoulos, A novel post-synthesis modification of CuO-CeO2 catalysts: effect on their activity for selective CO oxidation. ChemCatChem **10**, 2096–2106 (2018). https://doi.org/10.1002/cctc.201701968
76. P.P. Du, W.W. Wang, C.J. Jia, Q.S. Song, Y.Y. Huang, R. Si, Effect of strongly bound copper species in copper-ceria catalyst for preferential oxidation of carbon monoxide. Appl. Catal. A Gen. **518**, 87–101 (2016). https://doi.org/10.1016/j.apcata.2015.10.041
77. A. Martínez-Arias, A.B. Hungría, G. Munuera, D. Gamarra, Preferential oxidation of CO in rich H2 over CuO/CeO2: details of selectivity and deactivation under the reactant stream. Appl. Catal. B Environ. **65**, 207–216 (2006). https://doi.org/10.1016/j.apcatb.2006.02.003
78. H.C. Lee, D.H. Kim, Kinetics of CO and H2 oxidation over CuO-CeO2 catalyst in H2 mixtures with CO2 and H2O. Catal. Today **132**, 109–116 (2008). https://doi.org/10.1016/j.cattod.2007.12.028
79. X. Guo, J. Mao, R. Zhou, Influence of the copper coverage on the dispersion of copper oxide and the catalytic performance of CuO/CeO2(rod) catalysts in preferential oxidation of CO in excess hydrogen. J. Power Sources **371**, 119–128 (2017). https://doi.org/10.1016/j.jpowsour.2017.10.055
80. P.S. Barbato, S. Colussi, A. Di Benedetto, G. Landi, L. Lisi, J. Llorca, A. Trovarelli, CO preferential oxidation under H2-rich streams on copper oxide supported on Fe promoted CeO2. Appl. Catal. A Gen. **506**, 268–277 (2015). https://doi.org/10.1016/j.apcata.2015.09.018
81. A. Lopez Cámara, V. Cortés Corberán, L. Barrio, G. Zhou, R. Si, J.C. Hanson, M. Monte, J.C. Conesa, J.A. Rodriguez, A. Martínez-Arias, Improving the CO-PROX performance of inverse CeO2/CuO catalysts: doping of the CuO component with Zn. J. Phys. Chem. C **118**, 9030–9041 (2014). https://doi.org/10.1021/jp5009384
82. A. Kubacka, A. Martínez-Arias, M. Fernández-García, Role of the interface in base-metal ceria-based catalysts for hydrogen purification and production processes. ChemCatChem **7**, 3614–3624 (2015). https://doi.org/10.1002/cctc.201500593
83. A. Trovarelli, J. Llorca, Ceria catalysts at nanoscale: how do crystal shapes shape catalysis? ACS Catal. **7**, 4716–4735 (2017)
84. G.R. Kosmambetova, Structural organization of nanophase catalysts for preferential CO oxidation. Theor. Exp. Chem. **50**, 265–281 (2014). https://doi.org/10.1007/s11237-014-9376-4
85. I. López, T. Valdés-Solís, G. Marbán, An attempt to rank copper-based catalysts used in the CO-PROX reaction. Int. J. Hydrog. Energy **33**, 197–205 (2008). https://doi.org/10.1016/j.ijhydene.2007.09.011
86. C.G. MacIel, T.D.F. Silva, M.I. Hirooka, M.N. Belgacem, J.M. Assaf, Effect of nature of ceria support in CuO/CeO 2 catalyst for PROX-CO reaction. Fuel **97**, 245–252 (2012). https://doi.org/10.1016/j.fuel.2012.02.004
87. R. Zhang, J.T. Miller, C.D. Baertsch, Identifying the active redox oxygen sites in a mixed Cu and Ce oxide catalyst by in situ X-ray absorption spectroscopy and anaerobic reactions with CO in concentrated H2. J. Catal. **294**, 69–78 (2012). https://doi.org/10.1016/j.jcat.2012.07.005
88. V.D. Araújo, W. Avansi, A.J.S. Mascarenhas, H.M.C. Andrade, E. Longo, M.I.B. Bernardi, Ce0.97Cu0.03O2 nanocatalysts synthesized via microwave-assisted hydrothermal method:

characterization and CO-PROX catalytic efficiency. Earth Environ. Sci. Trans. R Soc. Edinburgh **1552**, 17–22 (2013). https://doi.org/10.1557/opl.2013.588
89. A. Arango-Díaz, E. Moretti, A. Talon, L. Storaro, M. Lenarda, P. Núñez, J. Marrero-Jerez, J. Jiménez-Jiménez, A. Jiménez-López, E. Rodríguez-Castellón, Preferential CO oxidation (CO-PROX) catalyzed by CuO supported on nanocrystalline CeO2 prepared by a freeze-drying method. Appl. Catal. A Gen. **477**, 54–63 (2014). https://doi.org/10.1016/j.apcata.2014.02.033
90. H. Yen, Y. Seo, S. Kaliaguine, F. Kleitz, Tailored mesostructured copper/ceria catalysts with enhanced performance for preferential oxidation of CO at low temperature. Angew Chemie Int. Ed. **51**, 12032–12035 (2012). https://doi.org/10.1002/anie.201206505
91. F. Zhang, C. Chen, W.M. Xiao, L. Xu, N. Zhang, CuO/CeO2 catalysts with well-dispersed active sites prepared from Cu 3 (BTC) 2 metal-organic framework precursor for preferential CO oxidation. Catal. Commun. **26**, 25–29 (2012). https://doi.org/10.1016/j.catcom.2012.04.028
92. S.H. Wang, L. Xu, P. Shen, C. Chen, N. Zhang, New approach to synthesis of CuO/CeO2 catalysts for preferential CO oxidation. Appl. Mech. Mater., 666–670 (2014)
93. X. Li, X.Y. Quek, D.A.J. Michel Ligthart, M. Guo, Y. Zhang, C. Li, Q. Yang, E.J.M. Hensen, CO-PROX reactions on copper cerium oxide catalysts prepared by melt infiltration. Appl. Catal. B Environ. **123–124**, 424–432 (2012). https://doi.org/10.1016/j.apcatb.2012.05.009
94. C. Tang, J. Sun, X. Yao, Y. Cao, L. Liu, C. Ge, F. Gao, L. Dong, Efficient fabrication of active CuO-CeO2/SBA-15 catalysts for preferential oxidation of CO by solid state impregnation. Appl. Catal. B Environ. **146**, 201–212 (2014). https://doi.org/10.1016/j.apcatb.2013.05.060
95. L. Gong, Z. Huang, L. Luo, N. Zhang, Promoting effect of MnOx on the performance of CuO/CeO2 catalysts for preferential oxidation of CO in H2-rich gases. React. Kinet. Mech. Catal. **111**, 489–504 (2014). https://doi.org/10.1007/s11144-013-0662-2
96. S. Zeng, T. Chen, K. Liu, H. Su, Promotion effect of metal oxides on inverse CeO2/CuO catalysts for preferential oxidation of CO. Catal. Commun. **45**, 16–20 (2014). https://doi.org/10.1016/j.catcom.2013.10.022
97. Z. Liu, J. Chen, R. Zhou, X. Zheng, Influence of ethanol washing in precursor on CuO-CeO2 catalysts in preferential oxidation of CO in excess hydrogen. Catal. Letters **123**, 102–106 (2008). https://doi.org/10.1007/s10562-008-9401-0
98. L.C. Chung, C.T. Yeh, Synthesis of highly active CuO-CeO2 nanocomposites for preferential oxidation of carbon monoxide at low temperatures. Catal. Commun. **9**, 670–674 (2008). https://doi.org/10.1016/j.catcom.2007.07.041
99. G. Marbán, I. López, T. Valdés-Solís, Preferential oxidation of CO by CuOx/CeO2 nanocatalysts prepared by SACOP. Mechanisms of deactivation under the reactant stream. Appl. Catal. A Gen. **361**, 160–169 (2009). https://doi.org/10.1016/j.apcata.2009.04.014
100. G. Sedmak, S. Hočevar, J. Levec, Kinetics of selective CO oxidation in excess of H2 over the nanostructured Cu0.1Ce0.9O2-y catalyst. J. Catal. **213**, 135–150 (2003). https://doi.org/10.1016/S0021-9517(02)00019-2
101. S. Xu, D. Sun, H. Liu, X. Wang, X. Yan, Fabrication of Cu-doped cerium oxide nanofibers via electrospinning for preferential CO oxidation. Catal. Commun. **12**, 514–518 (2011). https://doi.org/10.1016/j.catcom.2010.11.021
102. C. Borchers, M.L. Martin, G.A. Vorobjeva, O.S. Morozova, A.A. Firsova, A.V. Leonov, E.Z. Kurmaev, A.I. Kukharenko, I.S. Zhidkov, S.O. Cholakh, Cu–CeO2 nanocomposites: mechanochemical synthesis, physico-chemical properties, CO-PROX activity. J. Nanopart. Res. **18**, 344 (2016). https://doi.org/10.1007/s11051-016-3640-6
103. J. Han, H.J. Kim, S. Yoon, H. Lee, Shape effect of ceria in Cu/ceria catalysts for preferential CO oxidation. J. Mol. Catal. A Chem. **335**, 82–88 (2011). https://doi.org/10.1016/j.molcata.2010.11.017
104. M. Monte, D. Gamarra, A. López Cámara, S.B. Rasmussen, N. Gyorffy, Z. Schay, A. Martínez-Arias, J.C. Conesa, Preferential oxidation of CO in excess H2 over CuO/CeO2

catalysts: performance as a function of the copper coverage and exposed face present in the CeO2 support. Catal. Today **229**, 104–113 (2014)

105. D. Gamarra, A.L. Cámara, M. Monte, S.B. Rasmussen, L.E. Chinchilla, A.B. Hungría, G. Munuera, N. Gyorffy, Z. Schay, V.C. Corberán, J.C. Conesa, A. Martínez-Arias, Preferential oxidation of CO in excess H2 over CuO/CeO2 catalysts: characterization and performance as a function of the exposed face present in the CeO2 support. Appl. Catal. B Environ. **130–131**, 224–238 (2013). https://doi.org/10.1016/j.apcatb.2012.11.008
106. X. Guo, R. Zhou, A new insight into the morphology effect of ceria on CuO/CeO2catalysts for CO selective oxidation in hydrogen-rich gas. Cat. Sci. Technol. **6**, 3862–3871 (2016). https://doi.org/10.1039/c5cy01816a
107. Y. Xie, J. Wu, G. Jing, H. Zhang, S. Zeng, X. Tian, X. Zou, J. Wen, H. Su, C.J. Zhong, P. Cui, Structural origin of high catalytic activity for preferential CO oxidation over CuO/CeO2 nanocatalysts with different shapes. Appl. Catal. B Environ. **239**, 665–676 (2018). https://doi.org/10.1016/j.apcatb.2018.08.066
108. T. Kou, C. Si, J. Pinto, C. Ma, Z. Zhang, Dealloying assisted high-yield growth of surfactant-free <110> highly active Cu-doped CeO 2 nanowires for low-temperature CO oxidation. Nanoscale **9**, 8007–8014 (2017). https://doi.org/10.1039/c7nr02405c
109. X. Gong, B. Liu, B. Kang, G. Xu, Q. Wang, C. Jia, J. Zhang, Boosting Cu-Ce interaction in CuxO/CeO2 nanocube catalysts for enhanced catalytic performance of preferential oxidation of CO in H2-rich gases. Mol. Catal. **436**, 90–99 (2017). https://doi.org/10.1016/j.mcat.2017.04.013
110. A. Hornés, A.B. Hungría, P. Bera, A. López Cámara, M. Fernández-García, A. Martínez-Arias, L. Barrio, M. Estrella, G. Zhou, J.J. Fonseca, J.C. Hanson, J.A. Rodriguez, Inverse CeO2/CuO catalyst as an alternative to classical direct configurations for preferential oxidation of CO in hydrogen-rich stream. J. Am. Chem. Soc. **132**, 34–35 (2010). https://doi.org/10.1021/ja9089846
111. S. Zeng, K. Liu, T. Chen, H. Su, Influence of crystallite size and interface on the catalytic performance over the CeO2/CuO catalysts. Int. J. Hydrog. Energy **38**, 14542–14549 (2013). https://doi.org/10.1016/j.ijhydene.2013.09.023
112. S. Zeng, W. Zhang, M. Śliwa, H. Su, Comparative study of CeO2/CuO and CuO/CeO2 catalysts on catalytic performance for preferential CO oxidation. Int. J. Hydrog. Energy **38**, 3597–3605 (2013). https://doi.org/10.1016/j.ijhydene.2013.01.030
113. S.H. Zeng, T.J. Chen, H.Q. Su, Q. Man, W.L. Zhang, CeO2/CuO catalysts with Na2CO3 as precipitant for preferential oxidation of CO. Adv. Mater. Res **805-806**, 1297–1301 (2013)
114. S. Zeng, Y. Wang, B. Qin, X. Gu, H. Su, L. Li, K. Liu, Inverse CeO2/CuO catalysts prepared by different precipitants for preferential CO oxidation in hydrogen-rich streams. Cat. Sci. Technol. **3**, 3163–3172 (2013). https://doi.org/10.1039/c3cy00478c
115. S. Zeng, W. Zhang, N. Liu, H. Su, Inverse CeO2/CuO catalysts prepared by hydrothermal method for preferential CO oxidation. Catal. Letters **143**, 1018–1024 (2013). https://doi.org/10.1007/s10562-013-1065-8
116. S. Zeng, Y. Wang, K. Liu, F. Liu, H. Su, CeO 2 nanoparticles supported on CuO with petal-like and sphere-flower morphologies for preferential CO oxidation. Int. J. Hydrog. Energy **37**, 11640–11649 (2012). https://doi.org/10.1016/j.ijhydene.2012.05.086
117. S. Zeng, Y. Wang, S. Ding, J.J.H.B. Sattler, E. Borodina, L. Zhang, B.M. Weckhuysen, H. Su, Active sites over CuO/CeO2 and inverse CeO2/CuO catalysts for preferential CO oxidation. J. Power Sources **256**, 301–311 (2014). https://doi.org/10.1016/j.jpowsour.2014.01.098
118. S. Zeng, K. Liu, L. Zhang, B. Qin, T. Chen, Y. Yin, H. Su, Deactivation analyses of CeO2/CuO catalysts in the preferential oxidation of carbon monoxide. J. Power Sources **261**, 46–54 (2014). https://doi.org/10.1016/j.jpowsour.2014.03.043
119. A. López Cámara, V. Cortés Corberán, L. Barrio, G. Zhou, R. Si, J.C. Hanson, M. Monte, J.C. Conesa, J.A. Rodriguez, A. Martínez-Arias, Improving the CO-PROX performance of inverse CeO2/CuO catalysts: doping of the CuO component with Zn. J. Phys. Chem. C **118**, 9030–9041 (2014). https://doi.org/10.1021/jp5009384

120. S. Chen, L. Li, W. Hu, X. Huang, Q. Li, Y. Xu, Y. Zuo, G. Li, Anchoring high-concentration oxygen vacancies at interfaces of CeO2-x/Cu toward enhanced activity for preferential CO oxidation. ACS Appl. Mater. Interfaces **7**, 22999–23007 (2015). https://doi.org/10.1021/acsami.5b06302
121. C. Gu, R. Qi, Y. Wei, X. Zhang, Preparation and performances of nanorod-like inverse CeO2–CuO catalysts derived from Ce-1,3,5-Benzene tricarboxylic acid for CO preferential oxidation. React. Kinet. Mech. Catal. **124**, 651–667 (2018). https://doi.org/10.1007/s11144-018-1374-4
122. Y. Xie, Y. Yin, S. Zeng, M. Gao, H. Su, Coexistence of Cu+ and Cu2+ in star-shaped CeO2/CuxO catalyst for preferential CO oxidation. Catal. Commun. **99**, 110–114 (2017). https://doi.org/10.1016/j.catcom.2017.06.003
123. C. Chen, R. Wang, P. Shen, D. Zhao, N. Zhang, Inverse CeO2/CuO catalysts prepared from heterobimetallic metal-organic framework precursor for preferential CO oxidation in H2-rich stream. Int. J. Hydrog. Energy **40**, 4830–4839 (2015). https://doi.org/10.1016/j.ijhydene.2015.02.066
124. S. Zeng, L. Zhang, N. Jiang, M. Gao, X. Zhao, Y. Yin, H. Su, Multi-wall carbon nanotubes as support of copper-cerium composite for preferential oxidation of carbon monoxide. J. Power Sources **293**, 1016–1023 (2015). https://doi.org/10.1016/j.jpowsour.2015.04.115
125. Y. Gao, K. Xie, W. Wang, S. Mi, N. Liu, G. Pan, W. Huang, Structural features and catalytic performance in CO preferential oxidation of CuO-CeO2 supported on multi-walled carbon nanotubes. Cat. Sci. Technol. **5**, 1568–1579 (2015). https://doi.org/10.1039/c4cy01220h
126. A.B. Dongil, B. Bachiller-Baeza, E. Castillejos, N. Escalona, A. Guerrero-Ruiz, I. Rodríguez-Ramos, The promoter effect of potassium in CuO/CeO2 systems supported on carbon nanotubes and graphene for the CO-PROX reaction. Cat. Sci. Technol. **6**, 6118–6127 (2016). https://doi.org/10.1039/C6CY00304D
127. H. Zhang, C. Xu, J. Ding, H. Su, S. Zeng, RGO/MWCNTs/Cu x O-CeO2 ternary nanocomposites for preferential CO oxidation in hydrogen-rich streams. Appl. Surf. Sci. **426**, 50–55 (2017). https://doi.org/10.1016/j.apsusc.2017.07.154
128. C. Xu, S. Zeng, H. Zhang, Y. Xie, A. Zhang, G. Jing, H. Su, Facile hydrothermal procedure to synthesize sheet-on-sheet reduced graphene oxide (RGO)/Cu x O[sbnd]CeO2 nanocomposites for preferential oxidation of carbon monoxide. Int. J. Hydrog. Energy **42**, 14133–14143 (2017). https://doi.org/10.1016/j.ijhydene.2017.04.058
129. D. Zhang, J. Qin, D. Wei, S. Yang, S. Wang, C. Hu, Enhancing the CO preferential oxidation (CO-PROX) of CuO–CeO2/reduced graphene oxide (rGO) by conductive rGO-wrapping based on the interfacial charge transfer. Catal. Letters **148**, 3454–3466 (2018). https://doi.org/10.1007/s10562-018-2520-3
130. L. Shi, G. Zhang, Y. Wang, Tailoring catalytic performance of carbon nanotubes confined CuO–CeO2 catalysts for CO preferential oxidation. Int. J. Hydrog. Energy **43**, 18211–18219 (2018). https://doi.org/10.1016/j.ijhydene.2018.08.020
131. A. Martínez-Arias, M. Fernández-García, O. Gálvez, J.M. Coronado, J.A. Anderson, J.C. Conesa, J. Soria, G. Munuera, Comparative study on redox properties and catalytic behavior for CO oxidation of CuO/CeO2 and CuO/ZrCeO4 catalysts. J. Catal. **195**, 207–216 (2000). https://doi.org/10.1006/jcat.2000.2981
132. L.H. Reddy, G.K. Reddy, D. Devaiah, B.M. Reddy, A rapid microwave-assisted solution combustion synthesis of CuO promoted CeO2 -M x O y (M = Zr, La, Pr and Sm) catalysts for CO oxidation. Appl. Catal. A Gen. **445–446**, 297–305 (2012). https://doi.org/10.1016/j.apcata.2012.08.024
133. T. Caputo, R. Pirone, G. Russo, Supported CuO/Ce1-x Zr x O2 catalysts for the preferential oxidation of CO in H2-rich gases. Kinet. Catal. **47**, 756–764 (2006). https://doi.org/10.1134/S0023158406050156
134. J.L. Ayastuy, A. Gurbani, M.P. González-Marcos, M.A. Gutiérrez-Ortiz, Selective CO oxidation in H2 streams on CuO/Ce xZr 1-xO 2 catalysts: correlation between activity and low tem-

perature reducibility. Int. J. Hydrog. Energy **37**, 1993–2006 (2012). https://doi.org/10.1016/j.ijhydene.2011.04.178

135. J. Wang, L. Deng, D. He, J. Lu, S. He, S. He, Y. Luo, A facile and rapid route to synthesize CuO<inf>x</inf>/Ce<inf>0.8</inf>Zr<inf>0.2</inf>O<inf>2</inf> catalysts with high performance for CO preferential oxidation (CO-PROX). Int. J. Hydrog. Energy **40**, 12478–12488 (2015). https://doi.org/10.1016/j.ijhydene.2015.07.063
136. X. Guo, Z. Qiu, J. Mao, R. Zhou, Doping effect of transition metals (Zr, Mn, Ti and Ni) on well-shaped CuO/CeO 2 (rods): nano/micro structure and catalytic performance for selective oxidation of CO in excess H2. Phys. Chem. Chem. Phys. **20**, 25983–25994 (2018). https://doi.org/10.1039/c8cp03696a
137. H. Jin, R. You, S. Zhou, K. Ma, M. Meng, L. Zheng, J. Zhang, T. Hu, In-situ DRIFTS and XANES identification of copper species in the ternary composite oxide catalysts CuMnCeO during CO preferential oxidation. Int. J. Hydrog. Energy **40**, 3919–3931 (2015). https://doi.org/10.1016/j.ijhydene.2015.01.086
138. J.A. Cecilia, A. Arango-Díaz, V. Rico-Pérez, A. Bueno-López, E. Rodríguez-Castellón, The influence of promoters (Zr, La, Tb, Pr) on the catalytic performance of CuO-CeO<inf>2</inf> systems for the preferential oxidation of CO in the presence of CO<inf>2</inf> and H<inf>2</inf>O. Catal. Today **253**, 115–125 (2015). https://doi.org/10.1016/j.cattod.2015.02.012
139. J.E. De Oliveira, S. Rico-Francés, Z. Abdelouahab-Reddam, F. Coloma, J. Silvestre-Albero, A. Sepúlveda-Escribano, E.V. Ramos-Fernandez, High performance of Cu/CeO<inf>2</inf>-Nb<inf>2</inf>O<inf>5</inf> catalysts for preferential CO oxidation and total combustion of toluene. Appl. Catal. A Gen. **502**, 129–137 (2015). https://doi.org/10.1016/j.apcata.2015.05.033
140. J. Kugai, T. Moriya, S. Seino, T. Nakagawa, Y. Ohkubo, H. Nitani, Y. Mizukoshi, T.A. Yamamoto, Effect of support for PtCu bimetallic catalysts synthesized by electron beam irradiation method on preferential CO oxidation. Appl. Catal. B Environ. **126**, 306–314 (2012). https://doi.org/10.1016/j.apcatb.2012.07.028
141. J. Kugai, T. Moriya, S. Seino, T. Nakagawa, Y. Ohkubo, H. Nitani, H. Daimon, T.A. Yamamoto, CeO 2-supported Pt-Cu alloy nanoparticles synthesized by radiolytic process for highly selective CO oxidation. Int. J. Hydrog. Energy **37**, 4787–4797 (2012). https://doi.org/10.1016/j.ijhydene.2011.12.070
142. J. Kugai, T. Moriya, S. Seino, T. Nakagawa, Y. Ohkubo, H. Nitani, T. Akita, Y. Mizukoshi, T.A. Yamamoto, Effect of CeO2 support properties on structure of Pt-Cu nanoparticles synthesized by electron beam irradiation method for preferential CO oxidation. Chem. Eng. J. **223**, 347–355 (2013). https://doi.org/10.1016/j.cej.2013.02.116
143. S. Lang, M. Türk, B. Kraushaar-Czarnetzki, Novel PtCuO/CeO2/α-Al2O3 sponge catalysts for the preferential oxidation of CO (PROX) prepared by means of supercritical fluid reactive deposition (SFRD). J. Catal. **286**, 78–87 (2012). https://doi.org/10.1016/j.jcat.2011.10.017
144. E.S. Gnanakumar, J.M. Naik, M. Manikandan, T. Raja, C.S. Gopinath, Role of nanointerfaces in Cu- and Cu+ Au-based near-ambient-temperature CO oxidation catalysts. ChemCatChem **6**, 3116–3124 (2014). https://doi.org/10.1002/cctc.201402581
145. X. Liao, W. Chu, X. Dai, V. Pitchon, Bimetallic Au-Cu supported on ceria for PROX reaction: effects of Cu/Au atomic ratios and thermal pretreatments. Appl. Catal. B Environ. **142–143**, 25–37 (2013). https://doi.org/10.1016/j.apcatb.2013.05.010
146. L. Zhang, H.Y. Kim, G. Henkelman, CO oxidation at the Au-Cu interface of bimetallic nanoclusters supported on CeO2(111). J. Phys. Chem. Lett. **4**, 2943–2947 (2013). https://doi.org/10.1021/jz401524d
147. D.I. Potemkin, E.Y. Semitut, Y.V. Shubin, P.E. Plyusnin, P.V. Snytnikov, E.V. Makotchenko, D.Y. Osadchii, D.A. Svintsitskiy, S.A. Venyaminov, S.V. Korenev, V.A. Sobyanin, Silica, alumina and ceria supported Au-Cu nanoparticles prepared via the decomposition of [Au(en)2]2[Cu(C2O 4)2]3·8H2O single-source precursor: synthesis, characterization

and catalytic performance in CO PROX. Catal. Today **235**, 103–111 (2014). https://doi.org/10.1016/j.cattod.2014.04.026
148. J. Papavasiliou, Interaction of atomically dispersed gold with hydrothermally prepared copper-cerium oxide for preferential CO oxidation reaction. Catal. Today (2019). https://doi.org/10.1016/j.cattod.2019.02.026
149. A.A. Firsova, A.N. Il'ichev, T.I. Khomenko, L.V. Gorobinskii, Y.V. Maksimov, I.P. Suzdalev, V.N. Korchak, Selective oxidation of CO in the presence of hydrogen on CuO/CeO2 catalysts modified with Fe and Ni oxides. Kinet. Catal. **48**, 282–291 (2007). https://doi.org/10.1134/s0023158407020139
150. K. Sirichaiprasert, A. Luengnaruemitchai, S. Pongstabodee, Selective oxidation of CO to CO2 over Cu-Ce-Fe-O composite-oxide catalyst in hydrogen feed stream. Int. J. Hydrog. Energy **32**, 915–926 (2007). https://doi.org/10.1016/j.ijhydene.2006.10.060
151. S. Ma, G. Lu, Y. Shen, Y. Guo, Y. Wang, Y. Guo, Effect of Fe doping on the catalytic performance of CuO-CeO2 for low temperature CO oxidation. Cat. Sci. Technol. **1**, 669–674 (2011). https://doi.org/10.1039/c1cy00049g
152. Z. Lendzion-Bielun, M.M. Bettahar, S. Monteverdi, Fe-promoted CuO/CeO2 catalyst: structural characterization and CO oxidation activity. Catal. Commun. **11**, 1137–1142 (2010). https://doi.org/10.1016/j.catcom.2010.05.017
153. H. Bao, X. Chen, J. Fang, Z. Jiang, W. Huang, Structure-activity relation of Fe2O3-CeO2 composite catalysts in CO oxidation. Catal. Letters **125**, 160–167 (2008). https://doi.org/10.1007/s10562-008-9540-3
154. G. Landi, A. Di Benedetto, S. Colussi, P.S. Barbato, L. Lisi, Effect of carbon dioxide and water on the performances of an iron-promoted copper/ceria catalyst for CO preferential oxidation in H<inf>2</inf>–rich streams. Int. J. Hydrog. Energy **41**, 7332–7341 (2016). https://doi.org/10.1016/j.ijhydene.2016.03.141
155. V.D.B.C. Dasireddy, B. Likozar, J. Valand, Preferential oxidation of CO in H2/H2O/CO2 water–gas shift feedstocks over Cu-based carbon nanotubes-supported heterogeneous catalysts. Appl. Catal. B Environ. **237**, 1044–1058 (2018). https://doi.org/10.1016/j.apcatb.2018.06.069
156. J. Wang, C. Han, X. Gao, J. Lu, G. Wan, D. He, R. Chen, K. Chen, S. He, Y. Luo, Rapid synthesis of Fe-doped CuO-Ce0.8Zr0.2O2 catalysts for CO preferential oxidation in H2-rich streams: effect of iron source and the ratio of Fe/Cu. J. Power Sources **343**, 437–445 (2017). https://doi.org/10.1016/j.jpowsour.2017.01.084
157. Z. Liu, R. Zhou, X. Zheng, Influence of residual K+ on the catalytic performance of CuO - CeO2 catalysts in preferential oxidation of CO in excess hydrogen. Int. J. Hydrog. Energy **33**, 791–796 (2008). https://doi.org/10.1016/j.ijhydene.2007.10.011
158. A.B. Dongil, B. Bachiller-Baeza, E. Castillejos, N. Escalona, A. Guerrero-Ruiz, I. Rodríguez-Ramos, Promoter effect of alkalis on CuO/CeO2/carbon nanotubes systems for the PROx reaction. Catal. Today **301**, 141–146 (2018). https://doi.org/10.1016/j.cattod.2017.03.033

Chapter 4
Supported Nickel Nanocatalysts for the Dry Reforming of Methane: Effect of SBA-15's Pore Sizes on the Catalytic Performances of Nickel Nanoparticles

Oscar Daoura, Maya Boutros, Marie-Nour Kaydouh, Pascale Massiani, Franck Launay, and Nissrine El Hassan

4.1 Introduction

Catalysts made of transition metal (e.g., Ni, Co, Fe, Cu…) nanoparticles (NPs) have a wide range of applications such as in hydrogenation, hydrogenolysis or hydrodeoxygenation reactions; they can also be used in processes targeting the production of renewable energy resources such as the Fischer-Tropsch (FT) synthesis or the methane reforming and water–gas shift reaction [1–3]. In methane reforming, the size and consequently the dispersion of the involved metals play a key role in catalytic performances [4, 5]. As it is well-known, metallic particles with nanometric dimensions exhibit interesting physicochemical characteristics and usually, their catalytic activity is inversely proportional to their size [6]. Several oxide supports (alumina, silica, etc) have been tested in order to improve the dispersion of the active phase. Among them, siliceous carriers offer valuable features including their high surface areas, tunable nanoscale pore dimensions and sizes, variable accessible morphologies, and their rather low price [7–10]. However, due to diffusion limitation

O. Daoura
Laboratoire de Chimie Physique des Matériaux, LCPM/PR2N, Lebanese University, Faculty of Sciences II, Jdeideh, Lebanon

Laboratoire de Réactivité de Surface, LRS, Sorbonne Université, CNRS, Paris, France

M. Boutros
Laboratoire de Chimie Physique des Matériaux, LCPM/PR2N, Lebanese University, Faculty of Sciences II, Jdeideh, Lebanon

M.-N. Kaydouh · N. El Hassan (✉)
Department of Chemical Engineering, Faculty of Engineering, University of Balamand, Amioun El Koura, Lebanon
e-mail: nissrine.hassan@balamand.edu.lb

P. Massiani · F. Launay (✉)
Laboratoire de Réactivité de Surface, LRS, Sorbonne Université, CNRS, Paris, France
e-mail: franck.launay@sorbonne-universite.fr

M. Piumetti, S. Bensaid (eds.), *Nanostructured Catalysts for Environmental Applications*, https://doi.org/10.1007/978-3-030-58934-9_4

of the metallic precursors throughout the pore channels during the preparation step, the resulting metal nanoparticles are often not only dispersed inside the pores but also located on the outer surface of the grains, giving rise to aggregation. Moreover, examples of partial or even total loss of metallic active surface have been revealed during high temperature catalyst activation and/or operation as the result of NPs growth by migration, coalescence, and Ostwald ripening [11, 12].

SBA-15 type supports prepared by a self-assembly (S^0H^+)(X^-I^+) pathway (where S^0 stands for a neutral Surfactant (Pluronic P123), I^+ for cationic inorganic silica species, H^+ for protons, and X^- for chloride) are characterized by a two-dimensional hexagonal pore structure, a narrow pore size distribution, large specific surface area, and a rather high thermal stability (larger walls than MCM-41) [13, 14]. For these reasons, they are often considered as suitable carriers for the immobilization of metal NPs, especially nickel, affording notably heterogeneous catalysts for the dry reforming of methane (DRM) [15]. This reaction that consumes simultaneously methane and carbon dioxide to produce H_2 and CO (Eq. 4.1) in an equimolar ratio presents undoubtedly environmental and industrial interests.

$$CH_4 + CO_2 \rightarrow 2H_2 + 2CO \tag{4.1}$$

Indeed, it allows the conversion of two greenhouse gases into syngas with a H_2/CO ratio suitable for the Fischer-Tropsch synthesis (FTS) [15]. Owing to their good activity and the lower price of Ni compared to noble metals, Ni nanoparticles are preferred for DRM [16]. However, due to the endothermic nature of the reaction (high temperatures are needed in order to reach acceptable conversions) and to the existence of side reactions such as methane decomposition (Eq. 4.2) and Boudouard reaction (Eq. 4.3), nickel nanoparticles are suffering from sintering and coke deposition [17].

$$CH_4 \rightarrow 2H_2 + C_{(s)} \tag{4.2}$$

$$2CO \rightarrow CO_2 + C_{(s)} \tag{4.3}$$

In addition, the reverse water gas shift reaction (Eq. 4.4) can also take place, which leads to selectivity issues since it consumes the H_2 produced by the DRM process and generates more CO [18].

$$CO_2 + H_2 \rightarrow H_2O + CO \tag{4.4}$$

To improve their stability, the nickel nanoparticles can be deposited into the confined space of the pore channels of the SBA-15 support [18] whose diameter should play a key role, even though such effect was, to our knowledge, not specifically analyzed before. Hence, our target in the present work was to better understand confinement effects in SBA-15 and more precisely the impact of the pore sizes on the nickel nanoparticles incorporation in the channels, on their sizes, and on Ni dispersion. The "Two-Solvents" (TS) impregnation up 5 wt.% of nickel was applied on commercially available SBA-15 supports with 4, 5, and 7 nm pore diameters. The activities of those new catalysts were tested in the DRM reaction.

4.2 Experimental Part

4.2.1 Materials Preparation

Three commercially available SBA-15 supports (Sigma-Aldrich, CAS Number: 7631-86-9, ref.: 806803-5G, 806862-5G, 806854-5G) having different mean pore diameters (4, 5, and 7 nm), referred as S4, S5, and S7, respectively, were investigated here. Nickel (5 wt%) was incorporated by a postsynthesis method from an aqueous solution of $Ni(NO_3)_2.6H_2O$ affording Ni/S4, Ni/S5, and Ni/S7 samples. In details, the TS method used [19] consisted in suspending 1 g of the support in 35 mL of pentane (Sigma-Aldrich), then adding drop wise a volume of water equal to the silica pore volume (determined by N_2 physisorption and chosen in order to ensure a complete confinement of the Ni active phase in the pores of the support) and containing the appropriate amount of nickel precursor (0.25 g of $Ni(NO_3)_2.6H_2O$) (Sigma-Aldrich). The resulting solids were dried for three days at room temperature, then overnight at 60 °C, and they were finally calcined in air in an oven (static conditions) at 550 °C for 5 h (heating rate 1 °C min^{-1}).

4.2.2 Characterizations

Textural properties were determined from N_2 adsorption-desorption isotherms recorded on a Belsorp-max (MicrotracBEL) apparatus. Before measurements, the samples were degassed under vacuum for 2 h at 250 °C on BelprepII-vac. Specific surface areas were obtained using the BET equation. The mean pore diameters and specific pore volumes were determined using the BJH model.

X-ray Diffraction (XRD) data were recorded in the 2θ range between 5 and 90° on a Bruker D8 diffractometer apparatus using the Cu Kα radiation (λ = 1.5405 Å), 30 kV, and 10 mA conditions, a step size of 0.04° and 4 s per step. The average nickel particle size, D, for each sample was calculated (when possible) from the Scherrer equation (Eq. 4.5).

$$D = K\lambda / \beta\cos\theta \tag{4.5}$$

where K is a constant (K = 0.9), λ = 1.5405 Å, β is the full width at half maximum (FWHM) of the diffraction peak, and θ is the peak position.

Thermogravimetric analyses (TGA) were performed in order to quantify the carbon deposits on 10 mg of the spent catalysts using a TA SDT Q600 thermal analyzer instrument. Measurements were carried out from room temperature to 900 °C (heating rate 10 °C min^{-1}) under flowing air (100 mL min^{-1}).

Temperature Programmed Reduction (TPR), carried out on an Autochem 2910 (Micromeritics) apparatus equipped with a thermal conductivity detector, was used in order to study the nickel reducibility (on calcined materials). The samples (80 mg) were heated in a U-shaped quartz sample tube from room temperature to 800 °C at a rate of 10 °C min^{-1} using 5 vol% H_2/Ar (25 mL min^{-1}). A bath of isopropanol and liquid N_2 was used before the TCD detector in order to trap the water formed during NiO reduction. In addition, the nickel percentage (wt.%) was calculated from the quantitative H_2 consumption evaluated by integrating the reduction peak assuming the following reaction (Eq. 4.6)

$$NiO + H_2 \rightarrow Ni + H_2O \tag{4.6}$$

An Autochem 2910 apparatus was also used to perform reductions mimicking the in-situ reduction (detailed later, see Sect. 4.2.3) carried out before the catalytic test.

Transmission Electron Microscopy (TEM) analyses were performed on a JEOL-2020 (LaB) microscope operating at 200 kV. Prior to each measurement, the sample powder was embedded in a polymer matrix that was cut using an ultramicrotome, then slices were deposited on a copper grid coated with a porous carbon film. Hence, the observations made at room temperature could be done from the core to the surface of the grains of catalysts.

H_2 Chemisorption experiments were performed on a BELSORB-max equipment from BEL JAPAN. The samples (about 200 mg) were reduced in-situ under a flow of H_2 (50 mL min^{-1}) at 650 °C for 2 h using a ramp of 10 °C min^{-1}. The sample was then outgassed at 620 °C for 2 h under vacuum (about 5.10^{-1} Pa). A first, H_2 chemisorption was performed at 25 °C, the pressure at equilibrium being recorded when the pressure variation was below 0.02% per min. The sample was then outgassed for 2 h at 25 °C before a second H_2 chemisorption was done under the same conditions as above. Nickel particle size estimations were based on a truncated octahedron geometry, assuming a complete reduction, semispherical particles, and a H/Ni adsorption stoichiometry factor of 1. The experimental data were fitted with a Langmuir adsorption equation (Eq. 4.7) and the amount of surface nickel was calculated from the quantity adsorbed at saturation (Q_{max} in the model). The dispersion thus refers here to the molar percentage of surface Ni atoms compared to the total number of Ni atoms in the sample.

$$\theta = \frac{Q_{ads}}{Q_{max}} = \frac{\sqrt{KP}}{1+\sqrt{KP}} \tag{4.7}$$

θ: Fractional occupancy of the adsorption sites, Q_{ads}: Quantity adsorbed, Q_{max}: Quantity adsorbed on saturation, K: Equilibrium constant, P: Partial pressure of the adsorbate.

4.2.3 *Catalytic Measurements*

The catalysts were tested with a PID ENG & TECH Microactivity Effi Reactor. About 50 mg of powdered calcined material was loaded into the reactor (on quartz wool) and reduced in-situ at 650 °C for 2 h (10 °C min^{-1}) in a 5% H_2/Ar flow (30 mL min^{-1}) in order to ensure a complete conversion of nickel oxide into metallic nickel. After this pretreatment step, the temperature was decreased to 200 °C and the test of dry reforming of methane was started. This was conducted under atmospheric pressure, using a reactants ratio CH_4/CO_2 of 1:1 (each reactant being diluted at 10% in Ar) with a total flow of 60 mL min^{-1} (Gas Hourly Space Velocity (GHSV) = 72 L g^{-1} h^{-1}). The activity of the catalyst was measured by first increasing the temperature from 200 °C up to 650 °C (5 °C min^{-1}), then maintaining the temperature at 650 °C for 12 h for stability measurements. The gaseous products were analyzed online by an Agilent 490 micro-GC equipped with Plot-U (for the detection of CO_2) and molecular sieves (for the detection of H_2, CH_4, and CO) channels, and a thermal conductivity detector (TCD). The conversions of methane and carbon dioxide were calculated according to Eqs. (4.8) and (4.9), respectively, and the H_2/CO ratios were estimated using Eq. (4.10):

$$CH_4 \text{ conversion}, \% = \frac{\left(CH_{4(in)} - CH_{4(out)}\right)}{CH_{4(in)}} \times 100 \quad (4.8)$$

$$CO_2 \text{ conversion}, \% = \frac{\left(CO_{2(in)} - CO_{2(out)}\right)}{CO_{2(in)}} \times 100 \quad (4.9)$$

$$H_2 / CO = \frac{H_{2(out)}}{CO_{(out)}} \quad (4.10)$$

4.3 Results and Discussion

4.3.1 *Physicochemical Characterizations of the Supports and Calcined Ni-based SBA-15 Materials*

Figure 4.1 displays the nitrogen adsorption-desorption isotherms of the supports (as purchased) and of the corresponding impregnated samples. The corresponding textural parameters are listed in Table 4.1.

According to the IUPAC classification, the isotherms of the supports as well as their corresponding Ni-based solids are of type IV (a), in good agreement with what

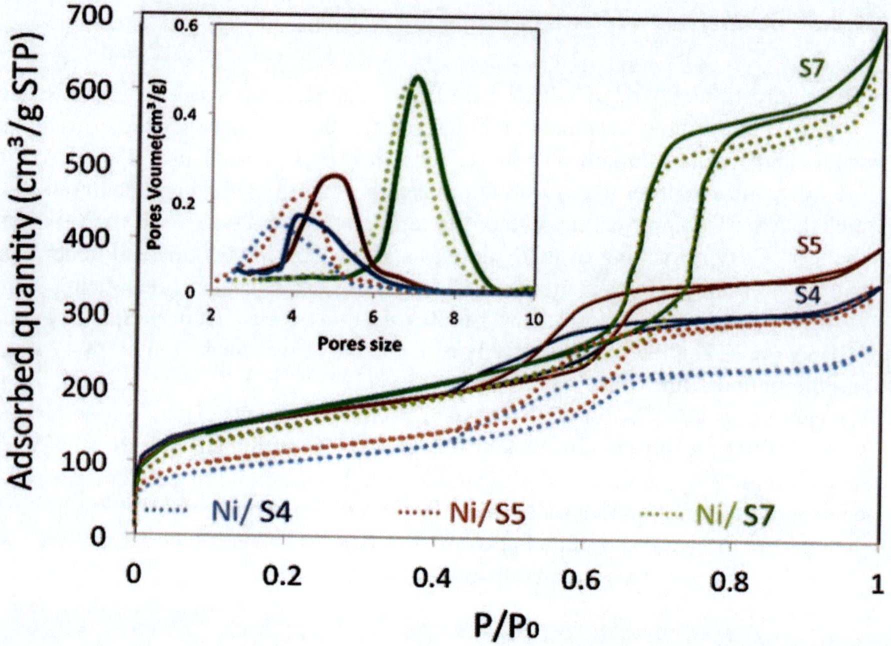

Fig. 4.1 N_2 adsorption-desorption isotherms at −196 °C and pores size distributions of the calcined Ni/S4, Ni/S5, and Ni/S7 materials and of their corresponding supports

Table 4.1 Textural properties of Ni/S materials

Materials	S_{BET} $m^2\ g^{-1}$	V_{Pores}[a] $cm^3\ g^{-1}$	$D_{Pores\ des.}$[b] (nm)
S4	573	0.42	4.0
Ni/S4	357	0.37	3.5
S5	571	0.64	5.0
Ni/S5	412	0.55	4.4
S7	593	1.02	7.0
Ni/S7	537	0.98	6.9

[a] Pores volume and [b] pores size estimated by BJH method from the desorption branch

is expected for mesoporous materials of the SBA-15 type [20], but with some variations between the samples.

The three supports (S4, S5, and S7) are characterized by very similar specific surface area of ~580 $m^2\ g^{-1}$ (Table 4.1). The value decreased after nickel impregnation by 37%, 28%, and 11% for Ni/S4, Ni/S5, and Ni/S7, respectively (Table 4.1). The pore volumes of the S4, S5, and S7 supports decreased as well after nickel impregnation by 0.05 (−12%), 0.09 (−14%), and 0.05 (−4%) $cm^3\ g^{-1}$, respectively (the weight of the support without nickel being taken into consideration). From those results, it can be emphasized that, whatever the carrier considered, the decrease of the pore volume was more important than expected assuming the incorporation of all Ni in the pores (5 wt.% and using a density of 8.9 g cm^{-3} for Ni(0)). This was specifically the case on the S4 and S5 supports, suggesting that total or partial pore blocking has occurred after nickel impregnation on these two supports. Such observation is in good agreement with the decrease of the mean pores diameter also observed for both S4- and S5-based materials after their impregnation and with their plugged hysteresis loops revealed by N_2 sorption in the P/P_0 region between 0.4 and 0.5 (Fig. 4.1).

The hystereses observed for S4 and S5 solids before and after impregnation are close to that of H5 type (especially for S4 solids), revealing the presence of both open and partially blocked mesopores (e.g., plugged hexagonal templated silicas) [20]. On the other hand, the S7 adsorption-desorption isotherms form an H1 hysteresis loop typical of well-organized hexagonal mesopores with a narrow pore size distribution (typical for SBA-15).

4.3.2 *Catalytic Activity*

The performances of the in-situ reduced samples toward the DRM reaction were evaluated in terms of CH_4 and CO_2 conversions under a GHSV of 72 L g^{-1} h^{-1}. The CH_4 and CO_2 conversions at $t = 0$ and after 12 h are presented in Fig. 4.2 as a function of the mean pores sizes in the catalyst. The three materials exhibited good catalytic performances, with conversions of both CH_4 and CO_2 close to those expected at thermodynamic equilibrium ones. Nevertheless, it can be seen that the activity tends to be improved while increasing the pores diameter of the SBA-15 support. Thus, Ni/S7 led to the highest conversion values of CH_4 and CO_2 (77% and 80%, respectively) and the deactivation after 12 h of catalytic test was the lowest (decrease

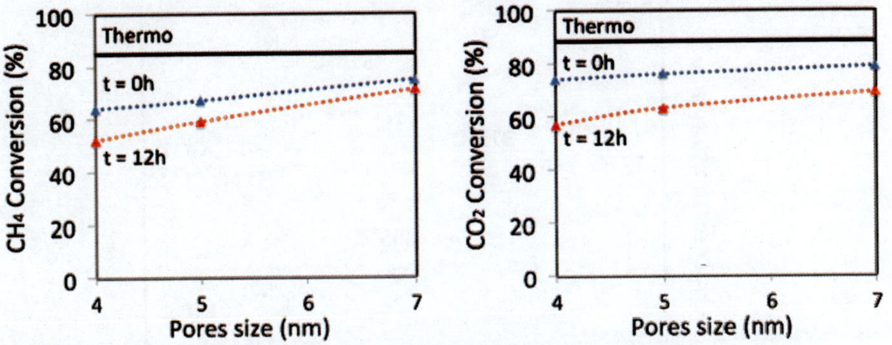

Fig. 4.2 Reactant conversions on the Ni-based SBA-15 catalysts at $t = 0$ (blue curve) and $t = 12$ h (orange curve) under a GHSV of 72 L g^{-1} h^{-1} (corresponding to a total flow of 60 mL min^{-1} and 50 mg of catalysts) as a function of the mean pores size of the supports; thermodynamic equilibrium values (black) calculated from the HSC (7.1) software (at 1 atm with an inlet feed ratio of $CO_2/CH_4 = 1$ and assuming no carbon formation occurring during the reaction) are also represented for comparison

of only 2% and 6% for CH_4 and CO_2, respectively). In contrast, Ni/S4 gave rise to the lowest conversion values (50% and 57%) and the highest deactivation after 12 h (decrease of 6% and 11% for CH_4 and CO_2, respectively).

In order to better understand those differences, a deeper investigation of the characteristics of the nickel-based species will be performed in the following part considering the Ni/S_4 and Ni/S7 samples having the smallest and largest pores of the series, respectively.

4.3.3 *NiO and Ni⁰ Characterization*

The first difference which could be clearly emphasized between the two solids arose from their nickel oxide reduction behavior.

The TPR profiles displayed in Fig. 4.3 show that the Ni/S4 and Ni/S7 samples are characterized by a "two-step" reduction profile. The H_2 consumption at lower temperature (375 °C) is usually attributed to the reduction of bulk NiO with low metal-support interaction and/or big aggregates [21]. Such component of the reduction profile was the major one in the case of Ni/S4. On the other hand, the second peak (centered at 455 or 510 °C for Ni/S4 and Ni/S7, respectively) can be tentatively assigned to the reduction of NiO species strongly interacting with the SiO_2 surface. From those results, we can clearly conclude that the Ni/S7 solid exhibits a more important fraction of nickel in good interaction with the support. Such strong Ni-support interaction should enhance the stability of nickel nanoparticles in the catalysts and hence contribute to the higher catalytic stability seen above for Ni/S7, thus confirming the crucial role of metal-support interaction in the stabilization of

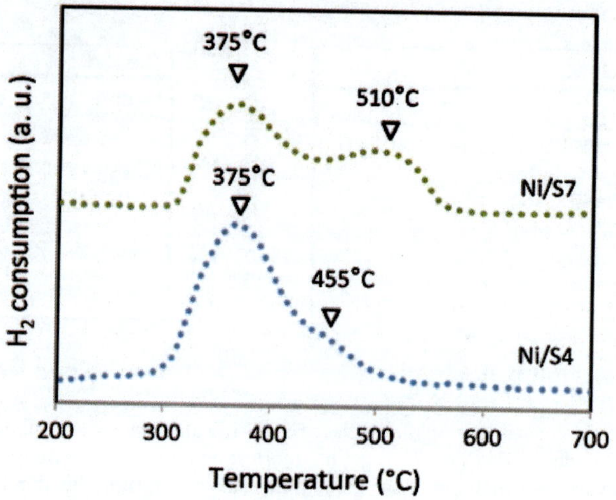

Fig. 4.3 H_2-TPR profiles of calcined Ni/S4 and Ni/S7 samples recorded with a H_2 5 Vol.%/Ar flow of 30 mL min^{-1} and a heating rate of 10 °C min^{-1}

Ni nanocatalysts [22, 23]. Last but not least, it is also important to note that the Ni wt.% estimated from the H_2 consumption during TPR analysis was 4.2 wt.% for both samples, which is close to the expected value of 5 wt.%.

The XRD patterns of the calcined and reduced Ni/S4 and Ni/S7 samples (Fig. 4.4) show a main wide peak in the 15° < 2θ°< 30° range typical of X-rays diffusion phenomena in amorphous silica walls. In addition, five intense diffraction peaks at 2θ° = 37.2°, 43.3°, 62.9°, 75.4°, and 77° corresponding to face-centered cubic lattice (Fm-3m) nickel oxide (JCPDS no. 89-7130) are observed for the calcined solids. After reduction under conditions very similar to those used just before the catalytic test (2 h at 650 °C), these peaks are substituted by three other ones at 2θ° = 44°, 52°, and 76°, characteristic of the face-centered cubic lattice (Fm-3m) of metallic Ni^0 (JCPDS no. 70-1849). The NiO particles sizes estimated from Scherrer equation (Eq. 4.5) were 7 nm and 5.4 nm for calcined Ni/S4 and Ni/S7, respectively (Table 4.2). In Ni/S7, this corresponds to a mean NiO size value smaller than the mean pores diameter of the S7 support, which would indicate that some particle size control by the pores occurred in this case. However, this was not the case for the S4-based material since the nickel oxide particles in Ni/S4 turned out to be larger than the pores size of S4. Such conclusions are in good agreement with above TPR data suggesting the presence of weakly interacting big NiO particles and with N_2 sorption results indicating some hindrance to pores access. Surprisingly, it is noteworthy that the XRD experiments performed after Ni/S4 and Ni/S7 reduction did not allow to show significant differences between NiO and Ni^0 particle size (Table 4.2).

In line with XRD results, clear differences between the two reduced solids were revealed by TEM analyses (Fig. 4.5).

In both cases, relatively good nickel dispersions could be emphasized from TEM images. However, some big Ni^0 aggregates were revealed on the external surface of the grains in the case of Ni/S4. Indeed, the particle sizes distribution extended

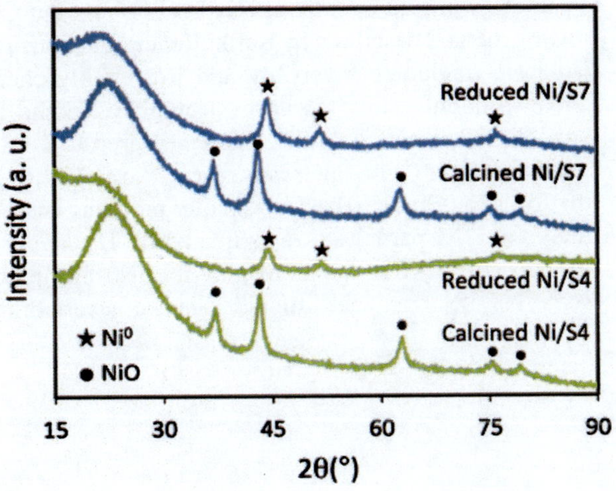

Fig. 4.4 XRD patterns of Ni/S4 and Ni/S7 samples after calcination or reduction (at 650 °C for 2 h (10 °C min^{-1}) in a 5% H_2/Ar flow (30 mL min^{-1}))

Table 4.2 H_2 chemisorption analysis at 20 °C after in-situ reduction under H_2 flow at 650 °C and nickel crystal size estimated from several methods for reduced and spent Ni/S materials

Materials	NiO size (nm)[a]	Ni^0 size (nm)[b]	Ni^0 size (nm)[c]	Ni dispersion (%)	Ni surface area ($m^2\ g^{-1}$)	Ni^0 size (nm) [d]	Ni^0 size (nm) [e]
Ni/S4	7.0	7.2	7.4	7.8	2.2	12	7.2
Ni/S7	5.4	5.1	5.8	11.6	3.2	8.7	5.8

Estimated from XRD results for [a] calcined and, [b] reduced solids; [c] Estimated from TEM images of the reduced solids; [d] Estimated from H_2 chemisorption; [e] Estimated from XRD results for spent solids

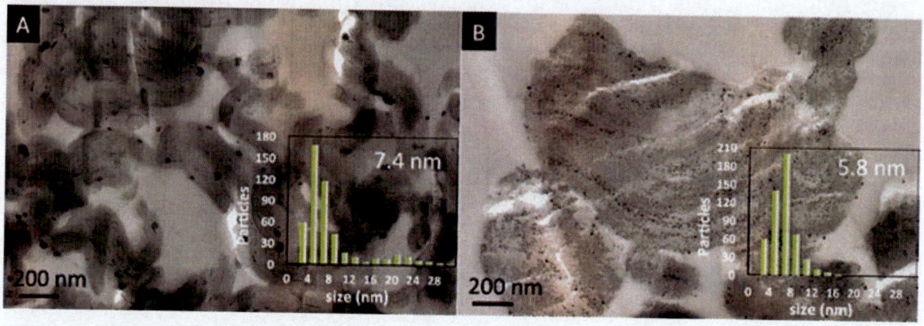

Fig. 4.5 TEM images of reduced A: Ni/S4 and B: Ni/S7 samples

towards 20–24 nm for Ni/S4 (Fig. 4.5a). This was not the case for the counts performed on the images of Ni/S7 (Fig. 4.5b), which reflects again, in good correlation with TPR results, a better incorporation and confinement of Ni in the pores of S7. In addition, the mean Ni particles size (reduced samples) estimated from the TEM images (on 400–500 particles) was 5.8 vs. 7.4 nm for Ni/S7 and Ni/S4, respectively. Such values are also in good agreement with the conclusion made from the discussion of the XRD results that emphasized indirectly a more successful confinement/incorporation effect for the support with the largest pores size (S7).

In this work, an attempt of Ni dispersion measurement was proposed by following the method described by Bartholomew and coworkers on the basis of the first H_2 isotherm (including reversibly and irreversibly chemisorbed hydrogen) [24]. H_2 chemisorption results (Table 4.2) emphasized the higher metal dispersion (and smaller nickel particles) in Ni/S7 compared to Ni/S4. These results are again in good agreement with the trends revealed by TEM and XRD, demonstrating further that Ni/S7 is characterized by smaller particles and better Ni dispersion. However, the greater Ni particles size estimated by H_2 chemisorption compared with those deduced from XRD and TEM may result from the weak bonding of hydrogen on some Ni sites or some difficulties in the accessibility of hydrogen to part of the Ni sites.

4.3.4 *Spent Solids Characterization*

After the catalytic test, the spent Ni/S catalysts were examined using TGA, XRD, and TEM in order to monitor the eventual modifications in their nickel dispersion as well as coke deposition, which could occur due to the side reactions. TGA analysis of the spent Ni/S catalysts (Fig. 4.6a) emphasizes the presence of carbon deposits of similar nature for both samples. The quantity of carbon was however, twice more for Ni/S4 compared to Ni/S7, being around 6 wt.%, which is however a value that remains below the quantities reported for other catalysts [25, 26]. The result for Ni/S7 is in good correlation with the stronger deactivation tendency exhibited by this catalyst during the DRM test (Fig. 4.2). Such observation also points out the excellent sintering resistance of nickel particles detected in majority inside the pores of S7. Indeed, it is well-known that the coke deposition is more favorable on big nickel aggregates [27]. In addition, no nanotubes were detected on the TEM images of Ni/S7 where all the nickel was incorporated in the porosity. In that case, the Ni^0 nanoparticles turned out to be still embedded into the porosity of S7 and no large

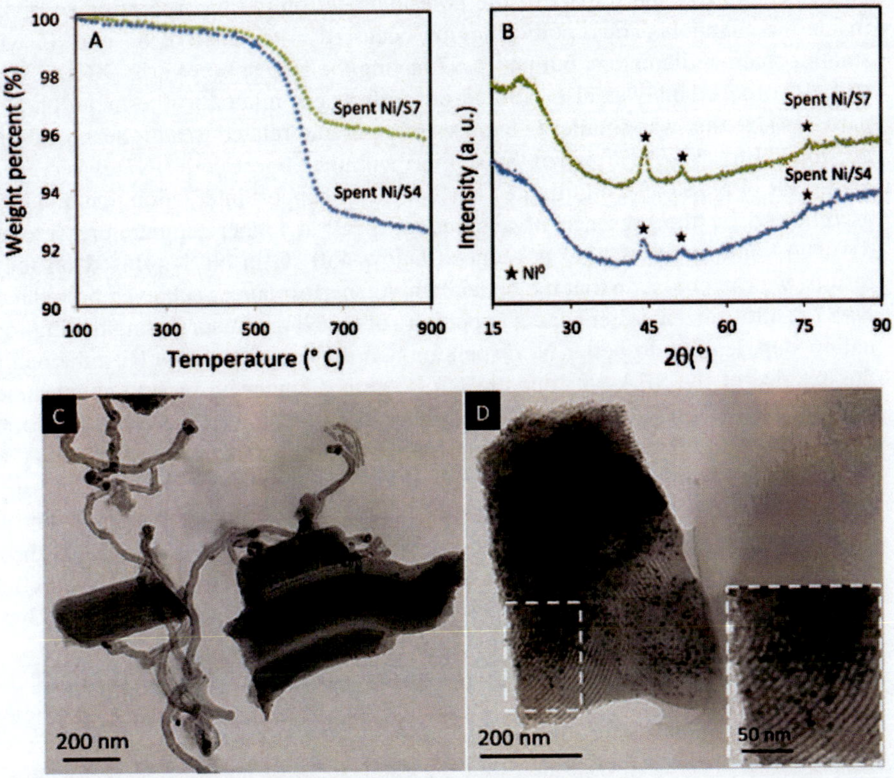

Fig. 4.6 (**a**) TGA profiles and (**b**) XRD patterns of spent Ni/S4 and Ni/S7 samples. (**c, d**) TEM images of spent Ni/S4 and Ni/S7 samples, respectively

nickel particles could be highlighted (Fig. 4.6d). It appears clearly that the big Ni aggregates detected on the external surface of the grains in the case of Ni/S4 (Fig. 4.6c) were less resistant to coke deposition contrary to those of Ni/S7. However, it is good to know that in the case of Ni/S7, carbon nanotubes were observed on some of the nickel particles, in minority, detected outside the grains. XRD results for both spent samples (Fig. 4.6b) exhibited three peaks corresponding to face centered cubic lattice of Ni^0 at $2\theta = 44°$, 52°, and 76° (JCPDS no. 70-1849) with no significant sintering (7.2 nm vs. 5.8 nm for Ni/S4 and Ni/S7, respectively) or carbon detected (due to the low quantity or to the presence of non ordered carbon that can be detected from TGA as well).

4.4 Conclusion

A systematic study of the effect of the size of pores in SBA-15 supports on the performances of nickel-SBA-15 catalysts in DRM reaction was carried out. N_2 sorption results obtained on the series of three catalysts with similar metal loading (4.2 wt.%) clearly emphasize the impact of the pore diameter on the incorporation of nickel inside the channels. Thus, pore plugging occurred in the case of S4 and S5 with smaller channel diameters, but not in S7 having the largest pores size. XRD, TEM, and H_2 sorption analysis also pointed out differences related to the mean Ni/NiO particles size that was smaller (~5 nm vs. ~7 nm) and related to a higher metal dispersion (11.6% vs. 7.8%) when the support with the largest pores (S7) was considered. TPR of Ni/S7 also highlighted a better metal-support interaction than in Ni/S4 as reflected by the presence of a reduction peak at higher temperature (around 550 °C) while most of NiO is reduced below 450 °C in Ni/S4. Based on these results, we can conclude that the better catalytic performances achieved on reduced Ni/S7 result from an easier access to porosity of the Ni precursor during the impregnation step, leading to better Ni dispersion. On the contrary, some Ni remained at the exterior of the SBA-15 grains in Ni/S4, giving bigger Ni-based nanoparticles poorly attached to the support and strongly sensitive to sintering. Such result looks like others reported in the literature. Indeed, in 2015, T. Shimizu and coworkers [28] demonstrated that Rh particles (4 wt.%) included in a mesoporous MCM-41 support used for hydrogenation of cinnamaldehyde tend to migrate out easily and to agglomerate on the external surface of the support during calcination, whereas those prepared in a mesostructured silica with larger pores size (MSU-H) remained quite unchanged. In addition, Yuan and coworkers [29] have shown that, for ordered hexagonal mesostructure ((Mo-Ni)-silica based catalysts with 13 wt.% of Mo and 3 wt.% of Ni), a support with a larger mean pore size gives rise to higher hydrodesulfurization performance efficiency. Also, Song and coworkers [30] who studied the pores size effect of siliceous supports on the performances of cobalt (15 wt.%) for the FT reaction demonstrated that the catalysts with pores diameter of about 6–10 nm displayed higher activity and good C5+ selectivity compared to those with smaller pores.

Acknowledgments This work was supported by the ANR (France) and CNRS-L (Lebanon) national agencies through their respective funding provided in the frame of the ERANETMED EU-FP7 initiative (Energy-065 SOL-CARE project, JC-ENERGY-2014 first call). The authors would like also to thank Dr. C. Thomas from LRS for the interpretation of the H_2 chemisorption experiments.

References

1. S. Chakraborty, H. Dai, P. Bhattacharya, N. Fairweather, M. Gibson, J. Krause, H. Guan, Iron-based catalysts for the hydrogenation of esters to alcohols. J. Am. Chem. Soc. **136**, 7869–7872 (2014)
2. K. Mallikarjuna, C. Bathula, K. Buruga, N. Shrestha, Y.Y. Noh, H. Kim, Green synthesis of palladium nanoparticles using fenugreek tea and their catalytic applications in organic reactions. Mater. Lett. **205**, 138–141 (2017)
3. H.Y. Zhao, D. Li, P. Bui, S.T. Oyama, Hydrodeoxygenation of guaiacol as model compound for pyrolysis oil on transition metal phosphide hydroprocessing catalysts. Appl. Catal. A Gen. **391**, 305–310 (2011)
4. H. Drobna, M. Kout, A. Sołtysek, V. González-Delacruz, A. Caballero, L. Capek, Analysis of Ni species formed on zeolites, mesoporous silica and alumina supports and their catalytic behavior in the dry reforming of methane. React. Kinet. Mech. Catal. **121**, 255–274 (2017)
5. X.Y. Gao, J. Ashok, S. Widjaja, K. Hidajat, S. Kawi, Ni/SiO_2 catalyst prepared via Ni-aliphatic amine compelxation for dry reforming of methane: effect of carbon chain number and amine concentration. Appl. Catal. A Gen. **503**, 34–42 (2017)
6. J. Woo Han, J.S. Park, M. Suk Choi, H. Lee, Uncoupling the size and support effects of Ni catalysts for dry reforming of methane. Appl. Catal. B Environ. **203**, 625–632 (2017)
7. X. Chen, S. Wang, J. Zhuang, M. Qiao, K. Fan, H. He, Mesoporous silica-supported NiB amorphous alloy catalysts for selective hydrogenation of 2-ethylanthraquinone. J. Catal. **227**, 419–427 (2004)
8. X. Yang, D. Chen, S. Liao, H. Song, Z. Li, Z. Fu, Y. Su, High-performance Pd-Au bimetallic catalyst with mesoporous silica nanoparticles as support and its catalysis of cinnamaldehyde hydrogenation. J. Catal. **291**, 36–43 (2012)
9. A.K. Prashar, S. Mayadevi, P.R. Rajamohanan, N. Devi, In situ encapsulation of Pt nanoparticles in mesoporous silica: synthesis, characterisation and effect of particle size on CO oxidation. Appl. Catal. A Gen. **403**, 91–97 (2011)
10. V. Gutiérrez, F. Nador, G. Radivoy, M.A. Volpe, Highly selective copper nanoparticles for the hydrogenation of α,β-unsaturated aldehydes in liquid phase. Appl. Catal. A Gen. **464-465**, 109–115 (2013)
11. T.W. Hansen, A.T. Delariva, S.R. Challa, A. Datye, Sintering of catalytic nanoparticles: particle migration or Ostwald ripening? Acc. Chem. Res. **46**, 1720–1730 (2013)
12. P. Munnik, P.E. de Jongh, K.P. de Jong, Recent developments in the synthesis of supported catalysts. Chem. Rev. **115**, 6687–6718 (2015)
13. L. Karam, S. Casale, H. Zakhem, N. El Hassan, Tuning the properties of nickel nanoparticles inside SBA-15 mesopores for enhanced stability in methane reforming. J. CO_2 Utiliz. **17**, 119–124 (2017)
14. M.N. Kaydouh, N. El Hassan, A. Davidson, S. Casale, H. El Zakhem, P. Massiani, Effect of the order of Ni and Ce addition in SBA-15 on the activity in dry reforming of methane. C. R. Chim. **18**, 293–301 (2015)
15. M.N. Kaydouh, *Nickel confinement effect into mesoporous silica supports based catalysts for the syngas production via the dry reforming of methane*. Dissertation, Sorbonne Université (2016)

16. O. Daoura, S. Daher, M.N. Kaydouh, N. El Hassan, P. Massiani, F. Launay, M. Boutros, Influence of the swelling agents of siliceous mesocellular foams on the performances of Ni-based methane dry reforming catalysts. Int. J. Hydrog. Energy **43**, 17205–17215 (2018)
17. O. Daoura, M.N. Kaydouh, N. El Hassan, P. Massiani, F. Launay, M. Boutros, Mesocellular silica foam-based Ni catalysts for dry reforming of CH_4 (by CO_2). J. CO_2 Utiliz. **24**, 112–119 (2018)
18. O. Daoura, G. El Chawich, M. Boutros, N. El Hassan, P. Massiani, O. Ersen, W. Baaziz, F. Launay, Aqueous nickel(II) hydroxycarbonate instead of nickel(0) colloids as precursors of stable Ni-silica based catalysts for the dry reforming of methane. Catal. Commun. **138**, 105953–105957 (2020)
19. J. Van der Meer, I. Bardez-Giboire, C. Mercier, B. Revel, A. Davidson, R. Denoyel, Mechanism of metal oxide nanoparticle loading in SBA-15 by the double solvent technique. J. Phys. Chem. C **114**, 3507–3515 (2010)
20. M. Thommes, K. Kaneko, A.V. Neimark, J. Oliver, F. Rodriguez-Reinoso, J. Rouquerol, K. Sing, Physisorption of gases, with special reference to the evaluation of surface area and pore size distribution (IUPAC Technical Report). Pure Appl. Chem., 1051–1060 (2015)
21. M.J.F.M. Verhaak, A.J. van Dillen, J.W. Geus, Measuring the acid-base properties of supported nickel catalysts using temperature-programmed desorption of ammonia. Appl. Catal. A Gen. **105**, 251–269 (1993)
22. Z. Taherian, M. Yousefpour, M. Tajally, B. Khoshandam, A comparative study of ZrO_2, Y_2O_3 and Sm_2O_3 promoted Ni/SBA-15 catalysts for evaluation of CO_2/methane reforming performance. Int. J. Hydrog. Energy **42**, 16408–16420 (2017)
23. S. Kawi, Y. Kathiraser, J. Ni, U. Oemar, Z. Li, T. Saw, Progress in synthesis of highly active and stable nickel-based catalysts for carbon dioxide reforming of methane. ChemSusChem **8**, 3556–3575 (2015)
24. C.H. Bartholomew, Hydrogen adsorption on supported cobalt, iron, and nickel. Catal. Lett. **7**, 27–51 (1990)
25. A. Albarazi, M.E. Gálvez, P. Da Costa, Synthesis strategies of ceria–zirconia doped Ni/SBA-15 catalysts for methane dry reforming. Catal. Commun. **59**, 108–112 (2015)
26. Z. Taherian, M. Yousefpour, M. Tajally, B. Khoshandam, Catalytic performance of Samaria-promoted Ni and Co/SBA-15 catalysts for dry reforming of methane. Int. J. Hydrog. Energy **42**, 24811–24822 (2017)
27. S. Arora, R. Prasad, An overview on dry reforming of methane: strategies to reduce carbonaceous deactivation of catalysts. RSC Adv. **6**, 108668–108688 (2016)
28. T. Shimizu, M. Ota, Y. Sato, H. Inomata, Effect of pore structure on catalytic properties of mesoporous silica supported rhodium catalysts for the hydrogenation of cinnamaldehyde. Chem. Eng. Res. Des. **104**, 174–179 (2015)
29. P. Yuan, J. Liu, Y. Li, Y. Fan, G. Shi, H. Liu, X. Bao, Effect of pore diameter and structure of mesoporous sieve supported catalysts on hydrodesulfurization performance. Chem. Eng. Sci. **111**, 381–389 (2014)
30. D. Song, J. Li, Effect of catalyst pore size on the catalytic performance of silica supported cobalt Fischer–Tropsch catalysts. J. Mol. Catal. A Chem. **247**, 206–212 (2006)

Chapter 5
Nanostructured Bimetallic Pd-based Catalysts for the Valorization of Lignocellulosic Biomasses

Emilia Paone and Francesco Mauriello

5.1 Introduction

Lignocellulosic biomasses, abundant and inedible resources, that can be easily obtainable also from agricultural residues and waste materials, represent an important opportunity for energy and alternative chemicals production, since one of the biggest challenges in the modern chemical industry is the overcoming of its historical dependence on fossil resources. The use of lignocellulosic biomasses represents also a great chance for the socio-economic development of marginal rural and mountain areas, helping, at the same time, to fight against the effect of climate changes.

The attention on the use of lignocellulosic biomasses to "transform" sustainably fossil resources to biofuels and bioderived chemicals can be explained considering the chemical composition of their key components (cellulose, hemicellulose and lignin) that have the potential for the sustainable production of several building block intermediates for modern biorefineries [1–8].

Cellulose and hemicellulose are mainly formed of C6- and C5-sugars, respectively, whilst lignin is mainly composed by phenolic units. Such a wealth of chemical functionalities represents, at present, the most promising alternative to petroleum resources, as shown in Fig. 5.1 [9–13].

Lignin, the most abundant natural phenolic biopolymer, is of particular interest for chemical industries since it is the only component that allows a sustainable production of aromatic compounds and intermediates [14–17]. In the last years, several strategies have been proposed for the lignin depolymerization into readily available

E. Paone (✉)
Dipartimento di Ingegneria Industriale, Università degli Studi di Firenze, Florence, Italy
e-mail: emilia.paone@unifi.it

F. Mauriello
Dipartimento DICEAM, Università Mediterranea di Reggio Calabria, Reggio Calabria, Italy

M. Piumetti, S. Bensaid (eds.), *Nanostructured Catalysts for Environmental Applications*, https://doi.org/10.1007/978-3-030-58934-9_5

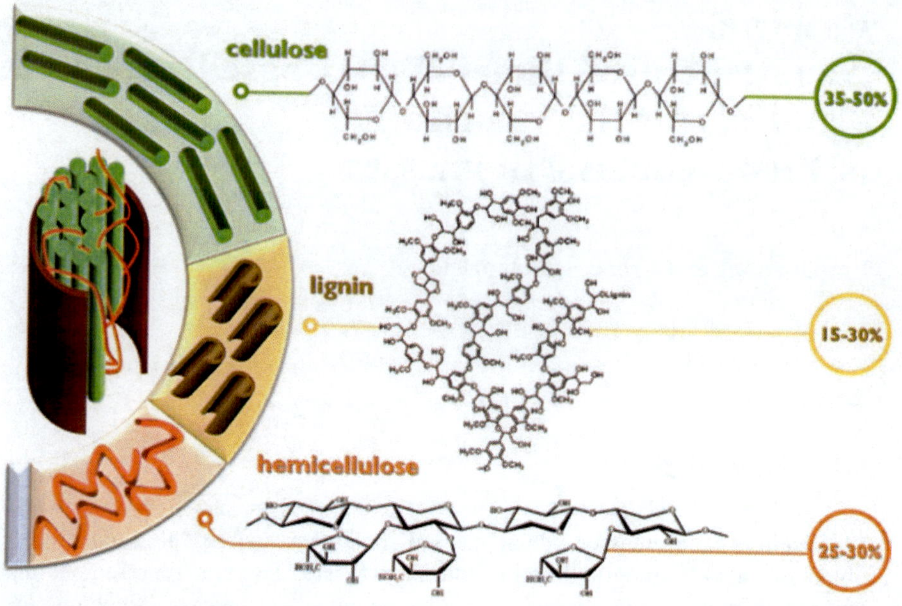

Fig. 5.1 Chemical structure of the lignocellulosic material and its constituents: cellulose, hemicellulose and lignin [14]

aromatic compounds. Since the lignin substructure is characterized by large amounts of etheric bonds, catalytic reductive technologies (hydrogenolysis) have received strong attention, allowing the C–O bond breaking by adding molecular hydrogen.

The hydrogenolysis of lignin generally requires harsh reaction conditions due to the high dissociation energies involved in the C–O bond cleavage [18]. Furthermore, the use of high-pressure molecular hydrogen leads to undesired aromatic ring hydrogenated products, thus decreasing the process efficiency and lowering the degree of lignin depolymerization (fully saturated products are less susceptible to a further hydrogenolysis process).

Therefore, in the last years, a significant growth towards an alternative catalytic approach avoiding the direct use of molecular hydrogen was considered. To this regard, the catalytic transfer hydrogenolysis (CTH) technique represents a valid green alternative to classic hydrogenation/hydrogenolysis reactions owing to the use of indirect H-source molecules [19]. Simple organic molecules, including short-chain alcohols (2-propanol, ethanol and methanol), can be efficiently used as H-donor for CTH reactions. At the same time, CTH molecules are generally easy to handle, potentially obtainable from renewable feedstocks and, due to their relatively lower tendency to release hydrogen, they may allow a higher production of aromatic compounds from lignin and its derived molecules [20, 21].

Transition metals have been successfully used in the catalytic valorization of biomass-derived molecules. Among them, palladium was less investigated, although it provides good results in the liquefaction and the hydrodeoxygenation of lignocellulosic biomasses [22, 23]. In particular, the commercial Pd/C has attracted a lot of research interest allowing an efficient transformation of lignin and of a variety of aromatic ethers using formic acid and 2-propanol as H-donors.

Moreover, palladium is characterized by a fundamental electronic configuration, unique among transition metals ($4d^{10}$, $5s^0$), that can be strongly influenced by the coordination environment. The presence of a second element, in fact, may influence the electronic structure of palladium leading to unusual catalytic properties different than those of monometallic Pd. Hence, heterogeneous catalysts based on palladium bimetallic systems have gained increasing attention for their enhanced activity and durability in different reactions including CTH reactions.

As a consequence, the preparation of bimetallic Pd-based catalysts, able to selective cleave the C–O bond in the presence of aromatic functionalities together with a deep insight on the basic chemistry of aromatic ethers, remains challenging to develop selective chemical processes to produce aromatics from lignin.

Therefore, the aim of this chapter is to present the synthesis and physico-chemical properties of nanostructured bimetallic Pd-based catalysts (Pd-M with M = Fe, Ni and Co) and their use in the transfer hydrogenolysis reactions of lignin-derived aromatic ethers by using 2-propanol as solvent/H-donor (Fig. 5.2). Several physico-chemical techniques (XRD, TEM, H_2-TPR, XPS and EXAFS) are included in order to highlight the key factors that may affect the catalytic properties of the bimetallic palladium catalysts.

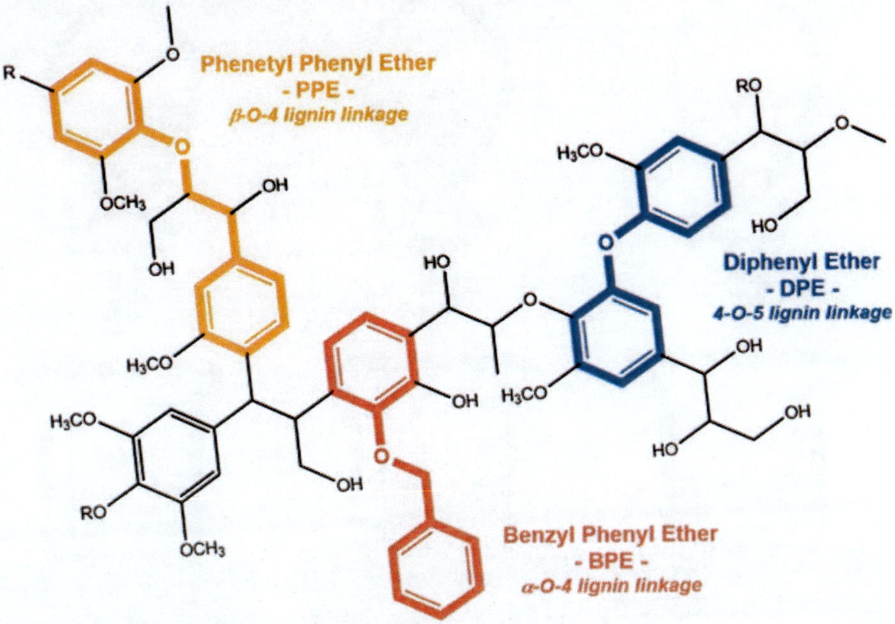

Fig. 5.2 Benzyl Phenyl Ether (BPE), 2-Phenethyl Phenyl Ether (PPE) and Diphenyl Ether (DPE) as simplest lignin-derived aromatic ethers

5.2 Heterogeneous Nanostructured Pd-based Bimetallic Catalysts

5.2.1 Preparation of Nanostructured Pd-based Bimetallic Catalysts

All catalysts presented in this chapter, designed with a nominal palladium loading of 3–5 wt%, were prepared by the co-precipitation technique (Fig. 5.3).

This technique implies the simultaneous precipitation of more than one precursor in a solution containing a precipitation agent (e.g., hydroxides, oxalates, formates and carbonates) [24, 25]. After formation of the co-precipitate, hydrothermal treatments (which transform amorphous precipitates into crystalline materials with improved thermal stability and surface acidity) may be carried out.

In catalysts with two or more metallic species, the final composition of the precipitate depends on the solubility difference between the components and the

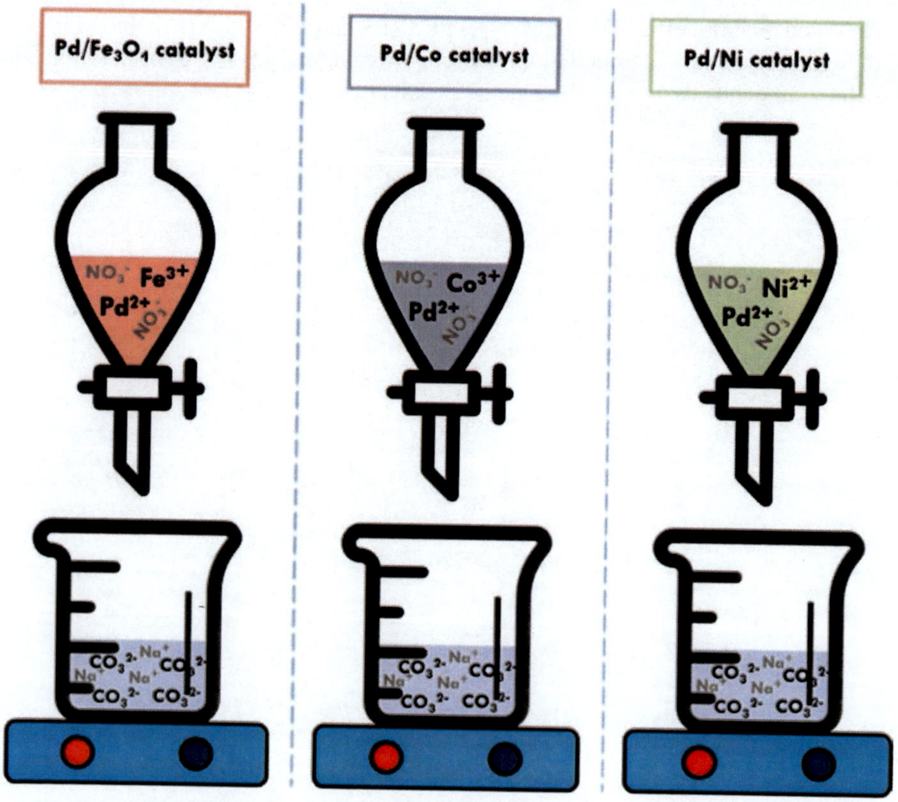

Fig. 5.3 Co-precipitation of the metallic precursors within a solution of carbonates for the preparation of bimetallic Pd-M (M = Fe, Ni and Co) catalysts

Table 5.1 Main characteristics of the presented bimetallic palladium-based catalysts

Catalyst	Support	Nominal Pd loading [wt%]	XRF	S.A. [$m^2\ g^{-1}$]	d_n [nm]
Pd/Fe	Fe_3O_4	5	8.7	170	2.4
Pd/Co	Co_3O_4	5	3.7	106	10.7
Pd/Ni	NiO	3	5.0	90	4.2

S.A. surface area, d_n mean particle size determined by TEM

chemistry taking place during the precipitation process. Subsequently, in order to obtain crystalline stoichiometric precipitates with an intimate inter-dispersion of the two metals, several parameters (e.g., pH, solutions' concentration, temperature and stirring speed) must be controlled.

Preparation of catalytic systems, discussed in this chapter, occurs by pouring a solution containing both nitrate precursors of palladium (II) and iron (III), cobalt (III) or nickel (II) into a solution containing sodium carbonate. The ensuing co-precipitate is washed, filtrated and then dried overnight at 120 °C and, before any further reaction, reduced with H_2 at 200 °C.

The main characteristics of the reduced Pd-based bimetallic catalysts are reported in Table 5.1.

5.2.2 Physico-Chemical Properties of Nanostructured Pd-M Catalysts

5.2.2.1 X-Ray Diffraction (XRD)

XRD patterns of co-precipitated bimetallic Pd-based catalysts are depicted in Fig. 5.4.

The pattern of Pd/Fe_3O_4, after the reduction at 200 °C under H_2 flow, exhibits only peaks related to Fe_3O_4 and no metallic palladium peaks are observed, indicating the presence of very small palladium particles highly dispersed [26–32]. Typical diffraction peaks related to the cubic crystal of magnetite, in agreement with the Joint Committee on Powder Diffraction Standards (JCPDS), are detected: 30.4° (220); 35.8° (311); 43.4° (400); 54.0° (422); 57.5° (511); 63.2° (440) [31–35].

The X-ray powder diffraction (XRD) pattern of the unreduced Pd/Co catalyst shows diffraction peaks typical of the pure cubic crystalline Co_3O_4 [36]. After reduction, diffraction peaks of metallic cobalt can be easily observed indicating the promoting effect of Pd in the support reduction. This point is worthy of remark since the complete reduction of Co_3O_4 into metallic Co in H_2 atmosphere generally occurs up to 400 °C [37]. Moreover, the absence of the Pd (111) diffraction peaks is indicative of highly dispersed palladium particles as confirmed also by SEM-EDX analysis [29].

The XRD pattern of the Pd/Ni catalyst, after the reduction at 250 °C under H_2 flow, shows peaks related to the reduction of NiO into metallic nickel. The typical

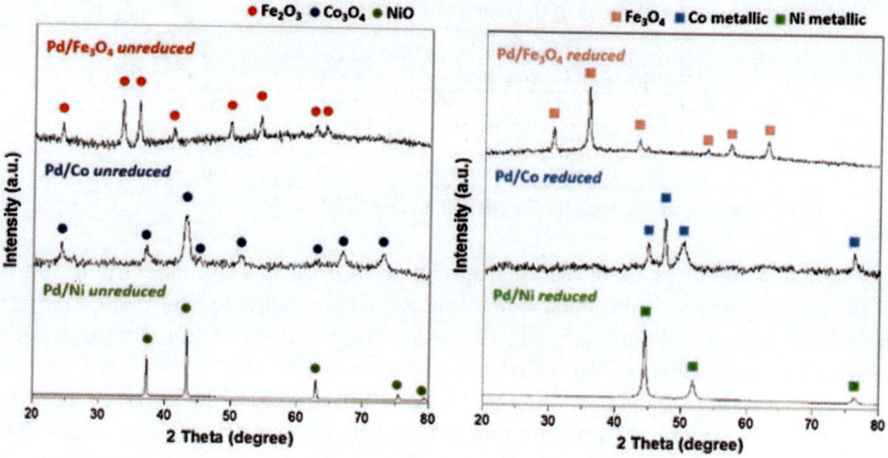

Fig. 5.4 XRD patterns of bimetallic Pd-based catalysts

diffraction peak of the metallic Pd (111) at $2\theta = 40.18°$ was not observed again suggesting the presence of extremely small and well-dispersed palladium particles. Diffraction lines at 2θ values of 44.4°, 51.8° and 76.2° were observed at slightly higher angles as compared to those of classic metallic nickel. These results indicate the presence of a Pd/Ni alloy phase, in good agreement with previous reports [30, 38, 39].

5.2.2.2 Transmission Electron Microscopy (TEM)

TEM images of investigated bimetallic Pd-based catalysts are reported in Fig. 5.5. The bimetallic Pd/Fe_3O_4 catalyst shows a predominance of extremely small palladium particles. A relatively narrow particle size distribution with diameter values ranging between 0.5 and 2.5 nm, with a majority of particles of 1.2 nm diameter, followed by a second population of slightly larger palladium particles, with 2.3 nm average diameter, was found [26]. TEM analyses are in perfect agreement with XRD results, which suggest the presence of highly dispersed palladium particles, extremely small, being the conventional limit of XRD instruments, for the detection of supported nanoparticles, usually in the range of 2–2.5 nm.

Representative TEM images of the reduced Pd/Co catalyst are also reported in Fig. 5.5 where distinctive lattice fringes with an interplanar distance of 0.19 nm, typical of the (111) plane of the metallic cobalt [40], can be easily detected confirming the complete reduction of the support. Smaller darker nanoparticles were identified, as bimetallic Pd-Co ensembles [41, 42] with a particle size distribution of about 8.7 nm.

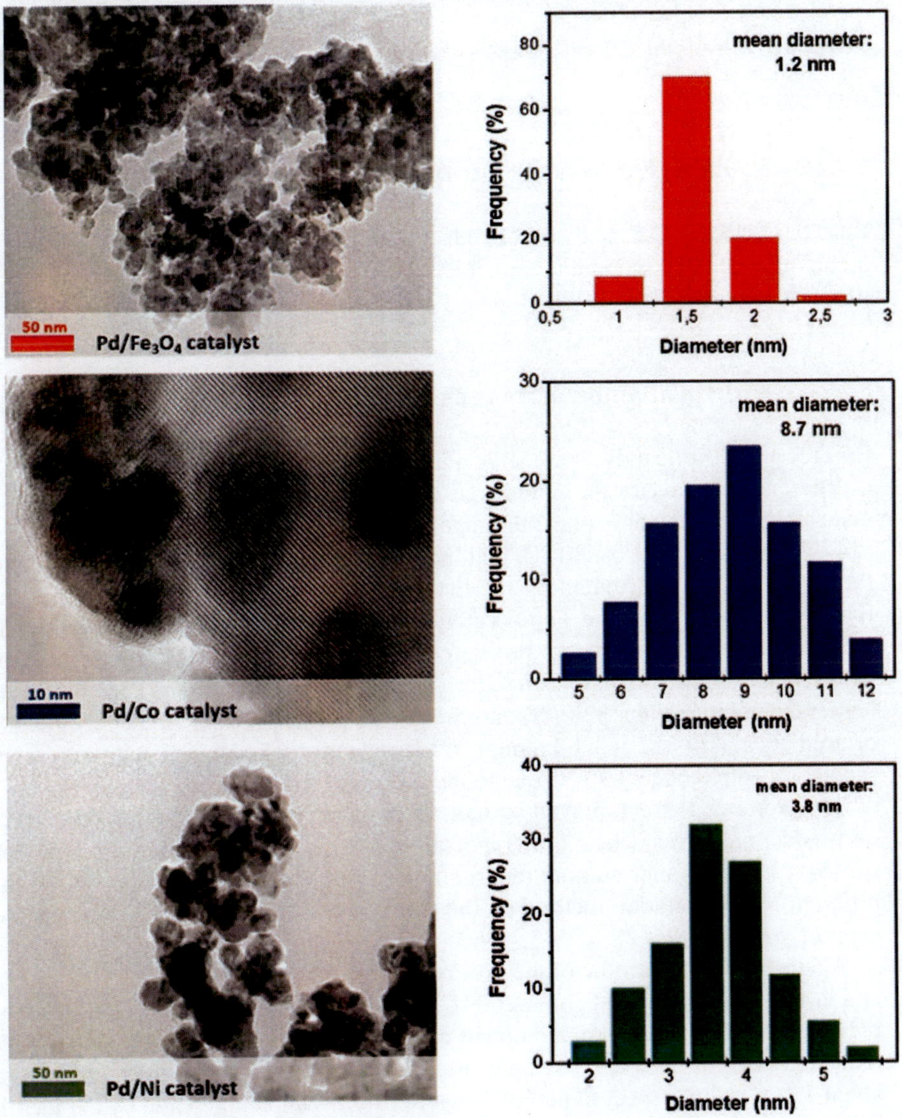

Fig. 5.5 TEM images of Pd-based catalysts and their relative particle size distribution

Conversely representative TEM analysis of the Pd/Ni system reveals faceted Ni metal particles and small palladium ensembles with a narrow particle size distribution.

5.2.2.3 Scanning Electron Microscopy (SEM)

Results of SEM-EDX analysis of reduced Pd/Fe_3O_4, Pd/Co and Pd/Ni catalysts, including elemental mapping, are depicted in Fig. 5.6. Highly dispersed palladium particles are easily detected in all investigated systems.

5.2.2.4 Hydrogen Temperature Programmed Reduction (H_2-TPR)

H_2-TPR profiles of catalysts are shown in Fig. 5.7.

The one referring to the unreduced co-precipitated Pd/Fe_3O_4 catalyst shows a main area containing only one intense peak of reduction at low temperature, starting at 44 °C and centred at ca. 80 °C, that includes reductions of either $Pd^{+2} \rightarrow Pd^0$ and $Fe^{+3} \rightarrow Fe_3O_4$, as determined by the hydrogen consumption calculation [26]. The H_2-TPR profile of the pure Fe_2O_3 catalyst is also included in the same square in order to compare the reductive profile of both the bimetallic catalyst and the iron oxide (Fig. 5.7). For the pure iron oxide (Fe_2O_3), the reduction starts at significantly higher temperatures (ca. 320 °C) and the profile presents two main peaks: the first centred at ca. 420 °C, which belongs to the reduction of $Fe^{+3} \rightarrow Fe_3O_4$, whilst the second is centred at ca. 690 °C due to the following reduction of $Fe_3O_4 \rightarrow FeO$.

The promoting effect, played by palladium nanoparticles on the iron oxide support reduction, is very clear and is indicative of a strong metal-support interaction (SMSI) [26, 28]. Other authors report similar results and underline the role of the different co-precipitation method on the effectiveness of palladium to promote the support reduction [26, 28, 29, 43, 44].

Again the H_2-TPR profile of the Pd/Co sample is characterized by only one broad and intense peak centred at about 260 °C, which can be related to the simultaneous reduction of both palladium and cobalt cations. Most important, the main H_2-TPR reduction peak of the Pd/Co sample (Fig. 5.7) is shifted to lower temperatures of about 150 °C with respect to that of the pure Co_3O_4 sample, prepared by the identical synthetic procedure (co-precipitation), and is characterized, in analogy with other reports, by a wide non-symmetric peak that includes both $Co_3O_4 \rightarrow CoO$ and $CoO \rightarrow Co$ reductions [45, 46].

This phenomenon is well known and is commonly ascribed to the promoting effect of well-dispersed palladium particles on the Co_3O_4 reduction indicating a strong interaction between Pd ions and the metal oxide support [47].

The H_2-TPR profile of NiO (Fig. 5.7) displays an intense reduction peak centred at ca. 350 °C corresponding to the reduction step of nickel oxide, with a shoulder at 310 °C, assigned to the reduction of oxygen species adsorbed on the surface [48, 49]. On the other hand, the H_2-TPR profile of the sample Pd/Ni, included in the

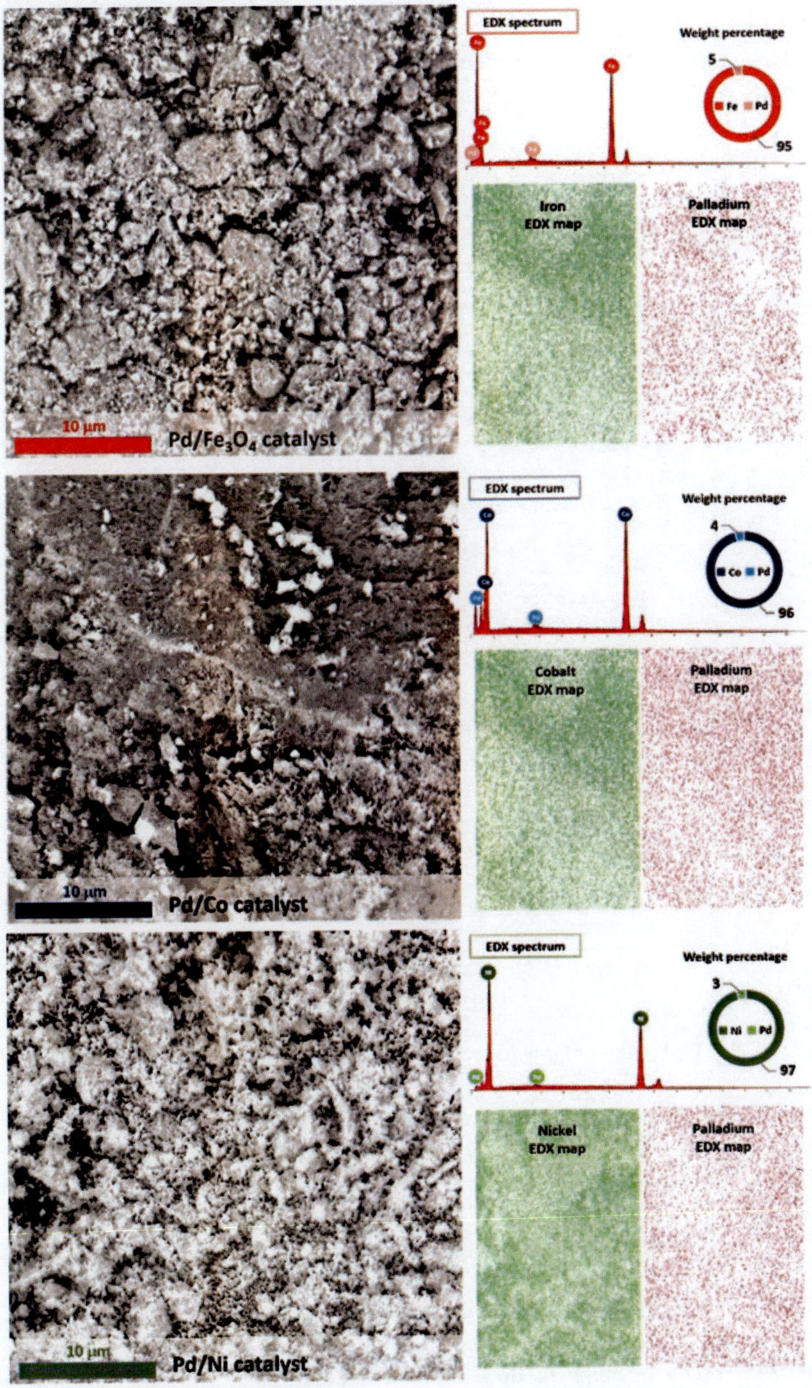

Fig. 5.6 SEM images of co-precipitated Pd-based catalysts

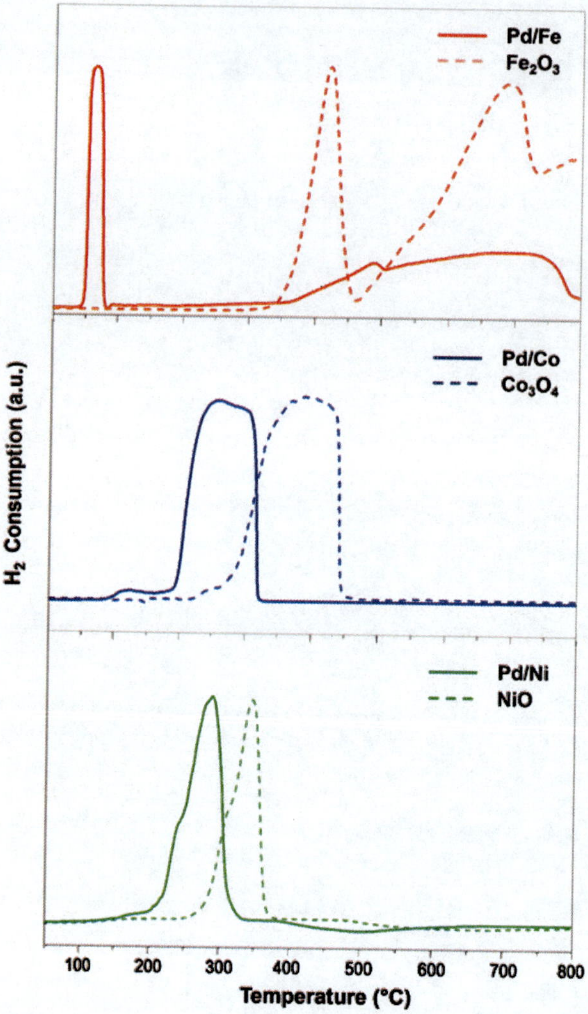

Fig. 5.7 H_2-TPR profiles of Pd-based catalysts and their corresponding metal oxide supports

same square, is analogous to that of the NiO sample although the centre of the main reduction peak is shifted to around 280 °C and displays the reduction of both Pd and Ni ions, indicative of a promoting effect of well-dispersed Pd particles on NiO reduction [27, 46].

5.2.2.5 X-Ray Photoelectron Spectroscopy (XPS)

The XPS analysis helps to understand the electronic properties of investigated bimetallic catalysts (Table 5.2).

Table 5.2 Binding energy values of Pd $3d_{5/2}$, Fe $2p_{3/2}$, Co $2p_{3/2}$, Ni $2p_{3/2}$ for *un*-reduced and reduced investigated Pd-based catalysts

Catalyst	Binding energy (eV)			
	Pd $3d_{5/2}$	Fe $2p_{3/2}$	Co $2p_{3/2}$	Ni $2p_{3/2}$
Pd/Fe_3O_4	335.3	710.7	–	–
Pd/Fe *unred*	336.6	710.9	–	–
Pd/Co	335.5	–	778.2	–
Pd/Co *unred*	336.8	–	779.9	–
Pd/Ni	335.6	–	–	852.2
Pd/Ni *unred*	336.8	–	–	853.7
Pd/C	335.0	–	–	–
Pd/C *unred*	336.6	–	–	–

XPS spectra of the reduced Pd/Fe sample are characterized from the absence of the satellite peak of Fe $2p_{3/2}$ at 718.5 eV, which indicates the absence of the Fe_2O_3 structure over the reduced surface (Table 5.2). Furthermore, the shift of ca. 0.3 eV of the Pd $3d_{5/2}$ binding energy to higher values, than that observed for metallic palladium, indicates the presence of partial positively charged metal Pd species ($Pd^{\delta+}$) (Table 5.2). That is attributed to an alloy formation in bimetallic-supported Pd catalysts [27].

The XPS spectrum of the reduced Pd/Co catalyst shows that the binding energy of the Pd $3d_{5/2}$ level appears at about 0.5 eV value higher than that of the binding energy of metallic palladium, again indicating the presence of partial positively charged Pd species, suggesting the formation of a Pd-Co alloy [50, 51]. At the same time, the presence of metallic cobalt is confirmed by its typical sharp peak at 778.2 eV [52].

XPS results of unreduced and reduced Pd/Ni samples were also achieved. The unreduced Pd/Ni sample spectrum shows, in the $Ni2p_{3/2}$ area, peaks corresponding to NiO and $Ni(OH)_2$ species [53]. After reduction at 200 °C, the binding energy of $Ni_2p_{3/2}$ is characterized by a shift to lower BE values of about 0.8 eV with respect to that of pure nickel. This can be ascribed to an electron density transfer from Pd to Ni species [54]. Accordingly, in the reduced spectrum of the PdNi system, the binding energy of the $Pd_3d_{5/2}$ level was found at 335.6 eV, about 0.6 eV higher than that reported for metallic palladium, confirming the formation of a bimetallic Pd/Ni ensemble that might be an alloy [55].

5.2.2.6 Extended X-Ray Absorption Fine Structure (EXAFS)

EXAFS analysis for all investigated Pd-based catalysts is reported in Table 5.3.

The EXAFS characterization of the reduced Pd/Fe_3O_4 catalyst reveals that some iron particles are alloyed into the palladium nanoparticles forming Pd–Fe bimetallic ensembles, since the Pd-Fe bond length is shorter than that of the Pd-Pd bond

Table 5.3 X-ray absorption fine structure characterization at the Pd K-edge for investigated Pd-based catalysts

Catalyst	Scattering pair	CN	R [Å]	DW [Å]	R [%]
Pd/Fe	Pd-Pd	2.59	2.71	0.091	2.0
	Pd-Fe	0.78	2.51	0.047	1.0
Pd/Co	Pd-Pd	0.3	2.69	0.021	1.9
	Pd-Co	3.92	2.51	0.047	0.8
Pd/Ni	Pd-Pd	–	–	–	–
	Pd-Ni	2.55	6.3	0.070	0.7
Pd/C	Pd-Pd	2.73	8.0	0.061	2.6

CN coordination number, *R* interatomic distance, *DW* Debye–Waller factor

(Table 5.3) and indicates that, on the Pd/Fe-reduced catalyst, strong interactions between palladium nanoparticles and the iron oxide support occur [26, 27].

On the other hand, the results of extended X-ray absorption fine structure (EXAFS) (Table 5.2) relative to the Pd/Co sample reveal also a shorter scattering Pd-Co path of 2.51 Å compared with that of the Pd–Pd distance of about 2.70 Å again confirming the formation of Pd-Co bimetallic ensembles [56, 57].

Also the Pd-Ni catalyst is characterized by a shorter Pd-Ni distance. However, the main peculiarity, in the PdNi sample, is the lack of any Pd-Pd scattering path implying the much higher concentration of Ni in the Pd-Ni alloy.

5.3 Conversion of Lignin-Derived Aromatic Ethers Promoted by Nanostructured Pd-M (M = Fe, Co, Ni) Catalysts

5.3.1 *Catalytic Valorization of Lignin-Derived Aromatic Ethers Promoted by the Bimetallic Pd/Fe System*

Catalytic systems based on carbon-supported bimetallic Pd–Fe sites were found to be active in the vapour phase hydrodeoxygenation of guaiacol and cresol, as well as in the cleavage of the C–O bond in benzyl phenyl ether (BPE) and 2-phenethyl phenyl ether (PPE) [58, 59]. Furthermore, in the last few years, the Pd/Fe_3O_4 catalyst (prepared by the co-precipitation technique) has shown an excellent performance in several catalytic reactions including the catalytic transfer hydrogenolysis (CTH) of polyols and aromatic/aliphatic esters [27, 39]. The unique catalytic ability of the Pd/Fe_3O_4 sample as well as of other analogous Pd-based co-precipitated systems was attributed to the strong metal/support interaction and to the formation of bimetallic Pd-M (M = Fe, Co, Ni) more than other important heterogeneous catalyst parameters such as surface area [27, 39].

Benzyl phenyl ether (BPE), in the presence of the heterogeneous Pd/Fe_3O_4 catalyst, at 240 °C and within 90 min time, is fully converted (100% conversion) to toluene and phenol as the only reaction products (100% aromatic yield).

Indeed, on using Pd/Fe_3O_4, as catalyst and 2-propanol as H-source, an appreciable BPE (0.1 M) conversion (19.7%) is achieved at temperature as low as 180 °C. The conversion of BPE increases on increasing the reaction temperature and it is fully converted into phenol and toluene at 240 °C. Cyclohexane and methyl cyclohexane are not detected, clearly indicating that the cleavage of the etheric C–O bond occurs as the primary reaction route under the reaction conditions adopted.

On the other hand, Pd/C is not active in the C–O bond breaking at 180 °C (0% conversion) under CTH conditions, whilst, at 210 °C and 240 °C, a very low conversion of BPE is achieved.

In order to verify the low tendency of the Pd/Fe_3O_4 catalyst to hydrogenate the aromatic ring, phenol and toluene were allowed to react in the same reaction conditions as those employed for BPE. In both cases, products deriving from aromatic ring hydrogenation are not observed.

The ability of the bimetallic Pd–Fe system in the C–O bond breaking without parallel hydrogenation of the aromatic ring is explained on the basis of a strong adsorption preference of the aromatic ring onto exposed Fe particles [60–63].

At 240 °C, the BPE conversion increases on increasing the reaction time, when passing from 30 to 90 min (Fig. 5.8). However, the selectivity to phenol slightly decreases after 12 h, as a result of further hydrogenation to cyclohexanol, in agreement with other studies that report a much higher hydrogenation rate of phenol than that of toluene [64].

Cross-check experiments, carried out under the same reaction conditions (240 °C, 90 min, 10 bar N_2), show also that both the presence of pure Fe_3O_4 as well

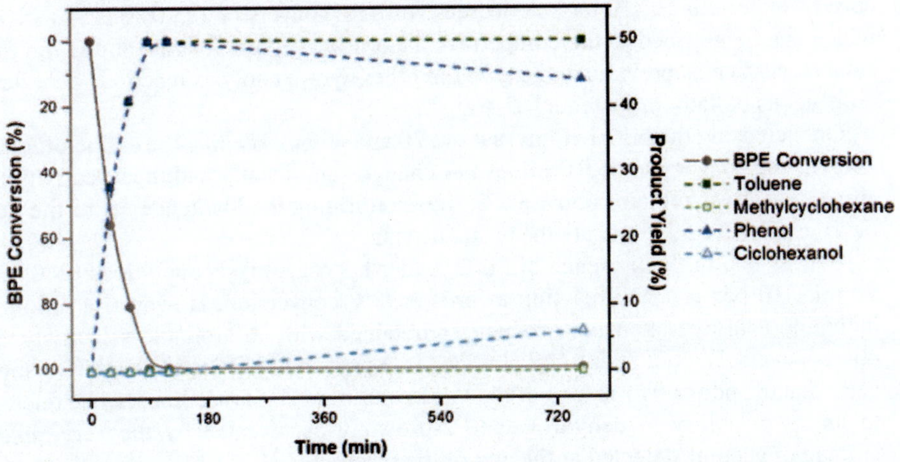

Fig. 5.8 Reaction time effect on the BPE conversion (%) and liquid products' selectivity (%) in the CTH at 240 °C

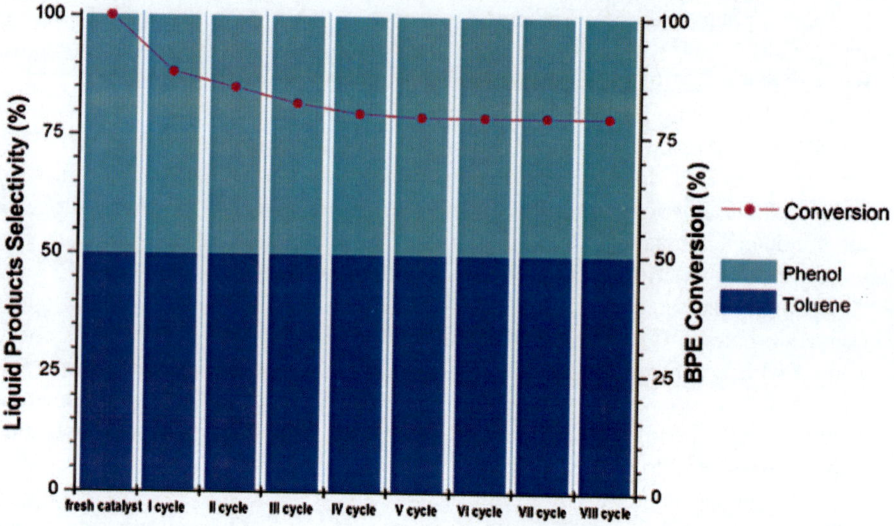

Fig. 5.9 Recycle of the Pd/Fe_3O_4 catalyst in the transfer hydrogenolysis of BPE (240 °C; 90 min; 0.25 g catalyst; 60 ml entry solution 0.1 M; 10 bar N_2 pressure)

as metallic Fe and the absence of any catalyst (non-catalyzed pyrolysis conditions) do not give any BPE conversion, implying that the presence of palladium is essential for the C–O bond breaking.

Catalyst recycle tests using the Pd/Fe_3O_4 catalyst at 240 °C were also performed and a BPE conversion up to 80% after eight recycling tests, was observed, highlighting the good stability of the catalyst (Fig. 5.9). Product's selectivity, at the same time, does not change appreciably.

For a comparison, the C–O bond breaking of BPE was also studied in the presence of molecular H_2 (10 bar). A decrease of BPE conversion (75.0%) is observed, which can be ascribed to the competitive adsorption of the substrate and H_2 on the catalyst surface as previously observed in other hydrogenolysis reactions over heterogeneous palladium systems [65–67].

On increasing the initial H_2 pressure to 20 and 40 bar, keeping the temperature at 240 °C, the conversion of BPE does not change significantly and the selectivity to arene derivatives remains above 98%, thus confirming the low tendency of the Pd/Fe_3O_4 catalyst to hydrogenate the aromatic ring.

Conversely, in the presence of Pd/C, under hydrogenolysis conditions (240 °C, 90 min, 10 bar H_2), the reaction affords a 98% conversion. However, a mixture including ring-hydrogenated products is obtained with an aromatic selectivity of 40%, conversely, poor results are obtained in 2-propanol CTH conditions. This suggests that the poor performance of the Pd/C catalyst in CTH reactions can be related to its lower ability to dehydrogenate 2-propanol as revealed by the very small amount of acetone detected at the end of the reaction [68].

The H-donor ability of simple primary (methanol, ethanol, 1-propanol, 1-butanol and 1-pentanol) and secondary (2-butanol, 2-pentanol and 3-pentanol) alcohols

Table 5.4 Transfer hydrogenolysis of BPE in the presence of the Pd/Fe_3O_4 catalyst (0.25 g of catalyst; 60 ml of entry solution; 240 °C; 10 bar N_2; 90 min) using simple primary and secondary alcohols as reaction solvents

Solvent	Conversion (mol/l)	Dehydrogenation (mol/l)	H-ability (mol/l)
Methanol	0.003	–	–
Ethanol	0.016	0.015	0.021
1-propanol	0.057	0.045	0.060
1-Butanol	0.052	0.050	0.057
1-Pentanol	0.055	0.050	0.058
2-Propanol	0.100	0.240	0.120
2-Butanol	0.035	0.090	0.038
2-Pentanol	0.019	0.080	0.019
3-Pentanol	0.002	0.075	0.002

with different carbon chain lengths was also tested on using the Pd/Fe_3O_4 catalyst (Table 5.4). A tight correlation between the moles/l of aldehyde or ketone formed during CTH reactions (H-donor ability) and the amount of BPE converted (moles/l) was found.

Noteworthily, with the exception of 2-propanol, secondary alcohols are generally less effective in the CTH of BPE than the primary ones. This is in remarkable contrast with the alcohol's dehydrogenation ability (calculated as the solvent dehydrogenated concentration obtained in the presence of the catalyst alone—moles/l) with the secondary alcohol that is generally more active than the primary one, due to the higher electron-releasing inductive effect. Accordingly, the reactivity of the same pure alcohols in the presence of the Pd/Fe_3O_4 catalyst has been investigated under the same reaction conditions (240 °C, 90 min, 10 bar N_2) in order to determine their dehydrogenation ability in the absence of the substrate (Table 5.4).

In the last case, the amount of aldehydes, formed from the primary alcohols, is slightly lower than that measured in the presence of BPE, suggesting that even though hydrogen can be transferred to the metal surface, the direct transfer of the donor solvent to BPE proceeds a little bit faster. Most important, however, the well-known higher tendency of secondary alcohols to release hydrogen is confirmed [69, 70]. Noteworthily, the amount of ketones formed is by far higher than that obtained in the transfer hydrogenolysis of BPE.

These results, considered all together, suggest that the CTH process of BPE is very sensitive to the steric hindrance of the H-donor molecule. Such a conclusion, coupled with the close relationship between the molar amount of carbonyl-derived compounds and that of the converted BPE, suggests that the H-transfer from the alcohol and the hydrogen promoting the C–O bond breaking occur in a unique chemical process.

In order to extend the substrate scope, the CTH of 2-phenethyl phenyl ether (PPE) and diphenyl ether (DPE) was finally investigated under the same reaction conditions as those of BPE.

The reaction of PPE at 240 °C shows a lower conversion (22%) and 100% yield of aromatics (ethyl benzene and phenol). Thus, the experimental data demonstrate again that the Pd/Fe_3O_4 catalyst has a higher selectivity towards the β-O-4 bond cleavage, whilst the phenyl ring is not involved.

On the contrary, under the reaction conditions adopted, DPE was not converted at all. These results can be explained taking into consideration that the 4-O-5 linkage is the most thermally stable bond having a higher bond dissociation energy (314 kJ mol^{-1}) with respect to the β-O-4 (289 kJ mol^{-1}) and the α-O-4 (218 kJ mol^{-1}) C–O bonds [71–73].

5.3.2 *Catalytic Valorization of Lignin-Derived Aromatic Ethers Promoted by the Bimetallic Pd/Co System*

In the presence of Pd/Co, after 3 h of reaction, the high BPE conversion at 210 °C (88%) and the total conversion (100%) at 240 °C evidence its excellent performance [74]. The product's distribution pattern changes within the investigated temperature range. Indeed, at the lower temperature (180 °C), the only products observed are toluene (Tol) and phenol (Phe). When the temperature increases, phenol is progressively converted into cyclohexanol (CXO) that becomes, at 240 °C, the main reaction product together with toluene. However, this is not to be considered a disfavouring result. It is worth highlighting, in fact, that cyclohexanol is an important feedstock in the industrial chemistry being used as a precursor to nylons, plastics, detergents and insecticides.

On the contrary, the commercial Pd/C exhibits a significant lower activity. Indeed, at 210 °C, the BPE conversion is less than 20% and the maximum is 45% at 240 °C, the highest temperature investigated. The modest performance of this monometallic Pd/C catalyst was already observed and attributed to its lower ability to dehydrogenate 2-propanol [27, 75].

Pure metallic cobalt, CoO and Co_3O_4 were also tested within the same temperature range and no BPE conversion was found, clearly indicating that the palladium presence in the catalyst is essential for the C–O bond breaking.

The lack of any appreciable reactivity, using either Pd/C- or Co-based catalysts, confirms that the marked activity, shown by the bimetallic Pd/Co catalyst, has to be attributed to the strong interaction between palladium and cobalt, as consequence of the preparation method (co-precipitation) in analogy with other reports attaining to glycerol hydrogenolysis [27].

The CTH of BPE, at different reaction times, at 210 °C (the best reaction temperature that maximizes the aromatic production) was also carried out. The BPE conversion is completed after 6 h of reaction, and the toluene yield progressively increases on increasing reaction times, reaching a maximum value at 360 min. However, no hydrogenation into methylcyclohexane is, in any case, registered. On the other hand, the highest yield of phenol is reached after 180 min and then it progressively decreases because of further hydrogenation to cyclohexanol.

The reusability of the Pd/Co catalyst was also evaluated under the harsh reaction conditions adopted (240 °C for 3 h). Pd/Co maintains its high activity after six consecutive recycling runs and no changes in product selectivity was found. This result clearly highlights the good stability of the catalyst.

The substrate scope was subsequently extended to additional aromatic ethers representative of lignin linkages, namely, PPE and DPE both under CTH conditions as well as under classical hydrogenolysis conditions. Catalytic tests show that: (1) improved conversions can be easily achieved in the presence of molecular hydrogen and (2) the cleavage of the C-O bond of PPE and DPE substrates is again less efficient due to the higher bond dissociation energies (β-O-4 = 289 kJ/mol and 4-O-5 = 314 kJ/mol) involved [76, 77].

5.3.3 *Catalytic Valorization of Lignin-Derived Aromatic Ethers Promoted by the Bimetallic Pd/Ni System*

Due to the good availability and the competitive market price of nickel precursors with respect to other transition metals, a significant interest was mainly devoted to address the design of homogeneous and heterogeneous Ni-based catalysts for hydrogenolysis of aromatic ethers [78]. Whilst homogeneous nickel catalysts were found highly efficient in the reductive C–O bond cleavage of arylalkyl ethers and diaryl ethers under mild reaction conditions, their application in lignin depolymerization has been partially limited for their high separation costs, reusability and handling [79, 80].

At the same time, in literature, a remarkable development of heterogeneous Ni-based catalysts for the hydrogenolysis of aromatic ethers as well as for the reductive valorization of the lignin was reported. Lercher and co-workers, for example, deeply investigated the conversion of aromatic ethers in the aqueous phase, elucidating kinetics and reaction pathways over several supported Ni catalysts showing the possibility to cleave C–O bonds via parallel hydrogenolysis, hydrolysis and hydrogenation/hydrodeoxygenation integrated steps [71, 81, 82]. In addition, a series of bimetallic Ni catalysts shows better performances towards aromatic products and stability in the hydrogenolysis of lignin model compounds if compared with analogous monometallic Ni systems [83–87]. Recently, the research group of Prof. Cai prepared and successfully applied some bimetallic Pd-Ni systems (BMNPs) in lignin hydrogenolysis as well as model compounds using low H_2 pressure as well as under self-hydrogenolysis conditions [88–90]. Among other heterogeneous bimetallic Ni-based catalysts, RANEY®Ni was proposed by the Rinaldi research group for the C–O bond cleavage of phenols, aromatic ethers and organosolv lignin under catalytic transfer hydrogenolysis (CTH) conditions [91–94].

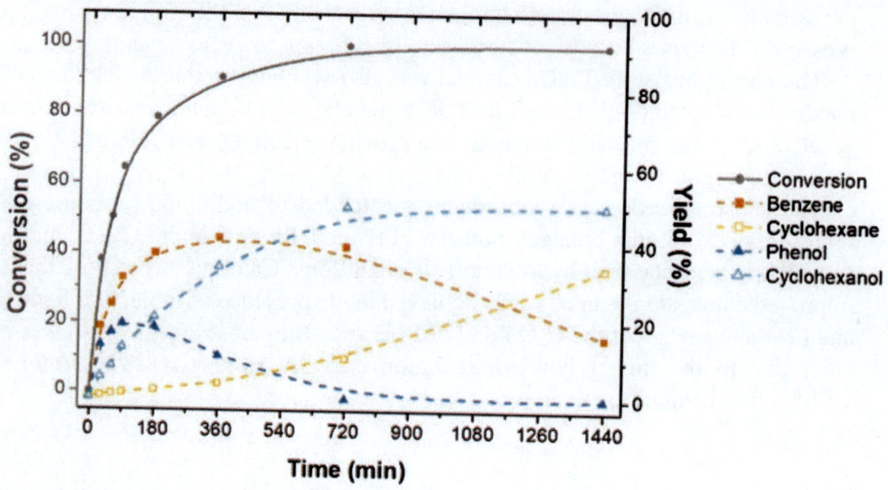

Fig. 5.10 Reaction time effect on the DPE conversion (%) and liquid products' selectivity (%) in the CTH at 210 °C

In the case of co-precipitated Pd/Ni catalysts, the attention was primarily focused on the CTH of DPE that has a dissociation C–O bond energy of 314 kJ/mol, the strongest structural link in lignin, to be broken and generally requires harsh reaction conditions.

In the presence of Pd/Ni, the transfer hydrogenolysis of DPE proceeds appreciably at 180 °C (39% conversion in 90 min) [95]. In addition to BEN and PHE, CXO is also obtained as the reaction product (14% selectivity). DPE conversion progressively rises (from 65% at 210 °C to 96% at 240 °C) on increasing the reaction temperature.

The highest production of aromatic products is observed at 210 °C where a selectivity of 48% and 30% to BEN and PHE, respectively, was measured. At 240 °C, a mixture of BEN (45%), CHX (6%), PHE (4%) and CXO (46%) is obtained.

With 210 °C, identified as the optimum reaction temperature to maximize the aromatic production, the CTH of DPE was also investigated at different reaction times (Fig. 5.10). As the reaction proceeds, in the presence of Pd/Ni, the BEN yield progressively increases, and only after 360 min, a discrete amount of CHX is registered. Comparatively, phenol reaches the highest yield value (19%, 90 min), then progressively decreases because of further hydrogenation into cyclohexanol.

A control experiment on the CTH of phenol (210 °C, 10 bar N_2, 90 min) gives a 97% yield of cyclohexanol under identical reaction conditions (see below). The larger phenol hydrogenation tendency observed, with respect to that of benzene, is in good agreement with previous literature reports [71, 81]. Analogous results are obtained at 180 and 240 °C.

The reusability of Pd/Ni, based on consecutive recycling tests at 210 °C for 90 min, was subsequently evaluated using the DPE conversion as model reaction.

Pd/Ni maintains its activity after eight consecutive runs and only slight changes in the product selectivity were found, highlighting the good stability of the catalyst.

Pd/Ni can also be magnetically recoverable and directly recycled (simple washing with 2-propanol is necessary) for the next reaction run.

At all reaction temperatures investigated, the monometallic Ni catalyst shows a markedly reduced activity. DPE conversion is ca. 10% at 180 °C and reaches a maximum value of 79% at 240 °C.

At the same time, a much higher tendency towards aromatic ring hydrogenation is always registered. In order to exclude the possibility that the activity of Pd/Ni may arise from the simple sum of both the monometallic Ni and Pd species, additional CTH reactions of DPE were carried out using Pd/C or unsupported metal palladium (resulting from the in situ reduction of PdO). In both cases, the molar amount of palladium was carefully checked to be equal to that of Pd/Ni catalysts used in all catalytic experiments (5.64 mol of Ni × 10^{-3} mol of Pd). The quantity of products observed, at all investigated temperatures, is indeed negligible, confirming that only Pd/Ni bimetallic species positively promote the CTH reaction of aromatic ethers.

The DPE C–O bond hydrogenolysis was also promoted using molecular hydrogen at 210 °C, and an increase in the production of aromatic ring hydrogenation derivatives (CXO and CHX) was noticed. On increasing the initial H_2 pressure, a slight decrease in DPE conversion could be observed although the aromatic yield remains as above, 40% (53%, 48% and 41% at 10, 20 and 40 bar of the initial H_2 pressure, respectively).

DPE is efficiently reduced also using a water/2-propanol mixture (50 wt %) with very similar conversions and chemoselectivities under both CTH and hydrogenolysis conditions. Unexpectedly, no DPE conversion was obtained by using only water as solvent suggesting that the Pd/Ni catalyst is not able to promote both hydrogenolysis and/or hydrolysis of the etheric C–O bond in pure water.

Having demonstrated the efficiency of the Pd/Ni catalyst in the hydrogenolysis and the CTH of diphenyl ether, the substrate scope was subsequently extended to the other aromatic ethers representative of lignin linkages, namely, PPE and BPE. CTH catalytic tests, carried out at 210 °C, show that improved conversions can be easily achieved because of lower dissociation energies for C–O bonds involved (β-O-4 = 289 kJ/mol and α-O-4 = 218 kJ/mol) with respect to that of DPE (4-O-5 = 314 kJ/mol).

In particular, BPE shows a conversion of 98%, with an aromatic yield of 97% whereas, in the case of PPE, a lower conversion (conversion, 67%; aromatic yield, 60%) was obtained.

For comparison, BPE and PPE were also tested using H_2 at an initial pressure of 10 bar. Under hydrogenolysis conditions, BPE is fully converted with an outstanding selectivity to aromatics (>99%) whilst PPE conversion slightly overcomes 75% with a high aromatic yield (70%).

The very high aromatic selectivity registered in the CTH and hydrogenolysis of BPE and PPE suggested also to extend the investigation to the hydrogenation of

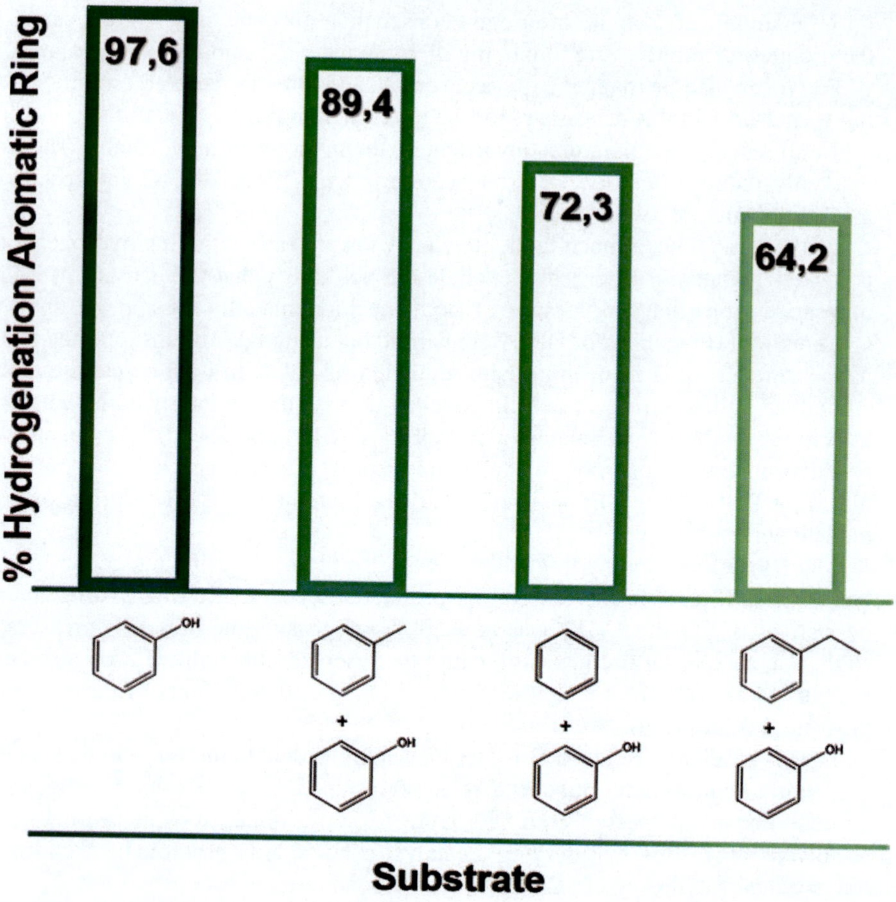

Fig. 5.11 Competitive catalytic hydrogenation of phenol in mixture with benzene, toluene and ethylbenzene at an equimolar feed composition at 210 °C under CTH conditions

phenol in the presence and in the absence of molecular hydrogen. Catalytic tests carried out at 210 °C clearly show:

1. that the hydrogenation of phenol is faster than the hydrogenolysis of DPE, PPE and BPE both under CTH as well as under hydrogenolysis conditions;
2. that the reaction proceeds through the intermediate cyclohexanone using 2-propanol; and
3. that a very low conversion into benzene is observed, in good agreement with other literature reports [91, 92].

These results, analysed together with those relative to aromatic ethers, imply that the hydrogenation process of phenol is very sensitive to the presence of other chemical species on the reaction system.

Therefore, a competitive catalytic hydrogenation reaction of phenol in mixture with benzene, toluene and ethylbenzene, added in equimolar feed composition at 210 °C under CTH conditions, was carried to mimic the reduction of phenol formed upon C–O bond cleavage. Results are reported in Fig. 5.11 and clearly show that hydrogenation of phenol progressively decreases in the presence of benzene (89%), toluene (72%) and ethylbenzene (64%).

Therefore, the addition in the reaction system of another aromatic molecule affords a twofold consequence on the hydrogenation of phenol: (1) a lower production of cyclohexanol and (2) a decrease of aromatic ring saturation as more as the arene derivative is substituted. The only possible conclusion is that a competitive adsorption between phenol and added aromatic hydrocarbons occurs on catalytic sites. The latter are more easily chemisorbed on the Pd/Ni surface with respect to phenol as the inductive effect of the substituent group at the ring increases (ethylbenzene > toluene > benzene). An analogous interpretation was already reported on the coordination chemistry of benzene and toluene on nickel (111), (110) and (100) surfaces [96].

Furthermore, no substrate hydrogenation was observed before the C-O bond cleavage, confirming that the first step, in the transfer hydrogenolysis of DPE, is the direct H-breaking of the C-O bond.

5.4 Conclusions

In this chapter, the preparation and physico-chemical properties of nanostructured bimetallic Pd-based catalysts as well as their use in the conversion of lignin-derivable molecules, starting from model aromatic ethers, such as Benzyl Phenyl Ether (BPE), 2-Phenethyl Phenyl Ether (PPE) and Diphenyl Ether (DPE) has been presented.

The physico-chemical and the structural properties of the co-precipitated Pd/Fe_3O_4, Pd/Co and Pd/Ni catalysts have been highlighted through XRD, SEM, TEM, H_2-TPR, XPS and EXAFS measurements. These physico-chemical techniques show the presence of well-dispersed palladium nanoparticles and reveal the existence of a strong metal–support interaction (in the case of the Pd/Fe_3O_4 catalysts) and alloys (in the case of Pd/Co and Pd/Ni catalysts) that enhance the ability of co-precipitated systems in the C-O bond cleavage of aromatic ethers.

The Pd/Fe_3O_4 catalyst is able to cleave the C–O bond of benzyl phenyl ether (BPE) under CTH conditions and, at the same time, to avoid parallel hydrogenation reactions of the aromatic ring, being one of the most selective heterogeneous catalysts in the production of arene derivatives. The experimental evidence shows that the direct hydrogen transfer from the α-H of the donor solvent to BPE occurs in a unique process.

The bimetallic Pd/Co catalyst, prepared by the co-precipitation technique, is able to effectively cleave the C–O bond of benzyl phenyl ether that, after 3 h at 240 °C, is selectively converted into toluene and cyclohexanol. It shows also a good stability and can be easily recoverable magnetically.

On the other hand, the CTH of the ethereal C–O bond of 2-phenethyl phenyl ether and diphenyl ether, in the presence of both Pd/Fe_3O_4 and Pd/Co catalysts, was found to be less efficient due to the higher bond dissociation energies involved.

The etheric C-O bond of lignin-derived aromatic ethers was efficiently cleaved also by the bimetallic Pd/Ni catalyst that can be efficiently used up to eight consecutive recycling tests and can be easily recoverable magnetically from the reaction medium. The catalytic tests clearly evidence that, in the presence of the Pd/Ni catalyst, cyclohexyl phenyl ether (CPE) is not a reaction intermediate revealing that the first step in the transfer hydrogenolysis of DPE is the direct cleavage of the C-O bond. Noteworthily, competitive catalytic hydrogenation reactions of phenol in mixture with benzene, toluene and ethylbenzene demonstrate that production of aromatics in the transfer hydrogenolysis of DPE, BPE and PPE is deeply influenced by the nature of the aryl groups, which form the aromatic ether structure, with a production of cyclohexanol that decreases as more as the arene derivative is substituted.

References

1. C.O. Tuck, E. Pérez, I.T. Horváth, R.A. Sheldon, M. Poliakoff, Valorization of biomass: deriving more value from waste. Science **337**, 695–699 (2012)
2. R.A. Sheldon, Green and sustainable manufacture of chemicals from biomass: state of the art. Green Chem. **16**, 950–963 (2014)
3. F.H. Isikgor, C.R. Becer, Lignocellulosic biomass: a sustainable platform for the production of bio-based chemicals and polymers. Polym. Chem. **6**, 4497–4559 (2015)
4. Q. Jin, L. Yang, N. Poe, H. Huang, Integrated processing of plant-derived waste to produce value-added products based on the biorefinery concept. Trends Food Sci. Technol. **74**, 119–131 (2018)
5. A.M. Ruppert, K. Weinberg, R. Palkovits, Hydrogenolysis goes bio: from carbohydrates and sugar alcohols to platform chemicals. Angew. Chem. Int. Ed. **51**, 2564–2601 (2012)
6. A. Fasolini, R. Cucciniello, E. Paone, F. Mauriello, T. Tabanelli, A short overview on the hydrogen production via aqueous phase reforming (APR) of cellulose, C6-C5 sugars and polyols. Catalysts **9**(11), 917 (2019)
7. C.-H. Zhou, X. Xia, C.-X. Lin, D.-S. Tong, J. Beltramini, Catalytic conversion of lignocellulosic biomass to fine chemicals and fuels. Chem. Soc. Rev. **40**, 5588–5617 (2011)
8. C. Xu, E. Paone, D. Rodríguez-Padròn, R. Luque, F. Mauriello, Recent catalytic routes for the preparation and the upgrading of biomass derived furfural and 5-hydroxymethylfurfural. Chem. Soc. Rev. **49**, 4273–4306 (2020)
9. H. Chen, Chemical composition and structure of natural lignocellulose, in *Biotechnology of Lignocellulose: Theory and Practice*, (Springer, Cham, 2014)
10. D. Klemm, B. Heublein, H.-P. Fink, A. Bohn, Cellulose: fascinating biopolymer and sustainable raw material. Angew. Chem. Int. Ed. **44**, 3358–3393 (2005)
11. C. Xu, R.A.D. Arancon, J. Labidi, R. Luque, Lignin depolymerisation strategies: towards valuable chemicals and fuels. Chem. Soc. Rev. **43**, 7485–7500 (2014)
12. M. Besson, P. Gallezot, C. Pinel, Conversion of biomass into chemicals over metal catalysts. Chem. Rev. **114**, 1827–1870 (2014)

13. A. Corma, S. Iborra, A. Velty, Chemical routes for the transformation of biomass into chemicals. Chem. Rev. **107**, 2411–2502 (2007)
14. J.B. Binder, R.T. Raines, Simple chemical transformation of lignocellulosic biomass into furans for fuels and chemicals. J. Am. Chem. Soc. **131**, 1979–1985 (2009)
15. P.J. Deuss, K. Barta, From models to lignin: transition metal catalysis for selective bond cleavage reactions. Coord. Chem. Rev. **306**, 510–532 (2016)
16. M. Zaheer, R. Kempe, Catalytic hydrogenolysis of aryl ethers: a key step in lignin valorization to valuable chemicals. ACS Catal. **5**, 1675–1684 (2015)
17. J. Zakzeski, P.C.A. Bruijnincx, A.L. Jongerius, B.M. Weckhuysen, The catalytic valorization of lignin for the production of renewable chemicals. Chem. Rev. **110**, 3552–3599 (2010)
18. M.V. Galkin, J.S.M. Samec, Lignin valorization through catalytic lignocellulose fractionation: a fundamental platform for the future biorefinery. ChemSusChem **9**, 1–16 (2016)
19. C. Espro, B. Gumina, T. Szumelda, E. Paone, F. Mauriello, Catalytic transfer hydrogenolysis as an effective tool for the reductive upgrading of cellulose, hemicellulose, lignin and their derived platform molecules. Catalysts **8**(8), 313 (2018)
20. E. Paone, T. Tabanelli, F. Mauriello, The rise of lignin biorefinery. Curr. Opin. Green Sustain. Chem. **24**, 1–6 (2020)
21. T. Tabanelli, E. Paone, P.B. Vàsquez, R. Pietropaolo, F. Cavani, F. Mauriello, Transfer hydrogenation of methyl and ethyl levulinate promoted by a ZrO2 catalyst: a comparison of batch vs continuous gas-flow conditions. ACS Sustain. Chem. Eng. **7**, 9937–9947 (2019)
22. M. Grilc, B. Likozar, J. Levec, Hydrodeoxygenation and hydrocracking of solvolysed lignocellulosic biomass by oxide, reduced and sulphide form of NiMo, Ni, Mo and Pd catalysts. Appl. Catal. B Environ. **150-151**, 275–287 (2014)
23. M. Grilc, B. Likozar, J. Levec, Simultaneous liquefaction and hydrodeoxygenation of lignocellulosic biomass over NiMo/Al_2O_3, Pd/Al_2O_3, and zeolite Y catalysts in hydrogen donor solvents. ChemCatChem **8**, 180–191 (2015)
24. Y. Zhang, H. Cao, J. Zhang, B. Xia, Synthesis of $LiNi_{0.6}Co_{0.2}Mn_{0.2}O_2$ cathode material by a carbonate co-precipitation method and its electrochemical characterization. Solid State Ionics **177**, 3303–3307 (2006)
25. T.H. Cho, Y. Shiosaki, H. Noguchi, Preparation and characterization of layered $LiMn_{1/3}Ni_{1/3}Co_{1/3}O_2$ as a cathode material by an oxalate co-precipitation method. J. Power Sources **159**, 1322–1327 (2006)
26. C. Espro, B. Gumina, E. Paone, F. Mauriello, Upgrading lignocellulosic biomasses: hydrogenolysis of platform derived molecules promoted by heterogeneous Pd-Fe catalysts. Catalysts **7**, 78 (2017)
27. F. Mauriello, H. Ariga, M.G. Musolino, R. Pietropaolo, S. Takakusagi, K. Asakura, Exploring the catalytic properties of supported palladium catalysts in the transfer hydrogenolysis of glycerol. Appl. Catal. B Environ. **166–167**, 121–131 (2015)
28. F. Liao, T.W.B. Lo, S.C.E. Tsang, Recent developments in palladium-based bimetallic catalysts. ChemCatChem **7**, 1998–2014 (2015)
29. J. Liu, B. Sun, J. Hu, Y. Pei, H. Li, M. Qiao, Aqueous-phase reforming of ethylene glycol to hydrogen on Pd/Fe_3O_4 catalyst prepared by co-precipitation: metal–support interaction and excellent intrinsic activity. J. Catal. **274**, 287–295 (2010)
30. M.G. Musolino, L.A. Scarpino, F. Mauriello, R. Pietropaolo, Glycerol hydrogenolysis promoted by supported Palladium catalysts. ChemSusChem **4**, 1143–1150 (2011)
31. L.S.F. Feio, C.E. Hori, L.V. Mattos, D. Zanchet, F.B. Noronha, J.M.C. Bueno, Partial oxidation and autothermal reforming of methane on Pd/CeO_2–Al_2O_3 catalysts. Appl. Catal. A Gen. **348**, 183–192 (2008)
32. G. Neri, G. Rizzo, L. De Luca, A. Donato, M.G. Musolino, R. Pietropaolo, Supported Pd catalysts for the hydrogenation of campholenic aldehyde: influence of support and preparation method. Appl. Catal. A Gen. **356**, 113–120 (2009)
33. J. Sun, S. Zhou, P. Hou, Y. Yang, J. Weng, X. Li, M. Li, Synthesis and characterization of biocompatible Fe_3O_4 nanoparticles. J. Biomed. Mater. Res. A **80**, 333–341 (2007)

34. H. Itoh, T. Sugimoto, Systematic control of size, shape, structure, and magnetic properties of uniform magnetite and maghemite particles. J. Colloid Interface Sci. **265**, 283–295 (2003)
35. W. Voit, D.K. Kim, W. Zapka, M. Muhammed, K.V. Rao, Magnetic behavior of coated superparamagnetic iron oxide nanoparticles in ferrofluids. Mater. Res. Soc. **676**, 781–786 (2001)
36. International Center for Diffraction Data, Powder Diffraction Database. Pennsylvania, PA (1997).
37. B.A. Sexton, A.E. Hughes, T.W. Turney, An XPS and TPR study of the reduction of promoted cobalt-kieselguhr Fischer-Tropsch catalysts. J. Catal. **97**, 390–406 (1986)
38. T. Jiang, Q. Huai, T. Geng, W. Ying, T. Xiao, F. Cao, Catalytic performance of Pd-Ni bimetallic catalyst for glycerol hydrogenolysis. Biomass Bioenergy **78**, 71–79 (2015)
39. M.G. Musolino, C. Busacca, F. Mauriello, R. Pietropaolo, Aliphatic carbonyl reduction promoted by palladium catalysts under mild conditions. Appl. Catal. A Gen. **379**, 77–86 (2010)
40. D. Li, L. Zhang, H. Chen, L.-X. Ding, S. Wanga, H. Wang, Nitrogen-doped bamboo-like carbon nanotubes: promising anode materials for sodium-ion batteries. Chem. Commun. **51**, 16045–16048 (2015)
41. S. Nandi, P. Patel, N.H. Khan, A.V. Biradar, R.I. Kureshy, Nitrogen-rich graphitic-carbon stabilized cobalt nanoparticles for chemoselective hydrogenation of nitroarenes at milder conditions. Inorgan. Chem. Front. **5**, 806–813 (2018)
42. Y. Yao, L.L. Gu, W. Jiang, H.C. Sun, Q. Su, J. Zhao, W.J. Ji, C.T. Au, Enhanced low temperature CO oxidation by pretreatment: specialty of the Au–Co_3O_4 oxide interfacial structures. Cat. Sci. Technol. **6**, 2349–2360 (2016)
43. D. Scholz, C. Aellig, I. Hermans, Catalytic transfer hydrogenation/hydrogenolysis for reductive upgrading of furfural and 5-(hydroxymethyl) furfural. ChemSusChem **7**, 268–275 (2014)
44. A.J.R. Hensley, Y. Hong, R. Zhang, H. Zhang, J. Sun, Y. Wang, J.S. McEwen, Enhanced Fe_2O_3 reducibility via surface modification with pd: characterizing the synergy within Pd/Fe catalysts for hydrodeoxygenation reactions. ACS Catal. **4**, 3381–3392 (2014)
45. Q. Liu, L.-C. Wang, M. Chen, Y. Cao, H.-Y. He, K.-N. Fan, Dry citrate-precursor synthesized nanocrystalline cobalt oxide as highly active catalyst for total oxidation of propane. J. Catal. **263**, 104–113 (2009)
46. F.B. Noronha, M. Schmal, C. Nicot, B. Moraweck, R. Frėty, Characterization of graphite-supported palladium–cobalt catalysts by temperature-programmed reduction and magnetic measurements. J. Catal. **168**, 42–50 (1997)
47. M.L. Cubeiro, J.L.G. Fierro, Selective production of hydrogen by partial oxidation of methanol over ZnO-supported palladium catalysts. J. Catal. **179**, 150–162 (1998)
48. X. Liu, Y. Zuo, L.P. Li, X.S. Huang, G.S. Li, Heterostructure NiO/Ce−xNixO_2: synthesis and synergistic effect of simultaneous surface modification and internal doping for superior catalytic performance. RSC Adv. **4**, 6397–6406 (2014)
49. C. Song, J. Zhao, H. Li, S. Luo, Y. Tang, D. Wang, Design, controlled synthesis, and properties of 2D CeO_2/NiO heterostructure assemblies. Cryst. Eng. Comm **19**, 7339–7346 (2017)
50. Y.-S. Fenga, X.-Y. Lina, J. Hao, H.-J. Xu, Pd–Co bimetallic nanoparticles supported on graphene as a highly active catalyst for Suzuki–Miyaura and Sonogashira cross-coupling reactions. Tetrahedron **70**, 5249–5253 (2014)
51. C.D. Wanger, W.M. Riggs, L.E. Davis, J.F. Moulder, G.E. Muilenberg, *Handbook of X-ray Photoelectron Spectroscopy* (Perkin-Elmer Corp, Minneapolis, MN, 1979)
52. B. Mierzwa, Z. Kaszkur, B. Moraweck, J. Pielaszek, In situ EXAFS study of the alloy catalyst Pd-Co (50%/50%)/SiO_2. J. Alloys Compd. **286**, 93–97 (1999)
53. M.C. Biesinger, B.P. Payne, L.W.M. Lau, A. Gerson, R.S.C. Smart, X-ray photoelectron spectroscopic chemical state quantification of mixed nickel metal, oxide and hydroxide systems. Surf. Interface Anal. **41**, 324–332 (2009)
54. N. Seshu Babu, N. Lingaiah, P.S. Sai Prasad, Characterization and reactivity of Al_2O_3 supported Pd-Ni bimetallic catalysts for hydrodechlorination of chlorobenzene. Appl. Catal. B Environ. **111–112**, 309–316 (2012)

55. L. Chen, H. Guo, T. Fujita, A. Hirata, W. Zhang, A. Inoue, M. Chen, Nanoporous PdNi bimetallic catalyst with enhanced electrocatalytic performances for electro-oxidation and oxygen reduction reactions. Adv. Funct. Mater. **21**, 4364–4370 (2011)
56. F. Liao, T.W.B. Lo, J. Qu, A. Kroner, A. Dent, S.C.E. Tsang, Tunability of catalytic properties of Pd-based catalysts by rational control of strong metal and support interaction (SMSI) for selective hydrogenolyic C–C and C–O bond cleavage of ethylene glycol units in biomass molecules. Cat. Sci. Technol. **5**, 3491–3495 (2015)
57. L. Nguyen, S. Zhang, L. Wang, Y. Li, H. Yoshida, A. Patlolla, A.I. Frenkel, S. Takeda, F.F. Tao, Reduction of nitric oxide with hydrogen on catalysts of singly dispersed bimetallic sites Pt1Com and Pd1Con. ACS Catal. **6**, 840–850 (2016)
58. J.K. Kim, J.K. Lee, K.H. Kang, J.W. Lee, I.K. Song, Catalytic decomposition of phenethyl phenyl ether to aromatics over Pd–Fe bimetallic catalysts supported on ordered mesoporous carbon. J. Mol. Catal. A Chem. **410**, 184–192 (2015)
59. J.K. Kim, J.K. Lee, K.H. Kang, J.W. Lee, I.K. Song, Selective cleavage of CO bond in benzyl phenyl ether to aromatics over Pd–Fe bimetallic catalyst supported on ordered mesoporous carbon. Appl. Catal. A Gen. **498**, 142–149 (2015)
60. B. Gumina, F. Mauriello, R. Pietropaolo, S. Galvagno, C. Espro, Hydrogenolysis of sorbitol into valuable C3-C2 alcohols at low H_2 pressure promoted by the heterogeneous Pd/Fe_3O_4 catalyst. Mol. Catal. **446**, 152–160 (2018)
61. J. Sun, A.M. Karim, H. Zhang, L. Kovarik, X.S. Li, A.J. Hensley, J.-S. McEwen, Y. Wang, Carbon-supported bimetallic Pd-Fe catalysts for vapor-phase hydrodeoxygenation of guaiacol. J. Catal. **306**, 47–57 (2013)
62. Y. Hong, H. Zhang, J. Sun, K.M. Ayman, A.J.R. Hensley, M. Gu, M.H. Engelhard, J.-S. McEwen, Y. Wang, Synergistic catalysis between Pd and Fe in gas phase dydrodeoxygenation of m-cresol. ACS Catal. **4**, 3335–3345 (2014)
63. A.J.R. Hensley, R. Zhang, Y. Wang, J.-S. McEwen, Tailoring the adsorption of benzene on PdFe surfaces: a density functional theory study. J. Phys. Chem. C **117**, 24317–24328 (2013)
64. N. Mahata, V. Vishwanathan, Gas phase hydrogenation of phenol over supported palladium catalysts. Catal. Today **49**, 65–69 (1999)
65. F. Mauriello, A. Vinci, C. Espro, B. Gumina, M.G. Musolino, R. Pietropaolo, Hydrogenolysis vs. aqueous phase reforming (APR) of glycerol promoted by a heterogeneous Pd/Fe catalyst. Cat. Sci. Technol. **5**, 4466–4473 (2015)
66. D. Cozzula, A. Vinci, F. Mauriello, R. Pietropaolo, T.E. Müller, Directing the cleavage of ester C–O bonds by controlling the hydrogen availability on the surface of coprecipitated Pd/Fe_3O_4. ChemCatChem **8**, 1515–1522 (2016)
67. H. Bernas, A. Taskinen, J. Wärnå, D.Y. Murzin, Describing the inverse dependence of hydrogen pressure by multi-site adsorption of the reactant: hydrogenolysis of hydroxymatairesinol on a Pd/C catalyst. J. Mol. Catal. A Chem. **306**, 33–39 (2009)
68. G.W. Huber, J.W. Shabaker, S.T. Evans, J.A. Dumesic, Aqueous-phase reforming of ethylene glycol over supported Pt and Pd bimetallic catalysts. Appl. Catal. B Environ. **62**, 226–235 (2006)
69. P. Panagiotopoulou, N. Martin, D.G. Vlachos, Effect of hydrogen donor on liquid phase catalytic transfer hydrogenation of furfural over a Ru/RuO_2/C catalyst. J. Mol. Catal. A Chem. **392**, 223–228 (2014)
70. N.S.M. Bertero, A.F. Trasarti, C.R. Apesteguía, A.J. Marchi, Solvent effect in the liquid-phase hydrogenation of acetophenone over Ni/SiO2: a comprehensive study of the phenomenon. Appl. Catal. A Gen. **394**, 228–238 (2011)
71. J. He, C. Zhao, J.A. Lercher, Ni-catalyzed cleavage of aryl ethers in the aqueous phase. J. Am. Chem. Soc. **134**, 20768–20775 (2012)
72. Y.R. Luo, *Comprehensive Handbook of Chemical Bond Energies* (CRC Press, Boca Raton, 2007)
73. Y.S. Choi, R. Singh, J. Zhang, G. Balasubramanian, M.R. Sturgeon, R. Katahira, G. Chupka, G.T. Beckham, B.H. Shanks, Pyrolysis reaction networks for lignin model compounds: unrav-

eling thermal deconstruction of β-O-4 and α-O-4 compounds. Green Chem. **18**, 1762–1773 (2016)
74. F. Mauriello, H. Ariga-Miwa, E. Paone, R. Pietropaolo, S. Takakusagi, K. Asakura, Transfer hydrogenolysis of aromatic ethers promoted by the bimetallic Pd/Co catalyst. Catal. Today (2019)
75. E. Paone, C. Espro, R. Pietropaolo, F. Mauriello, Selective arene production from transfer hydrogenolysis of benzyl phenyl ether promoted by a co-precipitated Pd/Fe_3O_4 catalyst. Cat. Sci. Technol. **6**, 7937–7941 (2016)
76. E. Dorrestijn, L.J.J. Laarhoven, I.W.C.E. Arends, P. Mulder, The occurrence and reactivity of phenoxyl linkages in lignin and low rank coal. J. Anal. Appl. Pyrolysis **54**, 153–192 (2000)
77. R. Parthasarathi, R.A. Romero, A. Redondo, S. Gnanakaran, Theoretical study of the remarkably diverse linkages in lignin. J. Phys. Chem. Lett. **2**, 2660–2666 (2011)
78. S. De, J. Zhang, R. Luque, N. Yan, Ni-based bimetallic heterogeneous catalysts for energy and environmental applications. Energy Environ. Sci. **9**, 3314–3347 (2016)
79. A.G. Sergeev, J.F. Hartwig, Selective, nickel-catalyzed hydrogenolysis of aryl ethers. Science **332**, 439–443 (2011)
80. A.G. Sergeev, J.D. Webb, J.F. Hartwig, A heterogeneous nickel catalyst for the hydrogenolysis of aryl ethers without arene hydrogenation. J. Am. Chem. Soc. **134**, 20226–20229 (2012)
81. J. He, L. Lu, C. Zhao, D. Mei, J.A. Lercher, Mechanisms of catalytic cleavage of benzyl phenyl ether in aqueous and apolar phases. J. Catal. **311**, 41–51 (2014)
82. J. He, L. Lu, C. Zhao, D. Mei, J.A. Lercher, Mechanisms of selective cleavage of C–O bonds in di-aryl ethers in aqueous phase. J. Catal. **309**, 280–290 (2014)
83. Q. Song, F. Wang, J. Xu, Hydrogenolysis of lignosulfonate into phenols over heterogeneous nickel catalysts. Chem. Commun. **48**, 7019–7021 (2012)
84. J. Zhang, H. Asakura, J. van Rijn, J. Yang, P. Duchesne, B. Zhang, X. Chen, P. Zhang, M. Saeys, N. Yan, Highly efficient, NiAu-catalyzed hydrogenolysis of lignin into phenolic chemicals. Green Chem. **16**, 2432–2437 (2014)
85. J. Zhang, J. Teo, X. Chen, H. Asakura, T. Tanaka, K. Teramura, N. Yan, A series of NiM (M = Ru, Rh, and Pd) bimetallic catalysts for effective lignin hydrogenolysis in water. ACS Catal. **4**, 1574–1583 (2014)
86. X. Cui, H. Yuan, K. Junge, C. Topf, M. Beller, F. Shi, A stable and practical nickel catalyst for the hydrogenolysis of C–O bonds. Green Chem. **19**, 305–310 (2017)
87. M.M. Ambursa, P. Sudarsanam, L. HweiVoon, S.B.A. Hamid, S.K. Bhargava, Bimetallic Cu-Ni catalysts supported on MCM-41 and Ti-MCM-41 porous materials for hydrodeoxygenation of lignin model compound into transportation fuels. Fuel Process. Technol. **162**, 87–97 (2017)
88. J.-W. Zhang, Y. Cai, G.-P. Lu, C. Cai, Facile and selective hydrogenolysis of β-O-4 linkages in lignin catalyzed by Pd-Ni bimetallic nanoparticles supported on ZrO_2. Green Chem. **18**, 6229–6235 (2016)
89. K.-K. Sun, G.-P. Lu, J.-W. Zhang, C. Cai, Selective hydrogenolysis of C-O bonds in lignin model compounds by Pd-Ni bimetallic nanoparticles in ionic liquids. Dalton Trans. **46**, 11884–11889 (2017)
90. J.-W. Zhang, G.-P. Lu, C. Cai, Self-hydrogen transfer hydrogenolysis of β-O-4 linkages in lignin catalyzed by MIL-100(Fe) supported Pd-Ni BMNPs. Green Chem. **19**, 4538–4543 (2017)
91. X. Wang, R. Rinaldi, Exploiting H-transfer reactions with RANEY®Ni for upgrade of phenolic and aromatic biorefinery feeds under unusual, low-severity conditions. Energy Environ. Sci. **5**, 8244–8260 (2012)
92. X. Wang, R. Rinaldi, A route for lignin and bio-oil conversion: dehydroxylation of phenols into arenes by catalytic tandem reactions. Angew. Chem. Int. Ed. **52**, 11499–11503 (2013)
93. X. Wang, R. Rinaldi, Solvent effects on the hydrogenolysis of diphenyl ether with raney nickel and their implications for the conversion of lignin. ChemSusChem **5**, 1455–1466 (2012)
94. X. Wang, R. Rinaldi, Bifunctional Ni catalysts for the one-pot conversion of Organosolv lignin into cycloalkanes. Catal. Today **269**, 48–55 (2016)

95. F. Mauriello, E. Paone, R. Pietropaolo, A.M. Balu, R. Luque, Catalytic transfer hydrogenolysis of lignin-derived aromatic ethers promoted by bimetallic Pd/Ni systems. ACS Sustain. Chem. Eng. **6**, 9269–9276 (2018)
96. C.M. Friend, E.L. Muetterties, Coordination chemistry of metal surfaces. 3. Benzene and toluene interactions with nickel surfaces. J. Am. Chem. Soc. **103**, 173–779 (1981)

Chapter 6
The Issue of Solid-Solid Contact in Catalytic Soot Oxidation and the Benefits of Catalyst Nanostructuring to Regeneration of Catalytic Diesel Particulate Filters

Gianluca Landi, Valeria Di Sarli, Almerinda Di Benedetto, and Luciana Lisi

6.1 Introduction

Diesel particulate filters (DPFs) represent the best option for removing particulate matter (PM), which is mostly composed of carbonaceous particles called "soot," from diesel engine exhaust. A DPF is typically made of a ceramic material, such as cordierite or silicon carbide (SiC). It consists of thousands of square-section parallel channels, with the opposite ends of adjacent channels being plugged (Fig. 6.1).

Figure 6.2 shows a schematic representation of the flow path in a channel of a "wall-flow" DPF.

The exhaust gas enters the inlet channel and passes through the porous walls to the adjacent outlet channels. This "wall-flow" path allows the soot particles to be retained in the inlet channels. More specifically, soot is accumulated inside the filter according to two distinct mechanisms of filtration [1]. The soot particles are first trapped inside the porous walls of the filter according to the mechanism of "deep-bed filtration." When the maximum packing density is reached, the walls become substantially impermeable to the particles, and a soot layer, also called "cake," is built up onto the porous filter surface. The formation of the cake layer is an important advantage for the filtration performance, given that the cake itself acts as a filter [1]. Indeed, most of the soot accumulated inside a DPF is trapped under this form, i.e., according to the mechanism of "cake filtration."

G. Landi · V. Di Sarli (✉) · L. Lisi
Istituto di Scienze e Tecnologie per l'Energia e la Mobilità Sostenibili (STEMS), Consiglio Nazionale delle Ricerche (CNR), Naples, Italy
e-mail: valeria.disarli@cnr.it

A. Di Benedetto
Dipartimento di Ingegneria Chimica, dei Materiali e della Produzione Industriale, Università di Napoli Federico II, Naples, Italy

M. Piumetti, S. Bensaid (eds.), *Nanostructured Catalysts for Environmental Applications*, https://doi.org/10.1007/978-3-030-58934-9_6

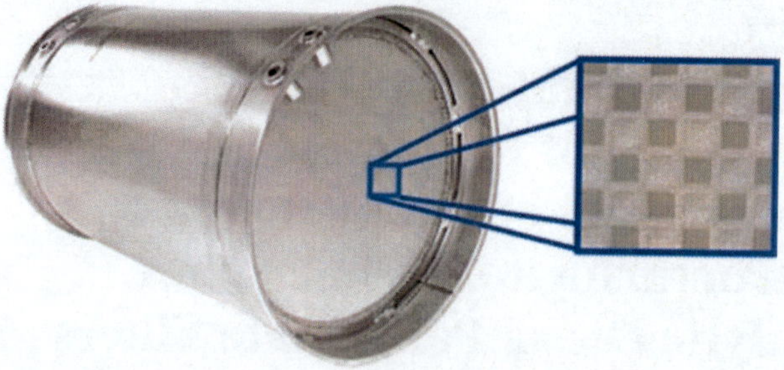

Fig. 6.1 A diesel particulate filter (DPF)—the magnified image shows the alternatively open and closed channels

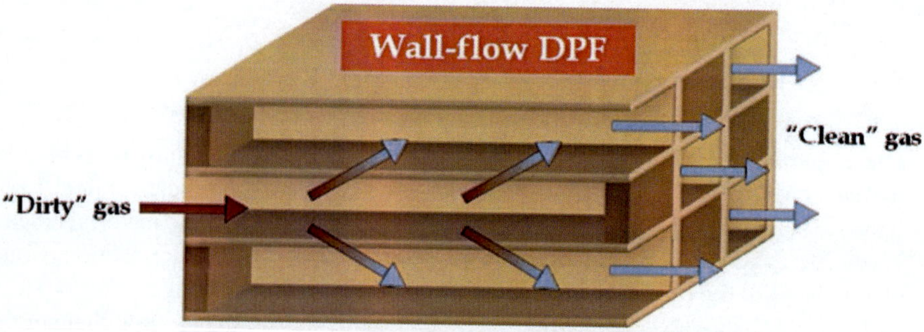

Fig. 6.2 Schematic representation of the flow path in a channel of a "wall-flow" DPF

The collected soot increases the backpressure of the diesel engine reducing its efficiency. Thus, the DPF needs to be periodically regenerated (cleaned) by soot combustion. Unfortunately, the temperature of the diesel exhaust gas (200–350 °C) is much lower than the temperature required to initiate and maintain soot oxidation. To raise the temperature up to the point that soot oxidation can be self-sustained in the filter at fast enough rates (>600 °C), external or engine (i.e., active) means (e.g., fuel burners, electric or microwave heating, injection of fuel in the exhaust, recirculation of exhaust gas, etc.) can be employed. Such an approach, also referred to as thermal regeneration, incurs additional energy costs and requires complex control techniques. Furthermore, in some cases, it can lead to the formation of excessively hot regions causing irreversible damage to the filter [2]. Finally, during thermal regeneration, soot is oxidized mostly to CO (values of selectivity to CO_2 around 40–50% were found [3, 4]).

Alternatively or complementarily, catalytic measures can be used to achieve soot oxidation at lower temperatures (250–550 °C) and/or to shorten the regeneration time period, thus allowing for energy saving (at an added cost of the catalyst).

Fuel-borne catalysts (FBCs), i.e., additives that are mixed with the fuel to lead to the formation of catalyst-doped soot during combustion in the engine, have been proposed as catalytic regeneration systems [5]. Additives are soluble compounds (octanoates, naphthenates, etc.) of metals (copper, iron, cerium, lead, manganese, etc.). In May 2000, PSA Peugeot Citroën launched the first series-production diesel passenger vehicle equipped with such a system. In particular, they used a Ce-based additive. By its sacrificial burning, particles of cerium oxide can be produced and embedded into soot particles, thus providing intimate soot-catalyst contact. This strategy has several critical issues. It requires the installation of a second tank for continuous additive supply in the fuel. The additive consumption leads to the accumulation of metal oxide "ash" inside the DPF, altering its structure. The fate of the catalyst particles after regeneration can be a further issue. Indeed, CeO_2 nanoparticles have a high long-term exposure toxicity [6]. To overcome these problems, the catalyst can be attached directly onto the walls of "catalytic" (i.e., catalyst-coated) DPFs.

Extensive research efforts have been focused on catalytic DPFs. However, there is still no general consensus regarding their ability to oxidize soot at low temperatures and under conditions relevant to practical applications. Indeed, catalytic soot oxidation involves a solid-solid catalysis, and thus, the regeneration performance of catalytic DPFs is dependent not only on the catalyst activity [3, 7], but also on the quality of the contact established between soot and catalyst particles [3, 4, 7–9]. As will be discussed in more detail in the following sections, in this framework, the development of effective catalytic DPFs is strictly linked to the development of nanostructured materials.

Nanoparticles are defined as particles having a diameter of less than 100 nm (0.1 μm) if spherical, while non-spherical nanoparticles are defined as particles with at least two dimensions smaller than 100 nm. When scaling down from micro to nano, physical, and chemical properties of particles significantly change both quantitatively and qualitatively. As expected, nanomaterials have a larger specific surface area than micromaterials. However, this is not the unique feature of nanoparticles with respect to microparticles. Nanoparticles exhibit significantly different properties with respect to microparticles, such as a greater reactivity, strength, fluorescence, conduction, etc. As a consequence, the scaling down from micro to nano does not trivially lead to an increase of catalyst activity related to a larger surface exposure, but it also changes the properties of the particles from an electronic point of view, thus changing their behavior and interaction with reactants and products. In other words, it also changes the intrinsic activity of the catalyst. For these reasons, nanoparticles are widely used in catalysis.

In heterogeneous catalysis, reaction occurs at the interface between the solid catalyst particle and reactants, which may be gaseous, liquid, and/or solid. The increase of the surface-to-volume ratio significantly affects the contact between reactants and active phase, thus enhancing the reactivity. But more importantly, it has been found that, when decreasing the particle size down to nanometers, the electronic state and the coordination environment of surface atoms undergo significant changes [10]. Goodman and coworkers [11] first demonstrated the role of nanosize in affecting the electronic character of Au nanoparticles supported on TiO_2 in CO oxidation. A gold nanoparticle at 1–2 nm may exhibit a molecule-like electronic state instead of a metallic state. This issue may strongly affect the catalytic performance.

Among all the catalysts investigated at the nanoscale, a prominent role is played by ceria. Several studies have shown that, due to the high surface area and highly reactive morphologies, ceria-based nanomaterials can be effectively used as catalysts for organic transformation reactions such as oxidation, reduction, hydrogenation, coupling reactions, and many more [12–14].

From the point of view of (nano)catalyst development and, more generally, chemical engineering, catalytic soot oxidation is a very interesting topic. Several issues are to be taken into consideration and linked to each other. First of all, solid-solid catalysis requires an intimate contact between reactant (soot) and active phase. The development of highly active catalysts is also a fundamental aspect requiring: (1) high mobility of oxygen and propensity to form surface active species; (2) nanostructure and/or special morphologies (in order to improve both contact and intrinsic activity); and (3) structural stability. In addition, there is a significant contribution of the interaction between solid phases and gaseous reactants to the catalytic performance, particularly in regard to the presence of gaseous oxygen carriers. Finally, engineering of particulate filters with nanostructured catalysts is not a trivial task. Most of these aspects are clearly related to each other, and thus, it is quite difficult to discuss each of them separately. In the following, we present sections devoted to specific topics and discuss the most relevant literature results in order to highlight how the issues mentioned above have been addressed and what solutions have been proposed. Obviously, the reader should consider that the topics of the different sections may partially overlap.

6.2 Type of Soot-Catalyst Contact

Soot typically consists of agglomerates with about 100 μm in size composed by smaller open particles in the range 0.1–1 μm, which in turn are made up of carbonaceous spherules of 10–50 nm (Fig. 6.3). The agglomerates can be easily disaggregated by treatment in ultrasound bath. The spherules are stuck together by shared carbon deposition forming the loose particles of 0.1–1 μm size. Although these particles have a high porosity (up to 0.95), they cannot be broken down into the individual spherules even by strong ultrasound treatment [15].

The surface of the spherules has adhering hydrocarbon material or soluble organic fraction (SOF) and inorganic material (mostly sulfates). Nevertheless, real soot collected from diesel engines under real conditions strongly depends on engine parameters, and consequently, Printex U is often used as model soot in studies reported in the literature [17].

The contact between the carbonaceous material and the grains of a solid catalyst is of paramount importance. Indeed, if the catalyst grains are not in intimate contact with the carbon surface, their effect will be lower. Ciambelli et al. [18] postulated two reactivity relationships, one for spherules within the field of action of the catalyst and another one for those initially outside. At lab-scale, Neeft et al. [19] developed two methods of mixing catalyst with soot, which they termed "loose" and

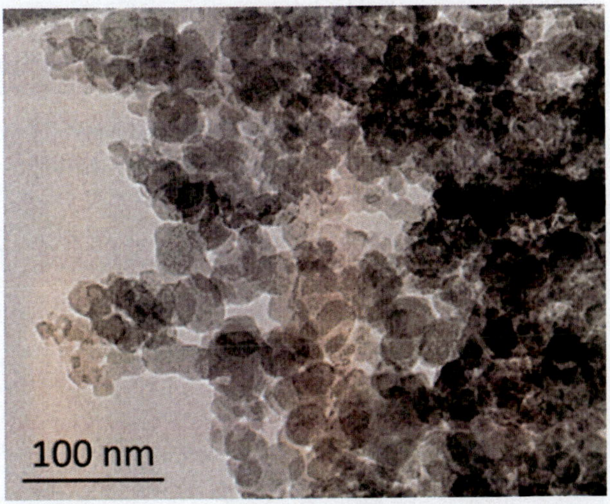

Fig. 6.3 Micrograph of typical diesel soot: particles consist of clusters of spherules [16]

"tight." The catalysts were prepared as particles of less than 125 μm size. In the loose mode, catalyst and carbon black (CB), with a 2:1 mass ratio of catalyst to CB, were simply mixed together with a spatula. In the tight mode, a mixture at the same ratio was mechanically milled for an hour. Since then, most of scientists who investigated new catalytic systems evaluated the catalytic performance using one or both types of contact between soot and catalyst particles, generally through temperature-programmed oxidation (TPO) experiments carried out in thermobalance consisting in an oxidation by increasing temperature of a soot-catalyst mixture.

The loose and tight methods allow a more reliable classification of the catalyst activity, although the catalytic performance can deeply depend on the contact type. Figure 6.4 shows the results obtained by Quiles-Díaz et al. [20] on a 2% CuO/ceria-zirconia catalyst and gives an example of the different behavior of the two contact systems.

Combustion of soot in tight contact with the catalyst takes place at significantly lower temperatures. Indeed, the tight mode forces soot and catalyst to be in intimate contact with each other. Generally, this analysis is used to discriminate catalysts in terms of intrinsic activity, because soot oxidation is not rate-limited by a poor soot-catalyst contact. On the other hand, the loose contact better reproduces the oxidation behavior of soot in the catalyzed (i.e., catalyst-coated) DPF.

Christensen et al. [21] evaluated the rate constant of various metal and metal oxide catalysts in loose and tight contact with soot. Regardless of the contact type, the rate constant outlines a volcano curve when plotted against the heat of oxygen chemisorption on the catalytic material. However, the maximum in the volcano curve corresponds to different values of heat of oxygen chemisorption for the two contact situations, and this result was attributed to mechanistic differences between oxidation in loose and tight contact. Among the tested catalysts, Co_3O_4 and CeO_2 are those closer to the maximum in tight contact, whereas Cr_2O_3 is closer to the maximum in loose contact. For both contacts, the activation energy for soot oxidation

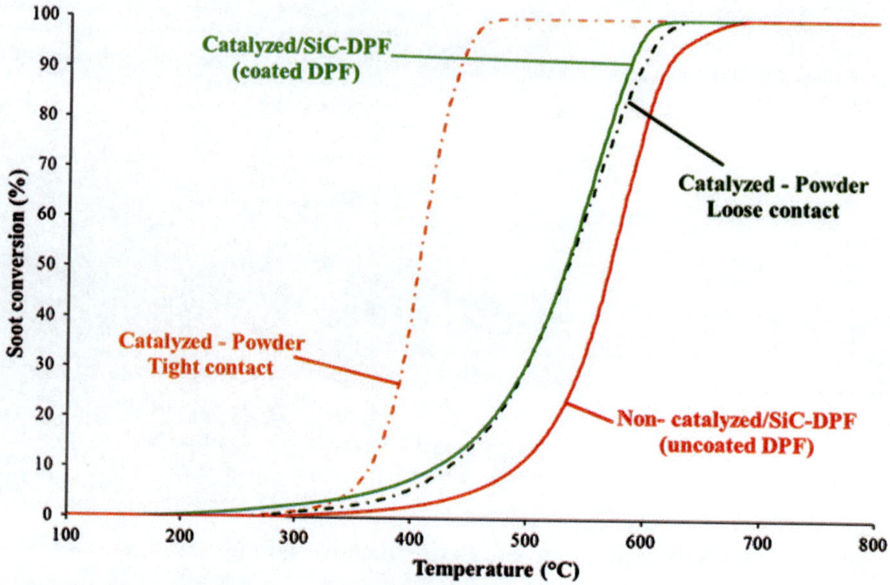

Fig. 6.4 Soot conversion curves versus temperature during heating-ramp experiments for different catalytic configurations [20]

follows a linear Brønsted-Evans-Polanyi relationship with the heat of oxygen chemisorption. However, the slope differs, being higher in loose contact. This suggests a transition state closer to the final state, namely a complete rupture of the oxygen-catalyst bond in loose contact.

Neeft et al. [19] found that the reaction rate is strongly influenced by the mass ratio of soot to catalyst, and the burn-off temperature significantly decreases as the ratio is decreased.

Liu et al. [22] reported that the difference between the temperature of ignition (T_{ig}) of soot combustion and that of the maximum (T_m) during TPO experiments can be taken as a measure of the nature of catalytic sites versus the number of contact points between these sites and soot. Low values of T_{ig} indicate the presence of very active sites which are capable of lowering the activation energy for soot oxidation. Therefore, T_{ig} reflects the nature of the active species regardless of the number of catalytic sites present. Once ignition takes place, the number of catalytic active sites in contact with soot becomes important, as it determines the further combustion rate. A marked difference between T_{ig} and T_m indicates a poor contact, while a small difference indicates a very good contact between soot and catalyst.

Besides the loose and tight contact, there exists another type of contact, the "pressure" contact. Hensgen and Stöwe [17] reproduced this contact type pressing a loose mixture of soot and (nano)ceria (with a 4:1 mass ratio of catalyst to soot) in a hydraulic press forming pellets that were subsequently crushed down to powder in a mortar. These authors compared the three types of contact (loose, tight, and pressure). They found that the T50 temperatures obtained with tight contact are much lower than in commercial DPF. On the other hand, the T50 temperatures of the loose and pressure contacts are similar to the temperatures in the real DPF systems, but the problem of these contacts is that no automation is possible. For five samples in loose contact, a standard deviation of T50 equal to 23.1 °C was reported. To reach a better reproducibility, a further contact type was developed by stirring the soot with the catalyst in acetone for 6 h. With this contact type, a better reproducibility, with a standard deviation of T50 of only 3.7 °C, was reached. This contact was called "wet" contact.

Hensgen and Stöwe [17] also studied the influence of the soot type on the contact mode. In particular, they investigated three different types of soot (fullerene soot (FS), Printex U (PU), and Printex 90 (P90)) in combination with the four different contact modes (loose, tight, wet, and pressure). In each contact mode, the T50 values of the FS are lower than those of the PU and P90 soot. For the loose, tight, and wet contacts, there exists a fixed order of the T50 value: tight < loose < pressure. The wet contact has not a fixed position within this order. While with PU, the wet contact is between the tight contact and the loose contact, with P90 and FS, the T50 value is even lower than in the case of the tight contact. Overall, in view of the standard deviations of the characteristic temperatures, the wet and tight contacts show similar T50 values, but the former is more reproducible.

Aneggi et al. [23] showed that a "supertight" contact can be obtained when ball-milling CeO_2-ZrO_2 powders and carbon soot for several hours. High-resolution transmission electron microscopy (HRTEM) revealed that the high-energy milling has the effect of progressively dissolving the large three-dimensional soot clamps and distributing more or less uniformly the carbon over the catalyst by wetting the catalyst surface with an almost two-dimensional thin layer of carbon. This result was also confirmed by the increasing metal/carbon atomic ratio detected by X-ray photoelectron spectroscopy (XPS). Accordingly, the number of contact points is significantly increased and the soot oxidation rate at low temperatures is boosted (a T50 value of 265 °C was found with 8 h milling). More recently, oxidation of soot in supertight contact with CeO_2-ZrO_2 was investigated by in situ environmental HRTEM from room temperature up to 550 °C, showing that the exceptional oxidation activity is related to the mobility of catalyst nanoparticles within the carbon globules during their consumption [24].

The results discussed above highlight the importance of the solid-solid contact in catalytic soot oxidation. The nanosize of the catalyst particles surely plays a key role in promoting this contact.

Lab-scale reaction conditions, especially if powdered soot-catalyst mixtures in tight (or similar) contact are used, can be significantly far from conditions of a real process of regeneration of catalytic DPF, thus limiting the possibility of extrapolation beyond the explored parameter ranges. On the other hand, studies in tight contact provide insight into the reaction mechanism and into the features related to the intrinsic activity, thus giving precious indications on the catalyst characteristics to be engineered in order to obtain improved catalytic systems.

6.3 Cerium Oxide (CeO_2)

CeO_2 is the most explored catalyst for soot oxidation [16]. This is also due to the fact that it is widely used in automotive three-way catalysts (TWCs) owing to its function of storage and release of oxygen under cyclic oxidizing and reducing environments. As a consequence, a lot of studies on the effect of the solid-solid contact on soot oxidation are focused on this material.

Simonsen et al. [25] studied CeO_2-catalyzed soot oxidation at the nanoscale by means of environmental transmission electron microscopy (ETEM). They found that catalytic oxidation involves processes that are confined to the region of the soot (carbon black, CB)-CeO_2 interface. Moreover, motions of CB agglomerates toward the catalyst surface act to re-establish the solid-solid interface and, thus, the CB-ceria contacts remain constant in the course of the oxidation process. Figure 6.5 shows the time-lapsed series of ETEM images recorded during CB oxidation over ceria.

It can be seen that agglomerates of CB particles move toward CeO_2 and vanish, whereas agglomerates of CB particles do not protrude and move away from the edges of the catalyst particles.

At the nanoscale, ambient pressure X-ray photoelectron spectroscopy (APXPS) revealed that soot oxidation over ceria involves two co-operative routes [26]. One occurs at the ceria-soot interface due to the formation of oxygen vacancies (related to Ce^{3+}); the other one occurs at the soot surface due to the formation of active superoxide species derived from the reaction between molecular O_2 and oxygen vacancies (Fig. 6.6).

Bassou et al. [27] evaluated the amount of oxygen transferred from ceria to soot via the soot-catalyst contacts using an experimental microkinetic approach. Consistently with the fact that the soot-ceria contacts do not change during oxidation

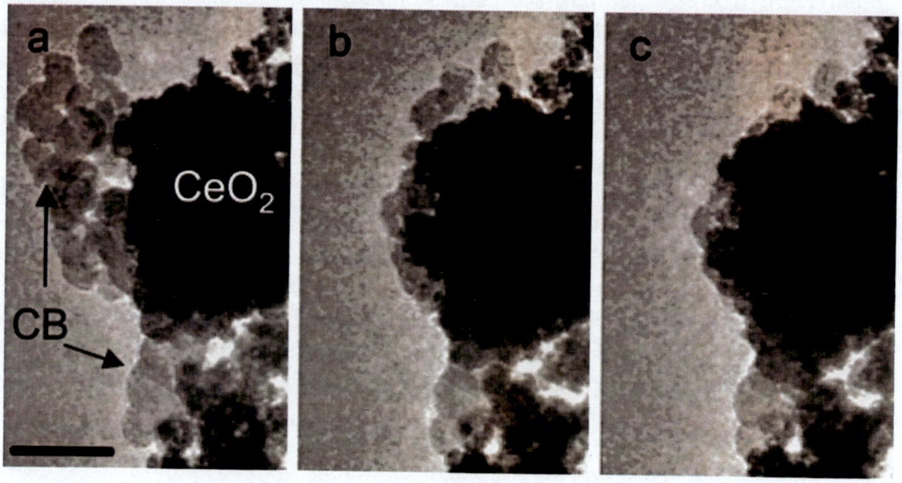

Fig. 6.5 Time-lapsed ETEM images during CB oxidation over ceria (the time interval between (**a**, **b**) images and (**b**, **c**) images is ~2 min; scale bar = 90 nm) [25]

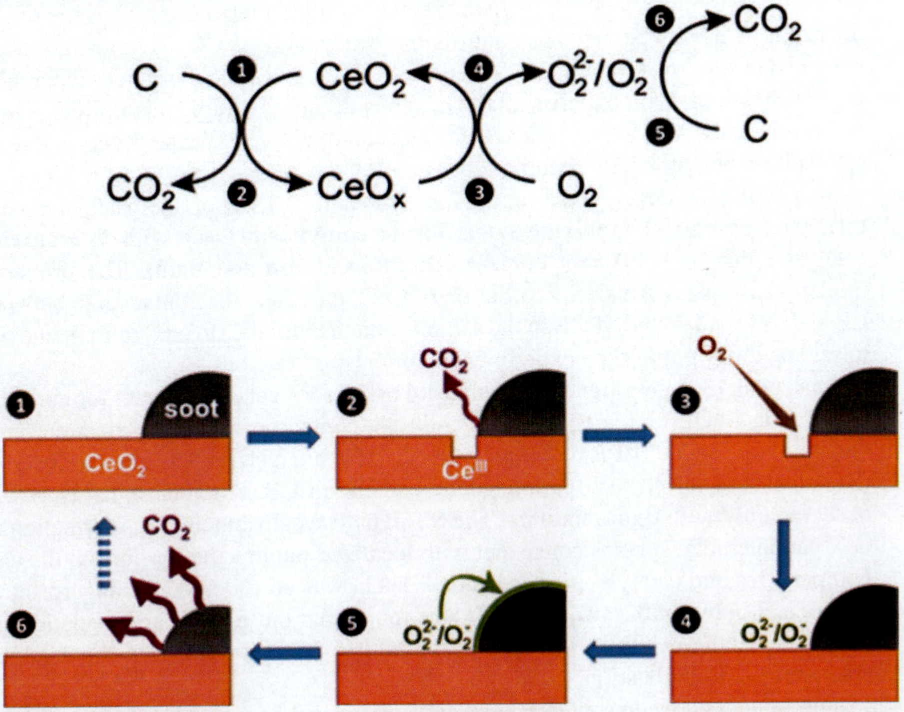

Fig. 6.6 The two co-operative routes involved in soot oxidation over ceria-based catalysts [26]

[25], they found that this amount (linked to the number of contacts) is constant with the progress of reaction, regardless of both type of soot-ceria contact (loose or tight) and ceria/soot weight ratio (10 or 1). This amount was evaluated after 2–3 cycles of temperature-programmed experiments with or without O_2, the first cycles being dominated by desorption and oxidation of the soluble organic fraction (SOF) preventing the evaluation of the oxygen transfer from ceria to soot from the productions of CO_2 and CO. However, the amount of oxygen available for soot oxidation (in μmol of O/g of soot) is dependent on the ceria-soot mixture and is correlated to the catalytic performance of ceria to decrease the temperature of soot oxidation by O_2. The authors also selected two plausible kinetic models (denoted as M1 and M2) of soot oxidation via surface defect sites, which are different by the implication (M2) or not (M1) of the diffusion of oxygen species on the ceria surface [28]. In the two models, the soot-ceria contacts are considered as a key kinetic parameter for soot oxidation. The active sites of soot (i.e., the carbon defect sites) and the active oxygen species of ceria are situated at the soot-ceria interface. The properties of this interface, such as the average number of contacts between a ceria particle and the soot particles, and the average surface area of a contact, are included in the mathematical formalism. The M1 model prevails at temperatures lower than 900 K, whereas the M2 model must be used at higher temperatures. This is due to the fact that surface oxygen species and, later, bulk oxygen species of ceria diffuse to the

interface, forming new active oxygen species. At high temperatures, it is the rate of diffusion of those species that controls the production of CO_2. The models were extended to soot oxidation in the presence of O_2. The M1 model provides theoretical curves consistent with experimental data, considering that rapid adsorption of oxygen on ceria allows for the surface concentration of active oxygen species to remain constant in the absence of external diffusion.

Issa et al. [29] developed a model for combustion of CB on the basis of experimental data obtained on mixtures of CB and a commercial CeO_2 with different CB/catalyst ratios and different contact conditions (loose and tight). The proposed model accounts for the size distribution of CeO_2 particles, the contact area between CB and CeO_2, the mass of CB in the sample, and the initial CB/CeO_2 ratio. It enables to satisfactorily predict the experimental combustion rate of CB.

Saab et al. [30] investigated the intensity of the soot-catalyst contact for mixtures of CB with CeO_2 (or Al_2O_3), in loose and tight contact, by means of electron paramagnetic resonance (EPR). The preliminary EPR analysis on CB showed a signal (S1) attributed to intrinsic paramagnetic centers on CB. Another signal (S2) was observed only with tight mixtures. The S2 signal was attributed to the formation of new paramagnetic species consistent with localized paramagnetic spins on the carbon particles and catalyst interface and, as such, was considered as a fingerprint of the contact between the two solids. It was found that these new paramagnetic species increase the reactivity of CB-CeO_2 mixtures in tight contact in the catalytic reaction of CB combustion.

The contact between soot (CB) and ceria was also investigated by SEM and EPR analyses [31]. Different procedures were followed to prepare loose and tight mixtures. CB, composed of primary particles in the 20–90 nm size range aggregated in structures with a foamy aspect, remains unchanged whatever the conditions of preparation of the mixture (mixing, grinding or sieving). CeO_2 alone is composed of aggregated grains in the 50–80 nm size range. Grinding CeO_2 samples provides particle disruption. SEM images showed a very limited solid-solid contact when CB and CeO_2 (with a 1:1 mass ratio) are ground together for 3 min (CeO_2 particles are sometimes surrounded by CB particles—loose sample), and an enhanced contact when CB and CeO_2 (at the same ratio) are ground together for 15 min (grinding promotes the formation of small catalyst particles and the increase of the contact points between the two solids—tight sample). In agreement with Saab et al. [30], the S2 EPR signal was identified as the paramagnetic print of the contact between CB and ceria. This signal was correlated to the number of soot-catalyst contact points, and the existence of a critical distance was suggested from which a liaison is established between the two solids communicating with each other through ultra-thin gas bolster (with or without physical touch). When this liaison is not established or does not exist, the S1 signal is only identified. Conversely, when this liaison is established, both EPR signals are identified. The two signals were observed for both loose and tight samples. However, in the case of the loose sample, the S2 intensity is low compared to the S1 intensity confirming that the contact between CB and CeO_2 is very poor. In addition, the number of paramagnetic species of the S2 signal is much lower than that found with the tight sample.

6.4 Addition of Other Elements to Ceria

In order to improve the activity of ceria toward soot oxidation, the addition of other elements has been proposed [16].

Sudarsanam et al. [32] investigated CeO_2 nanocubes decorated with CuO_x nanoparticles. Higher concentrations of Ce^{3+} ions and abundant oxygen vacancies, especially at the (nano)interface between CuO_x and CeO_2, improve the performance toward soot oxidation with respect to pristine ceria in both loose and tight conditions. Cui et al. [33] found that the doping of Cu in CeO_2 nanoflakes significantly promotes the formation of oxygen species that replenish the vacancies in both tight contact and NO_x-assisted loose contact (the NO_x assistance will be discussed in Sect. 6.6). The highest activity was observed for the sample of $Cu_{0.2}Ce_{0.8}O_2$ nanoflakes and was attributed to its moderate doping that induces both a large amount of surface-adsorbed oxygen species and a special flake morphology, thus providing effective contact area. Sudarsanam et al. [34] reported that the addition of iron, but mostly of copper, to ceria nanorods increases the number of active oxygen species.

The addition of La to CeO_2 also generates higher amounts of active oxygen species at lower temperatures, which are effective for soot oxidation in tight contact, when the oxygen transport from catalyst to soot is not limited [35]. Nevertheless, in loose contact, the active oxygen transfer is difficult and, before reacting with soot, the active oxygen species recombine with each other to form O_2. In this situation, the increase in specific surface area induced by the addition of La is the only factor that affects the activity.

Lim et al. [36] "simulated" diesel particulate matter (PM), consisting of primary soot particles and soluble organic fraction (SOF) on the surface, by liquid-phase adsorption of SOF (eicosane, $C_{20}H_{42}$) on soot (carbon black). They used Ag, Pt and Pd supported on CeO_2 or TiO_2 as catalysts and tested PM oxidation in both tight and loose contact. The catalytic performance for oxidation of SOF and soot in simulated PM depends on both metal species and supports. CeO_2 has an inherent activity for both SOF and soot oxidation, while TiO_2 is nearly inactive for both reactions. For SOF oxidation, which is independent of the contact conditions, Pt is the most active metal followed by Pd and Ag. On the contrary, Ag is more active than Pt and Pd toward soot oxidation under both tight and loose contact conditions, this reaction taking place in tight contact at a temperature lower than in loose contact, as expected.

Neeft et al. [37] suggested that the catalyst mobility (ability to move into deposited soot) is a major parameter determining the activity in loose contact, and this mobility correlates with the melting point or the partial pressure of the catalyst. In particular, catalysts with a high mobility can be low-melting-point compounds, which can melt and wet the soot particles, or materials with a high vapor-phase mobility. Actually, the catalyst mobility is a required feature to reduce the difference in activity between tight contact and loose contact. From this point of view, the work by Gardini et al. [38] is very interesting. These authors investigated soot oxidation catalyzed by very active silver nanoparticles exhibiting a limited difference in activity between loose mode and tight mode. The dynamic evolution of catalytic soot oxidation was visualized by means of in situ environmental transmission electron microscopy (ETEM) and bright-field TEM (BFTEM). The considerable activity of silver was at least in part ascribed to the significant mobility of the catalyst particles, responsible for ensuring the constant presence of a reactive soot-silver-oxygen interface. During soot oxidation, the reaction front moves, and attractive forces between metal and soot pull the silver particles along with the progressing front, thus causing a high mobility of the catalyst. The mobility is strongly influenced by the balance between attractive forces connecting silver agglomerates to the porous soot matrix and the size of the silver agglomerates themselves related to their internal cohesive energy. Figure 6.7 shows the BFTEM images of in situ oxidation of a soot-silver mixture in loose contact.

Mori et al. [39] investigated Ag/CeO_2-catalyzed soot oxidation at the nanoscale by means of ETEM. They found that reaction mainly occurs at the soot-catalyst interface (and not on soot itself), the soot particles gradually moving toward the catalyst and vanishing at the interface. They also observed that the silver-ceria interaction limits the formation of extremely large agglomerates of silver particles.

The addition of silver to ceria results in an enhanced soot oxidation activity with respect to pristine ceria [40–43]. Liu et al. [41] found that silver promotes the regeneration of O^-, present in large amount in spent CeO_2 catalysts, into highly active O_2^- leading to a ten-fold increase in soot oxidation activity over Ag/CeO_2 with respect to the unpromoted ceria catalysts. In agreement with these findings, Lee et al. [42] reported that the addition of silver to CeO_2 promotes the formation of the O_2^- superoxide, the amount of this active oxygen species being affected by the silver load. Zeng et al. [43] found that the presence of Ag nanoparticles induces a great number of oxygen vacancies in the CeO_2 lattice through the electronic transfer and that the Ag-assisted CeO_2 catalyst exhibits a better reduction performance. They also presented a reaction mechanism in which the O_2^- species is regarded as the determinant factor for the catalytic activity.

A positive effect of the silver addition was also reported by Deng et al. [44] for Ag/Ce-Zr catalysts. The catalysts were synthesized by precipitation in the presence

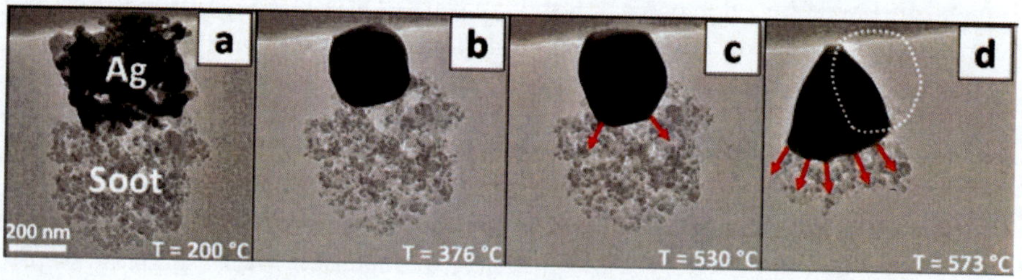

Fig. 6.7 Wetting and movement of a silver agglomerate during in situ oxidation of a soot-silver mixture in loose contact: BFTEM images showing (**a**) initial agglomeration and morphology of silver and soot, (**b**) coalescence of silver particles, (**c**) initial deformation of coalesced silver agglomerate due to capillary forces, and (**d**) movement of silver agglomerate. The red arrows indicate the direction of deformation of the silver agglomerate, and the dashed white line in sub-figure (**d**) highlights the previous position of the silver agglomerate as observed in sub-figure (**c**) [38]

of KOH and/or NH_3-H_2O, which allowed a good control of size and structure of the Ag particles. The very good performance in soot oxidation of the Ag-promoted catalysts, especially of those obtained using both precipitants, was attributed to the high capacity (associated with a suitable Ag^+/Ag^0 ratio) to activate and store oxygen species at a temperature much lower than unpromoted Ce-Zr. It was found that the high mobility of Ag contributes to the very good activity under loose contact conditions.

A good activity of Ag/CeO_2 was also reported by Corro et al. [45]. These authors did not use commercially available synthetic soot, but they generated soot by controlled combustion of several diesel-biodiesel blends. The addition of biodiesel to diesel modifies the soot composition, which becomes richer in aliphatic and oxygenated compounds. Thus, soot reacts faster over the Ag/CeO_2 catalyst (Fig. 6.8).

The addition of alkaline metals to CeO_2-based catalysts has also been proposed. Weng et al. [46] added potassium nitrate to copper-ceria catalysts. They found that the large amount of Cu-Ce interfacial sites is responsible for promoting the activation of oxygen, which is crucial for soot oxidation in tight contact. On the other hand, the low melting point of potassium nitrate improves the soot-catalyst contact in loose conditions. Alinezhadchamazketi et al. [47] found the same for K-added ceria-zirconia catalysts.

Sui et al. [48] prepared Cs-Ce-Zr catalysts by the sol-gel method. They reported that, under loose conditions, the solid-solid contact is improved and the soot oxidation rate speeds up when temperature is higher than that of melting of $CsNO_3$.

Shimokawa et al. [49] compared the addition of silver and potassium and found that K-promoted CeO_2 or TiO_2 samples oxidize catalytically a larger fraction of soot than Ag-promoted samples due to the higher mobility of K with respect to Ag in loose contact.

Zirconium has widely been used to partially substitute cerium forming mixed oxides. Ce-Zr mixed oxides have a higher specific surface area, a smaller crystal size, and enhanced redox properties, all features that determine a higher activity in

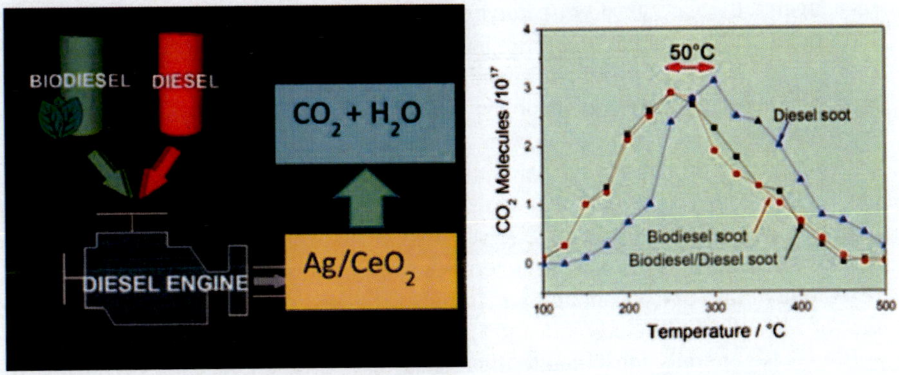

Fig. 6.8 The addition of biodiesel to diesel allows faster reaction of soot over Ag/CeO_2 catalyst [45]

both loose and tight contact. The substitution of Ce^{4+} for Zr^{4+} ions promotes the formation of defects (i.e., oxygen vacancies) in the ceria-zirconia lattice. Nevertheless, Piumetti et al. [50] showed that there exists a limit to the substitution of Ce^{4+} for Zr^{4+} because the surface density of the active redox element (Ce) decreases as the Zr content increases. They identified in $Ce_{0.9}Zr_{0.1}O_2$ the best composition of ceria-zirconia catalysts as nanopolyhedra. This morphology provides the best soot oxidation activity also compared to other catalysts with the same composition but different morphology, highlighting that the role of a suitable exposure of highly reactive planes prevails over the role of a large surface area.

Ce-Zr solid solutions were also studied by Trovarelli's group [23, 24, 26] with a focus on the soot-catalyst contact investigated at the nanoscale by means of advanced techniques, as already described in Sects. 6.2 and 6.3.

Trovarelli's group also investigated the activity of transition and alkaline-earth metal-doped catalysts supported on ceria or zirconia for NO_x-assisted soot oxidation [51]. A series of Cu- and Fe-impregnated catalysts over CeO_2 and ZrO_2 supports were prepared and characterized, and their catalytic activity was investigated by means of temperature-programmed oxidation (TPO) experiments. It was found that the copper-modified catalysts are more active. Moreover, the addition of Sr positively affects the performance of the materials, suggesting a synergic effect between transition metals and alkaline-earth metal. Copper is mainly involved in oxidation of NO to NO_2, which is a strong soot oxidizing agent, while strontium in the storage of NO_x species.

The role of copper as promoter of NO oxidation was also highlighted when this element was added to Ce-Pr solid solutions [52]. In particular, the benefit of mixing $Ce_{0.5}Pr_{0.5}O_{2-\delta}$ particles with and without copper in a single catalyst formulation was demonstrated. Indeed, the catalyst that combines $Ce_{0.5}Pr_{0.5}O_{2-\delta}$ particles with and without copper has two different types of active sites and optimizes the participation of two mechanisms of soot oxidation: the particles with copper mainly promote oxidation of NO to NO_2 (i.e., the NO_x-assisted mechanism), while those without copper are more effective in promoting the active oxygen mechanism.

In line with designing dual-functional catalysts, Castillo Marcano et al. [53] investigated the CeO_2/BaO/Pt catalytic system. The active oxides, CeO_2 and BaO, the former with oxidative properties and the latter with NO_x storage capability, were individually prepared and then mixed to form a "physical mixture," or they were simultaneously synthesized to obtain a solid solution through the so-called "co-synthesis" method. Pt was added in a subsequent step. The activity toward soot oxidation revealed that a physical mixture of CeO_2 and BaO allows to obtain more performing catalysts than the co-synthesis route, the former reaching a peak temperature of soot oxidation 25 °C lower than the latter. The Pt addition to the two catalysts further reduces their peak temperatures by around 30 °C in both cases.

Another class of Ce-containing materials proposed as soot oxidation catalysts is that of MnO_x-CeO_2 mixed oxides [54]. The good catalytic activity of Mn-Ce solid solutions is generally attributed to the formation of oxygen vacancies related to the generation of Ce^{3+}. Recently, Jampaiah et al. [55] prepared Mn_3O_4/CeO_2 catalysts in the form of dispersed Mn_3O_4 nanoparticles supported on CeO_2 microspheres. They varied the Mn_3O_4 load. The maximum soot oxidation activity, found with 20% Mn_3O_4 load, was attributed to the easiest generation of active oxygen species and to the best synergistic interaction at the interface between CeO_2 and Mn_3O_4, in turn due to the optimal deformation of the CeO_2 structure and the subsequent formation of oxygen vacancies. A high stability was also reported for these composite catalysts. In agreement, Sartoretti et al. [56] found that 5% Mn-doped CeO_2 shows a good and stable (after repeated cycles) activity for soot oxidation related to the ability of this catalyst to regenerate the most active defect sites.

To sum up, a wide variety of elements with different features (size, valence state, reducibility, mobility, melting point of their salts, etc.) has been proposed to promote the activity of CeO_2 toward soot oxidation. The main positive effects of their addition to ceria are as follows: enhanced formation of surface active species through the creation of oxygen vacancies/defect sites; improvement in the soot-catalyst contact conditions; introduction of an additional functionality cooperating in NO_x-assisted soot oxidation.

6.5 Other Catalysts

Although CeO_2, pure and in combination with other elements, is the most widely used catalyst for soot oxidation, several other compounds have been proposed as briefly reviewed in this section.

Wagloehner and Kureti [57] described a scheme of global reactions for the mechanism of soot oxidation on Fe_2O_3. According to this scheme, oxygen is transferred from the catalyst surface to soot by contact points. The resulting oxygen defects of the catalyst surface are refilled either by surface migration and final reoxidation by gas-phase oxygen or by diffusing bulk oxygen. The oxygen deficiency of the lattice is balanced by migration of oxygen from the surface or subsurface to the bulk of the catalyst. The authors also found that the local temperature in the catalyst-soot mixture strongly depends on the amount of catalyst (i.e., the catalyst acts as a temperature buffer). Thus, there exists an optimum catalyst/soot ratio reflecting a compromise between a high number of contact points (high catalyst mass) and a low heat capacity (low catalyst mass).

Perovskite-based catalysts were investigated for soot oxidation (see, e.g., [58–60]). Bensaid et al. [60] prepared several perovskites by the "solution combustion synthesis" (SCS) method described in Civera et al. [61]. They identified $Ce_{0.5}Pr_{0.3}La_{0.2}CrO_3$ as the most promising catalyst and demonstrated the feasibility of its deposition on a DPF.

As reported in Sect. 6.4, the addition of silver to ceria has positive effects on soot oxidation. Silver was also added to ZrO_2 by Haneda and Towata [62]. These authors attributed the outstanding performance of Ag nanoparticles supported on ZrO_2 in loose contact with soot to the good solid-solid contact and to the effective migration of active oxygen species from catalyst to soot. Ag was also supported on yttria-stabilized zirconia (YSZ) [63], and results showed a synergistic effect between the silver mobility and the availability of active oxygen species from the YSZ bulk, which leads to an improved activity with respect to pristine YSZ [64, 65].

Ag- and Ru-based catalysts were investigated by Castoldi et al. [66] for the simultaneous removal of soot and NO_x, and their behavior was compared with that of a model Pt-Ba/Al_2O_3 catalyst. It was found that both the Ag- and Ru-based formulations are active in soot oxidation, more than the traditional Pt-containing catalyst. In addition, the Ru-based sample showed remarkable performances in the deNO_x-desoot activity.

A catalyst for highly efficient NO_x capture and soot combustion was prepared by Dou et al. [67] in the form of a three-dimensional nanosheet array with small-sized active Co_3O_4 phase (5.7 nm) highly dispersed on a Mg/Al-oxide matrix.

The positive effect of alkali metals on catalytic soot oxidation, already discussed for CeO_2 in Sect. 6.4, has largely been demonstrated also for other materials. The promotion effect of potassium on the activity of transition metal (Mn, Fe, Co) spinels was reported by Legutko et al. [68]. These authors found that the location of the potassium promoter in the bulk, more than on the surface of the spinels, markedly lowers the onset temperature of soot oxidation.

Mul et al. [69] reported that the high soot oxidation activity in loose contact of catalytic systems containing an alkali metal chloride (KCl or CsCl or LiCl) and $CuMoO_4$ (or $CuWO_4$ or copper vanadates) can be ascribed to the formation of volatile copper chlorides. More specifically, it can be partially explained by the mobility and volatility of these compounds, resulting in an intimate contact between soot and catalyst. However, despite copper chlorides can be formed by reaction between KCl (which serves as a chlorine supplier) and $CuMoO_4$, the application of supported Cu/K/Mo/Cl catalysts was considered questionable due to the possible loss of activity caused by evaporation and decomposition of the active species.

Carrascull et al. [70] investigated KNO_3/ZrO_2 catalysts with different compositions, i.e., KNO_3 concentrations (in g of KNO_3/g of catalyst). They reported that the difference in temperature of the maximum soot combustion rate between loose contact and tight contact decreases with increasing KNO_3 concentration (in particular, from 95 °C with a concentration equal to 0.25% to 10 °C with 20%). This trend was attributed to the improved contact promoted by molten potassium nitrate.

Sui et al. [71] reported that KNO_3 greatly lowers the soot onset ignition temperature for Co-Sr catalysts. In agreement with Carascull et al. [70], they found that the soot-catalyst contact is strongly improved above the melting point of KNO_3.

An and McGinn [72] studied the wet contact and reported that the high activity they found for potassium-containing oxides is due to the intimate contact between soot and potassium cations caused by polar solvents.

Courcot and coworkers [73, 74] investigated the effect of potassium on both TiO_2 and ZrO_2 supports also in combination with copper. They reported that potassium not only favors the solid-solid contact, but also promotes the release of active oxygen species [74] and enhances the redox properties of copper [73, 74]. Thus, the role of the alkaline metal is not limited to improving the soot-catalyst contact conditions.

Galdeano et al. [75] explored cesium and other alkaline (Li and K) nitrates supported on hydrous zirconium. Under operative conditions similar to the catalyst behavior in the engine (loose contact) and in the presence of NO, the activity correlates with the electropositive character of the metal. The catalyst with cesium nitrate becomes active at a temperature lower than its melting point. It exhibits a combustion temperature of 364 °C within the values required for the catalyst to operate in the real case of an automobile. The catalyst with lithium nitrate becomes active at a temperature higher than its melting point, and the same occurs for the catalyst with potassium nitrate.

6.6 NO_x Assistance to Bridge the Tight-Loose Contact Gap

Diesel exhaust gases also contain nitrogen oxides which, in order to comply with the emission regulations, must be abated. However, as already mentioned in the previous two sections (mostly in Sect. 6.4), they positively affect soot oxidation.

Kaspera et al. [76] highlighted the role of NO_2 acting as an oxygen carrier from catalyst to soot lowering the temperature of 50% conversion in loose contact by ~100 °C, thus bridging the tight-loose contact temperature gap in NO_x-assisted catalytic oxidation of soot. This was observed on vanadium-doped cryptomelane $K_xMn_8O_{16}$ nanorods. It was found that the incorporation of vanadium into the cryptomelane framework substantially increases the desoot activity by promoting beneficial NO oxidation to NO_2. Two types of oxygen transport were distinguished: through space and across the surface. In the first case, NO_2 acts as an oxygen carrier from catalyst toward soot (through space transport). This situation corresponds to the experiments carried out in loose contact, especially in contactless modes. On the other hand, in tight contact, the oxygen reactive species are transported across the surface from the sites of oxygen activation and formation of oxygen reactive species toward the soot-catalyst contact points, where catalytic combustion actually takes place. This process is conventionally regarded as oxygen surface migration or spillover, as no auxiliary molecule is involved in the oxygen surface transport.

Likewise, Legutko et al. [68] found, for K-promoted transition metal spinels, that the difference in activity between tight contact and loose contact can be bridged in the presence of NO due to its transformation into NO_2, which acts as an oxygen carrier from the catalyst surface to the soot particles. The same was reported by Jakubek et al. [77] for nanostructured potassium-manganese oxides decorated with 1% Pd.

The effects of NO and NO_2 addition were investigated by Zhang et al. [78] for Pt/MnO_x-CeO_2 catalysts in tight and loose contact with soot. It was found that the promotion effects of NO_x are not exhibited in tight contact due to the inhibition of nitrate formation by the soot coverage on the catalyst surface. On the other hand, in loose contact, NO, more than NO_2, promotes the formation of surface nitrates which decompose providing surface active species able to accelerate soot oxidation.

Shen et al. [79] reported that, in tight contact, the "NO_2 assistance" to soot combustion on Fe-Ce-O catalysts is limited due to the inhibition of NO oxidation. Nevertheless, when soot is gradually consumed, the solid-solid contact becomes loose, thus resulting in an improvement of NO oxidation to NO_2.

Christensen et al. [80] found similar results for various metal and metal oxide catalysts. They observed a marked enhancement of the rate constant of soot oxidation in loose contact in the presence of NO_2. Among the tested catalysts, Cr_2O_3 is the most active one.

Zouaoui et al. [81] determined the kinetic parameters of soot oxidation in both loose and tight contact with CeO_2 under O_2 and NO/O_2. The activation energy was found to be strongly dependent on the type of contact in the presence of oxygen and to slightly depend when NO is also used.

6.7 Nanostructured Ceria Morphologies

In a real DPF, the catalytic activity is not the only important feature: an engineered morphology has to be designed to achieve better results [82]. It has been demonstrated that the catalytic activity of ceria toward soot oxidation depends on the contact points obtained at the interface between soot and catalyst and on the availability of active surface oxygen [83]. Both issues can be managed by modifying the morphology [82], thus also exposing more active crystalline planes [50, 84].

A CeO_2 morphology with fibrous structure has been proposed to maximize the contact between catalyst and soot particles [85, 86]. Despite their low specific surface area (SSA), these fibers have a filamentous structure that enhances the number of soot-fiber contact points and, in some cases, show better performances than foamy or higher SSA nanopowders. In other words, tailored morphologies can be achieved even with low specific areas. This proves that the specific surface area is not the only important factor in solid-solid catalysis. The fact that the sizes of soot particles and catalyst grains often have different orders of magnitude leads to a poor accessibility of the soot particles to the inner pores of the catalyst, and this aspect could be rate-limiting especially at low temperatures [85].

Miceli et al. [82] compared three CeO_2 morphologies, shown in the field-emission scanning electron microscopy (FESEM) images of Figs. 6.9, 6.10, and 6.11, in both

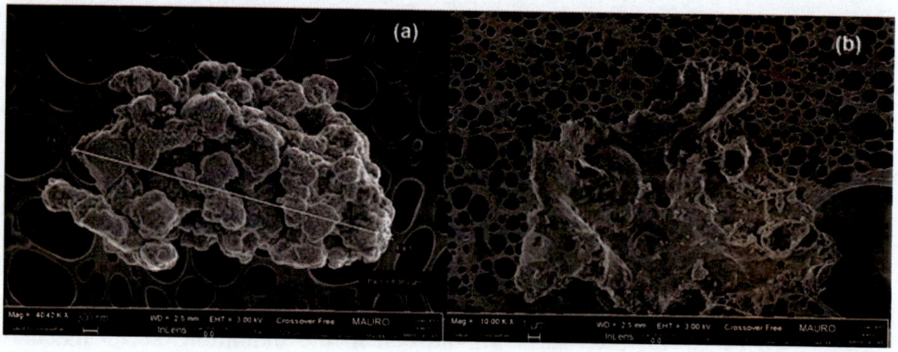

Fig. 6.9 FESEM images of ceria SCS nanopowders: **(a)** high and **(b)** low level of magnification [82]

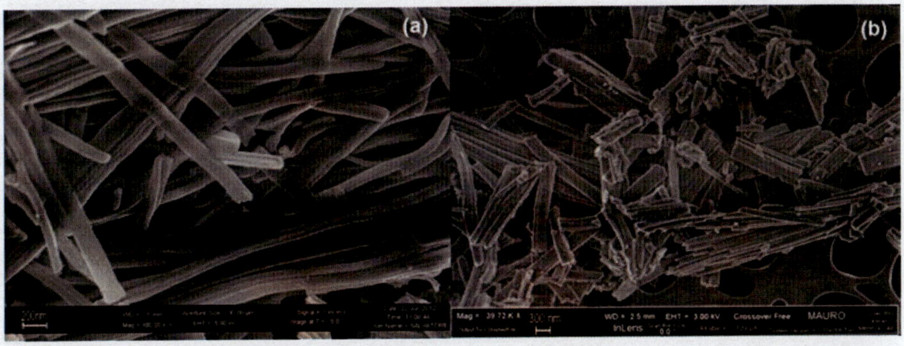

Fig. 6.10 FESEM images of ceria nanofibers: (**a**) high and (**b**) low level of magnification [82]

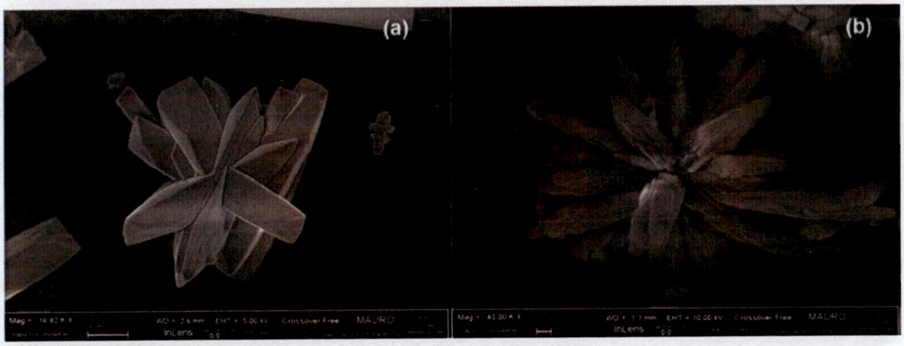

Fig. 6.11 FESEM images of ceria SA stars at different residence times inside the reactor in hydrothermal conditions: (**a**) 12 h and (**b**) 24 h [82]

tight and loose contact with soot: foamy structured CeO_2 prepared by the solution combustion synthesis (SCS) method (Fig. 6.9), nanofibers (Fig. 6.10), and self-assembled (SA) stars (Fig. 6.11).

They found that three-dimensional self-assembled stars, which have both a high specific surface area (105 m^2/g) and a high availability of contact points, give the best results especially in loose contact. Indeed, in tight contact, the mechanical force generates a particularly close contact between soot and catalyst, and thus, the advantages of the morphology are less important. Conversely, in loose contact, the morphology plays a more relevant role: nanofibers, despite the almost null surface area, exhibit an activity almost equivalent to that of SCS nanopowders. On the other hand, the high porosity of self-assembled stars provides more adsorbed oxygen to the contact points between soot and catalyst, which is likely to be in a sufficient amount to fully exploit this oxygen availability. Figure 6.12 shows the FESEM images representing a loose mixture of ceria SA stars and soot at different levels of magnification.

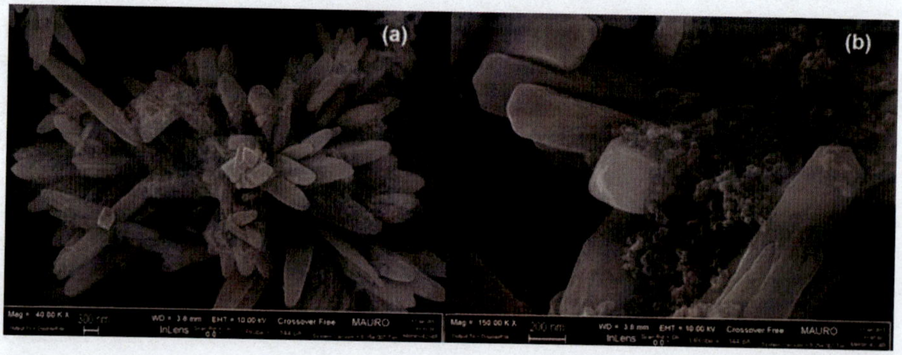

Fig. 6.12 FESEM images representing a loose mixture of ceria SA stars and soot: (a) low and (b) high level of magnification [82]

Miceli et al. [87] also found that self-assembled stars have a greater availability of oxygen vacancies and undergo more efficient redox cycles with a higher oxidative capability.

Zhang et al. [88] highlighted the role of the morphology of CeO_2 in loose contact, when the different physicochemical properties greatly influence the activity toward soot combustion also in the presence of NO. They found that nanorods perform better than nanoparticles and flakes due to a higher surface area and a higher adsorption of oxygen species on the surface.

On the contrary, Piumetti et al. [89] observed a "surface insensitivity" of CeO_2 with different textural and morphological properties, also dispersed on ZSM5, in both loose and tight contact at low temperatures, and they attributed this feature to the presence of adsorbed species (like hydroxyl groups, inert molecules, etc.) on the solid surface weakening the reactant-catalyst surface interaction through masking of the active sites. Nevertheless, above 410 or 370 °C (in loose or tight contact, respectively), a "surface sensitivity" was observed.

The same authors also investigated various CeO_2 morphologies and three contact conditions: tight contact, loose contact (with a 1–2 min mixing time), and a so-called "prolonged" loose contact (with a 30 min mixing time) [86]. In tight contact, nanofibers, having a low specific surface area, exhibit an onset temperature almost comparable to the one achieved with SCS nanopowders. An optimal contact between soot and catalyst is reached when the network of nanofibers has average pore diameters of the same order of magnitude as the soot particle ones. Conversely, sticks and flakes have excessive size, and a macroporous framework is formed, which poorly adapts to the soot particle average size. An important feature of SCS nanopowders is their better selectivity toward CO_2 than all the other morphologies due to the high specific surface area which provides a greater reactivity toward gaseous species, namely CO. The prolonged loose contact was adopted because fibers, sticks, and flakes are harder than SCS nanopowders, which are very soft. The latter are very easily and rapidly mixed up with the soot particles, while fibers require a longer time to be fully covered by the soot particles. In these conditions, even the selectivity to CO_2 of nanofibers is satisfactory and not much different from that of SCS nanopowders. In loose conditions (i.e., conditions which bring out the easier mixing of SCS nanoparticles with soot, due to their softness), no clear preeminence of either powders or fibers was observed. Even sticks and flakes approach the activity of SCS nanopowders. This encourages to better investigate the interaction between the morphology and the real contact conditions between soot and catalyst in DPFs and to tailor the catalytic support to enhance this contact while still minimizing the contribution of the catalytic layer to the pressure drop across the filter.

The effect of the ceria morphology was also reported by Aneggi et al. [90]. Nanocubes, which expose {100} surface, and nanorods, which expose {100}, {110}, and in part {111} surfaces, show a higher catalytic activity than conventional polycrystalline ceria, which exposes mainly {111} surface. Very interestingly, aged catalysts show irregular truncation of edges and corners and development of more reactive surface combinations in all crystal shapes at the expense of the specific surface area.

Cheng et al. [91] investigated a series of three-dimensionally ordered macroporous (3DOM) $Ce_{0.9-x}Fe_{0.1}Zr_xO_2$ catalysts with different Ce/Zr ratios. These catalysts were prepared by a colloidal crystal template method. The 3DOM unique structure promotes the contact of particulate matter (PM) with the active sites of the catalyst. The high Ce^{3+}/Ce^{4+} ratio, the amount of chemisorbed oxygen species, the good low-temperature reducibility, and the abundance of acidic sites were considered responsible for the excellent catalytic efficiency of the $Ce_{0.85}Fe_{0.1}Zr_{0.05}O_2$ sample for the simultaneous removal of PM and NO_x.

In conclusion, catalytic soot oxidation is a surface-dependent reaction. The use of catalysts with morphologies that maximize the extent of the soot-catalyst interface, even if this could be coupled to a low specific surface area, is recommended to achieve a good oxidation activity. An opportune choice of preferentially exposed crystal planes, more active than others, can further improve the catalytic performance.

6.8 Filters Coated with Nanostructured Catalysts

The importance of the soot-catalyst contact to an effective catalytic combustion of soot has also been highlighted by experiments carried out on both filters and "flow-through" monoliths coated with nanostructured catalysts. To examine what happens on the filter under as realistic as possible regeneration conditions is indeed crucial. As shown by Hinot et al. [92] with platinum nanoparticles, in going from catalyst-doped soot to a thin deposit (5–10 μm) of soot on a catalytic layer, the catalytic effectiveness may drastically drop. In particular, they found that the reduction in soot oxidation temperature with respect to thermal oxidation decreases from 140–250 °C (catalyst-doped soot) to 10–40 °C (soot deposited on a catalytic layer), and this was attributed to the increased "average" distance between soot and catalyst particles.

Fino's group prepared catalytic filters on the basis of a series of nanostructured perovskite catalysts [60, 93–95]. The preparation method employed in these works is the in situ solution combustion synthesis (SCS). A nanostructured and foamy catalyst coating over the DPF was obtained, and this issue was considered as an essential prerequisite not only for a good soot-catalyst contact, but also for a rather low-pressure drop throughout the channel walls.

More recently, perovskite-based flow-through monoliths were successfully tested for soot oxidation by Tang et al. [96]. In this work, monolithic catalysts were prepared by a two-step procedure. Cordierite monoliths were first washcoated with porous hollow γ-Al_2O_3 nanoparticles by dip coating. Active components, i.e., nanometric $LaKCoO_3$ or $LaCoO_3$ perovskite-type complex oxides, were then loaded by impregnation of the as-prepared monoliths with corresponding aqueous solution of metal salts. For comparison, the one-step method without the alumina coating was also investigated. It was found that the introduction of the γ-Al_2O_3 washcoat is crucial to greatly increase the surface area of the monolithic cordierite. This allows to increase the loading amount of active components and also to enhance their dispersion on the surface of the monolith, thus leading to improved conditions of soot-catalyst contact. The CO_2 concentration curves of Fig. 6.13, recorded during soot combustion over different monolithic catalysts, show that the presence of the γ-Al_2O_3 washcoat allows better performance.

The fact that the production of large-scale catalytic surfaces inside the DPF plays a key role in determining the efficiency of the soot-catalyst contact was also proven by Zhou et al. [97], who proposed a low-cost electroless coating approach that allows in situ growth of nanostructured metal crystals inside the pores of ceramic filters. Results from nanoscale imaging characterization and element-specific energy dispersive X-ray (EDX) spectra demonstrated the presence of a well-dispersed

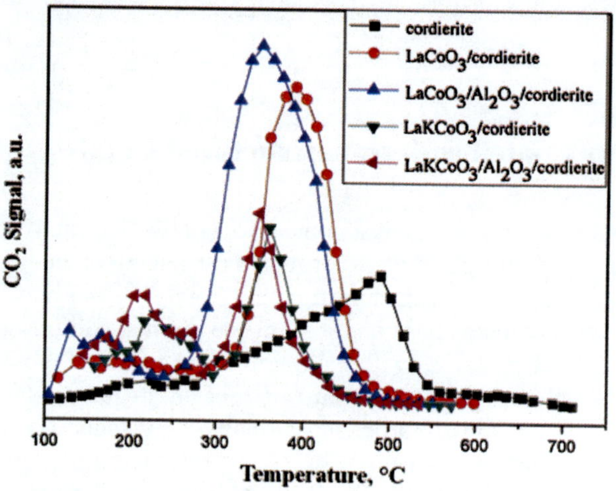

Fig. 6.13 CO_2 concentration curves recorded during soot combustion over different monolithic catalysts [96]

catalyst (Pt) throughout the filter, on the channel surfaces and inside the pores, resulting in a substrate that is highly active toward soot oxidation. According to the proposed method, the shape of the nanostructured catalyst can be tuned, from spheroid particle, rod-like wire to fern-like dendritic structure, by varying the starting reactant concentration and the length of the deposition time. The scanning electron microscopy (SEM) images of Fig. 6.14 show (a) the bare DPF sample and (b–d) Pt-decorated DPF samples, with Pt nanoparticles obtained starting from two different values of reactant concentration, (b) lower and (c) higher, and (d) Pt dendrites (at the higher value of reactant concentration).

The role of the catalyst shape in affecting the efficiency of the contact with soot was investigated, also at the filter level, by Kumar et al. [85], with a focus on nanostructured ceria catalysts. They found that nanofibers are very active with respect to other ceria morphologies, due to their arrangement in a network that enhances the number of soot-fiber contact points. The main advantage given by the nanofiber-catalyzed DPF, with respect to the other investigated morphologies, is related to the onset temperature. This temperature is 150 °C lower than the onset temperature of the un-coated DPF and more than 50 °C lower than the onset temperature of the DPF catalyzed with ceria through in situ SCS.

Filters and flow-through monoliths coated with nanostructured ceria or ceria-based materials were also investigated in more recent works [3, 4, 7, 8, 20, 98].

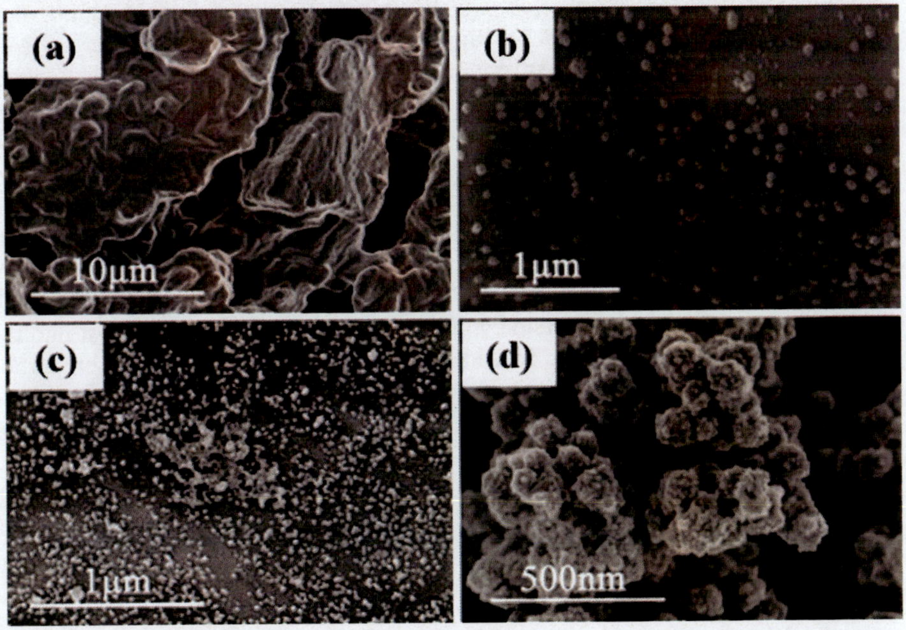

Fig. 6.14 SEM images of the Pt-decorated DPF samples: (**a**) bare filter; Pt nanoparticles obtained starting from two different values of reactant concentration, (**b**) lower and (**c**) higher; (**d**) Pt dendrites (at the higher value of reactant concentration) [97]

Nascimento et al. [7] used the sol-gel methodology to synthesize Ce-Fe binary mixed oxide catalysts (CeO_x/FeO_y). Cordierite monoliths were coated with the CeO_x/FeO_y catalysts. Samples of the cordierite coated with the binary mixed oxides were doped with Ag nanoparticles (addition of 5 wt.%). CeO_x/FeO_y- and $Ag/CeO_x/FeO_y$-coated ceramics were tested for their ability to reduce the soot emission during diesel combustion in a stationary engine. It was found that the silver-doped coating allows better performance as a result of both a relatively good soot-catalyst contact and an effective migration of active oxygen species from the catalyst to soot.

Quiles-Díaz et al. [20] deposited a nanosized 2% CuO/ceria-zirconia catalyst on ceramic DPFs using a simple and organic solvent-free procedure. The filter was simply dipped into an aqueous solution of the catalyst. The adopted procedure allowed to obtain a non-continuous deposit of catalyst on the channel walls, thus preventing the blocking of the filter pores. The regeneration performance of the catalyst-coated DPF is well simulated by the loose contact mode (see Fig. 6.4 in Sect. 6.2). This suggests that a rather weak soot-catalyst contact is established inside the filter under the conditions investigated.

To mimic the soot-catalyst contact achieved inside a real filter, Rico Pérez and Bueno-López [3] loaded nanoparticles of an optimized ceria-praseodymia active phase on silicon carbide (SiC) DPFs and designed an experimental set-up where a suspension of soot particles in air is forced to pass through the filter, thus simulating the filtering process in a real exhaust pipe. Results of regeneration tests performed at different soot loads (while keeping constant the catalyst load) allowed to identify a critical catalyst/soot ratio below which catalytic regeneration is hindered. This was attributed to the weakening of the soot-catalyst contact with increasing soot load. It was argued that the first soot particles loaded have more chances to come into contact with catalyst particles than particles loaded afterward. However, the link between the critical catalyst/soot ratio and the localization of contacted and non-contacted soot entities inside the filter was not elucidated.

A deep investigation of the effect of the catalyst/soot ratio on the "regime" of regeneration of SiC DPFs washcoated with nanometric ceria particles was performed in Di Sarli et al. [4]. The dip-coating procedure adopted for the deposition of ceria, along with the limited catalyst load and the nanometric size of the CeO_2 particles, ensured a high dispersion with a deep penetration of the catalyst into the (macro)pores of the filter walls. Indeed, in order to preserve the filtration properties of the bare support and, at the same time, promote the soot-catalyst contact inside the filter walls, the accumulation of a washcoat layer on top of the channel walls was prevented, thus ensuring the accessibility of soot to the pores of the filter. Results showed that, as the catalyst/soot ratio is decreased (i.e., the soot load is increased), a transition occurs from a regime of "almost purely catalytic regeneration" to a regime of "catalyst-assisted thermal regeneration." In the former regime, most of the soot is trapped inside the pores of the filter walls, thus coming into intimate contact with highly dispersed catalyst. As a consequence, regeneration occurs via the catalytic path at low temperatures. In the latter regime, most of the soot is accumulated in the form of a rather thick cake layer (15–20 μm) on top of the channel walls. The soot cake is substantially segregated from the catalyst and, as such, it

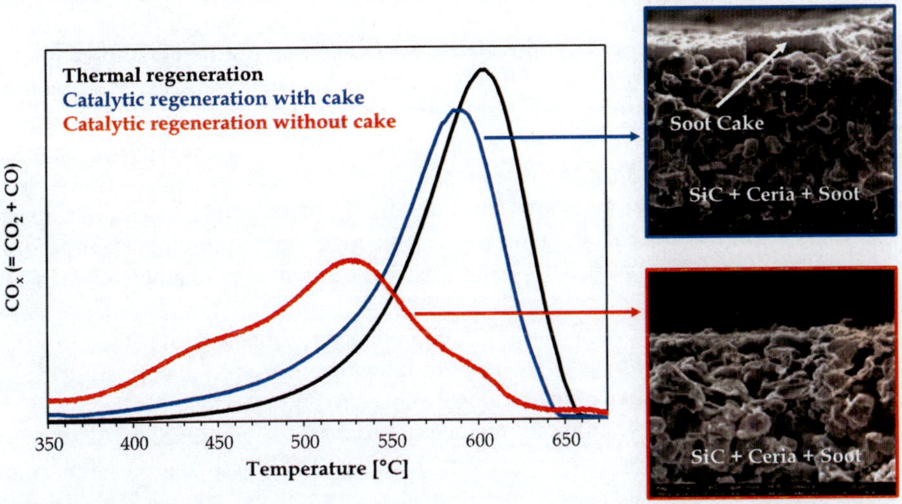

Fig. 6.15 Concentration of CO_x (=CO_2 + CO) versus filter temperature as recorded during three heating-ramp regeneration tests: thermal regeneration test, catalytic regeneration test with cake, and catalytic regeneration test without cake—the SEM images on the right-hand side show the cross section of a channel of the ceria-coated DPF with cake (top image) and without cake (bottom image) [4]

burns via the thermal path at high temperatures. Figure 6.15 is the graphical abstract of this work. It shows the plots of the concentration of CO_x (=CO_2 + CO) versus the filter temperature as recorded during three heating-ramp regeneration tests: thermal regeneration, catalytic regeneration with cake (high soot load and, thus, low catalyst/soot ratio), and catalytic regeneration without cake (low soot load and, thus, high catalyst/soot ratio). The images on the right-hand side are SEM images showing the cross section of a channel of the ceria-coated DPF with cake (top image) and without cake (bottom image).

This figure shows that, once assured a good catalyst dispersion inside the pores of the filter walls, in order to optimize the soot-catalyst contact, thus making regeneration of the DPF a truly catalytic process, it is essential to minimize/avoid the formation of the cake layer. This conclusion is supported further by the results of CFD-based simulations of soot combustion dynamics in a catalytic DPF showing that, once assumed all the soot trapped inside the filter walls to be in contact with the catalyst, fast and at the same time "safe" (i.e., low temperature) regeneration is not possible [99–101] unless the accumulation of soot as cake on top of the catalytic walls is prevented [101, 102].

As highlighted in Sect. 6.1, after the filtration stage, soot is trapped mostly in the form of cake. The results of Fig. 6.15, thus, pave the way to the passage to a continuous regeneration mode for catalytic DPFs, with combustion of soot occurring during (and not after) filtration, thus avoiding excessive soot storage inside the filter as cake.

In order to investigate the potential of (nano)ceria-coated DPFs for continuous regeneration, Di Sarli et al. [98] carried out isothermal regeneration tests—at temperature ranging from 200 to 600 °C—on filters loaded with a very low amount of soot suitably chosen to minimize the formation of the cake layer. Results showed that 475 °C is the minimum temperature at which the soot trapped inside the filter is burned off (and, thus, the filter is regenerated) via the catalytic path. At this temperature, the catalytic filter maintains substantially the same performance over repeated cycles of soot loading and regeneration, indicating that the thermal stability of the ceria catalyst is preserved. This issue was also confirmed by the outcomes of filter characterization.

The intrinsic activity of ceria can be enhanced through doping with a proper active metal, thus further decreasing the filter regeneration temperature to values included in the operative range of diesel exhausts (200–350 °C).

The potential for continuous regeneration was also envisaged in Wagloehner et al. [103] from temperature-programmed oxidation (TPO) studies on a filter coated with a nanosized Mn_3O_4 catalyst originated from flame spray pyrolysis (FSP-Mn_3O_4). Figure 6.16 was obtained from the graphical abstract of this work: the major part of the soot trapped inside the filter (around 80%) is in weak contact with the catalyst causing oxidation above 400 °C only, whereas the minor fraction of soot is in intimate contact with the catalyst evoking conversion already between 180 and 350 °C.

It is worth saying that structured configurations alternative to ceramic wall-flow filters have also been proposed. For instance, Bruneel et al. [104] prepared and tested a three-layer metallic soot filter coated with a CeO_2 buffer layer (used as

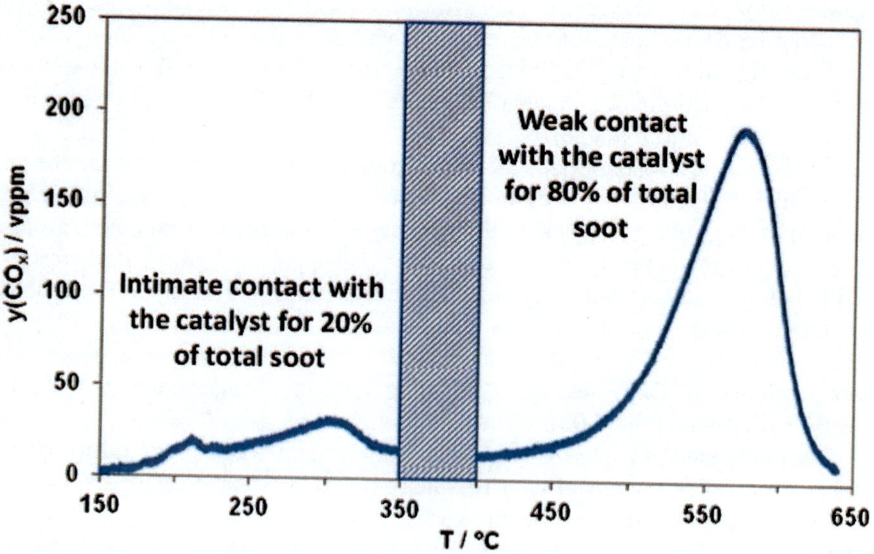

Fig. 6.16 TPO results obtained on a filter coated with a nanosized Mn_3O_4 catalyst originated from flame spray pyrolysis (FSP-Mn_3O_4) [103]

antioxidant) and a soot oxidation catalyst based on copper, molybdenum, and cerium oxides. More recently, Miró and coworkers developed novel structures based on ceramic papers [105, 106]. These are sheets of ceramic fibers (SiO_2-Al_2O_3) used as a support for the active phase. The best results were obtained with a mixed oxide containing Co, Ba, and K, showing a good dispersion onto the ceramic fibers [105].

6.9 Conclusions and Outlook

In this chapter, the main critical issue for regeneration of catalytic DPFs has been addressed, i.e., the efficiency of the contact in the solid-solid reaction between soot and catalyst, highlighting the close correlation of this issue with size and morphology of the catalyst particles.

The different types of soot-catalyst contact simulated at lab-scale have been analyzed in detail, ranging from the tight (or supertight) contact, which maximizes the contact points and provides information on the reaction mechanism, to the loose contact, which better reproduces the conditions of a real process of regeneration of catalytic DPF.

It has been highlighted that, even with a catalyst exhibiting an outstanding intrinsic activity, a poor contact with soot can strongly limit the performance. Indeed, when the soot-catalyst contact is weak, attempts to improve the oxygen mobility inside the catalyst lattice, thus increasing the soot oxidation rate, through suitable formulations and/or doping, may be in vain if the active oxygen is unable to reach the soot particle. On the other hand, an improved contact between the two solids allows enhancing all the phenomena occurring at the soot-catalyst interface. In this context, the nanosize of the catalyst particles plays a crucial role and, as such, represents a key factor for the production of highly performing soot oxidation catalysts.

Although the presence of nitrogen oxides in diesel exhaust gases can help to bridge the gap between loose and tight contact, it is essential to maximize the soot-catalyst contact inside the DPF, thus approaching tight conditions. In this regard, the increase of the number of contact points through the formation of special catalyst morphologies has been shown. A high exposure of active crystal planes of the catalyst particles provides better results than a high specific surface area related to a great catalyst porosity, since the soot particles cannot penetrate into the catalyst pores. The best results have been obtained when the size of the spaces created in the special catalyst morphology (for instance, a network of nanofibers) is of the same order of magnitude as the soot particle size.

As far as the catalyst deposition on the DPF is concerned, the importance of obtaining a high dispersion with a deep penetration of the catalyst nanoparticles into the pores of the filter walls has been highlighted. In addition, the critical issue of the amount of soot trapped inside the catalytic filter (i.e., the soot/catalyst ratio) has been addressed. The formation of a soot cake layer on top of the channel walls is not recommended, as it determines a poor contact with the catalyst located inside the pores of the walls. An almost totally catalytic regeneration of the DPF can more easily take place at lower temperatures if soot is mostly located inside the filter pores. If a reasonably low temperature of the catalytic process can be achieved, a continuous regeneration can be supposed: the newly filtered soot particles replace the quickly burning soot particles in the DPF pores, with an evident advantage also for the engine never working with a clogged filter.

The development of novel catalytic DPFs operating under continuous regeneration conditions is, thus, the most interesting challenge to be addressed in coming years. It involves the development of novel catalyst configurations showing improved contact conditions during filtration, a high activity at relatively low temperatures, and a high long-term stability.

References

1. P. Tandon, A. Heibel, J. Whitmore, N. Kekre, K. Chithapragada, Measurement and prediction of filtration efficiency evolution of soot loaded diesel particulate filters. Chem. Eng. Sci. **65**, 4751–4760 (2010)
2. K. Yang, J.T. Fox, R. Hunsicker, Characterizing diesel particulate filter failure during commercial fleet use due to pinholes, melting, cracking, and fouling. Emiss. Control Sci. Technol. **2**, 145–155 (2016)
3. V.R. Pérez, A. Bueno-López, Catalytic regeneration of diesel particulate filters: comparison of Pt and CePr active phases. Chem. Eng. J. **279**, 79–85 (2015)
4. V. Di Sarli, G. Landi, L. Lisi, A. Saliva, A. Di Benedetto, Catalytic diesel particulate filters with highly dispersed ceria: effect of the soot-catalyst contact on the regeneration performance. Appl. Catal. B Environ. **197**, 116–124 (2016)
5. L. Rocher, T. Seguelong, V. Harle, M. Lallemand, M. Pudlarz, M. Macduff, New generation fuel borne catalyst for reliable DPF operation in globally diverse fuels. SAE Technical Paper 2011-01-0297 (2011)
6. L. De Marzi, A. Monaco, J. De Lapuente, D. Ramos, M. Borras, M. Di Gioacchino, S. Santucci, A. Poma, Cytotoxicity and genotoxicity of ceria nanoparticles on different cell lines in vitro. Int. J. Mol. Sci. **14**, 3065–3077 (2013)
7. L.F. Nascimento, J.F. Lima, P.C. de Sousa Filho, O.A. Serra, Control of diesel particulate emission based on Ag/CeO_x/FeO_y catalysts supported on cordierite. Chem. Eng. J. **290**, 454–464 (2016)
8. V. Di Sarli, G. Landi, L. Lisi, A. Di Benedetto, Highly dispersed ceria for catalytic regeneration of diesel particulate filter. Adv. Sci. Lett. **23**, 5909–5911 (2017)
9. C. Su, Y. Wang, A. Kumar, P.J. McGinn, Simulating real world soot-catalyst contact conditions for lab-scale catalytic soot oxidation studies. Catalysts **8**, 247 (2018)
10. S. Cao, F.F. Tao, Y. Tang, Y. Li, J. Yu, Size- and shape-dependent catalytic performances of oxidation and reduction reactions on nanocatalysts. Chem. Soc. Rev. **45**, 4747–4765 (2016)
11. M. Valden, X. Lai, D.W. Goodman, Onset of catalytic activity of gold clusters on titania with the appearance of nonmetallic properties. Science **281**, 1647–1650 (1998)
12. F.Y. Kwong, A. Klapars, S.L. Buchwald, Copper-catalyzed coupling of alkylamines and aryl iodides: an efficient system even in an air atmosphere. Org. Lett. **4**, 581–584 (2002)
13. F.Y. Kwong, S.L. Buchwald, Mild and efficient copper-catalyzed amination of aryl bromides with primary alkylamines. Org. Lett. **5**, 793–796 (2003)
14. H. Rao, Y. Jin, H. Fu, Y. Jiang, Y. Zhao, A versatile and efficient ligand for copper-catalyzed formation of C-N, C-O, and P-C bonds: pyrrolidine-2-phosphonic acid phenyl monoester. Chem. Eur. J. **12**, 3636–3646 (2006)

15. B.R. Stanmore, J.F. Brilhac, P. Gilot, The oxidation of soot: a review of experiments, mechanisms and models. Carbon **39**, 2247–2268 (2001)
16. A. Bueno-López, Diesel soot combustion ceria catalysts. Appl. Catal. B Environ. **146**, 1–11 (2014)
17. L. Hensgen, K. Stöwe, Soot-catalyst contact studies in combustion processes using nano-scaled ceria as test material. Catal. Today **159**, 100–107 (2011)
18. P. Ciambelli, M. D'Amore, V. Palma, S. Vaccaro, Catalytic oxidation of an amorphous carbon black. Combust. Flame **99**, 413–421 (1994)
19. J.P.A. Neeft, M. Makkee, J.A. Moulijn, Catalytic oxidation of carbon black—I. Activity of catalysts and classification of oxidation profiles. Fuel **77**, 111–119 (1998)
20. S. Quiles-Díaz, J. Giménez-Mañogil, A. García-García, Catalytic performance of CuO/ $Ce_{0.8}Zr_{0.2}O_2$ loaded onto SiC-DPF in NO_x-assisted combustion of diesel soot. RSC Adv. **5**, 17018–17029 (2015)
21. J.M. Christensen, J.D. Grunwaldt, A.D. Jensen, Importance of the oxygen bond strength for catalytic activity in soot oxidation. Appl. Catal. B Environ. **188**, 235–244 (2016)
22. J. Liu, Z. Zhao, C.M. Xu, H. Wang, Study of the catalytic combustion of diesel soot over nanometric lanthanum-cobalt mixed oxide catalysts. React. Kinet. Catal. Lett. **87**, 107–114 (2005)
23. E. Aneggi, V. Rico-Perez, C. de Leitenburg, S. Maschio, L. Soler, J. Llorca, A. Trovarelli, Ceria-zirconia particles wrapped in a 2D carbon envelope: improved low-temperature oxygen transfer and oxidation activity. Angew. Chemie Int. Ed. **54**, 14040–14043 (2015)
24. E. Aneggi, J. Llorca, A. Trovarelli, M. Aouinea, P. Vernoux, In situ environmental HRTEM discloses low temperature carbon soot oxidation by ceria-zirconia at the nanoscale. Chem. Commun. **55**, 3876–3878 (2019)
25. S.B. Simonsen, S. Dahl, E. Johnson, S. Helveg, Ceria-catalyzed soot oxidation studied by environmental transmission electron microscopy. J. Catal. **255**, 1–5 (2008)
26. L. Soler, A. Casanovas, C. Escudero, V. Pérez-Dieste, E. Aneggi, A. Trovarelli, J. Llorca, Ambient pressure photoemission spectroscopy reveals the mechanism of carbon soot oxidation in ceria-based catalysts. ChemCatChem **8**, 2748–2751 (2016)
27. B. Bassou, N. Guilhaume, K. Lombaert, C. Mirodatos, D. Bianchi, Experimental microkinetic approach of the catalytic oxidation of diesel soot by ceria using temperature-programmed experiments. Part 1: impact and evolution of the ceria/soot contacts during soot oxidation. Energy Fuel **24**, 4766–4780 (2010)
28. B. Bassou, N. Guilhaume, K. Lombaert, C. Mirodatos, D. Bianchi, Experimental microkinetic approach of the catalytic oxidation of diesel soot by ceria using temperature-programmed experiments. Part 2: kinetic modeling of the impact of the ceria/soot contacts on the rate of oxidation. Energy Fuel **24**, 4781–4792 (2010)
29. M. Issa, C. Petit, A. Brillard, J.F. Brilhac, Oxidation of carbon by CeO_2: effect of the contact between carbon and catalyst particles. Fuel **87**, 740–750 (2008)
30. E. Saab, E. Abi-Aad, M.N. Bokova, E.A. Zhilinskaya, A. Aboukaïs, EPR characterisation of carbon black in loose and tight contact with Al_2O_3 and CeO_2 catalysts. Carbon **45**, 561–567 (2007)
31. M. Issa, C. Petit, H. Mahzoul, A. Aboukaïs, J.F. Brilhac, EPR and SEM characterizations of the contact between carbon black and cerium oxide. Top. Catal. **52**, 2063–2069 (2009)
32. P. Sudarsanam, B. Hillary, B. Mallesham, B.G. Rao, M.H. Amin, A. Nafady, A.M. Alsalme, B.M. Reddy, S.K. Bhargava, Designing CuO_x nanoparticle-decorated CeO_2 nanocubes for catalytic soot oxidation: role of the nanointerface in the catalytic performance of heterostructured nanomaterials. Langmuir **32**, 2208–2215 (2016)
33. B. Cui, S. Yan, Y. Xia, K. Li, S. Li, D. Wang, Y. Ye, Y.Q. Liu, $Cu_xCe_{1-x}O_2$ nanoflakes with improved catalytic activity and thermal stability for diesel soot combustion. Appl. Catal. A Gen. **578**, 20–29 (2019)

34. P. Sudarsanam, B. Hillary, M.H. Amin, N. Rockstroh, U. Bentrup, A. Brückner, S.K. Bhargava, Heterostructured copper-ceria and iron-ceria nanorods: role of morphology, redox, and acid properties in catalytic diesel soot combustion. Langmuir **34**, 2663–2673 (2018)
35. A. Bueno-López, K. Krishna, M. Makkee, J.A. Moulijn, Enhanced soot oxidation by lattice oxygen via La^{3+}-doped CeO_2. J. Catal. **230**, 237–248 (2005)
36. C.B. Lim, H. Kusaba, H. Einaga, Y. Teraoka, Catalytic performance of supported precious metal catalysts for the combustion of diesel particulate matter. Catal. Today **175**, 106–111 (2011)
37. J.P.A. Neeft, M. Makkee, J.A. Moulijn, Metal oxides as catalysts for the oxidation of soot. Chem. Eng. J. Biochem. Eng. J. **64**, 295–302 (1996)
38. D. Gardini, J.M. Christensen, C.D. Damsgaard, A.D. Jensen, J.B. Wagner, Visualizing the mobility of silver during catalytic soot oxidation. Appl. Catal. B Environ. **183**, 28–36 (2016)
39. K. Mori, K. Watanabe, T. Sato, H. Yamashita, Environmental transmission electron microscopy study of diesel carbon soot combustion under simulated catalytic-reaction conditions. ChemPhysChem **16**, 1347–1351 (2015)
40. E. Aneggi, J. Llorca, C. de Leitenburg, G. Dolcetti, A. Trovarelli, Soot combustion over silver-supported catalysts. Appl. Catal. B Environ. **91**, 489–498 (2009)
41. S. Liu, X. Wu, W. Liu, W. Chen, R. Ran, M. Li, D. Weng, Soot oxidation over CeO_2 and Ag/CeO_2: factors determining the catalyst activity and stability during reaction. J. Catal. **337**, 188–198 (2016)
42. J.H. Lee, S.H. Lee, J.W. Choung, C.H. Kim, K.Y. Lee, Ag-incorporated macroporous CeO_2 catalysts for soot oxidation: effects of Ag amount on the generation of active oxygen species. Appl. Catal. B Environ. **246**, 356–366 (2019)
43. L. Zeng, L. Cui, C. Wang, W. Guo, C. Gong, Ag-assisted CeO_2 catalyst for soot oxidation. Front. Mater. Sci. **13**, 288–295 (2019)
44. X. Deng, M. Li, J. Zhang, X. Hu, J. Zheng, N. Zhang, B.H. Chen, Constructing nano-structure on silver/ceria-zirconia towards highly active and stable catalyst for soot oxidation. Chem. Eng. J. **313**, 544–555 (2017)
45. G. Corro, A. Flores, F. Pacheco-Aguirre, U. Pal, F. Bañuelos, R. Araceli, A. Zehe, Biodiesel and fossil-fuel diesel soot oxidation activities of Ag/CeO_2 catalyst. Fuel **250**, 17–26 (2019)
46. D. Weng, J. Li, X. Wu, F. Lin, Promotional effect of potassium on soot oxidation activity and SO_2-poisoning resistance of Cu/CeO_2 catalyst. Catal. Commun. **9**, 1898–1901 (2008)
47. A. Alinezhadchamazketi, A.A. Khodadadi, Y. Mortazavi, A. Nemati, Catalytic evaluation of promoted CeO_2-ZrO_2 by transition, alkali, and alkaline-earth metal oxides for diesel soot oxidation. J. Environ. Sci. **25**, 2498–2506 (2013)
48. L. Sui, Y. Wang, H. Kang, H. Dong, L. Dong, L. Yu, Effect of Cs-Ce-Zr catalysts/soot contact conditions on diesel soot oxidation. ACS Omega **2**, 6984–6990 (2017)
49. H. Shimokawa, Y. Kurihara, H. Kusaba, H. Einaga, Y. Teraoka, Comparison of catalytic performance of Ag- and K-based catalysts for diesel soot combustion. Catal. Today **185**, 99–103 (2012)
50. M. Piumetti, S. Bensaid, N. Russo, D. Fino, Investigations into nanostructured ceria-zirconia catalysts for soot combustion. Appl. Catal. B Environ. **180**, 271–282 (2016)
51. V. Rico-Pérez, E. Aneggi, A. Trovarelli, The effect of Sr addition in Cu- and Fe-modified CeO_2 and ZrO_2 soot combustion catalysts. Catalysts **7**, 28 (2017)
52. V. Rico-Pérez, E. Aneggi, A. Bueno-López, A. Trovarelli, Synergic effect of Cu/$Ce_{0.5}Pr_{0.5}O_{2-\delta}$ and $Ce_{0.5}Pr_{0.5}O_{2-\delta}$ in soot combustion. Appl. Catal. B Environ. **197**, 95–104 (2016)
53. S.J. Castillo Marcano, S. Bensaid, F.A. Deorsola, N. Russo, D. Fino, Multifunctional catalyst based on BaO/Pt/CeO_2 for NO_2-assisted soot abatement and NO_x storage. Fuel **149**, 78–84 (2015)
54. Y. Gao, X. Wu, S. Liu, D. Weng, R. Ran, MnO_x-CeO_2 mixed oxides for diesel soot oxidation: a review. Catal. Surv. Jpn. **22**, 230–240 (2018)
55. D. Jampaiah, V.K. Velisoju, D. Devaiah, M. Singh, E.L.H. Mayes, V.E. Coyle, B.M. Reddy, V. Bansal, S.K. Bhargava, Flower-like Mn_3O_4/CeO_2 microspheres as an efficient catalyst for

diesel soot and CO oxidation: synergistic effects for enhanced catalytic performance. Appl. Surf. Sci. **473**, 209–221 (2019)
56. E. Sartoretti, C. Novara, F. Giorgis, M. Piumetti, S. Bensaid, N. Russo, D. Fino, In situ Raman analyses of the soot oxidation reaction over nanostructured ceria-based catalysts. Sci. Rep. **9**, 3875 (2019)
57. S. Wagloehner, S. Kureti, Study on the mechanism of the oxidation of soot on Fe_2O_3 catalyst. Appl. Catal. B Environ. **125**, 158–165 (2012)
58. Y. Teraoka, K. Nakano, S. Kagawa, W.F. Shangguan, Simultaneous removal of nitrogen oxides and diesel soot particulates catalyzed by perovskite-type oxides. Appl. Catal. B Environ. **5**, L181–L185 (1995)
59. A. Mishra, R. Prasad, Preparation and application of perovskite catalysts for diesel soot emissions control: an overview. Catal. Rev. **56**, 57–81 (2014)
60. S. Bensaid, G.A. Blengini, D. Fino, N. Russo, Diesel soot combustion with perovskite catalysts. Chem. Eng. Commun. **201**, 1327–1339 (2014)
61. A. Civera, M. Pavese, G. Saracco, V. Specchia, Combustion synthesis of perovskite-type catalysts for natural gas combustion. Catal. Today **83**, 199–211 (2003)
62. M. Haneda, A. Towata, Catalytic performance of supported Ag nano-particles prepared by liquid phase chemical reduction for soot oxidation. Catal. Today **242**, 351–356 (2015)
63. A. Serve, A. Boreave, B. Cartoixa, K. Pajot, P. Vernoux, Synergy between Ag nanoparticles and yttria-stabilized zirconia for soot oxidation. Appl. Catal. B Environ. **242**, 140–149 (2019)
64. E. Obeid, L. Lizarraga, M.N. Tsampas, A. Cordier, A. Boréave, M.C. Steil, G. Blanchard, K. Pajot, P. Vernoux, Continuously regenerating diesel particulate filters based on ionically conducting ceramics. J. Catal. **309**, 87–96 (2014)
65. E. Obeid, M.N. Tsampas, S. Jonet, A. Boréave, L. Burel, M.C. Steil, G. Blanchard, K. Pajot, P. Vernoux, Isothermal catalytic oxidation of diesel soot on yttria-stabilized zirconia. Solid State Ionics **262**, 253–256 (2014)
66. L. Castoldi, E. Aneggi, R. Matarrese, R. Bonzi, A. Trovarelli, L. Lietti, Simultaneous removal of soot and NO_x over silver and ruthenium-based catalysts. Top. Catal. **60**, 209–213 (2017)
67. L. Dou, T. Fan, H. Zhang, A novel 3D oxide nanosheet array catalyst derived from hierarchical structured array-like CoMgAl-LDH/graphene nanohybrid for highly efficient NO_x capture and catalytic soot combustion. Catal. Sci. Technol. **5**, 5153–5167 (2015)
68. P. Legutko, T. Jakubek, W. Kaspera, P. Stelmachowski, Z. Sojka, A. Kotarba, Strong enhancement of desoot activity of transition metal oxides by alkali doping: additive effects of potassium and nitric oxide. Top. Catal. **60**, 162–170 (2017)
69. G. Mul, J.P.A. Neeft, F. Kapteijn, M. Makkee, J.A. Moulijn, Soot oxidation catalyzed by a Cu/K/Mo/Cl catalyst: evaluation of the chemistry and performance of the catalyst. Appl. Catal. B Environ. **6**, 339–352 (1995)
70. A. Carrascull, I.D. Lick, E.N. Ponzi, M.I. Ponzi, Catalytic combustion of soot with a O_2/NO mixture. KNO_3/ZrO_2 catalysts. Catal. Commun. **4**, 124–128 (2003)
71. L. Sui, L. Yu, Y. Zhang, Catalytic combustion of diesel soot on Co-Sr-K catalysts. Energy Fuel **21**, 1420–1424 (2007)
72. H. An, P.J. McGinn, Catalytic behavior of potassium containing compounds for diesel soot combustion. Appl. Catal. B Environ. **62**, 46–56 (2006)
73. D. Courcot, C. Pruvost, E.A. Zhilinskaya, A. Aboukaïs, Potential of supported copper and potassium oxide catalysts in the combustion of carbonaceous particles. Kinet. Catal. **45**, 580–588 (2004)
74. H. Laversin, D. Courcot, E.A. Zhilinskaya, R. Cousin, A. Aboukaïs, Study of active species of Cu-K/ZrO_2 catalysts involved in the oxidation of soot. J. Catal. **241**, 456–464 (2006)
75. N.F. Galdeano, A.L. Carrascull, M.I. Ponzi, I.D. Lick, E.N. Ponzi, Catalytic combustion of particulate matter: catalysts of alkaline nitrates supported on hydrous zirconium. Thermochim. Acta **421**, 117–121 (2004)

76. W. Kaspera, P. Indyka, Z. Sojka, A. Kotarba, Bridging the gap between tight and loose contacts for soot oxidation by vanadium doping of cryptomelane nanorods catalyst using NO_2 as an oxygen carrier. Catal. Sci. Technol. **8**, 3183–3192 (2018)
77. T. Jakubek, K. Ralphs, A. Kotarba, H. Manyar, Nanostructured potassium-manganese oxides decorated with Pd nanoparticles as efficient catalysts for low-temperature soot oxidation. Catal. Lett. **149**, 100–106 (2019)
78. H. Zhang, S. Yuan, J.L. Wang, M. Gong, Y. Chen, Effects of contact model and NO_x on soot oxidation activity over Pt/MnO_x-CeO_2 and the reaction mechanisms. Chem. Eng. J. **327**, 1066–1076 (2017)
79. Q. Shen, G. Lu, C. Du, Y. Guo, Y. Wang, Y. Guo, X. Gong, Role and reduction of NO_x in the catalytic combustion of soot over iron-ceria mixed oxide catalyst. Chem. Eng. J. **218**, 164–172 (2013)
80. J.M. Christensen, J.D. Grunwaldt, A.D. Jensen, Effect of NO_2 and water on the catalytic oxidation of soot. Appl. Catal. B Environ. **205**, 182–188 (2017)
81. N. Zouaoui, M. Issa, D. Kehrli, M. Jeguirim, CeO_2 catalytic activity for soot oxidation under NO/O_2 in loose and tight contact. Catal. Today **189**, 65–69 (2012)
82. P. Miceli, S. Bensaid, N. Russo, D. Fino, CeO_2-based catalysts with engineered morphologies for soot oxidation to enhance soot-catalyst contact. Nanoscale Res. Lett. **9**, 254 (2014)
83. E. Aneggi, C. de Leitenburg, A. Trovarelli, On the role of lattice/surface oxygen in ceria-zirconia catalysts for diesel soot combustion. Catal. Today **181**, 108–115 (2012)
84. E. Aneggi, C. de Leitenburg, J. Llorca, A. Trovarelli, Higher activity of diesel soot oxidation over polycrystalline ceria and ceria-zirconia solid solutions from more reactive surface planes. Catal. Today **197**, 119–126 (2012)
85. P.A. Kumar, M.D. Tanwar, S. Bensaid, N. Russo, D. Fino, Soot combustion improvement in diesel particulate filters catalyzed with ceria nanofibers. Chem. Eng. J. **207–208**, 258–266 (2012)
86. S. Bensaid, N. Russo, D. Fino, CeO_2 catalysts with fibrous morphology for soot oxidation: the importance of the soot-catalyst contact conditions. Catal. Today **216**, 57–63 (2013)
87. P. Miceli, S. Bensaid, N. Russo, D. Fino, Effect of the morphological and surface properties of CeO_2-based catalysts on the soot oxidation activity. Chem. Eng. J. **278**, 190–198 (2015)
88. W. Zhang, X. Niu, L. Chen, F. Yuan, Y. Zhu, Soot combustion over nanostructured ceria with different morphologies. Sci. Rep. **6**, 29062 (2016)
89. M. Piumetti, S. Bensaid, N. Russo, D. Fino, Nanostructured ceria-based catalysts for soot combustion: investigations on the surface sensitivity. Appl. Catal. B Environ. **165**, 742–751 (2015)
90. E. Aneggi, D. Wiater, C. de Leitenburg, J. Llorca, A. Trovarelli, Shape-dependent activity of ceria in soot combustion. ACS Catal. **4**, 172–181 (2014)
91. Y. Cheng, J. Liu, Z. Zhao, Y. Wei, Y. Song, C. Xu, The simultaneous purification of PM and NO_x in diesel engine exhausts over a single 3DOM $Ce_{0.9-x}Fe_{0.1}Zr_xO_2$ catalyst. Environ. Sci. Nano **4**, 1168–1177 (2017)
92. K. Hinot, H. Burtscher, A.P. Weber, G. Kasper, The effect of the contact between platinum and soot particles on the catalytic oxidation of soot deposits on a diesel particle filter. Appl. Catal. B Environ. **71**, 271–278 (2007)
93. D. Fino, N. Russo, E. Cauda, G. Saracco, V. Specchia, La-Li-Cr perovskite catalysts for diesel particulate combustion. Catal. Today **114**, 31–39 (2006)
94. N. Russo, D. Fino, G. Saracco, V. Specchia, Promotion effect of Au on perovskite catalysts for the regeneration of diesel particulate filters. Catal. Today **137**, 306–311 (2008)
95. J. Caroca, G. Villata, D. Fino, N. Russo, Comparison of different diesel particulate filters. Top. Catal. **52**, 2076–2082 (2009)
96. L. Tang, Z. Zhao, K. Li, X. Yu, Y. Wei, J. Liu, Y. Peng, Y. Li, Y. Chen, Highly active monolith catalysts of $LaKCoO_3$ perovskite-type complex oxide on alumina-washcoated diesel particulate filter and the catalytic performances for the combustion of soot. Catal. Today **339**, 159–173 (2020)

97. Q. Zhou, K. Zhong, W. Fu, Q. Huang, Z. Wang, B. Nie, Nanostructured platinum catalyst coating on diesel particulate filter with a low-cost electroless deposition approach. Chem. Eng. J. **270**, 320–326 (2015)
98. V. Di Sarli, G. Landi, L. Lisi, A. Di Benedetto, Ceria-coated diesel particulate filters for continuous regeneration. AICHE J. **63**, 3442–3449 (2017)
99. V. Di Sarli, A. Di Benedetto, Modeling and simulation of soot combustion dynamics in a catalytic diesel particulate filter. Chem. Eng. Sci. **137**, 69–78 (2015)
100. V. Di Sarli, A. Di Benedetto, Operating map for regeneration of a catalytic diesel particulate filter. Ind. Eng. Chem. Res. **55**, 11052–11061 (2016)
101. V. Di Sarli, A. Di Benedetto, Combined effects of soot load and catalyst activity on the regeneration dynamics of catalytic diesel particulate filters. AICHE J. **64**, 1714–1722 (2018)
102. V. Di Sarli, A. Di Benedetto, Using CFD simulation as a tool to identify optimal operating conditions for regeneration of a catalytic diesel particulate filter. Appl. Sci. (Basel) **9**, 3453 (2019)
103. S. Wagloehner, M. Nitzer-Noski, S. Kureti, Oxidation of soot on manganese oxide catalysts. Chem. Eng. J. **259**, 492–504 (2015)
104. E. Bruneel, J. Van Brabant, M.T. Le, I. Van Driessche, Deposition of a Cu/Mo/Ce catalyst for diesel soot oxidation on a sintered metal fiber filter with a CeO_2 anti corrosion coating. Catal. Commun. **25**, 111–117 (2012)
105. S.A. Leonardi, F.E. Tuler, E.M. Gaigneaux, D.P. Debecker, E.E. Miró, V.G. Milt, Novel ceramic paper structures for diesel exhaust purification. Environ. Sci. Pollut. Res. **25**, 35276–35286 (2018)
106. N.A. Sacco, E.D. Banús, J.P. Bortolozzi, V.G. Milt, E.E. Miró, Ultrasound-assisted deposition of Co-CeO_2 onto ceramic microfibers to conform catalytic papers: their application in engine exhaust treatment. ACS Omega **3**, 18334–18342 (2018)

Chapter 7
Phosphorous-Based Titania Nanoparticles for the Photocatalytic Abatement of VOCs

Melodj Dosa, Marco Piumetti, Samir Bensaid, and Nunzio Russo

7.1 Introduction

A compound can be classified as Volatile Organic Compound (VOC) when it has a vapor pressure greater than 0.01 kPa at 20 °C [1]. The VOCs can evaporate at room temperature and they are carbon-based substances [2–4]. There are several compounds that are classified as VOCs, with specific physico-chemical and thermodynamics properties [5]. Alkanes, alcohols, halogenated hydrocarbons, aldehydes, ketones, aromatics, paraffins, olefins, and sulfur-based compounds are proper examples of VOCs [6–10].

As is known, VOCs are listed as the major contributors to air pollution. They are toxic substances for both environmental and human health. In the specific, they are responsible of the stratospheric ozone depletion and precursors of ground level smog [11–16].

The VOCs can be emitted by outdoor and indoor sources. Outdoor sources include those produced from industries (chemical and textile) and automotive field. On the other hand, household supplies (paints) and construction materials can be classified as indoor emission sources [10, 17–20].

In order to reduce the VOCs emissions, several technologies were proposed [21–23]. They can be divided into two groups: recovery or destruction. In the former group can be listed absorption, adsorption, biofiltration, condensation or separation by membrane. On the other hand, destruction technologies are characterized by processes able to convert VOCs [5].

Among the latter techniques, the most promising is the photocatalytic decomposition, namely the process in which a pollutant can be decomposed by a photocatalytic material (photocatalyst) in the presence of a light source [24–27].

M. Dosa · M. Piumetti (✉) · S. Bensaid · N. Russo
Department of Applied Science and Technology, Politecnico di Torino, Turin, Italy
e-mail: marco.piumetti@polito.it

M. Piumetti, S. Bensaid (eds.), *Nanostructured Catalysts for Environmental Applications*, https://doi.org/10.1007/978-3-030-58934-9_7

However, the photocatalytic phenomenon is characterized by a complex reaction route. The following parameters are the most important during the photocatalytic reaction: (1) photon absorption, (2) charge transport and trapping, (3) charge transfer, (4) molecular absorption, (5) reaction mechanism, (6) poisons and promoters, (7) surface and material structure [28].

1. **Photon absorption.** This phase is considered as a subsurface step. However, there are two major reasons that can influence what happen on the photocatalyst surface. If in the bulk there are lattice modification (i.e., truncations), on the surface, there are unsaturation that can create different mechanisms of charge transfer and, consequently, diverse process of photon absorption. As a second reason, the surface proton absorption can be influenced by the particles' dimensions: the lower nominal particles radius, the higher surface exposed, the higher optical absorption capacity.
2. **Charge transport and trapping.** The charge originated from the absorption in the photocatalyst subsurface region must reach the surface and be stabilized. Thus, the recombination phenomena with the electron/hole pair must be avoided. This is possible only if the photocatalyst structure is optimized (no recombination phenomena and easy charge transfer from the subsurface to the surface).
3. **Charge transfer.** This step is the core of the photocatalytic reaction. It is an interfacial charge exchange between the photocatalyst and the physi- or chemisorbed molecule. In the specific, the photocatalyst is the donor while the molecule on the surface is the acceptor. The charge transfer can be activated only by a light source. In general, a photocatalyst needs energy to promote its electron from the valence band (VB) to the conduction band (CB). During this electron promotion, holes are originated in the valence band. In these holes, OH groups or water molecules can react and originate $OH^{\bullet}$ radicals able to oxidize molecules in heterogeneous photocatalysis [29]. This mechanism can be activated by wavelength lower than 380 nm and this is a limitation in the possible utilization of UV-Visible light from the sun.
4. **Molecular absorption.** The photocatalytic phenomenon is influenced by the absorption state of the molecule. This aspect can influence both redox properties and electronic structure of the substance absorbate. Thus, as the molecule is absorbed on the surface, it can be more or less easily converted depending on how it is bonded with the photocatalyst surface.
5. **Reaction mechanism.** This step is complex, and it depends on the aspects analyzed into previous points.
6. **Promoters and poisons.** There are elements that can influence positively or negatively the photocatalytic phenomenon. In the specific, they can modify the reaction mechanism, block the sites, or change the energy (i.e., modification on the band gap energy). For example, the promoters can change the photoabsorption mechanism and they can also boost the charge transfer and the trapping phenomenon. Among the promoter there are platinum, silver, gold, palladium, copper, and rhodium. On the other hand, some elements create surface modifications (in terms of physico-chemical properties) that decrease the

reactivity during the photocatalytic reaction. In the latter case, they are considered as poisons.

7. **Surface and material structure.** The surface properties and structure of the photocatalyst can influence the photocatalytic phenomenon. The optimization of the synthesis procedure of the photocatalyst can arises the physico-chemical and electronic properties of the material. Therefore, the photocatalytic reaction is better performed.

A well-known photocatalyst used for the decomposition of VOCs under UV light is TiO_2. In the photocatalytic reaction, the electrons in the TiO_2 structure are promoted to the anatase conduction band (CB) and the corresponding holes are created in the valence band (VB). During this phenomenon, the $^{\bullet}OH$ and $^{\bullet}O^{2-}$ species are produced, according to the following Eqs. (7.1)–(7.4) [23]:

$$TiO_2 + h\nu \rightarrow h^+(VB) + e^-(CB) \tag{7.1}$$

$$H_2O + h^+(VB) \rightarrow OH^- + H^+ \tag{7.2}$$

$$h^+(VB) + OH^- \rightarrow \bullet OH \tag{7.3}$$

$$O_2 + e^-(CB) \rightarrow \bullet O^{2-} \tag{7.4}$$

These radical species additionally react with VOC molecules.

In this work, we considered ethylene as a model molecule of VOCs. In the specific, the ethylene is decomposed following Eqs. (7.5–7.8) to produce CO_2 and water vapor:

$$\bullet OH + C_2H_4 \rightarrow (C_2H_4)* \tag{7.5}$$

$$(C_2H_4OH)* + \bullet O_2 \rightarrow CO_2 + H_2O \tag{7.6}$$

$$Ti(IV) + e^- \rightarrow Ti(III) \tag{7.7}$$

$$Ti(III) + O_2 \rightarrow Ti(IV) + O^{2-} \tag{7.8}$$

During the last decades, it has been considered the possibility to introduce elements into the TiO_2 framework to improve whole activity with a special focus on the band gap energy. In fact, it has been observed that the presence of some elements, such as phosphorus, zirconium, nitrogen or iron [30–33], may render the TiO_2 active in the visible region.

In the present work, a set of doped TiO_2 systems, having different P-contents (0.6, 0.7, and 3 at.%), was studied. The samples were herein labeled as TiO_2-$P_{0.6}$, TiO_2-$P_{0.7}$, and TiO_2-P_3. For comparison purposes, a pure titania sample, labeled as "TiO_2" was synthesized.

The physico-chemical properties of the samples were investigated by complementary technique, such as X-ray Diffraction (XRD), N_2 physisorption at −196 °C, Field Emission Scanning Electron Microscopy (FESEM), Energy Dispersive X-ray (EDX) analysis, X-ray Photoelectron Spectroscopy (XPS), and (DR)UV-Vis spectroscopy. Then, the catalysts were tested for the total oxidation of ethylene under two different light sources (UVB and UVA radiations).

7.2 Experimental Part

7.2.1 Catalysts Preparation

7.2.1.1 Pure TiO_2

The titania-based catalysts were prepared using the sol-gel method [34, 35] as resumed in Fig. 7.1. Briefly, the pure titania sample, TiO_2, was prepared as follows:

Sixty milliliters of acetic acid (20 vol%, Sigma-Aldrich) was dropwised into 10 g of titanium (IV) butoxide. They were stirred together at room temperature for 4 h. Meanwhile, 6 g of Pluronic 123 (Sigma-Aldrich) was soften into 40 mL of ethanol (Sigma-Aldrich). Then, they were stirred at room temperature for 4 h.

After that, the solution containing Pluronic and ethanol was added to the first one (acetic acid and titanium butoxide) and stirred at room temperature for 24 h.

The obtained gel was placed in a Teflon autoclave and then heated in oven for 48 h at 85 °C. The solid was separated and dried at 80 °C. Finally, the dried powder was calcined at 450 °C for 4 h (heating rate of 1.8 °C min^{-1}).

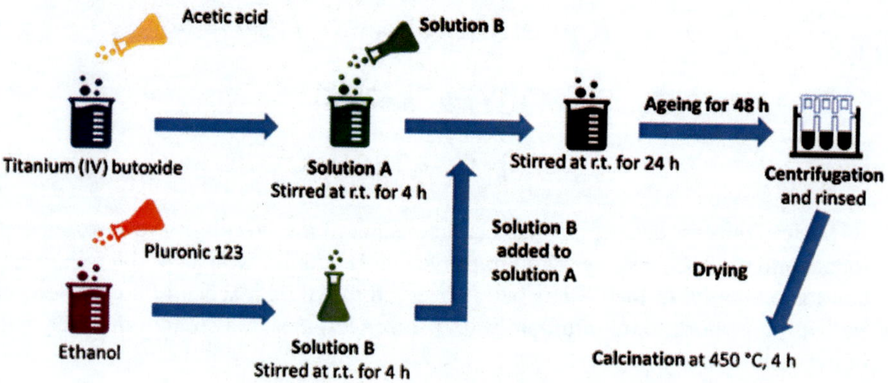

Fig. 7.1 Scheme of the TiO_2 synthesis procedure

7.2.1.2 Phosphorous-Doped TiO_2

Three phosphorous-doped titania samples were synthesized using the procedure described above (Sect. 7.2.1.1) slightly modified. In fact, during the preparation of the solution with acetic acid and titanium butoxide, different amounts of H_3PO_4 (85 wt%, Sigma-Aldrich) were added in order to have the total phosphorus content about 0.6, 0.7, and 3 at.%. The samples were labeled as TiO_2-$P_{0.6}$, TiO_2-$P_{0.7}$, and TiO_2-P_3.

7.2.2 *Catalysts Characterization Techniques*

The powder X-ray diffraction patterns were collected on an X'Pert Philips PW3040 diffractometer using Cu Ka radiation (2θ range = 10–70, step = 0.01θ, time per step = 0.2 s). The diffraction peaks were indexed according to the Powder Data File database (PDF-21999, International Centre of Diffraction Data). The crystallite size was calculated by Scherrer's equation (7.9):

$$D_c = \frac{K\lambda}{\beta \cos\theta} \tag{7.9}$$

where K is a constant (= 0.9), λ is the Cu Kα radiation (= 1.5405 Å), β is the full width at half maximum (FWHM) of the diffraction peak, and θ is the position of the peak.

The N_2 physisorption at −196 °C, performed using a Micromeritics Tristar II 3020 (v1.03, Micromeritics Instrument Corp.), was used to evaluate the total pore volume (V_p), the pore diameter (D_p), and the specific surface area (S_{BET}, calculated using the Brunauer–Emmett–Teller method) on samples previously outgassed at 200 °C for 4 h. The pore volume and pore diameter were estimated by the Barrett-Joyner Halenda (BJH) method, during the desorption phase.

The morphology of the samples was investigated using a field emission scanning electron microscopy (FESEM Zeiss MERLIN, Gemini-II column). Elemental analysis was carried out via energy dispersive X-ray (EDX) analysis (AZTec, Oxford Instruments).

The surface composition of the samples was analyzed via X-ray photoelectron spectroscopy (XPS), performed in a PHI Versa probe apparatus using a band-pass energy of 187.85 eV, a 45° take-off angle, and 100 μm diameter X-ray spot size.

The powder samples were analyzed via the (DR)UV-Vis spectroscopy. A UV-Vis double beam spectrophotometer (Varian Cary 500, Varian) was used. The spectra were collected in 200–450 nm regions, with a resolution of 2 nm.

7.2.3 *Photocatalytic Tests*

About 0.5 g of catalyst powder was well-spread inside a Pyrex reactor. The reactor was connected to mass flow controllers (Bronkhorst) that continuously feed, during the reaction, a mixture of 500 ppmv of ethylene and 10 vol.% of O_2 (volume balanced with N_2). The outflow from the reactor was analyzed using a nondispersive infrared analyzer (NDIR, Hartmann-Braun) and a gas chromatograph (GC, Varian CP-3800, ShinCarbon ST column, FID). The light sources used have two different wavelengths:

- UVB: wavelength = 312 nm, intensity = 12 W m^{-2}, 230 V, 50 Hz, 0.17 A MONTEPAONE SRL.
- UVA: wavelength = 365 nm, intensity = 8 W m^{-2}, 230 V, 50 Hz, 0.17 A MONTEPAONE SRL.

For each test, the corresponding lamp was located on the top of the reactor. A dark phase was performed before the test, in order to saturate the solid. The lamp was switched on when the system reached a steady-state condition (stable GC peak intensity), and the test ran for a time-on-stream (TOS) of 160 min.

7.3 Results and Discussion

7.3.1 *N_2 Physisorption at −196 °C and XRD*

Table 7.1 summarizes the textural properties as derived from the N_2 physisorption at −196 °C. All the synthesized samples were compared to P-25 (by Degussa), a commercial titania.

Noteworthy, all the synthesized samples exhibit a higher surface area compared to the commercial sample (P-25). Moreover, the doped samples show higher surface areas (TiO_2-$P_{0.6}$ = 144 $m^2\ g^{-1}$; TiO_2-$P_{0.7}$ = 119 $m^2\ g^{-1}$; TiO_2-P_3 = 168 $m^2\ g^{-1}$) and pore volume (TiO_2-$P_{0.6}$ = 0.50 $cm^3\ g^{-1}$; TiO_2-$P_{0.7}$ = 0.29 $cm^3\ g^{-1}$; TiO_2-P_3 = 0.58 $cm^3\ g^{-1}$) than the pure TiO_2 (S_{BET} = 91 $m^2\ g^{-1}$; V_p = 0.21 $cm^3\ g^{-1}$). This suggests that the

Table 7.1 Results as derived from the N_2 physisorption at −196 °C and XRD analysis

Samples	S_{BET}[a] ($m^2\ g^{-1}$)	V_p[b] ($cm^3\ g^{-1}$)	D_p[b] (nm)	D_c[c] (nm)
P-25	50	0.19	13	30
TiO_2	91	0.21	7	12
TiO_2-$P_{0.6}$	144	0.50	11	11
TiO_2-$P_{0.7}$	119	0.29	7	10
TiO_2-P_3	168	0.58	12	8

[a]Specific surface area calculated by the BET method
[b]Pore volume and pore diameter estimated by the BJH method, during the desorption phase
[c]Crystallite size calculated by the Scherrer's equation

incorporation of P into the TiO_2 framework promotes the textural and structural properties. The crystallite sizes of the samples follow the increasing order:

$$TiO_2 - P_3 (= 8\,nm) < TiO_2 - P_{0.7} < TiO_2 - P_{0.6} (= 11\,nm).$$

Then, it appears that particles with smaller D_c have more surface defects (i.e., edges and corners) compared to those with higher D_c values [36]. By a theoretical point of view, this effect can be explained considering the volume and area for a single particle. The volume can be expressed by the following equation (Eq. 7.10):

$$V = \frac{4}{3}\pi r^3 \tag{7.10}$$

On the other hand, the particle area can be expressed as follows (Eq. 7.11):

$$A = 4\pi r^2 \tag{7.11}$$

Then, the A-to-V ratio can be expressed by the following equation (Eq. 7.12):

$$\frac{A}{V} = \frac{3}{r} \tag{7.12}$$

Thus, if the particle radius decreases, the exposed surface area of the particles is higher and, therefore, higher probability to have surface defects for a proper amount (weight) of catalyst.

Since for smaller particles, there is a higher possibility to have more surface defects, there are more unsaturation that are able to create affinity between catalyst surface and the probe molecule (ethylene) [37].

Figure 7.2 shows the XRD diffractograms of the synthesized samples along with P-25 (by Degussa). All the prepared samples provide evidence of the anatase phase. In fact, the transition phase from anatase to rutile typically occurs at temperature above 500 °C [38]. On the other hand, the P-25 also exhibits the presence of rutile phase. The estimated ratio between the two TiO_2 phases for the latter sample is 70/30 (anatase/rutile) [39]. According to the literature, the anatase phase is photocatalytically more reactive than the rutile one due to the electron migration phenomenon that is faster compared to the same occurring in the rutile phase. In addition, the anatase has better photocatalytic activity due to its Fermi level that is higher by about 0.1 eV comparing with that of rutile [40]. Thanks to this better quality, the electron-hole recombination rate is very low [41].

In this work, the purpose was to obtain samples with only anatase phase (confirmed by XRD results, as shown in Fig. 7.2).

However, as known by the literature, if rutile and anatase are in strict contact, a synergistic effect may emerge. The electrons produced into rutile are promoted to anatase covalent band and this phenomenon reduces the recombination rate into the rutile phase [41–43]. In fact, the charge separation process can be achieved thanks

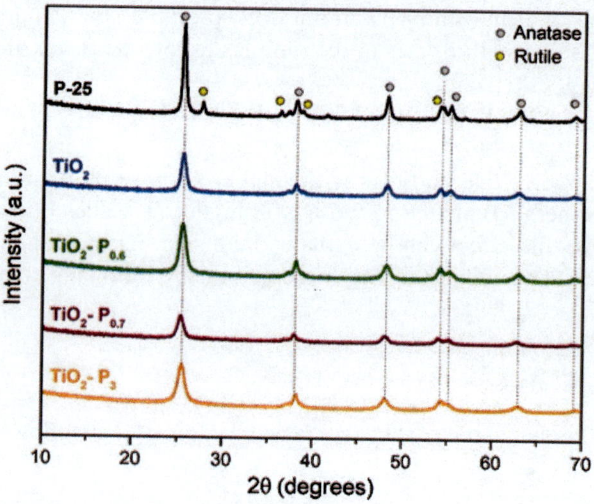

Fig. 7.2 XRD patterns of the synthesized samples along with commercial P-25 (by Degussa)

to this electron transfer to the rutile/anatase interface that is at lower energy state. The recombination phenomena decrease and electrons can easily move to TiO_2 conduction band [44]. As a whole, the presence of both anatase and rutile is beneficial in terms of photocatalytic activity due to the following issues [40]:

1. With the presence of both rutile and anatase phases, it is possible to extend the photocatalytic activity of titania at higher wavelength.
2. The charge separation from rutile to anatase is stabilized by the electron transfer and avoids the recombination phenomena of electron/hole pairs.
3. Rutile particles are characterized by smaller crystallite size, compared to anatase ones. This characteristic enhances the charge transfer because there are preferential points at the rutile/anatase interface (originated by the smaller rutile crystallites in close contact with the anatase phase) in which the photocatalytic phenomenon occurs.

7.3.2 *FESEM and EDX Analysis*

In Fig. 7.3 are reported the FESEM images for the prepared samples. As a whole, it appears that the prepared photocatalysts exhibit smaller particle sizes compared to the commercial material P-25.

To better investigate the particle sizes, the average diameters for each sample were analyzed by the ImageJ software [45]. The particle size distributions are reported in Fig. 7.4. The average diameter of the particles for the prepared samples

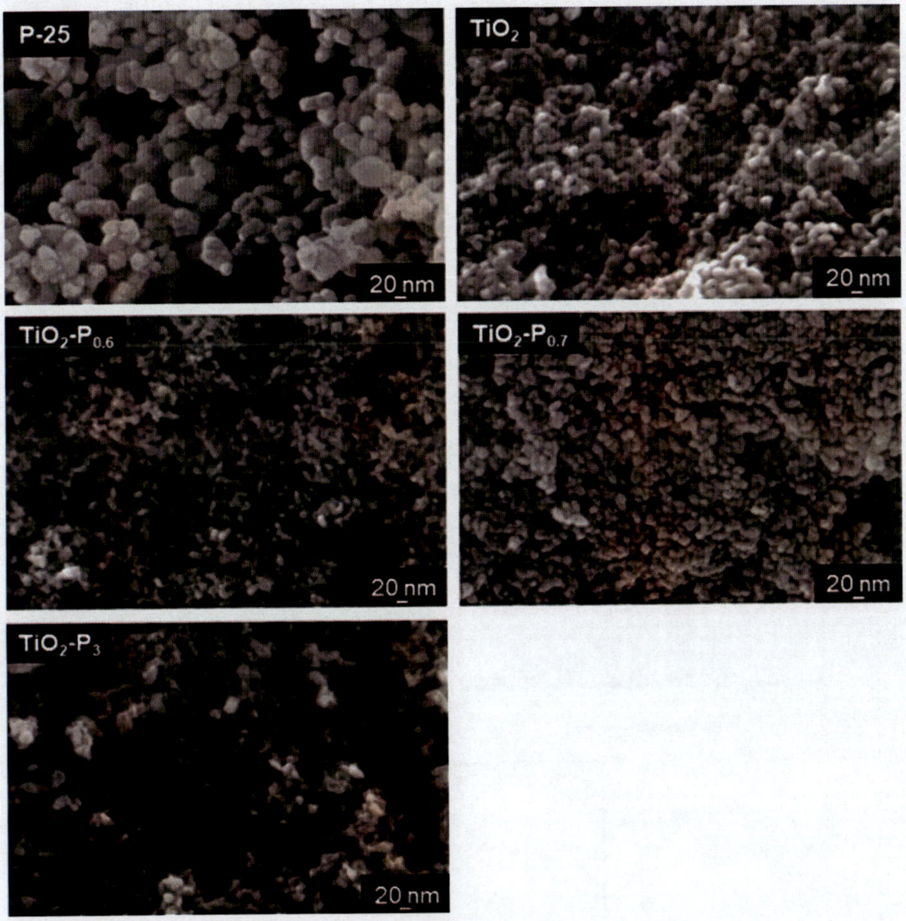

Fig. 7.3 FESEM images of the prepared samples along with the commercial P-25 (by Degussa)

is lower than 15 nm. In particular, the TiO_2-$P_{0.6}$ sample exhibits the lowest average particle diameter (about 9 nm) among the prepared photocatalysts. Moreover, these results are in agreement with the crystallite size: the higher the P-content, the smaller the crystallite, and the particle sizes.

The EDX analysis was performed over three different areas, in order to evaluate the P-content into doped-titania samples. The results are reported in Table 7.2. For all the samples, the average phosphorus content is close to the theoretical amount used during the synthesis procedure.

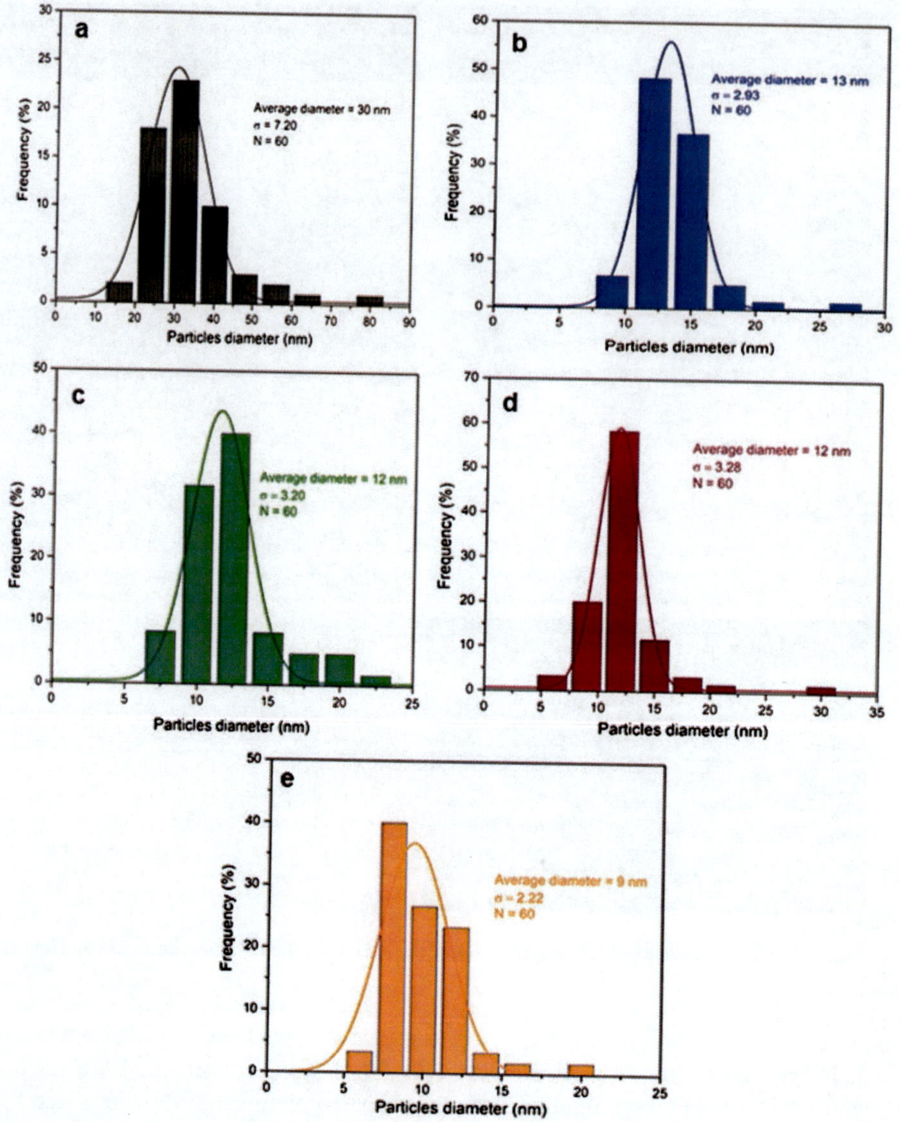

Fig. 7.4 Particles distribution analysis over (**a**) P-25, (**b**) TiO_2, (**c**) TiO_2-$P_{0.6}$, (**d**) TiO_2-$P_{0.7}$, and (**e**) TiO_2-P_3

7.3.3 XPS Analysis

Table 7.3 reports the XPS analysis of the prepared samples, as derived from the deconvolution peaks. As a whole, the synthesized catalysts have similar Ti and O contents (Ti/O ~ 1/3). This result is lower than the theoretical value of TiO_2 that is

Table 7.2 EDX analysis over the doped-titania samples investigated over three different areas

Elements	TiO_2-$P_{0.6}$	TiO_2-$P_{0.7}$	TiO_2-P_3
Ti	35.87	31.10	30.65
O	63.55	68.28	66.39
P	0.58	0.62	2.96
Tot.	100	100	100

Table 7.3 Elemental composition over the prepared samples derived from XPS analysis.

Catalysts	Ti (at.%)	O (at.%)	P (at.%)
TiO_2	26.8	73.2	–
TiO_2-$P_{0.6}$	26.9	71.6	1.5
TiO_2-$P_{0.7}$	27.0	71.6	1.4
TiO_2-P_3	22.7	73.2	4.1

about 1/2. As previously evidenced by the EDX analysis (*see* Table 7.2, Sect. 7.3.2), the bulk compositions of the doped samples exhibit the Ti-to-O ratio near to the theoretical value (1/2). On the other hand, according to the XPS analysis it appears that such materials present the Ti-to-O ratio near to 1/3.

These findings confirm the doped samples have higher amount of oxygen species on the surface rather than in the bulk.

The deconvoluted Ti 2*p* spectra are reported in Fig. 7.5. These spectra are characterized by a $2p_{3/2}$ spin-orbit peak (average binding energy (B.E.) = 458.8 eV) and a low-intensity peak in the $2p_{1/2}$ region. In particular, in the $2p_{3/2}$ region, two peaks appear: a signal at about 459 eV referred to Ti^{4+} and another component at about 457 eV assigned to Ti^{3+} [46]. The Ti^{4+} and Ti^{3+} abundancy was evaluated by the deconvolution of $2p_{3/2}$ region and the values (at.%) are reported in Table 7.4. The Ti^{3+}-to-Ti^{4+} ratio increases along with the P-content. Thus, the O-Ti-O bonding exhibits a weakening generated by the cation substitution into the titania framework.

The O 1*s* spectra are reported in Fig. 7.6. All the samples exhibit two characteristics peaks. One signal at lower B.E. (530 eV) related to bonded oxygens into the titania structure (Ti-O). Another one to higher B.E. (532–530 eV) associated to surface hydroxyl groups (O-H) [47–49]. For doped catalysts, an additional peak was observed at about 533 eV, herein labeled as "no lattice oxygen" (NLO*). According to the literature, this additional peak can be attributed to chemisorbed oxygen species on the catalysts surface [50, 51]. This NLO* peak is a characteristic only of the doped samples, since they are the samples with smaller crystallite sizes and with the higher probability to have more unsaturation on the catalyst's surfaces (*see* Sect. 7.3.1). These surface defects are responsible of the hydroxyl groups chemisorption [36].

Table 7.5 summarizes the relative abundancy of oxygen species evaluated by the deconvolution of O 1*s* XP spectra. As previously noticed by the O 1*s* spectra, only the doped samples exhibit the NLO* oxygen species. The presence of such species along with OH groups can promote the VOC abatement due to the formation of OH radicals, which play a key role during the VOC oxidation [52].

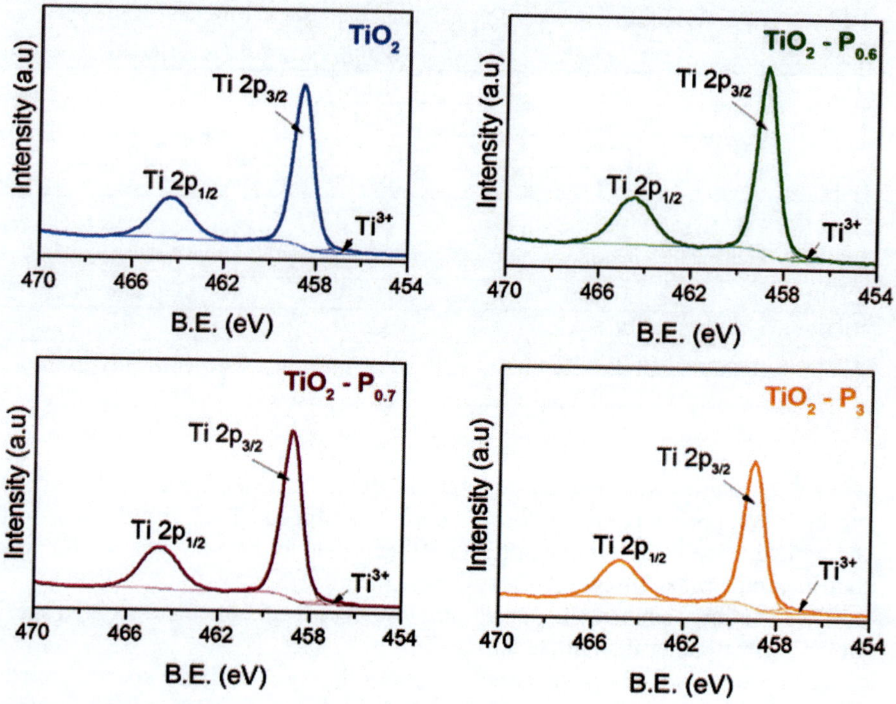

Fig. 7.5 Ti 2*p* X-ray photoelectron spectra of the samples

Table 7.4 Ti^{4+} and Ti^{3+} species observed on the catalyst surfaces evaluated by the deconvolution of Ti 2*p* XP spectra

	Ti 2*p*				
Catalysts	Ti^{4+} (at.%)	Position (B.E.)	Ti^{3+} (at.%)	Position (B.E.)	Ti^{3+}/Ti^{4+}
TiO_2	97.4	458.6	2.6	457.0	2.7
TiO_2-$P_{0.6}$	97.9	458.7	2.1	457.2	2.1
TiO_2-$P_{0.7}$	97.8	458.8	2.2	457.2	2.2
TiO_2-P_3	98.1	458.6	3.0	457.4	3.1

In Fig. 7.7 are reported the XP spectra for the P 2*p* core level to confirm the presence of P species on the catalyst surfaces. The position of the peaks occurs at about 134 eV corresponding to the P^{5+} species [46].

7.3.4 (DR)UV-Vis Spectroscopic Analysis

The (DR)UV-Vis spectra recorded for the synthesized catalysts are reported in Fig. 7.8. In particular, it appears that the absorption in the range 200–350 nm increases as a function of the P-content (Fig. 7.8 section a, TiO_2-P_3 is the sample that absorbs higher amount of light source in this range). The evaluation of the Band

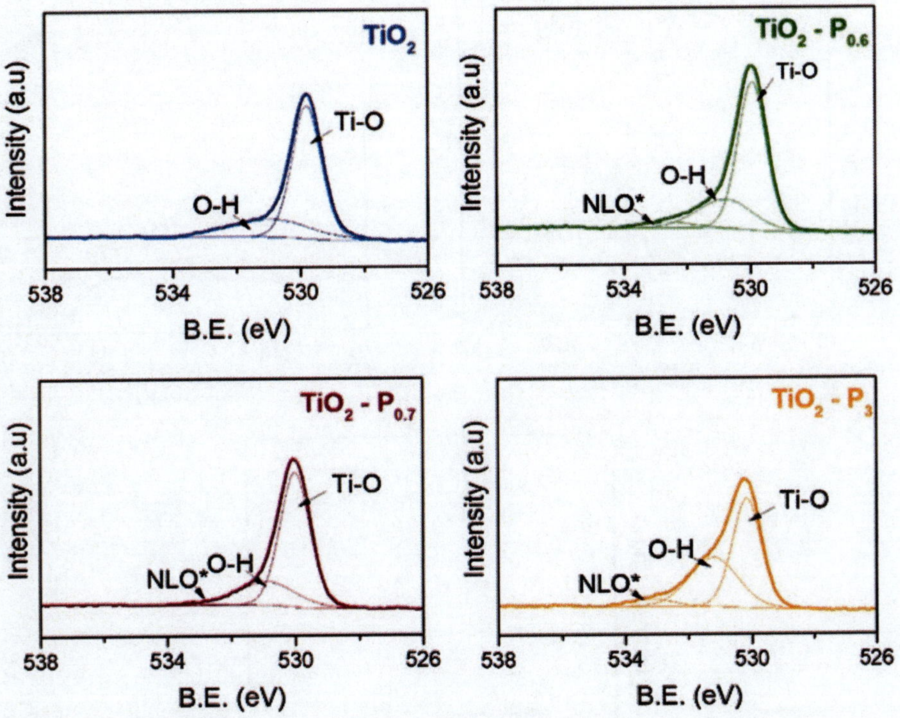

Fig. 7.6 O 1*s* X-ray photoelectron spectra of the samples

Table 7.5 Oxygen species on the catalyst surfaces as evaluated by the deconvolution of O 1*s* XP spectra

	O 1*s*					
Catalysts	Ti-O (at.%)	Position (B.E.)	O-H (at.%)	Position (B.E.)	NLO* (%)	Position (B.E.)
TiO_2	73.3	529.8	26.7	530.9	–	–
TiO_2-$P_{0.6}$	69.2	529.9	26.1	530.8	4.7	532.6
TiO_2-$P_{0.7}$	74.0	530.0	25.0	530.9	1.0	533.1
TiO_2-P_3	54.0	530.2	40.5	530.2	5.5	533.1

Gap energy was elaborated via the Tauc's plot (Fig. 7.8 sections b, c). The Band Gap Energies are reported in Table 7.6. Approximately, all the samples exhibit similar Band Gap Energies, corresponding to about 3.2 eV. During the photocatalytic phenomenon, this energy is required to promote the electrons from the valence band to the conduction band. In the present case, the catalyst with the lowest Band Gap Energy is the TiO_2-$P_{0.7}$ (3.13 eV). On the other hand, the highest Band Gap Energy can be observed for the TiO_2-P_3 sample (3.19 eV), along with TiO_2-$P_{0.6}$ (3.18 eV), thus suggesting higher absorption in the UV range [53].

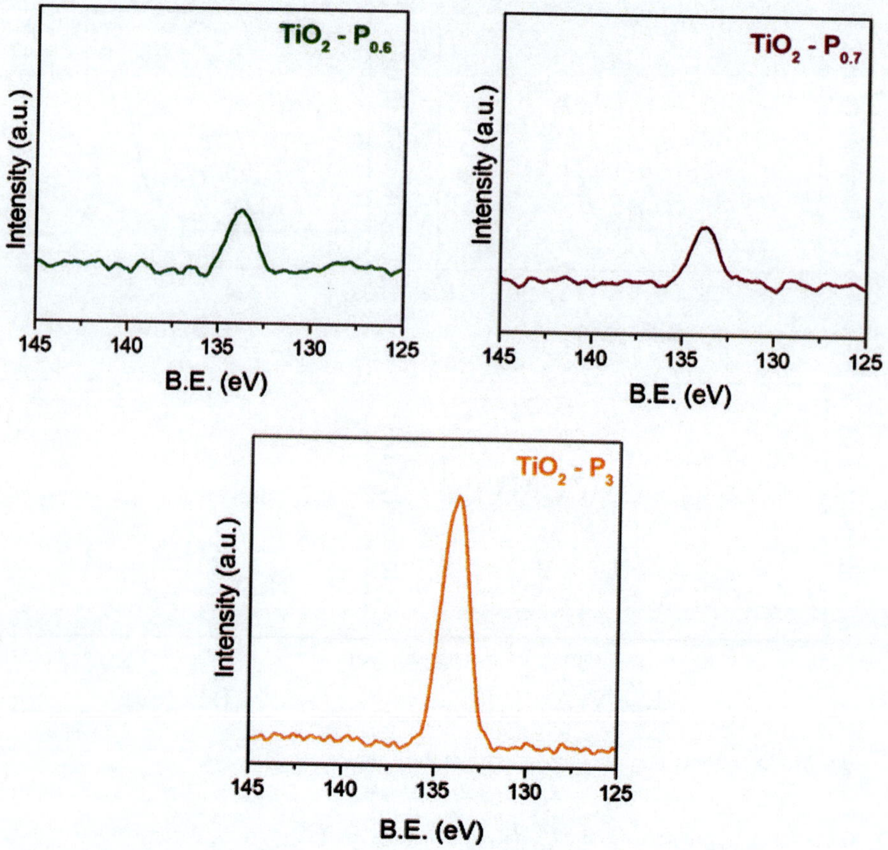

Fig. 7.7 P 2*p* X-ray photoelectron spectra of the doped samples

7.3.5 Catalytic Activity

The photocatalytic tests for the total oxidation of ethylene are reported in Fig. 7.9.

The ethylene conversion was studied for a TOS of 160 min. As a whole, the TiO_2 sample exhibited lower ethylene conversion compared to P-containing samples. In fact, as evidenced by our previous work [54] and other research studies [30, 55–61], the presence of phosphorus inside TiO_2 structure can enhance the photocatalytic activity as promoter element. This effect can be due to the possible modification of the crystalline and electronic structure of TiO_2 and then to the enhancedphotoabsorption mechanism (*see* Sect. 7.1).

Noteworthy, all the photocatalysts are more reactive under the UVA source rather than with the UVB range, although the lower intensity (namely, 8 and 12 W m^{-2}, respectively). This finding confirms that P-containing titania samples can be effective at higher wavelengths, in agreement with the literature [30, 55–61].

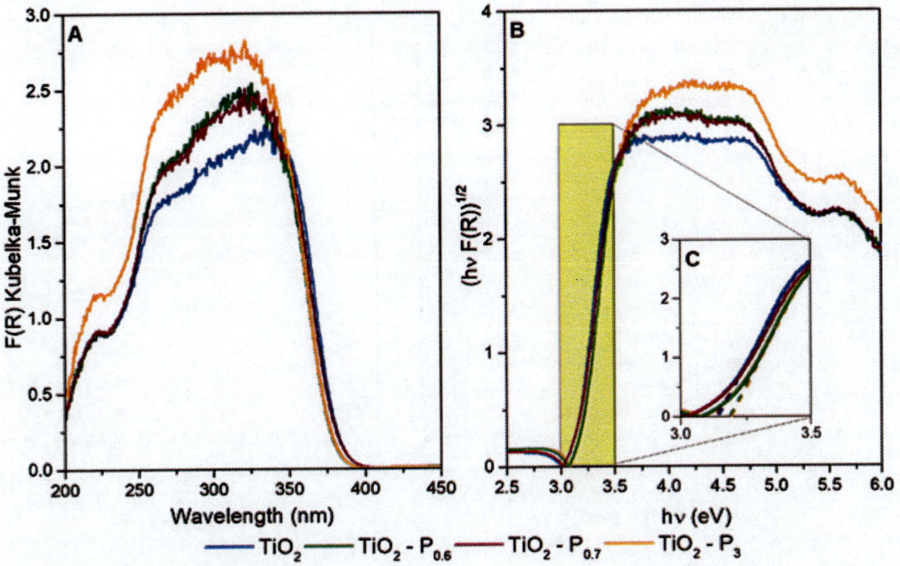

Fig. 7.8 (DR)UV-Vis spectra of the studied sample: (**a**) Kubelka-Munk plot, (**b**) Tauc's plot, and (**c**) Magnification of Tauc's plot

Table 7.6 Band Gap Energies (eV) evaluated over the studied catalysts.

Catalyst	Band gap energy (eV)
TiO_2	3.14
TiO_2-$P_{0.6}$	3.18
TiO_2-$P_{0.7}$	3.13
TiO_2-P_3	3.19

In order to better clarify the photocatalytic activity, the TOF was analyzed over the samples at two different reaction times (40 and 160 min). The results are reported in Fig. 7.10. The TOF values evaluated at 40 and 160 min show a positive trend as a function of the P-contents for the samples tested under UVB. The higher the P-content, the higher the reactivity. Thus, the best performing catalyst is the TiO_2-P_3, in agreement with the (DR)UV-Vis spectra. A similar trend, although less evident, can be observed for the samples tested under UVA. Nevertheless, in the latter case, it appears that samples with higher P-contents exhibited a lower reactivity at longer TOS, likely due to partial surface coverage by the presence of carbon-like molecules. Conversely, the optimum is reached with the TiO_2-$P_{0.6}$ sample (21.3 μmol h^{-1} g^{-1}).

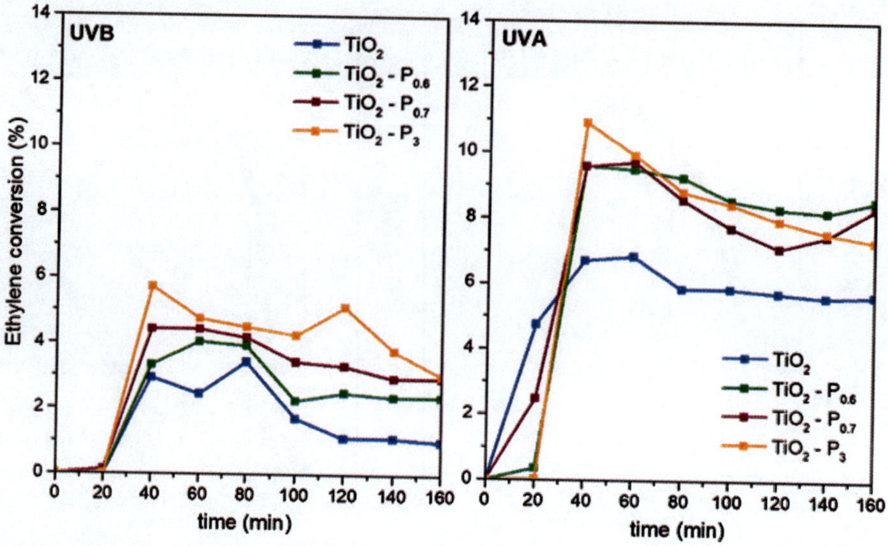

Fig. 7.9 Photocatalytic tests over the synthesized samples under two different sources: UVB and UVA

7.4 Conclusions

In this study, a set of phosphorous-doped titania were synthesized, at different P-contents, namely 0.6, 0.7, and 3 at.%. Pure TiO_2 was prepared for comparison purposes. All the synthesized samples had the presence of pure anatase and it was found that the incorporation of phosphorus into the solids renders the crystallite smaller.

Similarly, better textural properties were obtained with P-containing samples compared to pure TiO_2. UV-Vis spectroscopy confirmed a more intense absorption in the UV range for the P-containing samples. Catalytic results have shown a positive trend as a function of the P-contents for the samples tested under UVB. The higher the P-content, the higher the reactivity. A similar trend, although less evident, was for the samples tested under UVA. Nevertheless, in the latter case, it appears that samples with higher P-contents exhibited lower reactivity at longer TOS, likely due to the surface deposition of carbon-like molecules.

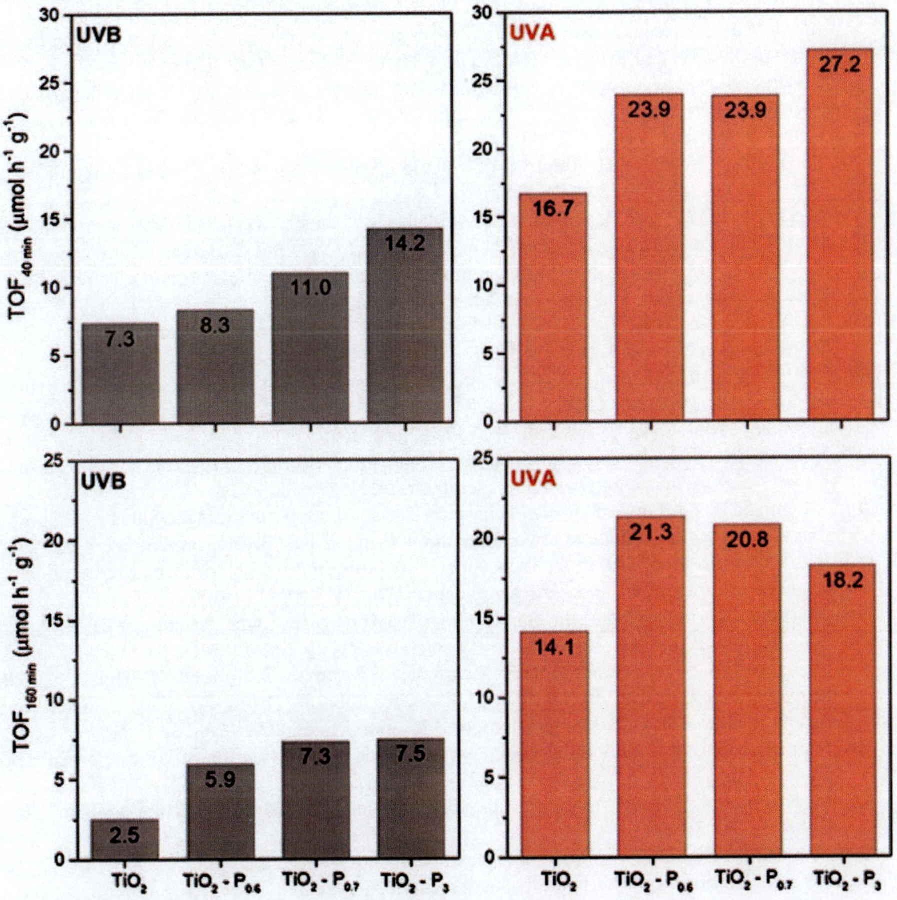

Fig. 7.10 TOF evaluated at 40 min (above) and 160 min (below) under different sources

References

1. Council of the European Parliament, Council Directive 1999/13/EC of 11 March 1999 on the limitation of emissions of volatile organic compounds due to the use of organic solvents in certain activities and installations, Off. J. Eur. Comm. L85/1–L85/22 (1999)
2. W.B. Li, J.X. Wang, H. Gong, Catalytic combustion of VOCs on non-noble metal catalysts. Catal. Today **148**, 81–87 (2010)
3. S. Ojala, S. Pitkäaho, T. Laitinen, N. NiskalaKoivikko, R. Brahmi, J. Gaálová, L. Matejova, A. Kucherov, S. Päivärinta, C. Hirschmann, T. Nevanperä, M. Riihimäki, M. Pirilä, R.L. Keiski, Catalysis in VOC abatement. Top. Catal. **54**, 1224–1256 (2011)
4. E. Olsen, F. Nielsen, Predicting vapour pressures of organic compounds from their chemical structure for classification according to the VOC-directive and risk assessment in general. Molecules **6**, 370–389 (2001)
5. M.S. Kamla, S.A. Razzak, M.M. Hossain, Catalytic oxidation of volatile organic compounds (VOCs)—a review. Atmos. Environ. **140**, 117–134 (2016)

6. P. Doggali, Y. Teraoka, P. Mungse, I.K. Shah, S. Rayalu, N. Labhsetwar, Combustion of volatile organic compounds over Cu-Mn based mixed oxide type catalysts supported on mesoporous Al_2O_3, TiO_2 and ZrO_2. J. Mol. Catal. A Chem. **358**, 23–30 (2012)
7. J. Carpentier, J.F. Lamonier, S. Siffert, E.A. Zhilinskaya, A. Aboukas, Characterisation of Mg/Al hydrotalcite with interlayer palladium complex for catalytic oxidation of toluene. Appl. Catal. A Gen. **234**, 91–101 (2002)
8. F.I. Khan, A.K. Ghoshal, Removal of volatile organic compounds from polluted air. J. Loss Prev. Process Ind. **13**, 527–545 (2000)
9. G.S.P. Soylu, Z. Özçelik, I. Boz, Total oxidation of toluene over metal oxides supported on a natural clinoptilolite-type zeolite. Chem. Eng. J. **162**, 380–387 (2010)
10. B. Ozturk, D. Yilmaz, Absorptive removal of volatile organic compounds from flue gas streams. Process Saf. Environ. Prot. **84**, 391–398 (2006)
11. M. Amann, M. Lutz, The revision of the air quality legislation in the European Union related to ground-level ozone. J. Hazard. Mater. **78**, 41–62 (2000)
12. P. Lakshmanan, L. Delannoy, V. Richard, C. Méthivier, C. Potvin, C. Louis, Total oxidation of propene over Au/xCeO_2-Al_2O_3 catalysts: influence of the CeO_2 loading and the activation treatment. Appl. Catal. B Environ. **96**, 117–125 (2010)
13. M.J. Molina, F.S. Rowland, *Stratospheric Sink for Chlorofluoromethanes: Chlorine Atom-Catalysed Destruction of Ozone* (Academic Press, New York, 1973)
14. J. Peng, S. Wang, Performance and characterization of supported metal catalysts for complete oxidation of formaldehyde at low temperatures. Appl. Catal. B Environ. **73**, 282–291 (2007)
15. B.J. Finlayson-Pitts, J.N. Pitts, Tropospheric air pollution: ozone, airborne toxics, polycyclic aromatic hydrocarbons, and particles. Science **276**, 1045–1052 (1997)
16. H. Rodhe, A comparison of the contribution of various gases to the greenhouse effect. Science **248**, 1217–1219 (1990)
17. M. Drobek, A. Figoli, S. Santoro, N. Navascués, J. Motuzas, S. Simone, C. Algieri, N. Gaeta, L. Querze, A. Trotta, G. Barbieri, R. Mallada, A. Julbe, E. Drioli, PVDF-MFI mixed matrix membranes as VOCs adsorbers. Micropor. Mesopor. Mater. **207**, 126–133 (2015)
18. S. Scirè, L.F. Liotta, Supported gold catalysts for the total oxidation of volatile organic compounds. Appl. Catal. B Environ. **125**, 222–246 (2012)
19. R. Koppmann, *Volatile Organic Compounds in the Atmosphere* (Wiley Online Library, Hoboken, 2007)
20. EPA, United Stated Environmental Protection Agency, Indoor Air Quality (IAQ), https://www.epa.gov/indoor-air-quality-iaq/introduction-indoor-air-quality
21. M. Magureanu, N.B. Mandache, P. Eloy, E.M. Gaigneaux, V.I. Parvulescu, Plasma-assisted catalysis for volatile organic compounds abatement. Appl. Catal. B Environ. **61**, 12–20 (2005)
22. A.H. Wani, R.M.R. Branion, A.K. Lau, Biofiltration: a promising and cost-effective control technology for odors, VOCs and air toxics. J. Environ. Sci. Heal. Part A Environ. Sci. Eng. Toxicol. **32**, 2027–2055 (1997)
23. M. Hussain, N. Russo, G. Saracco, Photocatalytic abatement of VOCs by novel optimized TiO_2 nanoparticles. Chem. Eng. J. **166**, 138–149 (2011)
24. M.A. Fox, M.T. Dulay, Heterogeneous photocatalysis. Chem. Rev. **93**, 341–357 (2002)
25. M.R. Hoffmann, S.T. Martin, W. Choi, D.W. Bahnemann, Environmental applications of semiconductor photocatalysis. Chem. Rev. **95**, 69–96 (1995)
26. A. Hagfeldt, M. Grätzel, Light-induced redox reactions in nanocrystalline systems. Chem. Rev. **95**, 49–68 (1995)
27. X. Chen, S.S. Mao, Titanium dioxide nanomaterials: synthesis, properties, modifications and applications. Chem. Rev. **107**, 2891–2959 (2007)
28. M.A. Henderson, A surface science perspective on TiO_2 photocatalysis. Surf. Sci. Rep. **66**, 185–297 (2011)
29. Z. Zhang, S. Brown, J.B.M. Goodall, X. Weng, K. Thompson, K. Gong, S. Kellici, R.J.H. Clark, J.R.G. Evans, J.A. Darr, Direct continuous hydrothermal synthesis of high surface area nanosized titania. J. Alloys Compd. **476**, 451–456 (2009)

30. P. Zhang, Y. Yu, E. Wang, J. Wang, J. Yao, Y. Cao, Structure of nitrogen and zirconium co-doped titania with enhanced visible-light photocatalytic activity. ACS Appl. Mater. Interfaces **6**, 4622–4629 (2014)
31. M.D. Hernández-Alonso, J.M. Coronado, B. Bachiller-Baeza, M. Fernández-García, J. Soria, Influence of structural and surface characteristics of Ti 1-xZrxO2 nanoparticles on the photocatalytic degradation of methylcyclohexane in the gas phase. Chem. Mater. **19**, 4283–4291 (2007)
32. W. Zhao, W. Ma, C. Chen, J. Zhao, Z. Shuai, Efficient degradation of toxic organic pollutants with Ni2O 3/TiO2-xBx under visible irradiation. J. Am. Chem. Soc. **126**, 4782–4783 (2004)
33. Y. Cong, J. Zhang, F. Chen, M. Anpo, D. He, Preparation, photocatalytic activity, and mechanism of nano-TiO2 Co-doped with nitrogen and iron (III). J. Phys. Chem. C **111**, 10618–10623 (2007)
34. S. Shamaila, A.K.L. Sajjad, F. Chen, J.L. Zhang, Synthesis and characterization of mesoporous-TiO_2 with enhanced photocatalytic activity for the degradation of chloro-phenol. Mater. Res. Bull. **45**, 1375–1382 (2010)
35. M. Piumetti, F.S. Freyria, M. Armandi, F. Geobaldo, E. Garrone, B. Bonelli, Fe- and V-doped mesoporous titania prepared by direct synthesis: characterization and role in the oxidation of AO7 by H_2O_2 in the dark. Catal. Today **227**, 71–79 (2014)
36. N. Musselwhite, G.A. Somorjai, Investigations of structure sensitivity in heterogeneous catalysis: from single crystals to monodisperse nanoparticles. Top. Catal. **56**, 1277–1283 (2013)
37. R.A. Van Santen, Complementary structure sensitive and insensitive catalytic relationships. Acc. Chem. Res. **42**, 57–66 (2009)
38. G. Li, L. Li, J. Boerio-Goates, B.F. Woodfield, High purity anatase TiO_2 nanocrystals: near room-temperature synthesis, grain growth kinetics, and surface hydration chemistry. J. Am. Chem. Soc. **127**, 8659–8666 (2005)
39. B. Ohtani, O.O. Prieto-Mahaney, D. Li, R. Abe, What is Degussa (Evonic) P25? Crystalline composition analysis, reconstruction from isolated pure particles and photocatalytic activity test. J. Photochem. Photobiol. A Chem. **216**, 179–182 (2010)
40. L.P. Childs, D.F. Ollis, Is photocatalysis catalytic? J. Catal. **66**, 383–390 (1980)
41. G. Li, C.P. Richter, R.L. Milot, L. Cai, C.A. Schmuttenmaer, R.H. Crabtree, G.W. Brudvig, V.S. Batista, Synergistic effect between anatase and rutile TiO2 nanoparticles in dye-sensitized solar cells. J. Chem. Soc. Dalton Trans. **45**, 10078–10085 (2009)
42. J. Zhang, P. Zhou, J. Liu, J. Yu, New understanding of the difference of photocatalytic activity among anatase, rutile and brookite TiO_2. Phys. Chem. Chem. Phys. **16**, 20382–20386 (2014)
43. D.S. Bhatkhande, V.G. Pangarkar, A.A. Beenackers, Photocatalytic degradation for environmental applications—a review. J. Chem. Technol. Biotechnol. **77**, 102–116 (2002)
44. R.R. Guimaraes, A.L.A. Parussulo, K. Araki, Impact of nanoparticles preparation method on the synergic effect in anatase/rutile mixtures. Electrochim. Acta **222**, 1378–1386 (2016)
45. M.D. Abràmoff, P.J. Magalhães, S.J. Ram, Image processing with imageJ. Biophoton. Int. **11**, 36–41 (2004)
46. J.F. Moulder, W.F. Stickle, P.E. Sobol, K.D. Bomben, *Handbook of Photoelectron Spectroscopy* (Perkin-Elmer Corporation, Eden Prairie, 1992)
47. M.M. Kumar, S. Badrinarayanan, M. Sastry, Nanocrystalline TiO_2 studied by optical, FTIR and X-ray photoelectron spectroscopy: correlation to presence of surface states. Thin Solid Films **358**, 122–130 (2000)
48. Y.D. Hou, X.Z. Fu, X.X. Wang, X.C. Wang, X.F. Chen, Z.X. Ding, L. Wu, N-doped SiO_2/TiO_2 mesoporous nanoparticles with enhanced photocatalytic activity under visible-light irradiation. Chemosphere **72**, 414–421 (2008)
49. S. Begin, G. Bertrand, F. Toma, O. Barres, C. Coddet, C. Meunier, D. Klein, Microstructure and environmental functionalities of TiO_2-supported photocatalysts obtained by suspension plasma spraying. Appl. Catal. B Environ. **68**, 74–84 (2006)
50. B. Bharti, S. Kumar, H.N. Lee, R. Kumar, Formation of oxygen vacancies and Ti^{3+} state in TiO_2 thin film and enhanced optical properties by air plasma treatment. Sci. Rep. **6**, 1–12 (2016)

51. S. Wang, L. Pan, J.J. Song, W. Mi, J.J. Zou, L. Wang, X. Zhang, Titanium-defected undoped anatase TiO2 with p-type conductivity, room-temperature ferromagnetism, and remarkable photocatalytic performance. J. Am. Chem. Soc. **137**, 2975–2983 (2015)
52. S. Futamura, H. Einaga, H. Kabashima, L.Y. Hwan, Synergistic effect of silent discharge plasma and catalysts on benzene decomposition. Catal. Today **89**, 89–95 (2004)
53. J.C. Yu, L. Zhang, Z. Zheng, J. Zhao, Synthesis and characterization of phosphated mesoporous titanium dioxide with high photocatalytic activity. Chem. Mater. **15**, 2280–2286 (2003)
54. M. Dosa, M. Piumetti, S. Bensaid, T. Andana, C. Galletti, D. Fino, N. Russo, Photocatalytic abatement of volatile organic compounds by TiO2 nanoparticles doped with either phosphorous or zirconium. Materials **12**, 2121 (2019)
55. E. Wang, T. He, L. Zhao, Y. Chen, Y. Cao, Improved visible light photocatalytic activity of titania doped with tin and nitrogen. J. Mater. Chem. **21**, 144–150 (2011)
56. F. Freyria, M. Compagnoni, N. Ditaranto, I. Rossetti, M. Piumetti, G. Ramis, B. Bonelli, Pure and Fe-doped mesoporous titania catalyse the oxidation of acid orange 7 by H_2O_2 under different illumination conditions: Fe doping improves photocatalytic activity under simulated solar light. Catalysts **7**, 213 (2017)
57. L. Lin, W. Lin, J.L. Xie, Y.X. Zhu, B.Y. Zhao, Y.C. Xie, Photocatalytic properties of phosphor-doped titania nanoparticles. Appl. Catal. B Environ. **75**, 52–58 (2007)
58. S. Guo, S. Han, M. Haifeng, C. Zeng, Y. Sun, B. Chi, J. Pu, J. Li, Synthesis of phosphorus-doped titania with mesoporous structure and excellent photocatalytic activity. Mater. Res. Bull. **48**, 3032–3036 (2013)
59. J.C.-S. Wu, C.H. Chen, A visible-light response vanadium-doped titania nanocatalyst by sol-gel method. J. Photochem. Photobiol. A Chem. **163**, 509–515 (2004)
60. K. Wilke, H.D. Breuer, The influence of transition metal doping on the physical and photocatalytic properties of titania. J. Photochem. Photobiol. A Chem. **121**, 49–53 (1999)
61. J. Reszczyńska, T. Grzyb, J.W. Sobczak, W. Lisowski, M. Gazda, B. Ohtani, A. Zaleska, Visible light activity of rare earth metal doped (Er^{3+}, Yb^{3+} or Er^{3+}/Yb^{3+}) titania photocatalysts. Appl. Catal. B Environ. **163**, 40–49 (2015)

Chapter 8
Colloidal Semiconductor Nanocrystals for Artificial Photosynthesis

Francesca S. Freyria

8.1 Introduction

During the 2018, the global average annual carbon dioxide concentration reached 407.4 (±0.1 ppm), the highest level of the last 800,000 years. Energy consumption is increasing every year, and currently, about 81% of world energy supply is from fossil fuels [1, 2]. Acceleration in energy consumption in 2018 was more than 2% driven by high growth in electricity and gas demand. In this scenario, solar energy conversion is a promising approach to solve this problem, especially because the energy from the Sun on the Earth's surface in 1 h (4.31020 J) almost matches the current world total energy consumption in a year. Obviously, the solar energy by itself could be able to largely satisfy the entire energy demand of the humankind [3, 4].

Chemical fuels can be promising candidates for energy storage from intermittent renewable energy sources such as sunlight. The formation of high energy C-C and C-H can be a tool to store energy which can release when they undergo combustion [5, 6]. Sunlight is the clean and the most abundant form of energy, but it is diffuse and intermittent so it is necessary to store it [3, 7–10].

An environmental-friendly way to produce simple fuels (e.g., H_2, CH_3OH, C_2H_5OH) is to use incident solar photon energy to drive energetically uphill reactions such as the water splitting and the reduction of CO_2. In this way, it is possible to obtain two great advantages simultaneously: (1) no additional CO_2 is released in the atmosphere; (2) the production of fuels is impressively accelerated. The current natural system takes millions of years to produce fossil fuels, which are quickly burnt by humans, finally emitting huge amount of CO_2 in the atmosphere with

F. S. Freyria (✉)
Department of Applied Science and Technology, Corso Duca degli Abruzzi, Turin, Italy

Department of Chemistry, Massachusetts Institute of Technology, Cambridge, MA, USA
e-mail: francesca.freyria@polito.it; freyria@mit.edu

M. Piumetti, S. Bensaid (eds.), *Nanostructured Catalysts for Environmental Applications*, https://doi.org/10.1007/978-3-030-58934-9_8

climate changes issues. The conversion of CO_2 already present in the atmosphere through a fast process could help to close the carbon cycle and to reach an equilibrium of CO_2 level concentration [4, 6, 8, 11].

Currently, there is a huge mismatch between the speed of which humankind emits greenhouse gases, especially CO_2, and the naturally pathway to capture them from the atmosphere. This disequilibrium is at the basis of the anthropogenic climate changes. Renewable energies could help not only to directly produce electricity, such as photovoltaic (PV) devices, hydrothermal, and geothermal systems, but also to produce fuels with the same characteristics of fossil fuels. These are called in general "renewable fuels" and when the renewable energy used for the process is the sunlight, they are called "solar fuels." Detz et al. [4] introduced four main categories for the methods to obtain renewable fuels based on how the energy is harvested: concentration, transformation, natural photosynthesis, and artificial photosynthesis. Under the last one, there are all the technologies to produce solar fuels through the exploitation of the solar light to drive reactions like water splitting or CO_2 reduction. A simple and cheap method can be accomplished by using particulate semiconductor (SC) materials where both absorption and catalysis occur in the single system. Recent economic calculations have estimated the average cost of H_2 is in the range of 1.6-3.5 $/kg with a solar to hydrogen conversion efficiency (STH) of 5–10% with a stability of 5 years [12, 13].

Colloidal nanocrystals (NCs), like QDs, are receiving great interest either as photosensitizer or as photocatalyst to drive water splitting and CO_2 reduction reactions. Semiconductor nanocrystals (SC-NCs) have a prodigious extinction coefficient in the solar spectrum [14], the possibility to tune the band gap up to the NIR-IR range of the solar irradiation [15, 16], the ability to generate multiple excitons [17], an active and crucial role in energy transfer in hybrid system [18–21]: all of these features provide the ground for excellent light-harvesting materials and photocatalysis [15, 22–25].

8.2 Photocatalysis/Photosynthesis with NCs: Mechanisms

To produce and commercialize solar fuels at a viable large scale in a near future, we need a low-cost and stable, durable photocatalytic colloid system. The process requires three main passages that all happen on the SC-NC-based system: (1) absorption of the photons and formation of excitons; (2) migration and charge separation of these photoexcited carriers; and (3) surface chemical reactions [12, 26, 27].

During these steps, there are some challenges to overcome to reach a desirable photocatalytic performance. In the following sections, we are going to describe some of them and to provide some strategies to address them.

8.2.1 *Thermodynamics and Kinetics of Photocatalytic Solar Fuels Production*

An exciton, i.e., an electron/hole (e^-/h^+) pair bound by Coulomb forces, can be generated upon absorption of incident photon by a semiconductor nanocrystal. To have a catalytic process is necessary that the exciton reaches the reaction centers at the surface of the semiconductor and dissociates into free charges which can react before the recombination [2, 12, 26, 27]. Excitons formation is the first fundamental process in photosynthesis. Photocatalysis and photosynthesis can be classified as light-driven systems, where chemical reactions are promoted by excited states. In the field of solar fuels production, photocatalysis and photosynthesis are often used without a clear distinction: in both processes, we have the absorption and conversion of photons into charge carriers which drive redox reactions at the surface [28], but the two processes are mainly different from a thermodynamic point of view (Fig. 8.1a). According to IUPAC definition, photocatalysis is: "*change in the rate of a chemical reaction or its initiation under the action of ultraviolet, visible or infrared radiation in the presence of a substance—the photocatalyst—that absorbs light and is involved in the chemical transformation of the reaction partners*" [29].

Whereas photosynthesis is: "*a metabolic process involving plants and some types of bacteria (e.g., Chromatiaceae, Rhodospirillaceae, Chlorobiaceae) in which light energy absorbed by chlorophyll and other photosynthetic pigments results in the reduction of CO_2 followed by the formation of organic compounds. In plants the overall process involves the conversion of CO_2 and H_2O to carbohydrates (and other plant material) and the release of O_2*" [29].

Based on these definitions, in a photocatalytic process light is only used to speed up a thermodynamically spontaneous reaction ($\Delta G < 0$) by reducing the kinetic barriers in the chemical conversion. In natural photosynthesis, instead, light is an essential ingredient of the reactions allowing processes that otherwise are thermodynamically forbidden ($\Delta G > 0$). In such a case, the reaction occurs only when a light absorber produces e^-/h^+ pairs whose energy is higher than ΔG [27].

Water splitting and CO_2 reduction are endergonic reactions, and the photon energy is necessary to reverse the sign of the associated Gibbs free energy. The rate of the spontaneous reverse reactions may be the limiting factor [27]. Therefore, it would be more appropriate to refer to these processes as photosynthetic processes rather than photocatalytic reactions. On the other hand, in literature, in the majority of the papers, it is more often used the term photocatalysis, especially in presence of reactions that involve sacrificial agents [2, 27, 30, 31].

When a photon with higher energy of the band gap is absorbed, an electron is promoted from the valence band (VB) to the conduction band (CB) with a subsequent loss of the energy in excess of the band gap due to thermalization process (Fig. 8.2) [28, 32]. Not every electron/hole in quasi-Fermi levels can be used to promote photocatalytic reactions (Fig. 8.1b). Indeed, not only energy band gap value needs to be larger than the difference in reduction and oxidation potentials of the two half reductions, but the band edge positions also need to be correctly aligned

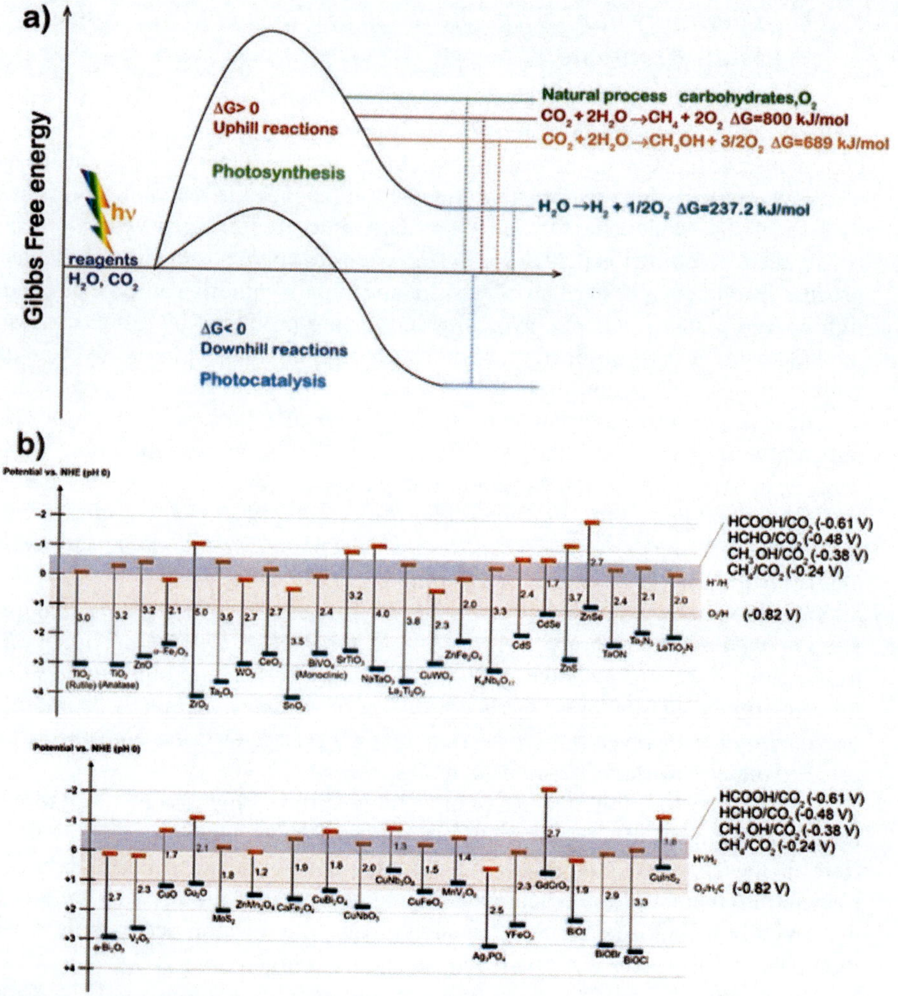

Fig. 8.1 (**a**) Energetics of photocatalysis and photosynthesis, inspired by [27] and [12]; (**b**) band edge positions of some typical semiconductors with respect to the reduction–oxidation potentials of CO_2 reduction and water splitting reactions. (Adapted with permission from Wang Q, et al. (2020) Particulate Photocatalysts for Light-Driven Water Splitting: Mechanisms, Challenges, and Design Strategies. Chem. Rev. 120:919–985). Copyright (2020) American Chemical Society [12]

with these redox potentials, having, in practice, the valence band below the oxidation potential whereas the conduction band above the reduction potential of the desired reaction (Fig. 8.2). To promote the electron or the hole transfer, it is necessary to guarantee overpotentials. To simplify, we can compare the band edge alignment between the conduction band and valence band of the semiconductor catalyst and the redox reaction to the band alignment on semiconductor junctions [2, 33, 34].

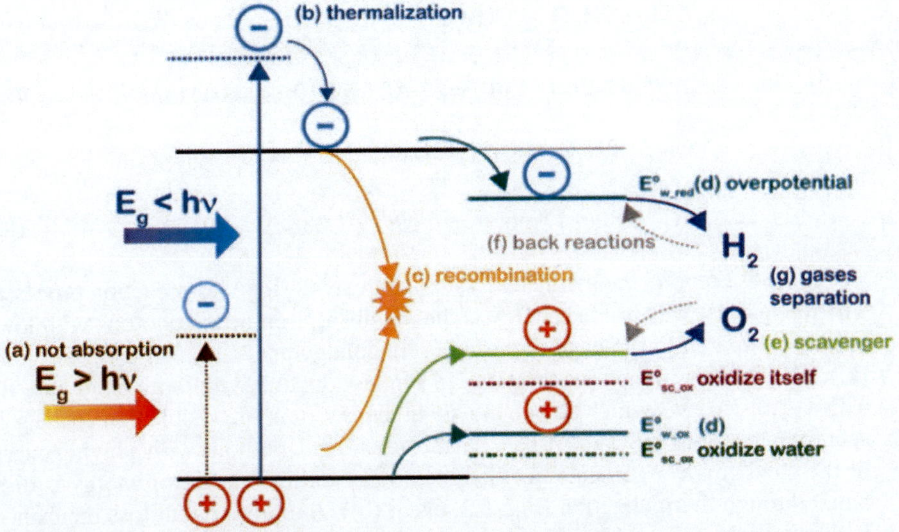

Fig. 8.2 Energy loss processes in SC-NCs during solar fuels reactions, based on [2, 43, 45]. Photons with lower energy (*a*) and with higher energy than the band gap (*b*); e^-/h^+ recombination (*c*); overpotential due to the band position of the NC and the redox levels of water (teal color) (*d*); photocorrosion of the SC is determined of the band position of its redox potentials with respect to the water redox potentials: e.g., for holes, if the oxidation potential of the water ($E°_{W\_OX}$, teal color) is lower than the oxidation potential of the semiconductor ($E°_{SC\_OX}$, purple), a hole scavenger is necessary (light green level, (*e*)), if $E°_{SC\_OX} < E°_{W\_OX}$ (dark green color), the water oxidation is instead favorited; possible back reactions (*f*) and gases separation (*g*)

When we have a Type I alignment (straddling configuration), the requirements of overpotentials are verified for both electrons and holes, whereas when we have a Type II alignment (staggered configuration), only one of the half-reactions can occur, and for the other half-reaction, we need to use a sacrificial agent.

For example, if we just consider the band alignment for water splitting, we would need a photocatalyst with the bottom of the conduction band more negative than the redox potential to reduce water molecules to hydrogen (H^+/H_2) and the top of the valence band more positive of the redox potential to oxidize water molecules to oxygen (O_2/H_2O) with an energy band gap between 1.23 and 3 eV [31] (Fig. 8.2, $E°_{W\_RED}$ and $E°_{W\_OX}$, in teal color).

Based on the Eqs. (8.1)–(8.3), the photons in the visible/NIR solar spectrum could have enough energy to carry out the reactions for the water splitting and CO_2 reduction.

$$H_2O \rightarrow H_2 + 1/2O_2 \quad \Delta G^0 = 237\,\text{kJ/mol} \overset{\Delta E^0 = \frac{\Delta G^0}{nF}}{\Rightarrow} \Delta E^0 = 1.23\,\text{eV} = 1008\,\text{nm} \qquad (8.1)$$

$$\begin{gathered} CO_2 + 2H_2O \rightarrow CH_3OH + 3/2O_2 \\ \Delta G^0 = 689\text{kJ/mol} \overset{\Delta E^0 = \frac{\Delta G^0}{nF}}{\Rightarrow} \Delta E^0 = 1.19\,\text{eV} = 1041\text{nm} \end{gathered} \tag{8.2}$$

$$\begin{gathered} CO_2 + 2H_2O \rightarrow CH_4 + 2O_2 \\ \Delta G^0 = 800\text{kJ/mol} \overset{\Delta E^0 = \frac{\Delta G^0}{nF}}{\Rightarrow} \Delta E^0 = 1.036\text{eV} = 1069\text{nm} \end{gathered} \tag{8.3}$$

Figure 8.2 reports the main energy loss pathway during photoreaction processes. The first easiest way to lose energy is the not absorption of the photons with lower energy compared to the band gap of the semiconductor [2, 12, 35] (Fig. 8.2, event (a)). This is a significant loss channel in semiconductors with large band gap, like TiO_2 (~3.2–3.0 eV), which absorb mostly in the UV or in the visible part of the solar spectrum nearer to the UV [36–39]. On the other hand, photons with a higher energy of the band gap can generate hot charge carriers and the excess of energy is dissipated through thermalization (Fig. 8.2, event (b)). Another crucial loss mechanism is the recombination of the electron/hole pair in the bulk or at the surface before their transfer to the catalytic center (Fig. 8.2, event (c)). Direct semiconductors with a suitable band gap can have a higher transition probability with high absorption coefficients as compared to the indirect semiconductors which result in lower absorption coefficients and a higher probability to generate nonradiative pathways for the e^-/h^+ pairs since phonons are involved [12]. When e^-/h^+ pairs are generated, it is important to avoid their recombination and, especially in direct semiconductors, Auger recombination could become an important channel [2]. The harvested light can be dissipated through radiative recombination (light emission) or through nonradiative recombination (heat generation). The long-lived photogenerated charges that do not undergo recombination can react with absorbed species at the surface. For efficient photocatalytic processes, it is necessary to transfer the charge on the surface and to have a good charge separation in order to run redox reactions with the absorbed species and to avoid back reactions or recombination [40]. Yet, it is also important to confine away the charge carrier from surface traps since a too strong confinement of a charge carrier can prevent reactions at the surface of the photocatalyst, especially the oxidation of holes which is considered the limiting step in the hydrogen generation [41]. In some cases, the photoluminescence quantum yield and photocatalytic performance might be competitive channels [40, 42]. Quasi-type II or type II core/shell semiconductors (Fig. 8.3a) can be good configurations to promote charge separation and decrease the charge carriers recombination. The thickness of the shell can positively impact on the quantum yield and the lifetime in the quantum dots due to a drop in nonradiative decay channels with the suppression of the surface traps. In NCs covered by a thick shell, the hole is strongly confined whereas the electron is more delocalized, consequently leading to a decrease of the overlap of the e^- and h^+ wavefunctions and therefore a more charge separation (below further descriptions about these configurations are provided). Another intrinsic loss pathway is the difference in the band alignment between the semicon-

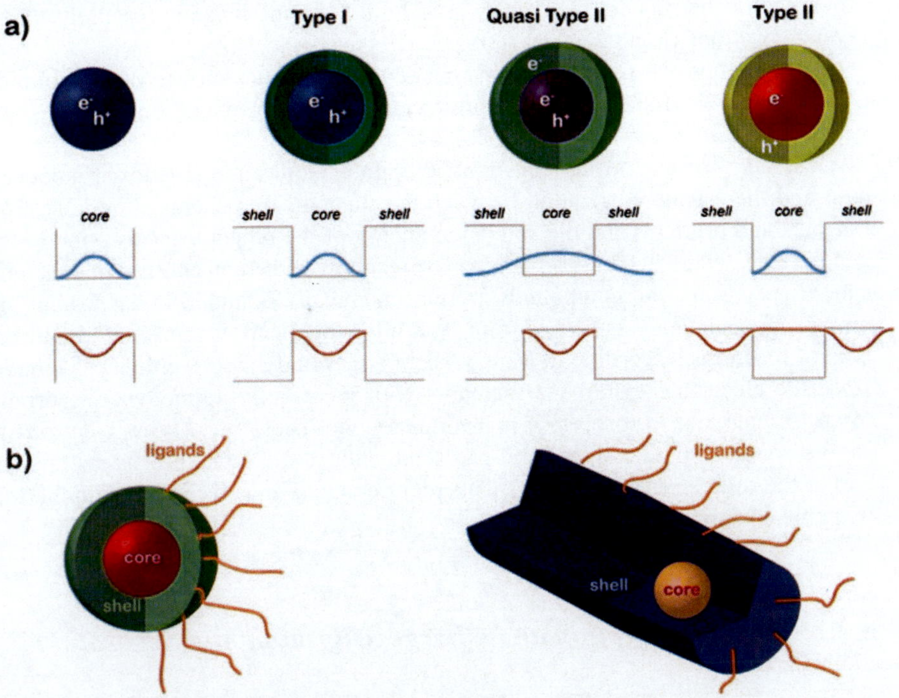

Fig. 8.3 (**a**) Schematic sketch of different band alignment possibilities in core/shell heterostructures, example of e⁻/h⁺ wavefunction are reported; (**b**) sketches of spherical core/shell QDs and core/shell nanorods (dot in rod)

ductor band positions and the redox potential levels of the reaction which is used to transfer the charges instead to run the photocatalytic reactions (Fig. 8.2, event (d)) [2]. A semiconductor NC is stable with respect to the electron reduction if its reduction potential is higher than either its CB or water reduction potential (H^+/H_2) and, with respect to the hole reduction, if its oxidation potential is lower than either its VB or the oxidation potential of water (O_2/H_2O) [43]. Photogenerated holes and (less likely) photogenerated electrons can also oxidize or reduce the semiconductor instead than to promote the surface reaction. This happens when the redox potentials of the photocatalyst (nanocrystal itself or ligands at the surface) have a better band alignment than the redox potential of the reaction. This is a common scenario in chalcogenide II–VI semiconductors where the oxidation of the lattice ions or thiol ligands, which then leads to the aggregation of the colloidal photocatalysts, can frequently happen. To avoid photooxidation, holes scavengers are used to prevent the damage of the photocatalyst by removing holes (Fig. 8.2, event (e)) [2]. Yu et al. added amorphous TiO_2 (Ti(IV)) as a hole cocatalyst on the CdS surface to make it stable and Ni(II) as cocatalyst to increase the photocatalytic reactivity. Ti(IV) acts as a h^+ sink and prevents the corrosion of CdS [44]. Moreover, the efficiency of the process is compromised by the thermodynamic feasibility of the reverse reactions (Fig.8.2, event (f)).

Finally, when the redox reactions happen on the same surface of the colloidal photocatalyst, the separation of the products (e.g., hydrogen and oxygen gases) can generate further energy losses (Fig.8.2, event (g)) [2].

Generally, for most of the reactions, thermodynamics and kinetics are independent and there is no correlation between the standard free energy difference from reagents and products and the activation energy of the others. An exception to this general rule occurs in "outer sphere" redox reactions, where in contrast to the "inner sphere" reactions, the compounds are not chemically bounded to each other and remain independently before, during, and after the electron transfer [46]. In this type of reactions, according to Marcus theory, thermodynamics influences kinetics since the electron transfer rate increases with increasing thermodynamic driving force, except when the process is in inverted regime. Therefore, a faster electron rate can be achieved with a wide band gap semiconductor, which absorbs just a small part of the solar spectrum [47, 48]. This requirement seems to conflict with thermodynamic constraints [2, 8].

8.2.2 *Light Absorption and Charge Migration and Separation*

Only ~5% of the solar light lies in the UV region (100–400 nm) [49], therefore, to run photocatalytic reactions is important to collect also the visible and the NIR part of the solar radiation. Several architectures have been proposed both to harvest more photons and to respect the thermodynamic and kinetic requirements.

One of the most used approaches to harvest light is to couple the catalyst to another material (sensitizer, upconverting material, plasmonic metal) which is able to absorb the light and deliver excitons/charged carriers to the reaction centers.

A method to do this is to couple a sensitizer (organic/inorganic materials) having a narrow band gap with a wide band gap catalyst (Fig. 8.4a). There are two main mechanisms for the sensitization process [45, 50]. Usually, the sensitizer has a LUMO (e.g., dye, Fig. 8.4a) or a conduction band edge (e.g., quantum dots, Fig. 8.4b) higher than the conduction band edge of the catalyst. Upon light, the photogenerated electrons are injected into the catalyst whereas the photogenerated holes react with holes scavengers present in the system. It is also possible to have a direct excitation of the electrons from the HOMO of the dye to the CB of the catalyst. Based on this approach, especially in the past, several oxides were coupled with dye sensitizers (this is the typical configuration of the dye-sensitized solar cell) [51]. The complex of the salt tris(bipyridine)ruthenium(II) chloride, $[Ru(bpy)_3]^{2+}$, has been largely applied as sensitizer either for the reduction or oxidation of the water [50, 52, 53]. Recently, new hybrid systems (organic/inorganic coupling) have been developed where the sensitizer is an inorganic semiconductor nanocrystal whereas the catalyst is an organic molecule. This latter configuration compared to the other one (dye as sensitizer and semiconductor NC) has shown several advantages

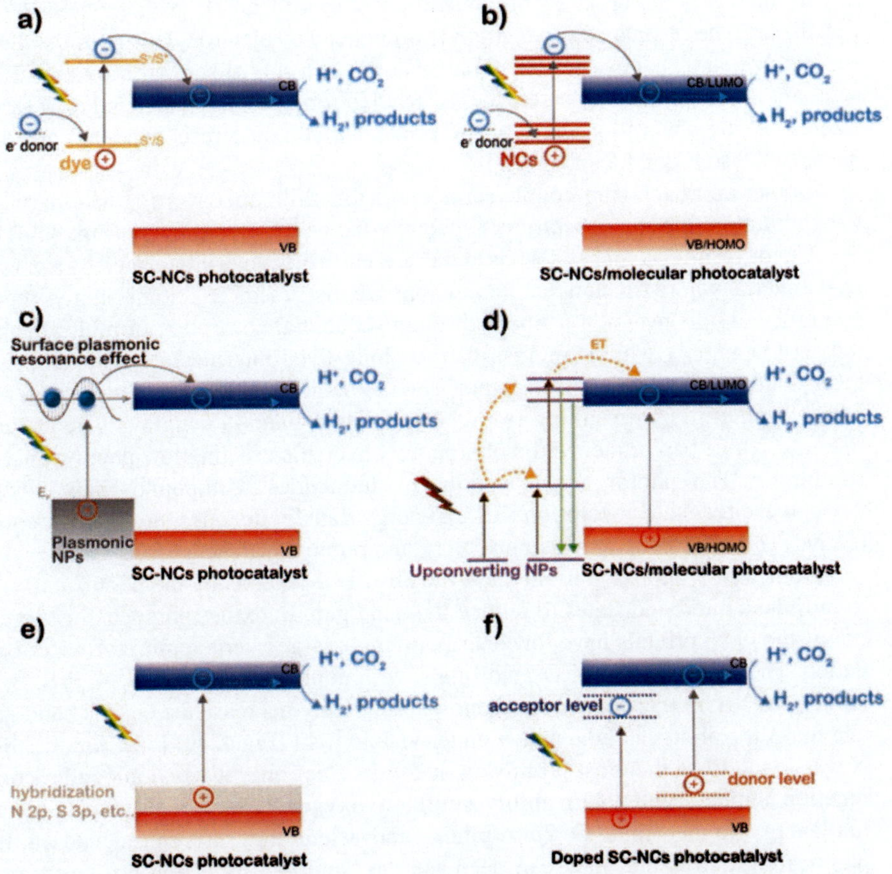

Fig. 8.4 Schematic illustration of different configurations to absorb light and to separate charge in SC-NCs systems for solar fuels reactions, inspired by [2, 12, 45]: (**a**) dye sensitizer; (**b**) SC-NC as sensitizer; (**c**) plasmonic metal semiconductor system; (**d**) energy transfer from upconverting NPs; (**e**) hybridization of orbitals; (**f**) introduction of interband through doping

since semiconductors, like quantum dots, can have (1) a very high extinction coefficient and a more tunable band gap and therefore can absorb more light compare to the dyes; (2) a higher stability again corrosion [22]. Moreover, comparing the active sites, molecular photocatalyst can be economically cheaper than semiconductor NCs [54].

Another method to inject more carries in the photocatalyst is to couple it with plasmonic noble metal nanoparticles (NPs) (Fig. 8.4c), especially silver and gold, in order to exploit the resonant electrons formed during the surface plasmonic resonance (SPR) excitation [55]. In plasmonic metal nanostructures, upon light excitation, when the frequency of photons is similar to the frequency of surface oscillating electrons, a resonant collective oscillation of electrons is formed (SPR), which largely increases the cross section absorbance, more than 10^5 as compared to dye sensitizers. The resonance wavelength is influenced by electron density of the material, its size, and shape. These nanoparticles act as a sort of sensitizer by absorbing resonant photons and transfer the SPR electrons to the semiconductor [55]. Therefore, the absorption band of the photocatalytic system is extended as compared to the semiconductor by itself.

Another approach is to couple semiconductors with upconverting nanoparticles to exploit more the solar spectrum, especially the near-infrared range (Fig. 8.3d) [2, 56]. Upconverting nanocrystals based on rare-earth elements are formed by a crystal host material where trivalent lanthanide ions are dispersed [56]. Some studies report that efficient upconversion happens through sequential absorption of multiple photons at low energy which are absorbed via long-lived intermediate states between the ground state and the emitting state. The lanthanide-based nanocrystals present a narrowband emission resulting from weak electron-phonon coupling. The crystalline host lattice can influence the photon upconversion creating low phonon energy conditions. The photon upconversion in lanthanides' compounds can happen through excited-state absorption (ESA), energy transfer upconversion (ETU), cooperative sensitization upconversion (CSU), and photon avalanche [57].

Another easy approach to increase the absorbed photons by the photocatalyst is to introduce interband states to reduce the band gap. In oxides photocatalysts, generally, the O 2*p* orbitals have low energy giving a large overpotential for water oxidation. The hybridization of O 2*p* orbitals, with nonmetals, such as S, N, F, S, B, P, and C, can lift up the valence band and consequently decrease the energy band gap and make the photocatalytic active under visible light (Fig. 8.4e) [58]. Among this, N was found to be the most promising due to its electronegativity, ionic radii, coordination numbers, and polarizability, similar to oxygen. N 2*p* orbitals display potential energy higher than O 2*p* orbitals, and when they are introduced in the photocatalyst structure, they can decrease the band gap by lifting up the valence band without perturbing the conduction band [32].

In d^0 oxides, the band gap can be narrowed by 1.1–1.5 eV by doping with Cu, Ag, Pb, Rh, Ga, Zn, Fe which can also lift the valence band (Fig. 8.4f). A same approach can be also used to lift down the CB [12, 59, 60]. In QDs, doping with ions as temporary charge carrier trap states can increase the lifetime of excited states and accelerate the rate of the reactions on the surface. QDs can be doped in situ or after, through a cation exchange between the surface and the lattice [61–63]. In core/shell ZnSe/ZnS, Mn^{2+} cations directly added during the synthesis can increase the excited lifetime up 1000 μm [64]. Cu^+ and Ag^+ doping can trap carriers in QDs, such as in InP and in CdSe [65, 66]. Wang et al. reported that the doping of CdS with Ni can increase the selectivity (up to 100%) of CO_2 photoreaction with water while efficiently trapping the photoexcited electrons at the surface and suppressing the hydrogen evolution [67].

Defects in the lattice can also create interband states and influence the band gap. For example, oxygen vacancies create interstitials $Ti^{4+}$and Ti^{3+} states decreasing the band gap to 1.54 eV and therefore boosting the photocatalytic properties of the

TiO_2 [2]. Tungsten oxides can also absorb more at higher wavelength when oxygen vacancies are present. In $CuInS_2$ QDs, possible point defects can cause exciton self-trapping with the delocalization of the holes [68].

8.2.3 Z-Scheme and Other Configuration Methods

The natural process of photosynthesis can be divided into three main steps: (1) solar light harvesting and charge separation in the photosystems I (PS I) and II (PS II); (2) electron transfer between redox parts; and (3) products formation (oxygen and carbohydrates) at the reaction sites. [26] The two systems PSI and PSII for each photon can generate an electron/hole pair although the water splitting process requires four electrons/holes whereas the NADP reduction two electrons/protons reaction [8, 69–71]. Therefore, a multielectron catalyst at the end of the charge separation is necessary as a bridge between the two reaction processes which require different e^-/h^+ pairs [8, 9, 71].

A method to mimic the natural photosynthetic process is to recreate the so-called Z-scheme configuration with, generally, two photoexcitation steps [9, 69, 71, 72].

The classic Z-scheme configuration (Fig. 8.5a) is two-step photoexcitation formed by two photocatalysts having narrow band gap which run separately the oxygen evolution reaction (OER) and the hydrogen evolution reaction (HER) [9, 69]. In such dual configuration, visible light is sufficient because the band gap of each photocatalyst has to satisfy the overpotential, kinetics, and thermodynamics requirements of only half-reaction for the water splitting. In Fig. 8.5a, under visible light e^-/h^+ pairs are generated in both OEP (oxygen evolution photocatalyst) and HEP (electron evolution photocatalyst). After the absorption of a photon in the OEP, the electrons will be excited in the CB by leaving holes in the VB which will oxidize the water oxygen. The electrons, in the CB of the OEP, will be transferred to the VB

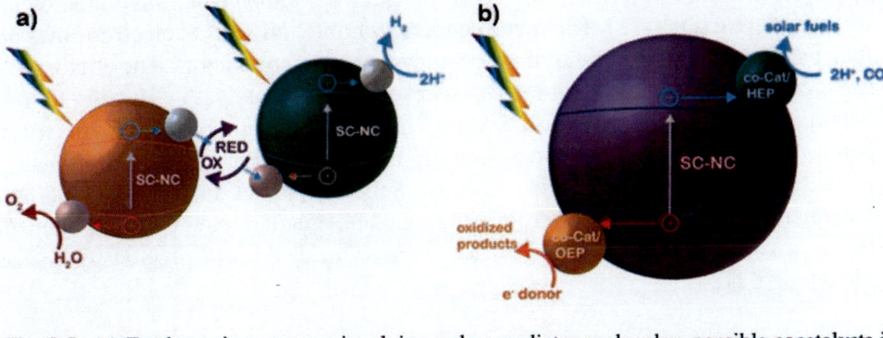

Fig. 8.5 (**a**) Z-scheme in two steps involving redox mediator molecules, possible cocatalysts in both the photocatalysts can be present; (**b**) in one step, sketch of semiconductor nanocrystal used as light antenna coupled with two cocatalysts for the reduction and oxidation, respectively

of the HEP through a shuttle redox couple (ions or molecules) in solution which will firstly reduce the oxidant (OX) to form the electron donors (RED), whereas the photoholes on the HEP surface will be oxidized by converting back the electrons. On the HEP side, the excited electrons in the CB will reduce the water to hydrogen. The main drawbacks of this system are the possible back reactions that can occur, e.g., the shuttle redox couple could react with the VB of the OEP or the CB of HEP [12, 69, 71, 73] A similar approach to separate the charge carriers and transfer them can be obtained with a Type II band alignment, which is similar an inverted Z-scheme without a redox mediator. This approach is commonly applied in photocatalytic systems, and it has good potential especially for just one-half reaction to produce solar fuels.

Another method to mimic the natural photosynthesis is to couple a main semiconductor with two cocatalysts where both the HER and the OER can occur (Fig. 8.5b). In this one-step photoexcited approach, solar light is harvested by the main photocatalyst which should have a suitable band gap to collect a large part of the solar spectrum and a good band alignment with the HE and OE cocatalysts [54, 71, 73]. Important drawbacks are due to the different timescales of the exciton generation and transportation, charge separation, and reaction. Typically, from a dynamic point of view, exciton generation has mainly a picosecond timescale, charge separation, and transfer to the reactive sites on the surface of the photocatalyst have a microsecond timescale while surface reaction of water splitting is much slower, from microsecond to second timescale, especially the OER. O_2 production occurs at the second timescale since it requires a transfer of four holes to the acceptor and a formation of an O-O double bond core [23, 26, 45, 74–76]. In contrast, H_2 production needs just two electrons and it is in millisecond timescale. This discrepancy of timescale of the two main reactions can create recombination of the charge carriers and energy losses. For example, in two of the most widely used photocatalysts, TiO_2 and CdS, the carrier lifetimes are on the nanosecond timescale, too short to run these reactions. Therefore, configurations to increase their carrier lifetime, charge separation, and the rate of the reactions are necessary. Heterojunction configurations can extend the carrier lifetime [77]. Coupling semiconductor/metal with appropriate band alignment can create barriers for either electrons or holes and avoid back reactions [78]. The metal (generally noble) NPs act as electron sinks and they usually show higher catalytic efficiency than semiconductors. The e^- storage in the metal NPs gives more reductive power to the photocatalyst by shifting to more negative value the apparent Fermi level in the metal-SC composite [77]. In a n-type heterojunction metal/semiconductor configuration, the Fermi level of the metal is lower than of the semiconductor, creating a Schottky barrier due to band bending at the interface. In a p-type heterojunction metal/semiconductor configuration, the barrier avoids the holes to be transferred, whereas the electrons can be easily injected in the metal NP [2].

8.3 Colloidal Semiconductors

Colloidal semiconductor nanocrystals, such as quantum dots and perovskite like QDs, can be excellent candidate to be applied in artificial photosystems thanks to their excellent properties to harvest solar light, to generate multiple excitons, and to be engineered in their surface properties [2, 12, 15, 23, 24, 79].

8.3.1 NCs (QDs and Perovskite-QDs) Coupled with Bulk Semiconductors

Semiconductor nanocrystals are demonstrating to be cutting-edge both as photosensitizers and as photocatalysts. As sensitizer, thanks the possibility to easily tune their band gap by changing the chemical composition or the size/morphology, they show high ability to harvest a broad part of the solar spectrum. Several examples are reported in literature of QDs coupled with wide band gap photocatalysts, especially for solar cell applications [80]. For solar fuels generation, PbS QDs coupled on TiO_2 showed to improve the CO_2 conversion up to five times since they can absorb light in the NIR range [81]. CdS QDs were also coupled to WO_3 photocatalyst to efficiently separate photoinduced e^-/h^+ pairs, as in Z-scheme without an electron shuttle [82]. Several examples are reported for photoelectrochemical applications (briefly described below) where TiO_2 or NiO_2 photoanodes are coupled with QDs as sensitizer, such as "giant" PbS/CdS [83], thick-shell CdSe/$CdSe_xS_{1-x}$/CdS QDs with pyramidal shape [84]. For example, CdSe/CdSeS alloy/CdS core/shell/shell QD sensitizer of mesoporous TiO_2 photoanode allowed to reach a very high photocurrent density to 17.5 mA/cm^2 under 1 sun illumination with a long stability [85]. More recently, perovskite type QDs or nanocystrals (PQDs or PNCs) [86] have been received an enormous interest thanks to their tunable properties and very easy and low-cost synthesis. Their main and important drawback is their instability in air and in water. New hybrid inorganic/organic shells and ligands are under study to make these new materials more stable in atmosphere. In nonaqueous medium, $CsPbBr_3$ QDs on graphene oxide were applied to convert CO_2 by improving electron extraction and transport [87]. They can be also stabilized on metal organic framework (MOF) such as UiO-66(NH_2) and without a sacrificial agent the rate of CO production reached 98.57 μmol/g [24]. Inorganic ligand-capped $CsPbBr_3$ NCs were also immobilized with metal complexes like $[Ni(terpy)_2]^{2+}$ (Ni(tpy)) for CO_2 reduction under visible light where Ni(tpy) suppressed the e^-/h^+ recombination by removing electrons [88].

Lead free cesium silver bismuth halide double perovskites, $Cs_2AgBiBr_6$, modified with sulfide showed good stability in moisture, light, and temperature and a good selectivity in the conversion of CO_2 [89]. Metal halide perovskites (MHPs) are at the forefront as promising class of new materials for solar applications. Huang et al. very recently highlighted some strategies to overcome their stability disadvantages, like the application of halogen acid solution for hydrogen production, the use of low polarity solvents for organic conversion and CO_2 reduction and their encapsulation during water-based processes [90].

8.3.2 Heterostructures

Especially, heterostructured nanocrystals with the epitaxial coupling of two or more semiconductors have gained attention thanks to the possibility to spatial manipulate the electron and hole wavefunctions through to their fine controllable synthesis [63, 91–93]. The type of band alignment can confine in different domains of the heterostructures photoexcited electrons and holes (Fig. 8.5a). In Type I band alignment, both the carriers are confined in the same domain whereas in Type II heterostructure, we have that the spatial separation of the photoexcited electron and holes occurs in the interface between two different SCs and this hampers the charge recombination although it increases the confinement and sometimes the unavailability of one charge carrier for surface reactions. In quasi-Type II configuration, one carrier (usually the electron) is delocalized over more than one domain whereas the hole is confined in a single domain [92] (Fig. 8.5a). The extended life times of the charge carriers and their spatial separation in quasi-Type II and Type II can be useful for photochemical production of fuels. CdSe/CdS and CdSe/ZnS core/shell nanocrystals having a small core and a thin shell show a Type I alignment with both electrons and holes in the core. This configuration can be more appropriated for a light harvest material to transport excitons. Instead, CdTe/CdSe and CdSe/ZnTe core/shell QDs show a Type II alignment where the hole is mostly in the core and the electron in the shell and vice versa, respectively [94]. Therefore, the separation of the carriers can be controlled by the morphology, the shell thickness, and core dimension [91, 95–98].

Anisotropic morphology, like nanorods [99], can further help to separate the holes and the electrons and facilitate the extraction of one charge carrier. Spherical QDs (Fig. 8.5b) can be considered 0D where the exciton is quantum confined in all the three dimensions. Elongated QDs in one direction are called nanorods (NR), and they have a diameter of few nanometers and lengths from 10 to 100 nm. In these 1D semiconductor nanostructures, the excitons are confined in the radial direction but not always along the axial direction. Thus, NR can present both properties typical of the QDs like the band gap tunability, strong carrier-carrier interactions, interfacial coupling, and bulk characteristics like long-distance charge transport and separation [100, 101]. Dukovic's group reported several studies on the carriers dynamics in NRs. In CdS and CdSe, NR with nonuniform width (dot in rod) trapped holes localized on undercoordinated chalcogen atoms diffuse with a random walk on the surface of the NRs hopping between trap states, whereas the electrons are more confined into wide-diameter regions of the NRs [92, 102]. Hamby et al. combining CdS NRs in a mixture with *Magnetococcus marinus* MC-1 (MmOGOR) have shown that NRs are a good sensitizer by providing electrons to enzymatically catalyze the formation of carbon–carbon bonds (2-oxoglutarate) via CO_2 reduction [103].

Heterostructured NRs composed by two different semiconductors have shown to further spatial separate the charge carriers. Dot in rod (core/shell) heterostructures, like ZnSe/CdS and CdSe/CdS, have also shown a fast hole trapping in the CdS rod and electron localization into the bulb, which can be manipulated by the chemical composition of the core [104]. More sophisticated heterostructures by creating core/shell semiconductors coupled with (noble) NPs as cocatalyst have gained consideration in photocatalysis, especially for solar fuel production. In platinum-tipped CdSe/CdS NRs, the holes are confined in the core while the electrons can be transferred to the metal NPs on the tip, significantly improving the hydrogen production, with an apparent quantum yield of 20% at 450 nm of excitation [105]. The length of the nanorod can also help in the charge separation as well as the position of the cocatalyst. In bare CdS, increasing the number of the cocatalysts along the rod seems to decrease the hydrogen production although the electron transfer is faster to the metal NP. This is probably because the electron transfer from the cocatalyst to the water reduction reaction is slower as compared to the recombination with the holes. Therefore, tipped CdS NRs showed a better photocatalytic performance [106] (Fig. 8.6).

In nickel-NPs decorating CdS NRs, Simon at al reported the use of a hydroxyl anion/radical couple as redox shuttle to relay the hole from the NR to the scavenger. This increases the H_2 generation and the photostability of the system [107].

Figure 8.7 shows that no-toxic Type II $CuAlS_2$/ZnS NRs can reduce aqueous bicarbonate ions to acetate and formate under visible light, which can be an interesting and encouraging starting point for these materials to be applied in CO_2 conversion under solar light [108]. Copper-based as well as tin-based NCs are a promising and environmental-friendly alternative to cadmium/lead based ones [109–111].

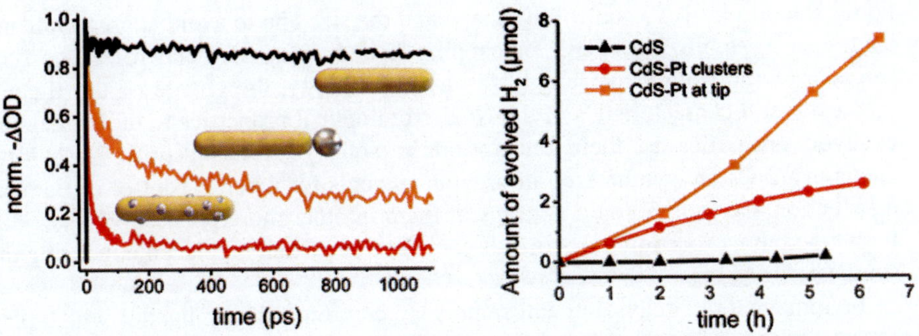

Fig. 8.6 Hydrogen generation by CdS NRs under illumination in neutral aqueous solution in the presence of sulfite ions. (Reprinted with permission from Simon T, et al. (2016) Electron Transfer Rate vs Recombination Losses in Photocatalytic H_2 Generation on Pt-Decorated CdS Nanorods. ACS Energy Lett, 1 (6), 1137–1142). Copyright (2016) American Chemical Society [106]

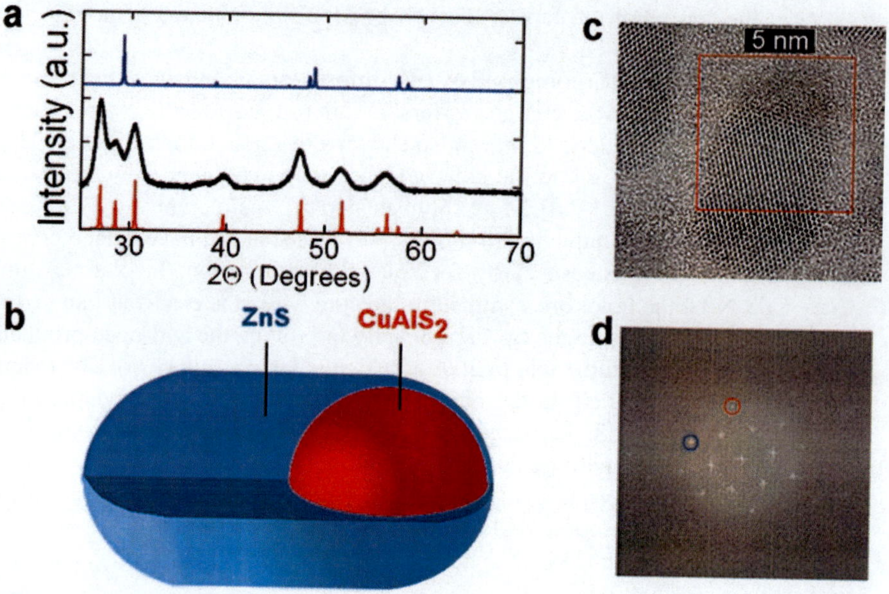

Fig. 8.7 Structural properties of $CuAlS_2/ZnS$ QDs. (**a**) XRD patterns, black line $CuAlS_2/ZnS$ QDs, blue line tetragonal $CuAlS_2$, red line hexagonal ZnS; (**b**) sketch of $CuAlS_2/ZnS$ QDs; (**c**) HRTEM image and (**d**) Fourier transform of the region marked in red in panel (**c**). (Reprinted with permission from Bhattacharyya B, et al. (2018) Efficient Photosynthesis of Organics from Aqueous Bicarbonate Ions by Quantum Dots Using Visible Light. ACS Energy Lett, 3(7): 1508–1514. Copyright (2018) American Chemical Society [108]

8.3.3 Surface Ligands

The best synthesis route to obtain homogenous in size, morphology, physico-chemical properties nanocrystals is a colloidal approach where NCs are formed in liquid solution [14]. To control the shape and the size and to avoid aggregation and coalescence, stabilizing agents or capping ligands play a crucial role. The type (organic or inorganic), the length, the hydrophobicity, the charge of the ligands dictate several characteristics of the NCs, from optical properties to biological and catalytic properties to their integration in solid devices/substrate [112–116]. Generally, NCs are synthesized in organic solvents for a better control of the size distribution and morphology. To transfer them in other more polar solvents or different medium, procedures are ligands exchange or partially replacement during the synthesis [18, 114].

In homogenous catalysis, metal centers are coordinated with ligands which play an important part creating electrostatic and steric effects that influence the catalytic performance, especially the selectivity [117]. They also have a similar role in heterogeneous catalysis with NCs. Capping ligands in NCs have a delicate role since they can either enhance the selectivity and the reaction rate or can have a detrimental effect by blocking adsorption of molecules at the surface [118]. Therefore, to have just positive roles, the coverage of ligands on the surface of the NCs catalysts should not block the active site and should facilitate the interaction between the reactants and the surface of the catalyst and mediate the energy transfer amid NCs and molecules [113, 119, 120]. Their positive effects can be the selectivity control, the chiral modification, the improvement of the adsorption, the surface crowding regulation, the increase of photocatalyst stability as well as the rate and the type of charge transfer [119]. Yu et al. reported an efficient catalytic H_2 production by engineering the ligand of InP and InP/ZnS QDs. Surface S^{2-} ligands showed to increase the photocatalytic performance of the QDs since they can act as hole traps and they can improve the charge separation efficiency of the system. They reported TON values up to 128,000 per InP/ZnS QDs and an internal quantum yield of 31% under green light [121] (Fig. 8.8).

InP-based QDs are promising candidate as photocatalyst since their band gap can be tuned in the NIR region, and they have a pretty high quantum yield and are Cd and Pb free, which can make them more environmentally compatible. InP/ZnS QDs were applied without any sacrificial agent to photocatalyze under visible light a standard C-C reaction [122]. They can also be covered by an oxide shell, like ZnO, to increase their photostability and to decrease reabsorption losses [123].

Ternary Cu-based QDs have also shown to be a good no-toxic photocatalyst [109, 110]. Huang et al. reported the application of $CuInS_2$ (CIS) as sensitizer for a classic rhenium bipyridine complex (Re catalyst) for CO_2 reduction [124]. Through a click chemistry reaction, the CIS QDs were covalently linked to the Re catalyst

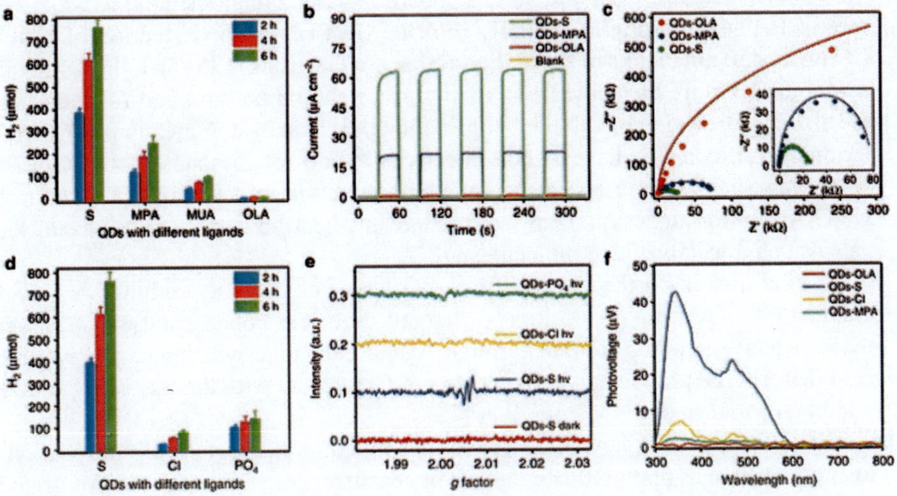

Fig. 8.8 Influence of different ligands on the hydrogen evolution performance and photophysical properties of InP/ZnS QDs (525 nm, 15 min). (**a**) Hydrogen evolution for InP/ZnS QDs capped with different organic ligands; (**b**) transient photocurrent; (**c**) electrochemical impedance spectra; (**d**) hydrogen evolution for InP/ZnS QDs capped with different inorganic ligands. (**e**) EPR spectra; (**f**) SPV spectra (Reprinted under the terms of the Creative Commons CC BY [121])

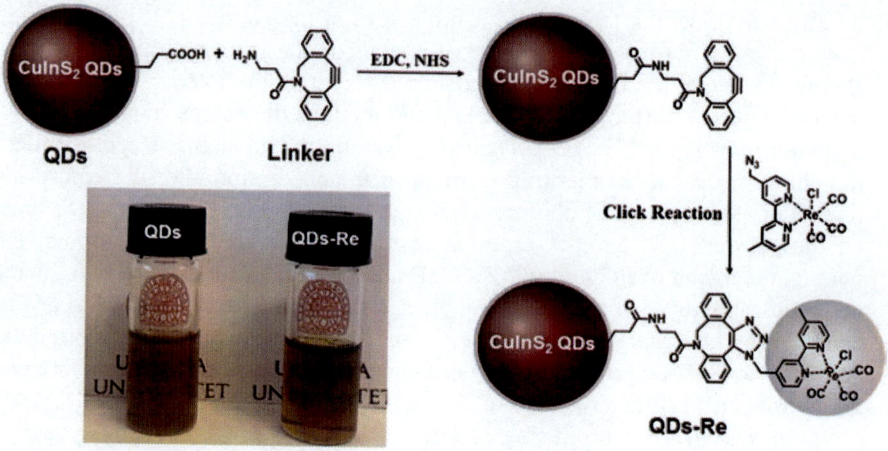

Fig. 8.9 Schematic routes of the covalently linked CuInS2 QD–Re hybrid system. In the photograph $CuInS_2$ QDs and the $CuInS_2$ QD–Re hybrid system in DMSO. (Reproduced under a Creative Commons Attribution 3.0 Unported Licence. Huang et al. (2018) covalently linking $CuInS_2$ quantum dots with a Re catalyst by click reaction for photocatalytic CO_2 reduction. Dalt Trans 47:10775–10783. Published by The Royal Society of Chemistry [124])

and this favored a fast electron transfer which promoted formation of reduced Re species, boosting the photocatalytic performance of the hybrid catalyst [124] (Fig. 8.9).

3-Mercapto-1-propanol (MPO) capped CIS/ZnS QDs have also shown to be 18 times better sensitizer than the more standard Ir-based organometallic sensitizer for meso-tetraphenylporphyrin iron(III) chloride (FeTTP) in the reduction of CO_2 to CO under 450 nm thanks to the ability to transfer electrons [125]. In hybrid systems, where the NCs are the sensitizers, their surface should be modified for increasing the affinity adsorption of CO_2 and eliminating the cocatalyst. For example, hydride-terminated silicon NCs have a good selectivity to convert gaseous CO_2 to CO [126]. Hydrogen-Si QDs were also used for photoreduction in a 0.01 M CO_2-saturated Na_2CO_3 solution under halogen lamp whereas when the surface is oxidized, they can also photocatalyze dye molecules [127].

Bi et al. increased the selectivity of CO_2 conversion by assembling negatively charged CdS QDs with a positively charged dinuclear cobalt catalyst. QDs were made negatively water soluble through ligand exchange with mercaptopropionic acid (MPA). The photocatalytic reduction of CO_2 to CO with this assembly showed good performance under visible light [128].

Weiss outlined four key strategies to be applied on the surface of colloidal NCs to endorse them as photoredox catalysts or sensitizers of redox catalysts for organic transformations: (1) increasing the affinity interaction between the NC and the substrate or cocatalyst in absence of a covalent linkage by tuning surface charge density; (2) optimization of the active sites of the NCs surface thanks to ligand exchange; (3) removing oxidative equivalents from the NC through "hole shuttle" ligands; (4)

decreasing the kinetic barrier for proton coupled electron transfer reactions by controlling the protons concentration on the NC surface [74].

8.3.4 Assembly and Energy Transfer

A remarkable feature of natural PS is the ability to work with the low photon flux of the sunlight, whose transport to the catalytic centers is efficiency-limiting. Indeed, this issue is overcome by the energy transfer [25, 26, 75, 129]. At the catalytic center, then, the excitons can dissociate into free charges. In natural leaves, the pigments in the antenna complexes of the PSII absorb and transfer the excitation energy to the reaction center with a diffusion length of some tens of nanometers. The high energy transfer, more than 80% yield in PSII, is able to deliver excitons in the reaction center every 10 ms which allows oxidizing water with 4 holes at very high quantum efficiency [130].

So far, the majority of sensitized photocatalytic systems is based on the approach 1: sensitizer = 1: catalyst or when multiple sensitizers per catalyst are used, for example by coupling multiple sensitizers to single catalysts, like TiO_2, they do not transfer energy among themselves but (1) directly to the photocatalyst or (2) they funnel charges instead of energy [22, 131]. Self-assembly of CdSe/CdS QDs binding through polyacrylate ions as surface ligands to platinum NPs as cocatalyst showed to increase the hydrogen evolution initial internal quantum yield of ~65% by facilitating the electron transfer among particles upon irradiation with visible light [132].

Excitons transfer is more energy-conservative than charge transfer in highly polar solvents, such as water, since excitons can be considered as "neutral particles" that hop from site to site without provoking a reorganization of the molecules typically of the charging and discharging processes [131]. Recently, Weiss' group has shown that energy migration among QDs assemblies can boost photocatalytic reactions for solar fuels [131, 133]. Kodaimati et al. demonstrated the feasibility of energy transfer sensitization by coupling together through electrostatic interaction two sets of core/shell QDs, CdSe/ZnS, and CdSe/CdS as sensitizer and photocatalyst, respectively. By funneling energy from multiple sensitizer QDs to catalytic QDs with a ratio of 4:1, they increased 13 times the reduction efficiency of protons to H_2 compared to the system where energy transfer was absent [131]. Lian et al. applied negatively charged assemblies of core/shell CIS/ZnS QDs to sensitize positively charged trimethylaminofunctionalized iron tetraphenylporphyrin catalysts (FeTMA) to photoconvert CO_2 to CO in water with 450 nm light. The turnover of CO with a selectivity of 99% was 450 after 30 h of illumination, and the sensitization efficiency was 11 times higher [133] (Fig. 8.10).

Self-assembly in hybrid heterostructures can help either to faster transfer charged carriers or, even better, to transfer excitons to the catalytic center and allows multi-electron reactions to occur also under low photon flux.

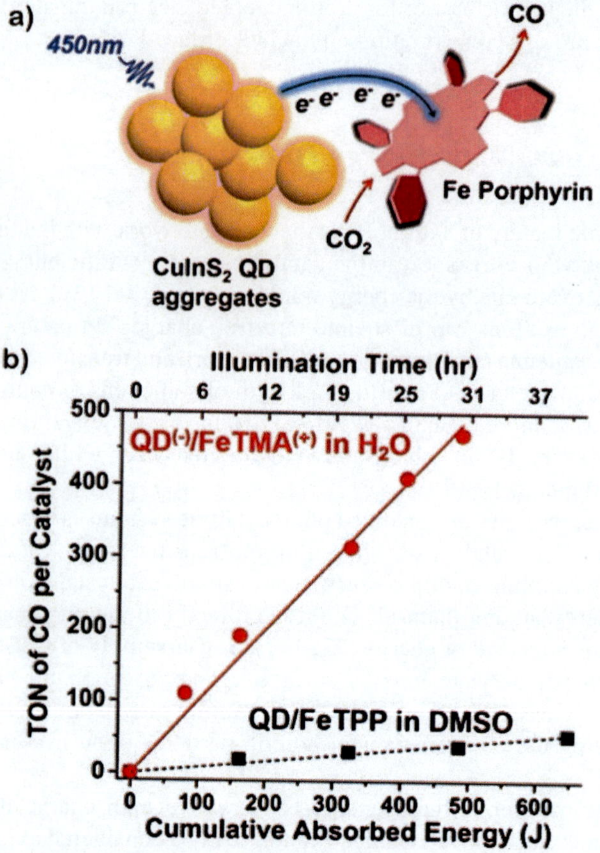

Fig. 8.10 **(a)** QDs assembly sensitizes Fe-porphyrin catalyst for CO_2 reduction; **(b)** plot of TON vs absorbed photon energy. (Reprinted with permission from Lian S, et al. (2018) Photocatalytically Active Superstructures of Quantum Dots and Iron Porphyrins for Reduction of CO_2 to CO in Water. ACS Nano). Copyright (2018) American Chemical Society [133])

8.4 Device Architectures

8.4.1 Current Solar-Based Methods to Produce Solar Fuels

To produce solar fuels or chemicals through solar light and CO_2, several strategies are being applied, such as biocatalytic and thermo-catalytic approaches, and photoelectrochemical techniques. Currently, the three most applied schemes are (Fig. 8.11): (1) photovoltaic–electrolysis (PV–EC) cell; (2) photoelectrochemical (PEC) cell; (3) photocatalytic (PC) cell due to their comparatively low cost and more simplicity architecture. In these methods, it is possible to apply either homogeneous or heterogeneous catalysts [6, 70]. In the homogenous system, photocatalyst, sensitizer, and possible scavengers are in the same phase of the reactants. In the

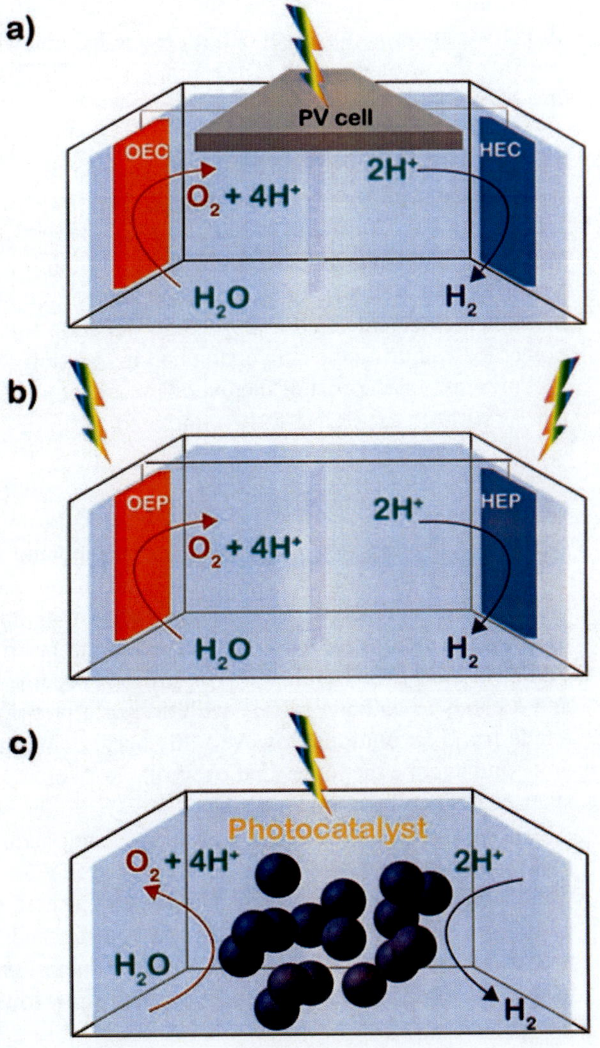

Fig. 8.11 Main devices configurations, inspired by [6]: (**a**) PV-EC cell; (**b**) PEC cell; (**c**) PC cell

past, the most molecular (photo)catalysts studied for these applications were based on rare metal center complexes, like Re, Ru, and Pd, whereas more recently earth-abundant transition metals like Fe, Co, Ni, Mn have been acquiring more attention. Often, the durability of this organometallic complexes-based system is still an issue [70, 134]. Semiconductors as photocatalyst in heterogeneous system can be a good alternative, either applied directly in solution or to build photoanode/ photocathode.

Among the different strategies (Fig. 8.11), PV-EC design has the highest technology readiness level and some commercialization devices are already available on the market. PV-EC configuration uses an independent solar cell (but also other renewable energy sources like geothermal or wind) to produce electricity which is directly transferred as direct current to the EC cell where the anode and the cathode perform the water splitting [135] (Fig. 8.11a). This part could be also in the dark and the solar light could only be used by the PV to produce electricity to "help" the water splitting reaction. Usually, the PV panel is connected in series to the electrolyzer to have the same current and voltage from the two independent devices. To increase the stability and avoid degradation, the PV is generally out of the electrolytic liquid [6]. To design efficient PV-EC, several parameters have to be considered, like exchange of current density, system resistance, electrode area, light intensity, solar cell efficiency, and Tafel slope. Pretty high efficiency, more the 30%, has recently been reached by applying 3-junction solar cells [136]. However, several power losses during the conversion and integration of electrical energy into fuel are still present and the cost of the overall architecture still remains pretty high making the H_2 still too expensive to be compared to fossil fuels and therefore limiting its application on large scale [136].

Another, less complex method to produce solar fuels is based on PEC cell, wherein the light is absorbed by one or both of the photoelectrodes (Fig. 8.11b). The (photo)anode and (photo)cathode are connected and usually only the light absorber side is in contact with the liquid, while the charge collector is isolated. In this system, the two gases, H_2 and O_2, are produced in separate compartments of the vessel keeping the charge balance with a membrane [135]. Different configurations for PEC cells have been developed. The simplest scheme is either an n-type photoanode and a catalytic cathode for H_2 evolution or a p-type photocathode and a catalytic anode for O_2 evolution. However, this basic configuration needs a bias voltage to work and to cope the losses and overvoltage. More complicated configurations consider tandem cells or the PV-PEC.

As photoanode, various types of n-type semiconductors have been studied, especially metal-oxides, like TiO_2, Fe_2O_3, WO_3, $BiVO_4$, since they are very stable in water. On the other hand, usually they absorb a small part of the solar light having a wide band gap and a low charge carrier transfer and fast recombination. The most promising are $BiVO_4$, and Fe_2O_3. Semiconductors suitable for photocathode can have a narrower band gap but most of them show lower performances compared to the photoanode electrode.

The simplest and the cheapest solar fuel production configuration is the PC cell (Fig. 8.11c), where the semiconductor nanoparticles are dispersed in the liquid. The simplicity of the cells makes them very attractive for a possible large-scale application but so far, they show low production efficiency, especially compared to the PV-EC.

The stability of the sensitizer/photocatalyst hybrid system and the drawback reactions, which are more favorable since the evolution of H_2 and O_2 occurs on the same NCs, are the two main issues that need to be overcome to increase the performance of this system. New architectures for new hybrid heterostructured colloidal NCs have been proposed in the previous sections of this chapter.

8.4.2 More Recent Solar-Based Methods to Produce Solar Fuels

Two remarkable factors in the natural photosynthesis are the light harvesting and the ability to run multielectron reactions under low photon flux. From a technical design point of view for the cell, it is possible to couple devices that harvest the solar light to transfer more photons to devices that run the reaction. Luminescent solar concentrators, waveguide polymeric slabs where luminophores are embedded, can collect, down convert, and concentrate solar photons. Liu et al. have reported an enhancement of 420% compared to a PEC device alone when a multishell/core QDS LSC was coupled to the photoanode. The better performance is mainly due to a higher stability of the photoanode thanks to the light down-conversion operated by the LSC [137]. Cambié et al. have also reported that photomicroreactors channels for photoredox reactions can be integrated in the LSC [138–140].

8.5 Summary and Outlook

Colloidal photocatalysis has great potential for solar fuel production and other redox reactions (Table 8.1), showing desirable features either of the homogenous or of the classical heterogeneous photocatalysis. Yet it presents also particular challenges, which can be overcome and controlled by tuning several physico-chemical properties of the nanocrystals. Photocatalytic reactions based on multielectron chemistry are very dependent on active sites, defect type and density, surface properties, and spatial migration of excitons. Colloidal NCs show excellent light absorption, multiple exciton generation, surface functionalization, and charge separation properties which can be easily tuned and controlled by synthesis procedures. By combining a deeper understanding of multielectron chemistry and advanced synthesis strategies, it is possible to design better photocatalytic systems through hybrid coupling, heterostructures, morphologies, surface engineering, and self-assembly techniques. For example, it is possible to dope, to create heterojunction through morphological architectures, to couple with metallic NPs, and to control assembly to further engineer the energy transfer, spatial charge separation, and reactions on the surface.

Colloidal PC system shows a very simple architecture and a low cost since the photocatalyst is directly dispersed in a medium. Some issues need to be overcome in order to increase the efficiency and to make it ready for widespread commercial applications.

Crucial aspects for an ideal photocatalyst are high activity, high stability, low cost, and scalability in its production. For a practical use of NCs, further researches should focus on the developing new cadmium and lead free QDs, such as InP, ternary Cu-based QDs, Ag- and Zn-based QDs. Moreover, to increase the environmental stability, new surface ligands, shell, and functionalization should be investigated

Table 8.1 Comparison of organic dyes, bulk semiconductors, and QDs for solar fuel production, inspired by [24]

Photocatalyst	Advantages	Drawbacks	Examples
Molecular compound	Straightforward design, synthesis, and characterization	Expensive, sometimes scarcity of metal complex, difficult synthesis, purity	Ru- and Ir-based complexes Fe/Cu-based system Earth-Abundant First-Row Transition based systems
Bulk SC	Inexpensive, generally water-based synthesis, easy synthesis, high reaction yield, low toxicity	Wide band gap, mainly absorption in the UV-vis range, low exciton kinetics, fast carrier recombination	TiO_2, Fe_2O_3, WO_3, $BiVO_4$
NC-SC	Properties similar either to homogenous or heterogeneous photocatalysts, high light absorption from the UV-vis to the NIR range, simple synthesis, easy tunable photochemical properties, tunable exciton dynamics, multiple exciton generation, fast and long diffusion length of carriers	Possible toxicity, discrete stability	QDs, NRs, etc.., perovskite NCs

as well as to enhance the surface selectivity for solar fuel conversion. The interesting and keep going progresses in this field push colloidal NCs to be promising candidate either for photosensitizer or photocatalyst, which will allow solar to fuel conversion to be applied on a large scale in the near future.

Acknowledgments The LuSH Art project has received funding from the European Union's Horizon 2020 research and innovation programme under the Marie Skłodowska–Curie grant agreement No 843439. The author thanks Prof. Barbara Bonelli for the constructive comments to improve the readability of the paper.

Bibliography

1. J. Blunden, D.S. Arndt, State of the climate in 2018. Bull. Am. Meteor. Soc. **100**(9), Si–S305 (2019)
2. J.K. Stolarczyk, S. Bhattacharyya, L. Polavarapu, J. Feldmann, Challenges and Prospects in Solar Water Splitting and CO_2 Reduction with Inorganic and Hybrid Nanostructures. ACS Catal. **8**(4), 3602–3635 (2018). https://doi.org/10.1021/acscatal.8b00791
3. X.-B. Li, C.-H. Tung, L.-Z. Wu, Semiconducting quantum dots for artificial photosynthesis. Nat. Rev. Chem. **2**, 160–173 (2018). https://doi.org/10.1038/s41570-018-0024-8

4. R.J. Detz, J.N.H. Reek, B.C.C. van der Zwaan, The future of solar fuels: when could they become competitive? Energy Environ. Sci. **11**, 1653–1669 (2018). https://doi.org/10.1039/C8EE00111A
5. T. Kunene, L. Xiong, J. Rosenthal, Solar-powered synthesis of hydrocarbons from carbon dioxide and water. Proc. Natl. Acad. Sci. U. S. A. **116**, 9693–9695 (2019). https://doi.org/10.1073/pnas.1904856116
6. J.H. Kim, D. Hansora, P. Sharma, et al., Toward practical solar hydrogen production – an artificial photosynthetic leaf-to-farm challenge. Chem. Soc. Rev. **48**, 1908–1971 (2019). https://doi.org/10.1039/C8CS00699G
7. L. Fan, C. Xia, F. Yang, et al., Strategies in catalysts and electrolyzer design for electrochemical CO_2 reduction toward C_{2+} products. Sci. Adv. **6**, eaay3111 (2020). https://doi.org/10.1126/sciadv.aay3111
8. J. Barber, Photosynthetic energy conversion: natural and artificial. Chem. Soc. Rev. **38**, 185–196 (2009). https://doi.org/10.1039/b802262n
9. D.G. Nocera, The artificial leaf. Acc. Chem. Res. **45**, 767–776 (2012). https://doi.org/10.1021/ar2003013
10. H. Zhou, R. Yan, D. Zhang, T. Fan, Challenges and perspectives in designing artificial photosynthetic systems. Chem. A Eur. J. **22**, 9870–9885 (2016). https://doi.org/10.1002/chem.201600289
11. J. Artz, T.E. Müller, K. Thenert, et al., Sustainable conversion of carbon dioxide: an integrated review of catalysis and life cycle assessment. Chem. Rev. **118**, 434–504 (2018). https://doi.org/10.1021/acs.chemrev.7b00435
12. Q. Wang, K. Domen, Particulate photocatalysts for light-driven water splitting: mechanisms, challenges, and design strategies. Chem. Rev. **120**, 919–985 (2020)
13. M.R. Shaner, H.A. Atwater, N.S. Lewis, E.W. McFarland, A comparative technoeconomic analysis of renewable hydrogen production using solar energy. Energy Environ. Sci. **9**, 2354–2371 (2016). https://doi.org/10.1039/C5EE02573G
14. C.B. Murray, D.J. Norris, M.G. Bawendi, Synthesis and characterization of nearly monodisperse CdE (E = S, Se, Te) semiconductor nanocrystallites. J. Am. Chem. Soc. **115**, 8706–8715 (1993). https://doi.org/10.1021/ja00072a025
15. H. Zhao, F. Rosei, Colloidal quantum dots for solar technologies. Chemistry **3**, 229–258 (2017). https://doi.org/10.1016/j.chempr.2017.07.007
16. E.H. Sargent, Infrared quantum dots. Adv. Mater. **17**, 515–522 (2005). https://doi.org/10.1002/adma.200401552
17. M.C. Beard, Multiple exciton generation in semiconductor quantum dots. J. Phys. Chem. Lett. **2**, 1282–1288 (2011). https://doi.org/10.1021/jz200166y
18. F.S. Freyria, J.M. Cordero, J.R. Caram, et al., Near-infrared quantum dot emission enhanced by stabilized self-assembled J-aggregate antennas. Nano Lett. **17**, 7665–7674 (2017). https://doi.org/10.1021/acs.nanolett.7b03735
19. B.J. Walker, V. Bulović, M.G. Bawendi, Quantum dot/J-aggregate blended films for light harvesting and energy transfer. Nano Lett. **10**, 3995–3999 (2010). https://doi.org/10.1021/nl1018639
20. P. Moroz, L. Royo Romero, M. Zamkov, Colloidal semiconductor nanocrystals in energy transfer reactions. Chem. Commun. **55**, 3033–3048 (2019). https://doi.org/10.1039/C9CC00162J
21. B.J. Walker, G.P. Nair, L.F. Marshall, et al., Narrow-band absorption-enhanced quantum dot/J-aggregate conjugates. J. Am. Chem. Soc. **131**, 9624–9625 (2009). https://doi.org/10.1021/ja902813q
22. X.-B. Li, C.-H. Tung, L.-Z. Wu, Quantum dot assembly for light-driven multielectron redox reactions, such as hydrogen evolution and CO_2 reduction. Angew. Chemie Int. Ed. **58**, 10804–10811 (2019). https://doi.org/10.1002/anie.201901267

23. M.S. Kodaimati, K.P. McClelland, C. He, et al., Viewpoint: challenges in colloidal photocatalysis and some strategies for addressing them. Inorg. Chem. **57**, 3659–3670 (2018). https://doi.org/10.1021/acs.inorgchem.7b03182
24. H. Wu, X. Li, C. Tung, L. Wu, Semiconductor quantum dots: an emerging candidate for CO_2 photoreduction. Adv. Mater. **31**, 1900709 (2019). https://doi.org/10.1002/adma.201900709
25. J.R. Caram, S. Doria, D.M. Eisele, et al., Room-temperature micron-scale exciton migration in a stabilized emissive molecular aggregate. Nano Lett. **16**, 6808–6815 (2016). https://doi.org/10.1021/acs.nanolett.6b02529
26. T. Mirkovic, E.E. Ostroumov, J.M. Anna, et al., Light absorption and energy transfer in the antenna complexes of photosynthetic organisms. Chem. Rev. **117**, 249–293 (2017). https://doi.org/10.1021/acs.chemrev.6b00002
27. F.E. Osterloh, Photocatalysis versus photosynthesis: a sensitivity analysis of devices for solar energy conversion and chemical transformations. ACS Energy Lett. **2**, 445–453 (2017). https://doi.org/10.1021/acsenergylett.6b00665
28. A. Mills, S. Le Hunte, An overview of semiconductor photocatalysis. J. Photochem. Photobiol. A Chem. **108**, 1–35 (1997). https://doi.org/10.1016/S1010-6030(97)00118-4
29. Book G IUPAC Gold Book
30. S.C. Roy, O.K. Varghese, M. Paulose, C.A. Grimes, Toward solar fuels: photocatalytic conversion of carbon dioxide to hydrocarbons. ACS Nano **4**, 1259–1278 (2010). https://doi.org/10.1021/nn9015423
31. N. Serpone, A.V. Emeline, V.K. Ryabchuk, et al., Why do hydrogen and oxygen yields from semiconductor-based photocatalyzed water splitting remain disappointingly low? Intrinsic and extrinsic factors impacting surface redox reactions. ACS Energy Lett. **1**, 931–948 (2016). https://doi.org/10.1021/acsenergylett.6b00391
32. W. Wang, M.O. Tadé, Z. Shao, Nitrogen-doped simple and complex oxides for photocatalysis: a review. Prog. Mater. Sci. **92**, 33–63 (2018). https://doi.org/10.1016/j.pmatsci.2017.09.002
33. T. Hisatomi, J. Kubota, K. Domen, Recent advances in semiconductors for photocatalytic and photoelectrochemical water splitting. Chem. Soc. Rev. **43**, 7520–7535 (2014). https://doi.org/10.1039/C3CS60378D
34. X. Wang, C. Li, Interfacial charge transfer in semiconductor-molecular photocatalyst systems for proton reduction. J. Photochem. Photobiol. C: Photochem. Rev. **33**, 165–179 (2017). https://doi.org/10.1016/j.jphotochemrev.2017.10.003
35. X. Chen, S. Shen, L. Guo, S.S. Mao, Semiconductor-based photocatalytic hydrogen generation. Chem. Rev. **110**, 6503–6570 (2010). https://doi.org/10.1021/cr1001645
36. K. Mori, H. Yamashita, M. Anpo, Photocatalytic reduction of CO_2 with H_2O on various titanium oxide photocatalysts. RSC Adv. **2**, 3165 (2012). https://doi.org/10.1039/c2ra01332k
37. B. Bonelli, S. Esposito, F.S. Freyria, Mesoporous titania: synthesis, properties and comparison with non-porous titania, in *Titanium Dioxide*, ed. by M. Janus, (Intech, Rijeka, 2017)
38. F. Freyria, M. Compagnoni, N. Ditaranto, et al., Pure and Fe-doped mesoporous titania catalyse the oxidation of acid orange 7 by H_2O_2 under different illumination conditions: Fe doping improves photocatalytic activity under simulated solar light. Catalysts **7**, 213 (2017). https://doi.org/10.3390/catal7070213
39. R. Nasi, S. Esposito, F.S. Freyria, et al., Application of reverse micelle sol – gel synthesis for bulk doping and heteroatoms surface enrichment. Materials (Basel) **12**(6), 937 (2019). https://doi.org/10.3390/ma12060937
40. F. Ehrat, T. Simon, J.K. Stolarczyk, J. Feldmann, Size effects on photocatalytic H_2 generation with CdSe/CdS core-shell nanocrystals. Z. Phys. Chem. **229**, 205–219 (2015). https://doi.org/10.1515/zpch-2014-0635
41. H. Inoue, T. Shimada, Y. Kou, et al., The water oxidation bottleneck in artificial photosynthesis: how can we get through it? An alternative route involving a two-electron process. ChemSusChem **4**(2), 173–179 (2011). https://doi.org/10.1002/cssc.201000385
42. L. Buzzetti, G.E.M. Crisenza, P. Melchiorre, Mechanistic studies in photocatalysis. Angew. Chemie Int. Ed. **58**, 3730–3747 (2019). https://doi.org/10.1002/anie.201809984

43. S. Chen, L.-W. Wang, Thermodynamic oxidation and reduction potentials of photocatalytic semiconductors in aqueous solution. Chem. Mater. **24**, 3659–3666 (2012). https://doi.org/10.1021/cm302533s
44. H. Yu, X. Huang, P. Wang, J. Yu, Enhanced photoinduced-stability and photocatalytic activity of CdS by dual amorphous cocatalysts: synergistic effect of Ti(IV)-hole cocatalyst and Ni(II)-electron cocatalyst. J. Phys. Chem. C **120**, 3722–3730 (2016). https://doi.org/10.1021/acs.jpcc.6b00126
45. X. Li, J. Yu, M. Jaroniec, X. Chen, Cocatalysts for selective photoreduction of CO_2 into solar fuels. Chem. Rev. **119**, 3962–4179 (2019). https://doi.org/10.1021/acs.chemrev.8b00400
46. T.P. Silverstein, Marcus theory: thermodynamics CAN control the kinetics of electron transfer reactions. J. Chem. Educ. **89**, 1159–1167 (2012). https://doi.org/10.1021/ed1007712
47. R.A. Marcus, N. Sutin, Electron transfers in chemistry and biology. BBA Rev. Bioenerg. **811**(3), 265–322 (1985)
48. R.A. Marcus, On the theory of oxidation-reduction reactions involving electron transfer. I. J. Chem. Phys. **24**, 966–978 (1956). https://doi.org/10.1063/1.1742723
49. IARC Working Group, Solar and ultraviolet radiation, in *Radiation Volume 100 D A Review of Human Carcinogens*, ed. by L. Galiche, (International Agency for Research on Cancer, World Health Organization, Lyon, 2012)
50. X. Zhang, T. Peng, S. Song, Recent advances in dye-sensitized semiconductor systems for photocatalytic hydrogen production. J. Mater. Chem. A **4**, 2365–2402 (2016). https://doi.org/10.1039/C5TA08939E
51. B. O'Regan, M. Grätzel, A low-cost, high-efficiency solar cell based on dye-sensitized colloidal TiO_2 films. Nature **353**, 737–740 (1991). https://doi.org/10.1038/353737a0
52. S. Bensaid, C. Ottone Melis, S. Hernández, et al., A simple model for a complex system: kinetics of water oxidation with the $[Ru(bpy)_3]^{2+}/S_2O_8^{2-}$photosystem as catalyzed by Mn_2O_3 under different illumination conditions Samir. Chem. Eng. J. **311**, 143–152 (2017). https://doi.org/10.1016/j.cej.2016.11.087
53. M. Armandi, S. Hernandez, S. Vankova, et al., Visible-light driven oxidation of water as catalyzed by Co-APO-5 in the presence of Ru sensitizer. ACS Catal. **3**, 1272–1278 (2013). https://doi.org/10.1021/cs400067m
54. Y. Xu, A. Li, T. Yao, et al., Strategies for efficient charge separation and transfer in artificial photosynthesis of solar fuels. ChemSusChem **10**, 4277–4305 (2017). https://doi.org/10.1002/cssc.201701598
55. S. Linic, P. Christopher, D.B. Ingram, Plasmonic-metal nanostructures for efficient conversion of solar to chemical energy. Nat. Mater. **10**, 911–921 (2011). https://doi.org/10.1038/nmat3151
56. Q. Zhang, F. Yang, Z. Xu, et al., Are lanthanide-doped upconversion materials good candidates for photocatalysis? Nanoscale Horiz. **4**, 579–591 (2019). https://doi.org/10.1039/c8nh00373d
57. B. Zhou, B. Shi, D. Jin, X. Liu, Controlling upconversion nanocrystals for emerging applications. Nat. Nanotechnol. **10**, 924–936 (2015). https://doi.org/10.1038/nnano.2015.251
58. A. Lin, D. Qi, H. Ding, et al., Carbon-doped titanum dioxide nanocrystals for highly efficient dye-sensitized solar cells. Catal. Today **281**, 636–641 (2017). https://doi.org/10.1016/j.cattod.2016.06.033
59. K. Watanabe, A. Iwase, S. Nozawa, et al., Effects of coapplication of Rh-doping and Ag-substitution on the band structure of Li_2TiO_3 and the photocatalytic property. ACS Sustain. Chem. Eng. **7**, 9881–9887 (2019). https://doi.org/10.1021/acssuschemeng.9b00513
60. M. Etzi Coller Pascuzzi, E. Selinger, A. Sacco, et al., Beneficial effect of Fe addition on the catalytic activity of electrodeposited MnO_x films in the water oxidation reaction. Electrochim. Acta **284**, 294–302 (2018). https://doi.org/10.1016/j.electacta.2018.07.148
61. K. Miszta, D. Dorfs, A. Genovese, et al., Cation exchange reactions in colloidal branched nanocrystals. ACS Nano. **5**, 7176–7183 (2011). https://doi.org/10.1021/nn201988w

62. H. Li, M. Zanella, A. Genovese, et al., Sequential cation exchange in nanocrystals: preservation of crystal phase and formation of metastable phases. Nano Lett. **11**, 4964–4970 (2011). https://doi.org/10.1021/nl202927a
63. L. De Trizio, L. Manna, Forging colloidal nanostructures via cation exchange reactions. Chem. Rev. **116**, 10852–10887 (2016)
64. C. Pu, J. Ma, H. Qin, et al., Doped semiconductor-nanocrystal emitters with optimal photoluminescence decay dynamics in microsecond to millisecond range: synthesis and applications. ACS Cent. Sci. **2**, 32–39 (2016). https://doi.org/10.1021/acscentsci.5b00327
65. H.D. Nelson, S.O.M. Hinterding, R. Fainblat, et al., Mid-gap states and normal vs inverted bonding in luminescent Cu^+- and Ag^+-doped CdSe nanocrystals. J. Am. Chem. Soc. **139**, 6411–6421 (2017). https://doi.org/10.1021/jacs.7b01924
66. K.E. Hughes, S.R. Ostheller, H.D. Nelson, D.R. Gamelin, Copper's role in the photoluminescence of $Ag_{1-x}Cu_x\ InS_2$ nanocrystals, from copper-doped $AgInS_2$ (x~0) to $CuInS_2$ (x=1). Nano Lett. **19**, 1318–1325 (2019). https://doi.org/10.1021/acs.nanolett.8b04905
67. J. Wang, T. Xia, L. Wang, et al., Enabling visible-light-driven selective CO_2 reduction by doping quantum dots: trapping electrons and suppressing H_2 evolution. Angew. Chemie Int. Ed. **57**, 16447–16451 (2018). https://doi.org/10.1002/anie.201810550
68. K.E. Knowles, H.D. Nelson, T.B. Kilburn, D.R. Gamelin, Singlet–triplet splittings in the luminescent excited states of colloidal Cu^+ :CdSe, Cu^+:InP, and $CuInS_2$ nanocrystals: charge-transfer configurations and self-trapped excitons. J. Am. Chem. Soc. **137**, 13138–13147 (2015). https://doi.org/10.1021/jacs.5b08547
69. Y. Wang, H. Suzuki, J. Xie, et al., Mimicking natural photosynthesis: solar to renewable H_2 fuel synthesis by Z-scheme water splitting systems. Chem. Rev. **118**, 5201–5241 (2018). https://doi.org/10.1021/acs.chemrev.7b00286
70. A.U. Pawar, C.W. Kim, M.T. Nguyen-Le, Y.S. Kang, General review on the components and parameters of photoelectrochemical system for CO_2 reduction with in situ analysis. ACS Sustain. Chem. Eng. **7**, 7431–7455 (2019)
71. F. Wen, C. Li, Hybrid artificial photosynthetic systems comprising semiconductors as light harvesters and biomimetic complexes as molecular cocatalysts. Acc. Chem. Res. **46**, 2355–2364 (2013). https://doi.org/10.1021/ar300224u
72. S. Sato, T. Arai, T. Morikawa, et al., Selective CO_2 conversion to formate conjugated with H_2O oxidation utilizing semiconductor/complex hybrid photocatalysts. J. Am. Chem. Soc. **133**, 15240–15243 (2011). https://doi.org/10.1021/ja204881d
73. K. Maeda, Z-scheme water splitting using two different semiconductor photocatalysts. ACS Catal. **3**, 1486–1503 (2013). https://doi.org/10.1021/cs4002089
74. E.A. Weiss, Designing the surfaces of semiconductor quantum dots for colloidal photocatalysis. ACS Energy Lett. **2**(5), 1005–1013 (2017). https://doi.org/10.1021/acsenergylett.7b00061
75. G.D. Scholes, G.R. Fleming, A. Olaya-Castro, R. Van Grondelle, Lessons from nature about solar light harvesting. Nat. Chem. **3**, 763–774 (2011)
76. T. Kondo, J.B. Gordon, A. Pinnola, et al., Microsecond and millisecond dynamics in the photosynthetic protein LHCSR1 observed by single-molecule correlation spectroscopy. Proc. Natl. Acad. Sci. U. S. A. **166**, 11247–11252 (2019). https://doi.org/10.1073/pnas.1821207116
77. P.V. Kamat, Manipulation of charge transfer across semiconductor interface. A criterion that cannot be ignored in photocatalyst design. J. Phys. Chem. Lett. **3**, 663–672 (2012). https://doi.org/10.1021/jz201629p
78. H. Kisch, Semiconductor photocatalysis - mechanistic and synthetic aspects. Angew. Chemie Int. Ed. **52**, 812–847 (2013). https://doi.org/10.1002/anie.201201200
79. N.T.K. Thanh, N. Maclean, S. Mahiddine, Mechanisms of nucleation and growth of nanoparticles in solution. Chem. Rev. **114**, 7610–7630 (2014). https://doi.org/10.1021/cr400544s
80. D. Wang, F. Yin, Z. Du, et al., Recent progress in quantum dot-sensitized solar cells employing metal chalcogenides. J. Mater. Chem. A **7**, 26205–26226 (2019). https://doi.org/10.1039/c9ta10557c

81. C. Wang, R.L. Thompson, P. Ohodnicki, et al., Size-dependent photocatalytic reduction of CO_2 with PbS quantum dot sensitized TiO_2 heterostructured photocatalysts. J. Mater. Chem. **21**, 13452 (2011). https://doi.org/10.1039/c1jm12367j
82. J. Jin, J. Yu, D. Guo, et al., A hierarchical Z-scheme CdS-WO_3 photocatalyst with enhanced CO_2 reduction activity. Small **11**, 5262–5271 (2015). https://doi.org/10.1002/smll.201500926
83. L. Jin, G. Sirigu, X. Tong, et al., Engineering interfacial structure in "Giant" PbS/CdS quantum dots for photoelectrochemical solar energy conversion. Nano Energy **30**, 531–541 (2016). https://doi.org/10.1016/j.nanoen.2016.10.029
84. H. Zhao, G. Liu, F. Vidal, et al., Colloidal thick-shell pyramidal quantum dots for efficient hydrogen production. Nano Energy **53**, 116–124 (2018). https://doi.org/10.1016/j.nanoen.2018.08.042
85. K. Wang, X. Tong, Y. Zhou, et al., Efficient solar-driven hydrogen generation using colloidal heterostructured quantum dots. J. Mater. Chem. A **7**, 14079–14088 (2019). https://doi.org/10.1039/c9ta03026c
86. H. Utzat, K.E. Shulenberger, O.B. Achorn, et al., Probing linewidths and biexciton quantum yields of single cesium lead halide nanocrystals in solution. Nano Lett. **17**, 6838–6846 (2017). https://doi.org/10.1021/acs.nanolett.7b03120
87. Y.-F. Xu, M.-Z. Yang, B.-X. Chen, et al., A $CsPbBr_3$ perovskite quantum dot/graphene oxide composite for photocatalytic CO_2 reduction. J. Am. Chem. Soc. **139**, 5660–5663 (2017). https://doi.org/10.1021/jacs.7b00489
88. Z. Chen, Y. Hu, J. Wang, et al., Boosting photocatalytic CO_2 reduction on $CsPbBr_3$ perovskite nanocrystals by immobilizing metal complexes. Chem. Mater. **32**, 1517–1525 (2020). https://doi.org/10.1021/acs.chemmater.9b04582
89. N. Pai, J. Lu, M. Wang, et al., Enhancement of the intrinsic light harvesting capacity of $Cs_2AgBiBr_6$ double perovskite via modification with sulphide. J. Mater. Chem. A **8**, 2008–2020 (2020). https://doi.org/10.1039/C9TA10422D
90. H. Huang, B. Pradhan, J. Hofkens, et al., Solar-driven metal halide perovskite photocatalysis: design, stability, and performance. ACS Energy Lett., 1107–1123 (2020). https://doi.org/10.1021/acsenergylett.0c00058
91. G.S. Selopal, H. Zhao, Z.M. Wang, F. Rosei, Core/shell quantum dots solar cells. Adv. Funct. Mater. **27**(30), 1701468 (2020). https://doi.org/10.1002/adfm.201908762
92. J.K. Utterback, H. Hamby, O.M. Pearce, et al., Trapped-hole diffusion in photoexcited CdSe nanorods. J. Phys. Chem. C **122**, 16974–16982 (2018). https://doi.org/10.1021/acs.jpcc.8b05031
93. H. Li, A.G. Kanaras, L. Manna, Colloidal branched semiconductor nanocrystals: state of the art and perspectives. Acc. Chem. Res. **46**, 1387–1396 (2013)
94. S. Kim, B. Fisher, H.-J. Eisler, M. Bawendi, Type-II quantum dots: CdTe/CdSe(core/shell) and CdSe/ZnTe(core/shell) heterostructures. J. Am. Chem. Soc. **125**, 11466–11467 (2003). https://doi.org/10.1021/ja0361749
95. L. Wang, K. Nonaka, T. Okuhata, et al., Quasi-type II carrier distribution in CdSe/CdS core/shell quantum dots with type I band alignment. J. Phys. Chem. C **122**, 12038–12046 (2018). https://doi.org/10.1021/acs.jpcc.7b11684
96. S.S. Lo, T. Mirkovic, C.H. Chuang, et al., Emergent properties resulting from type-II band alignment in semiconductor nanoheterostructures. Adv. Mater. **23**, 180–197 (2011). https://doi.org/10.1002/adma.201002290
97. I. Coropceanu, M.G. Bawendi, Core/shell quantum dot based luminescent solar concentrators with reduced reabsorption and enhanced efficiency. Nano Lett. **14**, 4097–4101 (2014). https://doi.org/10.1021/nl501627e
98. S. Brovelli, R.D. Schaller, S.A. Crooker, et al., Nano-engineered electron–hole exchange interaction controls exciton dynamics in core–shell semiconductor nanocrystals. Nat. Commun. **2**, 280 (2011). https://doi.org/10.1038/ncomms1281

99. I. Coropceanu, A. Rossinelli, J.R. Caram, et al., Slow-injection growth of seeded CdSe/CdS nanorods with unity fluorescence quantum yield and complete shell to core energy transfer. ACS Nano. **10**, 3295–3301 (2016). https://doi.org/10.1021/acsnano.5b06772
100. K. Wu, T. Lian, Quantum confined colloidal nanorod heterostructures for solar-to-fuel conversion. Chem. Soc. Rev. **45**, 3781–3810 (2016). https://doi.org/10.1039/C5CS00472A
101. K. Wu, Z. Chen, H. Lv, et al., Hole removal rate limits photodriven H_2 generation efficiency in CdS-Pt and CdSe/CdS-Pt semiconductor nanorod–metal tip heterostructures. J. Am. Chem. Soc. **136**, 7708–7716 (2014). https://doi.org/10.1021/ja5023893
102. J.K. Utterback, A.N. Grennell, M.B. Wilker, et al., Observation of trapped-hole diffusion on the surfaces of CdS nanorods. Nat. Chem. **8**, 1061–1066 (2016). https://doi.org/10.1038/nchem.2566
103. H. Hamby, B. Li, K.E. Shinopoulos, et al., Light-driven carbon−carbon bond formation via CO_2 reduction catalyzed by complexes of CdS nanorods and a 2-oxoacid oxidoreductase. Proc. Natl. Acad. Sci. U. S. A. **117**, 135–140 (2020). https://doi.org/10.1073/pnas.1903948116
104. A.N. Grennell, J.K. Utterback, O.M. Pearce, et al., Relationships between exciton dissociation and slow recombination within ZnSe/CdS and CdSe/CdS dot-in-rod heterostructures. Nano Lett. **17**, 3764–3774 (2017). https://doi.org/10.1021/acs.nanolett.7b01101
105. L. Amirav, A.P. Alivisatos, Photocatalytic hydrogen production with tunable nanorod heterostructures. J. Phys. Chem. Lett. **1**, 1051–1054 (2010). https://doi.org/10.1021/jz100075c
106. T. Simon, M.T. Carlson, J.K. Stolarczyk, J. Feldmann, Electron transfer rate vs recombination losses in photocatalytic H_2 generation on Pt-decorated CdS nanorods. ACS Energy Lett. **1**, 1137–1142 (2016). https://doi.org/10.1021/acsenergylett.6b00468
107. T. Simon, N. Bouchonville, M.J. Berr, et al., Redox shuttle mechanism enhances photocatalytic H_2 generation on Ni-decorated CdS nanorods. Nat. Mater. **13**, 1013–1018 (2014). https://doi.org/10.1038/nmat4049
108. B. Bhattacharyya, A.K. Simlandy, A. Chakraborty, et al., Efficient photosynthesis of organics from aqueous bicarbonate ions by quantum dots using visible light. ACS Energy Lett. **3**, 1508–1514 (2018). https://doi.org/10.1021/acsenergylett.8b00886
109. E.C. Hansen, S.N. Bertram, J.J. Yoo, M.G. Bawendi, Zinc thiolate enables bright Cu-deficient Cu-In-S/ZnS quantum dots. Small **1901462**, 1–6 (2019). https://doi.org/10.1002/smll.201901462
110. E.C. Hansen, Y. Liu, H. Utzat, et al., Blue light emitting defective nanocrystals composed of earth-abundant elements. Angew. Chemie **59**(2), 860–867 (2019). https://doi.org/10.1002/ange.201911436
111. H. Zhao, C. Wang, G. Liu, et al., Efficient and stable hydrogen evolution based on earth-abundant SnSe nanocrystals. Appl. Catal. B Environ. **264**, 118526 (2020). https://doi.org/10.1016/j.apcatb.2019.118526
112. L.M. Rossi, J.L. Fiorio, M.A.S. Garcia, C.P. Ferraz, The role and fate of capping ligands in colloidally prepared metal nanoparticle catalysts. Dalton Trans. **47**, 5889–5915 (2018). https://doi.org/10.1039/C7DT04728B
113. P. Liu, R. Qin, G. Fu, N. Zheng, Surface coordination chemistry of metal nanomaterials. J. Am. Chem. Soc. **139**, 2122–2131 (2017). https://doi.org/10.1021/jacs.6b10978
114. Y. Chen, J.M. Cordero, H. Wang, et al., A ligand system for the flexible functionalization of quantum dots via click chemistry. Angew. Chemie Int. Ed. **130**, 4742–4746 (2018). https://doi.org/10.1002/anie.201801113
115. D.R. Kauffman, P.R. Ohodnicki, B.W. Kail, C. Matranga, Selective electrocatalytic activity of ligand stabilized copper oxide nanoparticles. J. Phys. Chem. Lett. **2**, 2038–2043 (2011). https://doi.org/10.1021/jz200850y
116. H. Fujiwara, H. Hosokawa, K. Murakoshi, et al., Effect of surface structures on photocatalytic CO_2 reduction using quantized CdS nanocrystallites. J. Phys. Chem. B **101**, 8270–8278 (1997). https://doi.org/10.1021/jp971621q

117. M.D. Kärkäs, O. Verho, E.V. Johnston, B. Åkermark, Artificial photosynthesis: molecular systems for catalytic water oxidation. Chem. Rev. **114**, 11863–12001 (2014). https://doi.org/10.1021/cr400572f
118. S. Kunz, Supported, ligand-functionalized nanoparticles: an attempt to rationalize the application and potential of ligands in heterogeneous catalysis. Top. Catal. **59**, 1671–1685 (2016). https://doi.org/10.1007/s11244-016-0687-7
119. Z. Niu, Y. Li, Removal and utilization of capping agents in nanocatalysis. Chem. Mater. **26**, 72–83 (2014). https://doi.org/10.1021/cm4022479
120. Z. Huang, M.L. Tang, Designing transmitter ligands that mediate energy transfer between semiconductor nanocrystals and molecules. J. Am. Chem. Soc. **139**, 9412–9418 (2017). https://doi.org/10.1021/jacs.6b08783
121. S. Yu, X.B. Fan, X. Wang, et al., Efficient photocatalytic hydrogen evolution with ligand engineered all-inorganic InP and InP/ZnS colloidal quantum dots. Nat. Commun. **9**, 1–10 (2018). https://doi.org/10.1038/s41467-018-06294-y
122. I.N. Chakraborty, S. Roy, G. Devatha, et al., InP/ZnS quantum dots as efficient visible-light photocatalysts for redox and carbon–carbon coupling reactions. Chem. Mater. **31**, 2258–2262 (2019). https://doi.org/10.1021/acs.chemmater.9b00086
123. S. Sadeghi, H. Bahmani Jalali, R. Melikov, et al., Stokes-shift-engineered indium phosphide quantum dots for efficient luminescent solar concentrators. ACS Appl. Mater. Interfaces **10**(15), 12975–12982 (2018). https://doi.org/10.1021/acsami.7b19144
124. J. Huang, M.G. Gatty, B. Xu, et al., Covalently linking $CuInS_2$ quantum dots with a Re catalyst by click reaction for photocatalytic CO_2 reduction. Dalton Trans. **47**, 10775–10783 (2018). https://doi.org/10.1039/C8DT01631C
125. S. Lian, M.S. Kodaimati, D.S. Dolzhnikov, et al., Powering a CO_2 reduction catalyst with visible light through multiple sub-picosecond electron transfers from a quantum dot. J. Am. Chem. Soc. **139**, 8931–8938 (2017). https://doi.org/10.1021/jacs.7b03134
126. W. Sun, C. Qian, L. He, et al., Heterogeneous reduction of carbon dioxide by hydride-terminated silicon nanocrystals. Nat. Commun. **7**, 12553 (2016). https://doi.org/10.1038/ncomms12553
127. Z. Kang, C.H.A. Tsang, N.-B. Wong, et al., Silicon quantum dots: a general photocatalyst for reduction, decomposition, and selective oxidation reactions. J. Am. Chem. Soc. **129**, 12090–12091 (2007). https://doi.org/10.1021/ja075184x
128. Q.-Q. Bi, J.-W. Wang, J.-X. Lv, et al., Selective photocatalytic CO_2 reduction in water by electrostatic assembly of CdS nanocrystals with a dinuclear cobalt catalyst. ACS Catal. **8**, 11815–11821 (2018). https://doi.org/10.1021/acscatal.8b03457
129. S. Doria, T. Sinclair, N. Klein, et al., Photochemical control of exciton superradiance in light-harvesting nanotubes. ACS Nano. **12**, 4556–4564. https://doi.org/10.1021/acsnano.8b00911
130. K. Amarnath, D.I.G. Bennett, A.R. Schneider, G.R. Fleming, Multiscale model of light harvesting by photosystem II in plants. Proc. Natl. Acad. Sci. U. S. A. **113**, 1156–1161 (2016). https://doi.org/10.1073/pnas.1524999113
131. M.S. Kodaimati, S. Lian, G.C. Schatz, E.A. Weiss, Energy transfer-enhanced photocatalytic reduction of protons within quantum dot light-harvesting–catalyst assemblies. Proc. Natl. Acad. Sci. **115**, 8290–8295 (2018). https://doi.org/10.1073/pnas.1805625115
132. X.-B. Li, Y.-J. Gao, Y. Wang, et al., Self-assembled framework enhances electronic communication of ultrasmall-sized nanoparticles for exceptional solar hydrogen evolution. J. Am. Chem. Soc. **139**, 4789–4796 (2017). https://doi.org/10.1021/jacs.6b12976
133. S. Lian, M.S. Kodaimati, E.A. Weiss, Photocatalytically active superstructures of quantum dots and iron porphyrins for reduction of CO_2 to CO in water. ACS Nano. **12**(1), 568–575 (2018). https://doi.org/10.1021/acsnano.7b07377
134. R. Francke, B. Schille, M. Roemelt, Homogeneously catalyzed electroreduction of carbon dioxide—methods, mechanisms, and catalysts. Chem. Rev. **118**, 4631–4701 (2018). https://doi.org/10.1021/acs.chemrev.7b00459

135. C. Ottone, S. Hernández, M. Armandi, B. Bonell, *Testing Novel Water Oxidation Catalysts for Solar Fuels Production* (Springer International Publishing, Cham, 2019)
136. J. Jia, L.C. Seitz, J.D. Benck, et al., Solar water splitting by photovoltaic-electrolysis with a solar-to-hydrogen efficiency over 30%. Nat. Commun. **7**, 13237 (2016). https://doi.org/10.1038/ncomms13237
137. G. Liu, B. Sun, H. Li, et al., Integration of photoelectrochemical devices and luminescent solar concentrators based on giant quantum dots for highly stable hydrogen generation. J. Mater. Chem. A **7**, 18529–18537 (2019). https://doi.org/10.1039/C9TA06437K
138. D. Cambié, F. Zhao, V. Hessel, et al., A leaf-inspired luminescent solar concentrator for energy-efficient continuous-flow photochemistry. Angew. Chemie Int. Ed. **56**, 1050–1054 (2017). https://doi.org/10.1002/anie.201611101
139. D. Cambié, J. Dobbelaar, P. Riente, et al., Energy-efficient solar photochemistry with luminescent solar concentrator based photomicroreactors. Angew. Chemie Int. Ed. **58**, 14374–14378 (2019). https://doi.org/10.1002/anie.201908553
140. F. Zhao, D. Cambié, J. Janse, et al., Scale-up of a luminescent solar concentrator-based photomicroreactor via numbering-up. ACS Sustain. Chem. Eng. **6**, 422–429 (2018). https://doi.org/10.1021/acssuschemeng.7b02687

Chapter 9
Photocatalysis with Nanoparticles for Environmental Applications: Reactor Design Issues

Ilenia Rossetti, Francesco Conte, Antonio Tripodi, and Gianguido Ramis

9.1 Introduction

3×10^{24} J is the annual solar irradiance, ca. 10,000 times the worldwide energy consumption [1, 2]. This strongly supports the use of solar power to drive chemical processes, either for the photocatalytic/photochemical synthesis of fuels or chemicals or for the photoconversion of noxious compounds.

The photodegradation of various pollutants has been proposed through photocatalytic processes thanks to the in situ formation of powerful oxidising species upon irradiation. Different approaches can involve homogeneous processes, such as the Photo-Fenton reaction [3] or photooxidation through UVC irradiation of H_2O_2 or heterogeneous photocatalysis. The heterogeneous photocatalytic approach has the advantage of separating and recovering the solid catalyst or using it immobilised on a separate phase with respect to the reacting mixture, limiting its impact, especially when treating wastewaters [2, 4–6].

By far, the most used heterogeneous photocatalyst is TiO_2, which can be photoexcited upon absorption of a radiation with energy larger than its bandgap. An electron is thus promoted to the conduction band of the semiconductor, leaving a hole in the valence band,

$$TiO_2 \xrightarrow{h\nu \geq E_g} h^+ + e^-.$$

I. Rossetti (✉) · F. Conte · A. Tripodi
Chemical Plants and Industrial Chemistry Group, Dipartimento di Chimica, Università degli Studi di Milano and INSTM Unit Milano-Università, Milan, Italy
e-mail: ilenia.rossetti@unimi.it

G. Ramis
DICCA, Università degli Studi di Genova and INSTM Unit-Genova, Genova, Italy

M. Piumetti, S. Bensaid (eds.), *Nanostructured Catalysts for Environmental Applications*, https://doi.org/10.1007/978-3-030-58934-9_9

The hole h^+ is a strong oxidant, with a Relative Oxidation Power ROP = 2.35 [7] and can oxidise directly an organic pollutant in water or interact with water forming hydroxyl radicals, characterised by slightly lower ROP = 1.25,

$$h^+ + R \rightarrow R^{+\cdot},$$

$$h^+ + H_2O \rightarrow OH^{\cdot} + H^+,$$

$$OH^{\cdot} + RH \rightarrow R^{\cdot} + H_2O$$

with R being a generic organic moiety. On the other hand, O_2 can act as electron scavenger forming a superoxide radical anion, which in turn acts as oxidant,

$$e^- + O_2 \rightarrow O_2^{\cdot -},$$

$$O_2^{\cdot -} + H^+ \rightleftarrows HO_2^{\cdot},$$

$$2\,HO_2^{\cdot} \rightarrow H_2O_2,$$

$$H_2O_2 + e^- \rightarrow OH^{\cdot} + OH^-.$$

All the radicals so formed contribute to the oxidation of the target pollutant.

Various applications of photocatalytic oxidation processes can be found in either liquid or gas phase [8]. However, the research is mainly focussed on the development of photoactive materials, often neglecting the careful design of the photoreactor and of the relative process. This is a key point, since the reactor geometry and size, the radiation source and its distribution across the reacting medium hugely impact the photocatalytic performance, determining the success or failure of the process and posing important issues for the scale-up. Therefore, this chapter presents some different photoreactor configurations, predominantly focussing on water treatment, but dealing also in some cases with different reactions, e.g. H_2 production or CO_2 photoreduction if they present interesting reactor layouts that may suggest improvements also for water decontamination. A further section deals with modelling of different photoreactors and presents the main descriptors and mathematical approaches for their design.

9.2 Photoreactor Configurations

The main features that an efficient photoreactor should exhibit are: (1) high surface exposure of the photocatalyst per unit volume; (2) appropriate mixing and turbulence regime to ensure fast mass transfer of the reactants and efficient suspension of the catalyst in the case of slurry configuration; (3) good transparency of the solution/

suspension to radiation and (4) effective and continuous oxygen supply for oxidation reactions.

Different types of photoreactors are described in the literature. Some examples are reported in Fig. 9.1.

The most used configuration for photoreactors is the slurry one, at least in a research stage, where the catalyst is suspended in the aqueous medium containing the pollutants and irradiated either from the external surface of the reactor or through immersed lamps. This configuration has the advantage of exploiting the intrinsic activity of the catalytic material and providing optimal active site exposure to light and to the contaminant. Provided that efficient mixing conditions are achieved, it might ensure that mass transfer limitations are negligible and that statistically all the solid particles can be reached by radiation. However, a portion of the whole solid volume is shadowed by the absorption and scattering effects of the solid itself, depending on the solid particle concentration. To limit this drawback, typically small reactor volume is adopted, obtaining a more uniform radiation pattern penetrating the whole reacting volume. Furthermore, issues for scale-up and industrial exploitation arise from the need to separate and recover the catalyst after use.

A UV Free-Surface Reactor (UV-FSR) has been used for the degradation of dyes in a pilot 10 L scale [10]. It is basically a stirred tank reactor, working in continuous or batch mode, with part of its surface irradiated by one or several UV lamps. Intensive stirring throughout the process provides almost ideally turbulent conditions in the aqueous reactor content. This results in a constantly renewed surface, allowing water or contaminants to meet statistically the photon source, independent of colour and turbidity of the suspension or solution. It was tested with or without a heterogeneous photocatalyst as AOP for water treatment.

A fully new concept of high pressure slurry photoreactor, operating up to 20 bar pressure, has been developed. This prototype, with ca. 1.5 L capacity, has been used at the moment to improve the CO_2 solubility in water, boostening the reaction rate for CO_2 photoreduction and for the photoreforming of organic molecules for H_2 production [11–18]. However, the possibility to increase significantly O_2 concentration in water makes it very promising for also oxidation reactions in water treatment.

The choice between the use of suspended catalysts in a slurry arrangement or immobilised materials is hard. On one hand, a slurry configuration is preferable to maintain unaltered the intrinsic activity of the photocatalyst, since after deposition, some decrease of performance is often noticed. However, in order to achieve continuous operation when using slurry type reactors, as required for water treatment, catalyst separation, recovery and recycle are needed. A possible option is to use a selective membrane in a hybrid membrane photoreactor, as recently reviewed [5, 19]. In this arrangement, a slurry type photoreactor is used, with suitable irradiation, and a selective membrane is immersed in it to collect the clarified solution as permeate, retaining the photocatalyst in the reactor [20]. A fouling problem may arise for the membrane, since, depending on the surface charge on the photocatalyst particles, agglomeration may occur, altering the membrane properties [21]. A possible solution consists in adapting the solution pH based on surface charge measurements.

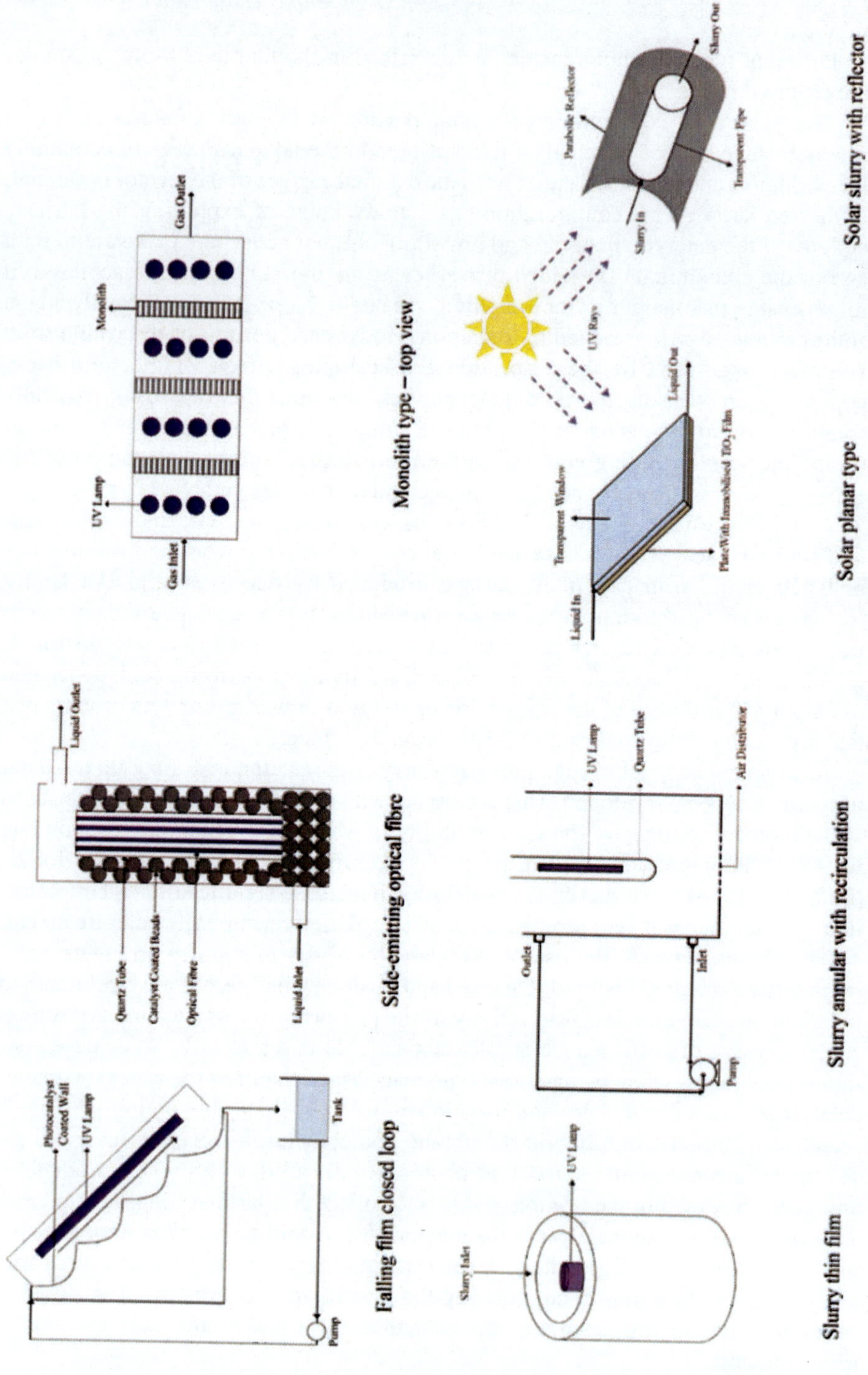

Fig. 9.1 Sketch of different photoreactor types. (Adapted and reproduced from [9] by kind permission. Copyright (2013) of Elsevier)

Different photoreactor configurations have been reviewed recently, for application to Advanced Oxidation processes (AOPs) [2]. A very simple option to overcome the limitations of slurry reactors and improve the throughput is to adopt a fixed bed configuration, which is possible by immobilisation of the active phase over a transparent material, such as glass beads [1]. However, transparency is limited and the surface interaction with the liquid is less efficient than slurry.

The possibility to use photomicroreactors, where the catalyst is immobilised in different forms over a solid substrate, allows us to overcome the downstream separation problems of the catalyst powder and to irradiate a clear and transparent solution, but also implies some limits for the decrease of active site exposure with respect to slurries.

Photocatalytic microreactors have been developed with significant surface to volume ratio (>10,000 m^2/m^3) to overcome the latter limitation. Another very interesting feature that underpins the use of microreactors is that when optimised flow patterns are achieved, heat and mass transfer limitations are negligible and a uniform and predictable irradiation can be achieved [2]. Different flow patterns in three-phase microreactors are reported for the fluid flow through the microchannels, such as bubble flow [22], slug flow [23] or annular flow [24]. Microreactors can assume different shape, such as microcapillary [25], single- [26] or multi-microchannel reactors [27] or planar structure [28].

An example of capillary microreactor has been proposed by Eskandarloo et al. [26] by coating a stainless steel plate including microchannels with the photocatalyst to accomplish the oxidation of terephthalic acid under UV irradiation. Additionally, Castedo et al. [29] reported the photoproduction of H_2 with Au/TiO_2 immobilised in a silicon microreactor, which guarantees higher versatility.

Monolith-type microreactors are an interesting configuration, which consists of a regular array of channels, usually straight and coated with a thin layer of catalyst, irradiated efficiently though optical fibres located inside the channels. The fibres should not be coated to distribute uniformly the radiation across the monolith (Fig. 9.2) [30].

Different configurations have been developed. For instance, a photomicroreactor containing TiO_2 nanoparticles coated over ZnO nanorod arrays has been obtained by growing the active phase in the capillaries [31]. The assembly was used to photodegrade methylene blue. In addition, the mass and photon transfer have been improved for the reduction of Cr(VI) to Cr(III) comparing two photoreactors: a monolithic tubular photoreactor and a micro-meso-structured photoreactor, achieving in both cases a satisfactory irradiated surface to volume ratio [32]. To reduce photon transfer limitations, the tubular photoreactor was packed with transparent cellulose acetate monoliths loaded with the catalyst by dip-coating. For the other layout, a thin film was uniformly deposited on a glass slab or on a network of channels and chambers printed in the back stainless steel slab, achieving much better performance than the monolith reactor.

A completely different configuration is achieved in a twin reactor, used, e.g. to achieve the photoreduction of CO_2 by CO, which is a stronger reducing agent than water [33]. It consists of two compartments separated by an ion exchange membrane,

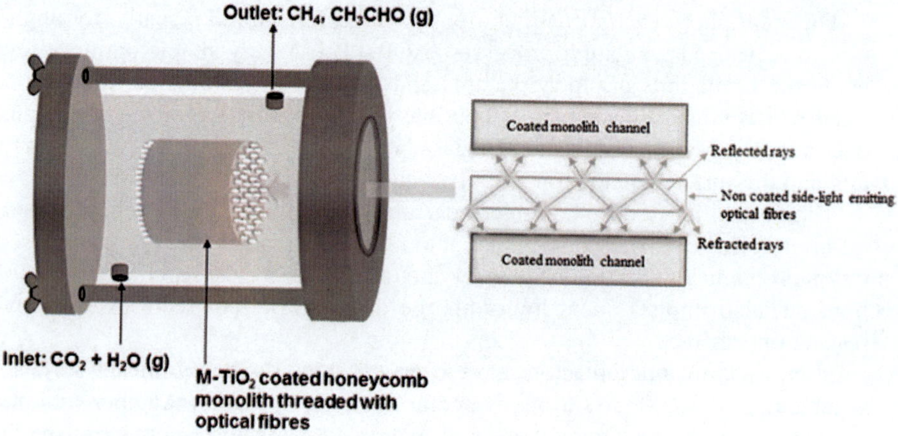

Fig. 9.2 Example of honeycomb-type micro-photoreactor, irradiated through optical fibres. (Reproduced from [30] by kind permission. Copyright (2016) Elsevier)

in the specific case allowing H^+ transfer between the two reacting sections, e.g. Nafion. The same concept has been applied to the CO_2 photoreduction assisted by H_2 evolution, where the latter is accomplished by photoreforming organic pollutants in water [1]. The same twin reactor configuration has also been applied by Baniasadi et al. [34] for the photoproduction of H_2 under visible light. Two different catalysts, one for H_2 evolution (Pt/$SrTiO_3$:Rh) and the other for oxygen reaction (WO_3), were separately loaded in two compartments, separated by a Nafion membrane, and using the Fe(II)–Fe(III) couple as electron transfer agent. A Compound Parabolic Concentrator (CPC) is also included. The reactor performance is not only dependent on the flow regime, needing high turbulence, but also sensitive to the high pressure drop under that conditions.

The conversion of the Direct Red 23 (DR23) dye has been successfully achieved using solar light with Fe_3O_4/TiO_2, coated on a glass tube and inserted in an annular shaped photoreactor. This assembly is also mounted in a CPC configuration and allows almost quantitative abatement of the dye [35].

A solar Offset Multi Tubular Photoreactor (OMTP) is a variant of the CPC photoreactor layout, including additional tubes in the space occupied by the axes of intersection of the CPC reflective involutes. This allows a considerable increase of the irradiated reactor volume (by 79%) and of the fluid residence time (up to 1.8 factor) with respect to a conventional CPC with the same footprint. With this assembly, the degradation efficiencies significantly increased for various pollutants [36].

A combined solar-electro-Fenton approach has been applied for the pilot scale removal of different pesticides and emerging contaminants [37]. The selected reaction is based on a redox cycle mediated by the Fe^{2+}/Fe^{3+} couple, where the ferrous ions react with H_2O_2 to provide $^{\bullet}OH$ radicals as Reactive Oxygen Species (ROS). The electrochemical formation of H_2O_2 in situ avoids the handling, storage and transportation of this reactant. Furthermore, the irradiation with UVA radiation, or better, with solar light, promotes the regeneration of Fe^{2+} with concomitant production of additional $^{\bullet}OH$ through photoreduction of Fe^{3+}, the photodegradation of possible Fe(III)-carboxylate complexes formed as intermediates and the additional direct photolysis of the pollutants or intermediates. The pilot reactor was supplied with a CPC concentrating the solar radiation flux to the solution. The pilot scale reactor (Fig. 9.3), the largest available for this application, allowed 100 L capacity with natural solar irradiation in the facility of the Plataforma Solar de Almería (Spain). The electrochemical section of the device was constituted of four plate-and-frame electrochemical reactors, while the irradiated area was 2 m^2 for a 23 L volume of solution.

A similar reactor configuration on a pilot scale has been used for the photo-Fenton degradation of a solution of tetracyclines, which are the second largest group of antibiotics worldwide, largely excreted in water reservoirs through hospital or

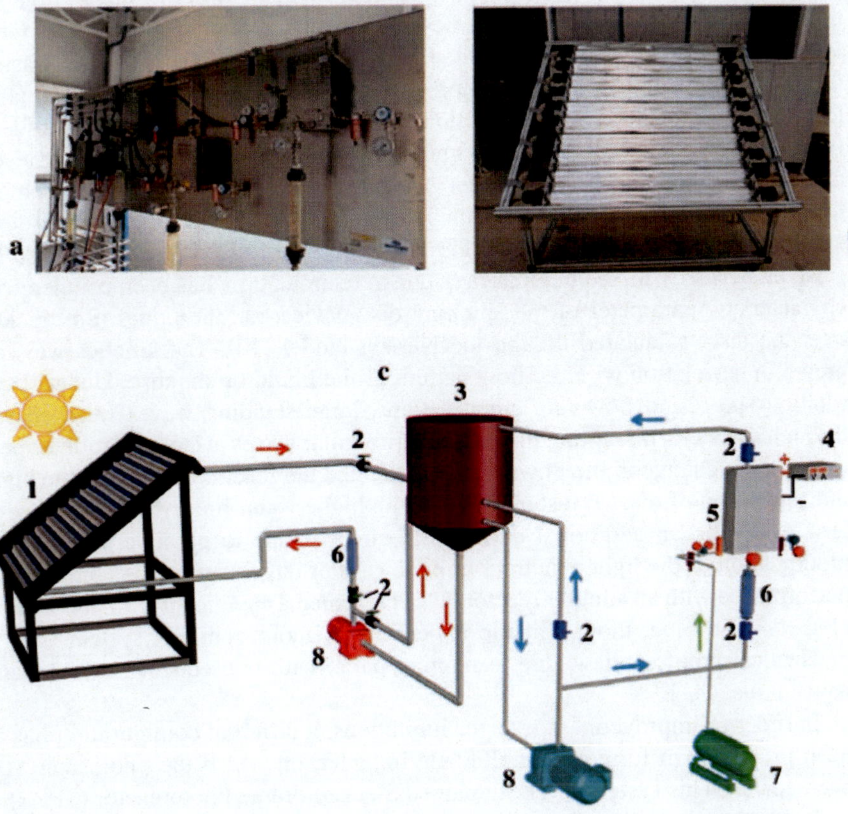

Fig. 9.3 Front view of **(a)** the four filter-press type electrochemical cells of the pilot unit and **(b)** the CPC photoreactor. In **(c)**, schematic diagram of the pilot unit equipped with one cell (examined in this work), showing (*1*) CPC photoreactor, (*2*) valve, (*3*) feed tank, (*4*) power supply, (*5*) electrochemical reactor, (*6*) liquid flowmeter, (*7*) air compressor and (*8*) magnetic pump. (Reproduced from [37] by kind permission. Copyright (2019) by Elsevier)

domestic wastes, besides through industrial residue. Given the ineffective removal through conventional biological treatments, to avoid the insurgence of bacterial resistance, a pilot-scale photoreactor constituted of a CPC and 6 borosilicate glass tubes in series with a batch recirculating circuit was used to treat 35 L of a tetracycline solution, achieving almost 90% conversion in 2 h under different operating conditions [38]. Also, a CPC system for hydrogen production under solar light irradiation has been designed and modelled by Cao et al. [39].

A further different configuration is constituted by spinning disc reactors, which are proposed as tools for process intensification [40–43]. The reactor is composed of a spinning disc placed in the horizontal plane, which is let to rotate at different spinning velocity. The reacting mixture is injected from a central nozzle and moves towards the periphery due to centrifugal draft. The photocatalyst may be immobilised on the disc and irradiation can be achieved from top. The intensification of the process mainly occurs through a more uniform and effective penetration of light through the thin liquid layer formed over the disc and by improvement of the mass transfer coefficient. This device has been applied to the photodegradation of methylene blue and of dehydroabietic acid, a resin acid found in pulp and paper wastewater, which acts as endocrine disruptor, with adverse ecological effects even at very low concentration. The overall reaction rate has been measured at different feeding flow rates and spinning regimes, obtaining widely different results and even a change of the apparent reaction order. The latter parameter is expected as first order for most photocatalytic reactions, but it turned out to be second order for the selected applications in very specific cases. The mass balance of a Continuously Stirred Tank Reactor, perfectly mixed but unsteady due to recirculation, has been coupled with a characteristic parameter of the spinning disc device, i.e. the liquid film thickness over the disc, calculated though the Nusselt model [40]. The kinetics was interpreted in correlation with the flow pattern of the liquid on the disc. Under specific rotation speed and flow rate combinations, large standing waves were observed through a camera, disrupting in an array of irregular waves at large rotational speeds. Overall, the change in the flow structure affected the reaction rate, with the biggest intensification of mass transfer rate with the highest non-linearity of the wave pattern. However, an additional detrimental effect is due to possible scattering and modification of the light penetration in the case of large waves. The comparison of performance with an annular reactor layout revealed a significant gain of conversion of methylene blue, though at the expense of a more complex system from the mechanical point of view, due to moving parts, with consequent heavier maintenance issues [41].

In order to improve mass transfer limitations, a different configuration has also been proposed, in form of a multiblade impeller, on which the photocatalyst has been coated. This system has been named Stacked Frame Photoreactor (SFPR) and is sketched in Fig. 9.4 [44]. In this photoreactor, the propeller rotation helps mixing and thus a more efficient mass transfer with respect to immobilised photocatalytic surfaces.

An impinging jet stream photoreactor was set up with a TiO_2-coated disc for the abatement of phenol in water. Different parameters were optimised, such as liquid

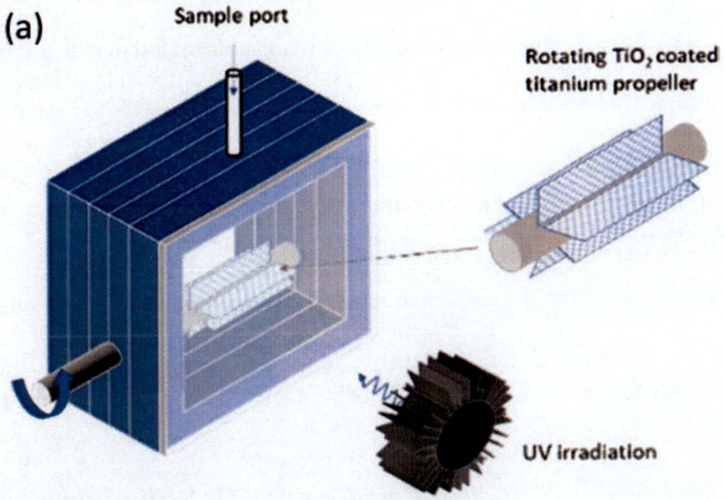

Fig. 9.4 Schematic of the SFPR with the rotating titanium propeller housed within the reactor. (Reprinted from [44] by kind permission. Copyright (2019) of American Chemical Society)

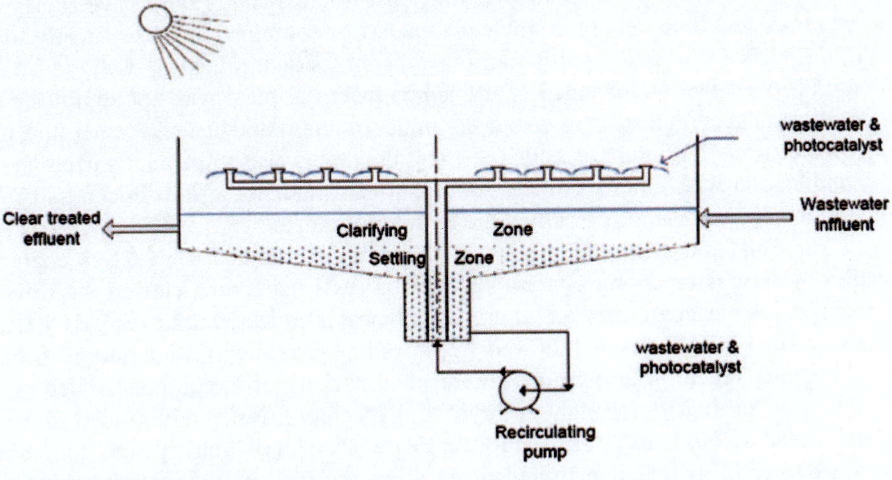

Fig. 9.5 Schematic drawing of TiO_2 "water-bell" photoreactor. (Reproduced from [48] by kind permission. Copyright (2018) by Elsevier)

flow rate, disc diameter, nozzle-to-disc distance and initial pollutant concentration. A newly developed hydrodynamic model, Langmuir-Hinshelwood kinetics and mass transfer phenomena were used to rate the system and it was compared with a spinning disc reactor, showing better performance [45–47].

A "water-bell" TiO_2-based photoreactor (Fig. 9.5) has been designed to achieve a thin water/photocatalyst film efficiently irradiated through UVA lamps, envisaging a future application with solar light [48]. The thin film has the advantage of exchanging oxygen efficiently with surrounding air and of allowing high photocatalyst concentration, since high light penetration is not needed due to small film thickness. This also allows loading the catalyst on coarse supporting particles, to ensure easy settling and recovery of the catalyst in a so-called clarifying zone. The system is designed at a large scale as a modular set of nozzles that spray the water/photocatalyst suspension as a film to a basin for recirculation. The catalyst loading proved to be a more sensitive parameter than irradiance and the presence of an efficient electron scavenger also demonstrated important, as well as the pH.

A falling film reactor has been tested for the photodegradation of methylene blue. Ca. 4 g of TiO_2 were deposited in a four step configuration exposing ca. 0.2 m^2 surface to UVA irradiation. The mechanism of oxidation has been investigated, together with the optimisation of the reaction conditions [49]. Furthermore, falling film microreactors irradiated with LED lamps were constituted by parallel microchannels for photocatalytic-assisted synthesis [2, 50].

Similarly, Bahbani Ghatar et al. [51] investigated the same reaction in a UVA-irradiated plane falling film photoreactor based on ZnO. Also, in this case, the optimisation of the operating conditions was the objective of the work, using an experimental design over different variables: the slope of the plate, the number of lamps, their distance from the plate and the flow rate of wastewater. The combined effect of all these parameters over the observable result (dye conversion) was expressed as a function of the single parameters or combined in couples. Only some of the variables revealed particularly significant, as determined by ANOVA analysis. For instance, the distance of the lamps from the plate was not significant by itself, but the combination of distance/number of lamps and distance/water flow rate revealed important. Furthermore, the water flow rate was significant by itself and in combination with other parameters, since it can modify the film thickness of the solution and turbulence can affect the light absorbance.

The FluHelik photoreactor was used for a photochemical UVC/H_2O_2 degradation of various emerging contaminants [52]. Tests have been carried out in both ultrapure water matrix and a real urban wastewater collected after secondary treatment. The FluHelik reactor showed 1.3 times higher mineralisation with respect to a reference reactor. Furthermore, the residual toxicity of the photodegraded products was evaluated through embryo toxicity bioassays. Furthermore, a mili-photoreactor (NETmix) was used for the degradation of oxytetracycline from urban wastewater [53]. It is constituted of an array of small cylindrical chambers and prismatic transport channels and the system is assembled with a quartz slab. The optimisation of the radiation source, ensuring uniform light distribution, and the H_2O_2 concentration was performed. The same type of reactor was also used with a heterogeneous catalytic system, based on TiO_2 irradiated with a UVA source [54] and for bromate reduction [55].

The photodegradation of VOCs in air has been tested on a semi-pilot scale in a 420 L photoreactor consisting of a photocatalytic chamber that included a set of 9 W UVA fluorescent lamps and a photocatalytic filter. The latter was assembled on a cellulose-based material on which a titania-silica coating has been immobilised.

The air flow containing the pollutant was fed and recirculated to a first larger chamber, where sampling points allowed us to measure the concentration of the target molecules vs. time. The flow rate was measured before entering the photocatalytic module and air circulation was provided by a fan [56].

A summary of the different possible configurations is reported in Table 9.1.

Table 9.1 Main reactor configurations

Reactor type	Features	Advantages	Disadvantages
Slurry	Catalyst in suspension; intense mixing needed	Good surface exposure; limited mass transfer if efficient mixing is achieved	Scale-up issues; inefficient light penetration due to scattering or opaque solutions; recovery of catalyst powder
Fixed bed	Catalyst immobilised over beads	Immobilised catalyst; no recovery issues	Light penetration is inefficient
Microreactors/ capillary reactors	Catalyst immobilised in microchannels exposed to light	Good catalyst exposure to light and reactants; limited size of the channels and so limited mass transfer issues	Needed high surface to volume ratio
Monoliths	Catalyst immobilised in microchannels of a monolith exposed to light	Good surface to volume ratio; effective irradiation with LED or optical fibres in each channel	Need of complex arrangement for irradiation; complex coating procedures
Twin reactor	Separate compartments for different half reactions	Separate production of different products	Applications limited by needs of transport through a membrane; rates possibly limited for the same reason
Single or multitubular with compound parabolic collectors (CPC)	Effective collection and concentration of radiation on the tubes	Enhanced irradiation; relatively compact and simple design for multitubular systems	Sizing dependent on reaction rate
Spinning disc	Solution/suspension spread in thin film irradiated on top	Short optical path; limited mass transfer	Possible irregular flow shape; mechanically more complex due to moving parts
Stacked frame/ rotating impeller	Catalyst immobilised over rotating blades	Improved mass transfer	Mechanically more complex due to moving parts; irradiation spreading and averaging over the surface
Water bell/falling film	Solution/suspension recirculated in thin layers flowing	Improved mass transfer; effective irradiation	Need of high recirculation; intrinsically unsteady operation

9.3 Photoreactor Scale-Up and Modelling Issues

One of the key points for modelling photochemical or photocatalytic reactors is the need to include the radiation as a "pseudo-reactant" in the kinetics of the process. This allows us to integrate the characteristic equations for the reactor used (mass, momentum and heat balances) to size or rate the photoreactor. Computational methods in photocatalysis have been very recently summarised and include methods for the solution of the radiation pattern [57].

The first issue is to estimate the Local Volumetric Rate of Photon Absorption (LVRPA, e^a), i.e. a sort of "concentration of photons". LVRPA expresses the useful photon concentration as they were reactants and thus the useful number of photons absorbed by the surface of the catalyst in the unit volume of the reactor [58].

The radiative transport equation (RTE) [59, 60] represents the most rigorous approach,

$$\frac{dI_\lambda(s,\Omega)}{ds} = -k_\lambda I_\lambda(s,\Omega) - \sigma_\lambda I_\lambda(s,\Omega) + \frac{1}{4\pi}\sigma_\lambda \int_0^{4\pi} p(\Omega' \rightarrow \Omega) I_\lambda(s,\Omega') d\Omega'.$$

I_λ represents the irradiance at wavelength λ, s is a spatial coordinate and Ω represents a directional solid angle. The properties of the photocatalytic material are instead represented by an absorption coefficient k_λ and a scattering coefficient σ_λ, both depending on the wavelength. The term to be integrated $p(\Omega' \rightarrow \Omega)$ represents a scattering phase function, i.e. the probability that around the position s, a photon can be redirected from the direction Ω' to Ω. Therefore, the first term on the right represents the absorption, the second the out-scattering and the third the in-scattering of photons.

Different approaches were implemented to solve the RTE, where a rigorous Discrete Ordinate Method (DO) discretises the volume coupling and conventional mass and energy balances of Computational Fluid Dynamic (CFD) methods, with this additional radiation balance. More precisely, the knowledge of the hydrodynamics is needed to calculate the catalyst distribution in slurry reactors. CFD simulation typically solves the flow pattern, from which the time-averaged or steady state results of catalyst distribution are used to calculate the radiation field through the RTE [9].

In the case of homogeneous photoreactors, the Navier-Stokes equations are used, while for multiphase flow, an Eulerian-Eulerian (E-E) or Eulerian-Lagrangian (E-L) approach can be applied. The E-E model solves the mass and momentum balance equations for the different phases, considered as interpenetrated continua. Additional relations are needed in the case of solid phases. The E-E approach has been more frequently implemented for multiphase flows in this application [61–66]. The E-L approach simulates the trajectory of the dispersed phase particles through the equations of motion, by applying a force balance, while the continuous phase is solved through an Eulerian approach. Different turbulence models have been used for bubble columns and photocatalytic reactors, among which the κ-ε one, in case accounting for bubble induced turbulence, is a good compromise between accuracy and computational demand [9]. On the contrary, for immobilised catalysts, the near-wall zone is computed by discretisation under laminar conditions.

The Discrete Ordinate (DO) model transforms the RTE, which is an integro-differential equation, into a system of algebraic equations [67, 68]. The radiation field is divided into a number of discrete directions for which the RTE is solved. Alternatively, a Finite Volume (FV) method is more flexible with respect to geometries, while the DO can span the whole optical thickness, has dedicated non-grey models and has been successfully applied to both immobilised and slurry photoreactors. A very nice example of application of CFD modelling to photoreactor design is proposed by Boyjoo et al. [66], where a DO approach is used, also implemented as a FV method in a commercial software for CFD, such as Fluent®. The reaction order with respect to the LVRPA showed variable depending on light intensity, i.e. 0.5 at high irradiance (due to more frequent parasitic electron-hole recombination events) and 1 for low irradiance. Following an E-E model, the phases are modelled as interpenetrating continua for the solution of the mass and momentum conservation balances. The rate of reaction is expressed as

$$-r_{\mathrm{p}} = kf\left(c_{\mathrm{p}}\right)\mathrm{LVRPA}^{m},$$

where $m = 0.5–1$ depending on the regimes and $f(c_{\mathrm{p}})$ was a function of the concentration of pollutant to be treated corresponding to a first order kinetics or a Langmuir-Hinshelwood expression. The radiation modelling was carried out through DO modelling, by dividing the radiation field into discrete directions and solving the RTE separately for each of them. The FV tool in Fluent® was also used, by discretising each control volume (e.g. 72 solid angles). However, the simultaneous solution of the radiation field and the fluid dynamics is not currently possible in the commercial tool. So, at first, the simulation was made on the phases, getting the mean catalyst concentration and velocity vectors. Then, the radiation was modelled, accessing the catalyst local concentration data previously stored, to calculate the optical properties of the medium and LVRPA. The latter was then used to model the reaction rate and, consequently, the reactor balances. Different photoreactor layouts were tested, holding 1, 2 or 4 lamps, with optimised, different grids for the fluid dynamic simulation and the radiation modelling [66].

CFD modelling was also applied to the SFPR reactor depicted in Fig. 9.4 using a rotating machinery turbulent flow κ-ε model [44]. The rays emitted by a LED lamp were modelled as deriving from a point source and so, depending on the distance from the TiO_2-coated propeller, the amount of radiation reaching the surface was calculated. As reasonable, the latter parameter decreased upon increasing the distance from the lamp. However, another important feature was highlighted through the model. It was previously reported that rapidly switching LED sources can improve the activity due to more controlled excitation, which would decrease the recombination of the photogenerated charges [44, 69–71]. This effect can be achieved through the rotation of the propeller, which alternates the surfaces exposed to light. The calculated irradiance at different lamp-reactor distance and with different positions of the propeller is exemplified in Fig. 9.6. The effect of the lamp distance from the photocatalyst layer is immediately evident between 4 cm (Fig. 9.6a, b) and 7 cm (Fig. 9.6c, d). The shadowing zones are also evidenced during the rotation of the photocatalyst.

Similarly, a CFD approach allowed us to optimise the configuration of a UV LED array, tested for the photodegradation of cinnamic acid as model for agro-industrial waste water [72]. LEDs may be usually represented as point sources and can now reach high intensity irradiance, which makes the reaction rate order with respect to radiation intensity non-linear. Thus, the accurate prediction of light distribution (again with fluid dynamic assessment of catalyst and reactant distribution across the reactor) is a fundamental task for optimisation. Different reactor geometries were also tested [73].

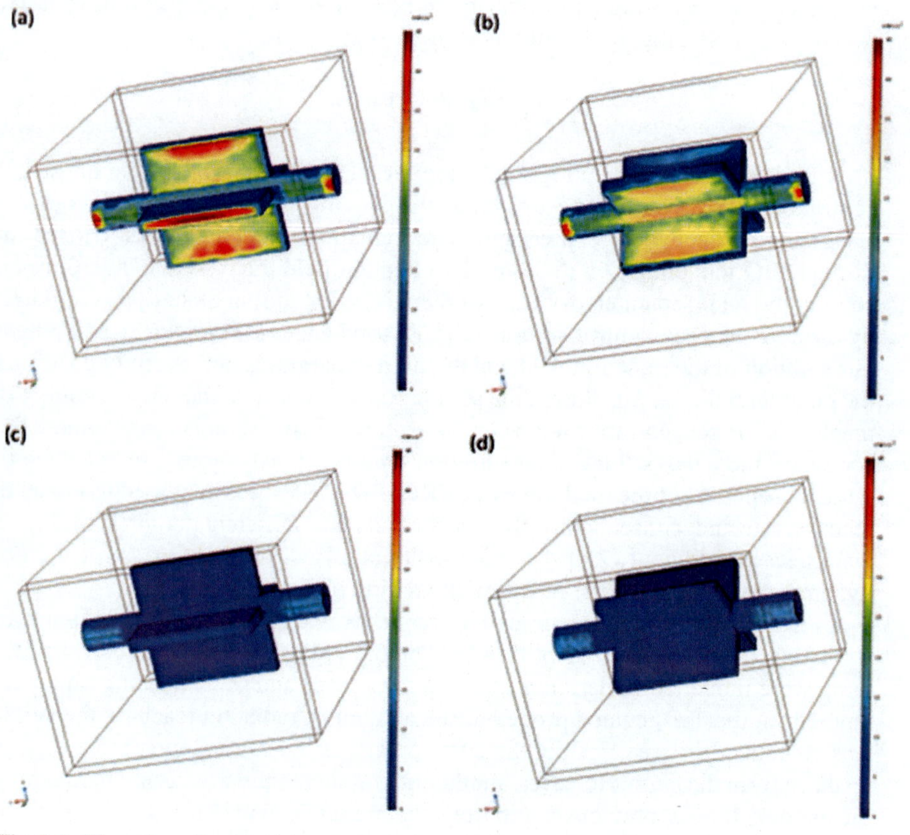

Fig. 9.6 3D surface plots of the irradiance on the catalyst surface at 4 cm lamp/reactor distance in both the initial configuration (**a**) and the rotated propeller position (**b**) or at 7 cm distance in the initial static position (**c**) and the rotated propeller one (**d**). (Reproduced from [44] by kind permission. Copyright (2019) by American Chemical Society)

The best piping arrangement in a pilot annular photoreactor has been modelled, considering a FluHelik reactor and a conventional Jet reactor. The residence time and radiation distribution have been calculated by CFD for the degradation of 3-amino-5-methylisoxazole with H_2O_2/UVC and UVC processes [74].

Various reactor layouts were modelled through CFD. For instance, as a non-exhaustive list, a channelled optical fibre reactor was applied to ethylene photodegradation in gas phase [75] and for its oxidation in non-ideal fixed bed flow reactors [76]; a baffled flat plate photoreactor irradiated with UVA-LED was used for the degradation of $Fe(CN)_6^-$, a recalcitrant compound present in mining wastewaters [77]; the FV module implemented in Fluent® was used to model a single phase annular photoreactor [78]; an annular photocatalytic reactor to degrade Rhodamine B was simulated using Fluent® [79]; the photodegradation of acetaldehyde in air has been modelled in a multi-tubular reactor [80].

Alternatively, a Monte Carlo (MC) stochastic approach can be used to solve the RTE. For both the DO and MC approaches, the rigorous solution is obtained at the expense of computational cost, so that typically, these methods are employed for constant irradiance sources, such as lamps, avoiding solar photoreactors for which the variation of the input source would be scarcely manageable.

The MC method [58] is based on the random scattering and absorption phenomena (Fig. 9.7, right), which can be modelled in a slab reactor through the definition of a grid of thickness dx and by correlating the trajectory of each photon in the slab through an array R of numbers, which assumes random values between 0 and 1. A mean free path (l) of the photon is defined as the free pathway between two consecutive absorption or scattering events. This is correlated with the extinction coefficient of the suspension (β) as

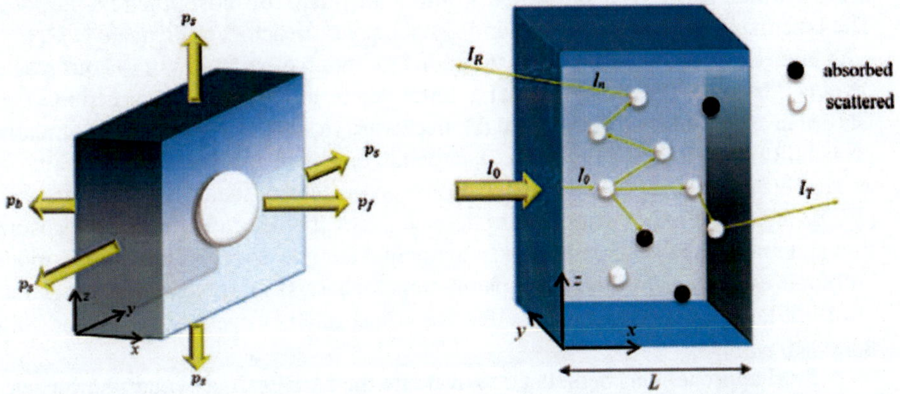

Fig. 9.7 *Left:* Scattering directions considered by the SFM in a differential volume of liquid containing a particle. *Right:* schematic diagram of the slab photoreactor with absorbing and scattering particles suspended in the inside. I_0, I_R and I_T are the incident, reflected and transmitted intensity of light, L is the reactor length and l the free path of photons. (Reproduced from [58] by kind permission. Copyright (2016) by Elsevier)

$$l = -\frac{1}{\beta}\ln R,$$

l and the direction θ_1 where the photon is emitted allow the calculation of the new position (x_i) of the photon, which should lie between 0 and L; otherwise, the photon exits the slab and it is no more useful,

$$x_i = x_{i-1} + l\cos\theta_i.$$

When the free path is passed, the photon is absorbed or scattered based on the value of R: the photon is absorbed and its location recorded if $R \geq \omega$, where ω is defined as the scattering albedo, i.e. the ratio between the scattering and the extinction coefficients, defining the scattering probability. If $R < \omega$, the photon is scattered towards a new random direction, whose longitude and latitude angles (azimuth α and zenith φ, respectively) are found randomly in the range [0; 2π]. φ is expressed as a function of R and of the phase function, while the azimuth as

$$\alpha_{i+1} = 2\pi R.$$

Finally, the scattering angle is

$$\cos\theta_{i+1} = \cos\varphi_{i+1}\cos\theta_i + \sin\varphi_{i+1}\cos\alpha_{i+1}\sin\theta_i.$$

The radiative heat transfer in a multi-channel solar reactor was performed through MC on a single channel level [81]. A coating of $ZnFe_2O_4$ was considered to optimise the channel geometry and film thickness.

Another alternative approach is the Six Flux Absorption-Scattering Model (SFM), which provides a simplified solution adequate for most practical purposes. The latter has been proposed for homogeneous photoreactors by Brucato et al. [82], with pioneering simplified models to predict radiation distribution in various reactor layouts [83–87]. The SFM model admits scattering in the six directions of a Cartesian space, defined as forward (f), backward (b) and sideward (s), with relative probability p_i (i = f,b,s) (Fig. 9.7). p_i depends on the phase function for scattering and has been variously calculated depending on the type of scattering that is assumed [58, 82, 88, 89]. For heterogeneous photocatalysts, the determination of the absorption and scattering coefficients is a fundamental step for the reliability of the model and some estimates of the relevant parameters for TiO_2 is reported by Acosta-Herazo et al. [58], together with a very effective comparison between the MC and SFM approaches.

A SFM approach has been used to evaluate the LVRPA in a planar photoreactor, distinguishing the radiation fraction in UVA and Vis regions when coupling materials able to absorb different spectral ranges and therefore, they contribute differently to absorption and scattering. In the case of polychromatic sources, such as solar radiation, the photoactive materials should be carefully analysed to predict their optical properties and to calculate the relevant parameters. In the example reported by Acosta-Herazo et al. [90], TiO_2 absorbs and scatters light in the UVA region, while goethite absorbs both UVA and part of the Visible spectrum. The SFM is therefore applied to polychromatic radiation and the net radiation absorbed by the goethite photocatalyst was calculated summing the contributions from the UVA and the Visible radiation. So the net radiation absorbed by goethite was calculated summing both contributions,

$$e^a_{\mathrm{TiO_2}}(x) = e^a_{\mathrm{UVA}}(x)\left(\frac{\mathrm{W}}{\mathrm{m}^3}\right),$$

$$e^a_{\mathrm{goethite}}(x) = e^a_{\mathrm{UVA}}(x) + e^a_{\mathrm{Vis}}(x)\left(\frac{\mathrm{W}}{\mathrm{m}^3}\right).$$

The optical properties of both materials were averaged in the relevant spectral range and depend on catalyst concentration, e.g.

$$\kappa^* = \frac{\int_{\lambda_{\min}}^{\lambda_{\max}} \kappa^*_\lambda I_\lambda \,\mathrm{d}\lambda}{\int_{\lambda_{\min}}^{\lambda_{\max}} I_\lambda \,\mathrm{d}\lambda}\left(\frac{\mathrm{m}^2}{\mathrm{kg}}\right),$$

$$\sigma^* = \frac{\int_{\lambda_{\min}}^{\lambda_{\max}} \sigma^*_\lambda I_\lambda \,\mathrm{d}\lambda}{\int_{\lambda_{\min}}^{\lambda_{\max}} I_\lambda \,\mathrm{d}\lambda}\left(\frac{\mathrm{m}^2}{\mathrm{kg}}\right),$$

$$g = \frac{\int_{\lambda_{\min}}^{\lambda_{\max}} g_\lambda I_\lambda \,\mathrm{d}\lambda}{\int_{\lambda_{\min}}^{\lambda_{\max}} I_\lambda \,\mathrm{d}\lambda}\,(-),$$

where κ is the absorption coefficient, σ is the scattering coefficient, while g is a parameter used to compute the phase function (e.g. defined according to Henyey-Greenstein, but different approaches are also reviewed in [9]). The limits for integration depend on the absorption edge of each material.

The optical thickness of an absorbing medium can also be calculated, assuming a characteristic length L of the system (e.g. slab length or annulus thickness), as [9]

$$\tau_{\mathrm{abs}} = \kappa_\lambda L.$$

The particle agglomeration, if occurring near the isoelectric point of the material, can also be an issue, which is accounted for by correcting the extinction coefficient (β) with respect to aggregation,

$$\beta^*_\lambda = \kappa^*_\lambda + \sigma^*_\lambda,$$

$$\beta^*_{\text{aggr},\lambda} = \beta^*_\lambda \frac{\ln T(t, W_{\text{cat}})}{\ln T(t_1, W_{\text{cat}})},$$

$$\sigma^*_{\text{aggr},\lambda} = \beta^*_{\text{aggr},\lambda} - \kappa^*_\lambda,$$

where t_1 is the time in which the suspension is formed and T is the transmission, depending on catalyst concentration (W_{cat}) and, in the case of particles aggregation, on time t,

$$T(t, W_{\text{cat}}) = \frac{\text{Transmitted radiation flux at time } t \text{ and catalyst loading } W_{\text{cat}}}{\text{Transmitted radiation flux at time } t \text{ and pure water}}.$$

Different parameters have been introduced to analyse and compare the performance of photoreactors, such as the Initial Rate of Photon Absorption (IRPA),

$$\text{IRPA} = -\left[\frac{\text{d}}{\text{d}x}\left(\text{e}^a(x)\right)\right]_{x=0},$$

which may result in the following equation:

$$\text{IRPA} = -I_0 c_{\text{cat}}^2 \psi \quad \left(\frac{\text{W}}{\text{m}^4}\right),$$

$$\psi = (\beta^*)^2 \frac{a^2\left(1-\omega_{\text{corr}}^2\right)}{(1-\gamma)\omega_{\text{corr}}}\left[\gamma\left(\omega_{\text{corr}} - 1 - \sqrt{1-\omega_{\text{corr}}^2}\right) - \left(\omega_{\text{corr}} - 1 + \sqrt{1-\omega_{\text{corr}}^2}\right)\right]$$

with $\beta^* = \sigma^* + \kappa^*$ being specific extinction coefficient. The IRPA is very sensitive to catalyst concentration and thus, the optimal catalyst amount has to be calculated depending on its optical properties and be in turn related to the reactor thickness L to maximise the absorption of the incoming radiation.

Another useful indicator is the Total Rate of Photon Absorption (TRPA), which quantifies the total radiation absorbed in the whole volume,

$$\frac{\text{TPRA}}{A} = \int_0^L \text{e}^a(x)\,\text{d}x.$$

Also, in this case, useful parameters relative to TiO_2 are reported by Acosta-Herazo et al. [90].

The SFM approach has also been used to model the LVRPA in a CPC photoreactor for the abatement of acetaminophen, a hugely used drug [91], by using a Langmuir-Hinshelwood kinetic expression. The reactor mass balance has been solved as a series of plug-flow and continuous stirred reactors to treat 5000 L of wastewater (daily amount of a medium-size Colombian hospital).

Different lamp models were proposed, as line source, surface or volume emitters [9], whose equations are summarised in the cited reference.

In the case of immobilised photocatalysts (planar coatings, monoliths, etc.), there is no scattering and absorption at difference with slurry systems (if a clear and transparent medium is treated). In this case, the RTE is calculated through the Beer-Lambert equation where the incident light on the photocatalyst surface (I_0) is calculated by subtracting the reflected part from the inlet radiation intensity ($I_0 = I_{in} - I_r$). Possibly, some scattering may be due to air bubbles, if present [9].

Once the model for the source and the integration of the RTE gives quantitative indications on the useful radiation amount available as a "pseudo-reactant", the kinetic equation can be derived. At first, intrinsic kinetics is considered, i.e. neglecting the possible physical limitations due to mass transport in the fluid phase or in the porous solid particles. Subsequently, these factors should be added.

In general, the kinetic equation can be written as a function of the concentration of the reactants, usually the pollutant concentration in the case of water treatment. The function is commonly first order or expressed in Langmuir-Hinshelwood form,

$$-r = kf\left(c_i\right),$$

$$k = f\left(\text{pH}, T, c_{\text{O}_2}, W_{\text{cat}}, \text{LVRPA}\right).$$

The dependence on the radiation intensity is often expressed through a "reaction order" of the LVRPA, which depends on different parameters, as reviewed elsewhere [9]. Furthermore, photon absorption with varying catalyst concentration has been modelled by Cao et al. [92].

A generalised model was developed to rate solar CPCs with different arrangement for degradation of water pollutants. A modified Langmuir–Hinshelwood kinetics has been developed, together with a new model to quantify the "effective" quantum yield. It is computed from the optical properties of the catalyst and the incident photon flux, without dependence on the composition, the operating conditions, the geometry and the scale of the reactor [93].

The RTE has been integrated with a Dirichlet boundary condition on the film surface exposed to irradiation, [60]

$$\frac{\mathrm{d}I_\lambda\left(s\right)}{\mathrm{d}s} = -k_\lambda I_\lambda\left(s\right).$$

$$I_{\lambda,s} = I_{\lambda,0}.$$

It can be integrated to yield

$$I_{\lambda,s} = I_{\lambda,0}\mathrm{e}^{-k_\lambda s}.$$

The mass transport of each species has been modelled under steady state conditions according to Fick's first law, as diffusive flux N_i,

$$N_i = -D_{\text{eff}} \nabla c_i,$$

$$D_{\text{eff}} = \frac{D_{ij}\varepsilon}{\tau}.$$

Take into account the effective diffusivity D_{eff} as a function of the diffusion coefficient of species i in the reaction medium j ($D_{i,j}$), the porosity (ε) and the tortuosity factor (τ) of pores constituting the photocatalyst.

Accordingly, the rate of conversion of species i was defined assuming a first order dependence on concentration and an order γ on light intensity, as

$$r_i = -\eta k_{r,\text{app}} c_i I_{\lambda,s}^{\gamma} (1 - \varepsilon).$$

The effectiveness of the catalyst (η) has been conventionally defined as

$$\eta = \frac{\tanh(\phi)}{\phi},$$

where the Thiele modulus for a first order reaction was formulated as

$$\phi = (\delta - y)\sqrt{\frac{k_{r,\text{app}} I_{\lambda,0}^{\gamma}}{D_{\text{eff}}}},$$

representing as usual the ratio between the potential rate of reaction and that of diffusion inside the pores of the solid. A discussion on the suggested values for the various parameters can be found in [60] and references therein. The adopted model neglects diffusional limitations in the liquid phase, which is reasonable for monolith or microchannel photoreactors operating at high Reynolds number. The normalised reaction rate (r/r_{max}) was modelled depending on the thickness of the photocatalyst coating for the two front side (FSI) and back side (BSI) irradiation (Fig. 9.8b, c). The study returned the optimal value of the thickness for both cases and the attenuation of light intensity depending on the latter parameter. A correlation for film thickness optimisation depending on the various operating parameters was also proposed [60].

Additionally, CO_2 photoreduction has been investigated in the presence of a more powerful reducing agent than H_2O, studying the effect of different operating parameters in a twin reactor to predict the productivity of methanol under natural irradiance exposure [33]. A model for the reactor has been derived considering the reaction rate as a function of irradiance, where higher m exponents mean a better utilisation of photons in the given reactor geometry,

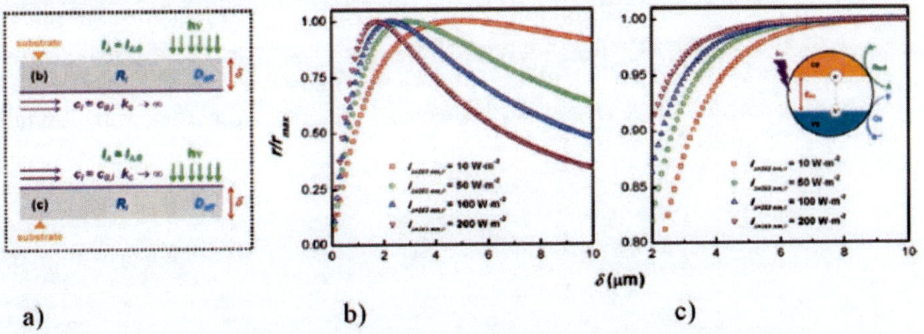

Fig. 9.8 (**a**) Top, Back Side (BSI) and bottom, Front Side (FSI) Irradiation of photocatalyst coating. (**b**, **c**) Normalised reaction rate vs. photocatalyst film thickness (δ) for BSI and FSI arrangement, respectively. (Reproduced from [60] by kind permission. Copyright (2017) by Elsevier)

$$r_j = I^m \left(k_j \prod_{i=1} c_i^{v_i} \right).$$

Mass transfer was computed according to the two-film theory and the solubility following Henry's coefficient, calculated keeping into account the effect of the ionic strength of the solution. Sunlight was modelled using dedicated software. It was concluded that temperature and pressure were determinant factors to improve the mass transfer coefficients and gas solubility in water, respectively [13, 16, 17].

The optimal photocatalyst layer thickness has been calculated for the degradation of methylene blue as model reaction in microphotoreactors [60]. The comparison between two irradiation modes in immobilised catalyst layers, i.e. BSI or FSI, was compared, developing a radiation model to be included in the reaction kinetics. The fluid dynamic properties of the fluid in the channel were also considered. In the BSI arrangement, the maximum incident radiation and the maximum pollutant concentration occur on the opposite side of the assembly, while in FSI, they occur on the same side of the photocatalytic film (as they were in a counter- or co-current arrangement, respectively) (Fig. 9.8).

A similar comparison was done for the gas phase photodegradation of n-decane as model compound for VOCs. In such a case, however, no significant difference between FSI or BSI illumination was noticed, thanks to the high porosity of the catalytic layer and high diffusivity of the reactant [94].

Kinetic modelling has been considered by Charles et al. [95] in a microreactor assuming a Langmuir-Hinshelwood expression to represent the degradation of salicylic acid, with K being the adsorption constant of the compound to oxidise and C_s its surface concentration,

$$r = \frac{kKC_s}{1 + KC_s}.$$

The residence time distribution was evaluated in the microreactor with the auxilium of an inert tracer and compared with a plug-flow reactor. The mass transfer coefficient $k_{m,ext}$ has been calculated through the Sherwood number (Sh), thanks to an existing correlation developed by Ergu et al. [95, 96] as a function of the Reynolds (Re) and Schmidt (Sc) numbers,

$$\mathrm{Sh} = \frac{k_{\mathrm{m},ext} d_{\mathrm{h}}}{D},$$

$$\mathrm{Re} = \frac{\rho u d_{\mathrm{h}}}{\mu},$$

$$\mathrm{Sc} = \frac{\mu}{\rho D},$$

$$\mathrm{Sh} = 2.076 \times \mathrm{Re}^{0.28} \left(\frac{L}{d_{\mathrm{h}}} \right)^{-0.12} \times \mathrm{Sc}^{0.33},$$

where μ is the fluid viscosity, ρ its density, u the linear velocity, D the diffusion coefficient and d_h the characteristic length of the system, e.g. the channel hydraulic diameter.

Different arrangements of micro- and meso-photoreactors have been proposed, either for scale-up or numbering-up purposes [97] for the degradation of dyes or organic pollutants. The flow systems have been characterised according to the flow rate, Reynolds, Damköhler and Peclet numbers, also identifying the definition of pertinent efficiency quantifiers. At first, the irradiance (I) has been simply defined as dependent on the optical absorbance of the medium, obtained multiplying the extinction coefficient ε, the concentration of the absorbing species c and the optical path (l). The incident radiation I_0 can be measured through a radiometer,

$$I = I_0 \mathrm{e}^{-\varepsilon c l}.$$

The authors considered strictly laminar flow regimes. The second Damköhler ($\mathrm{Da_{II}}$) number was calculated considering the ratio between the reaction and the mass transfer rates as

$$\mathrm{Da_{II}} = \frac{\text{reaction rate}}{\text{diffusive mass transfer rate}} = \frac{k'' H}{D}$$

with H being the channel height and $k'' = k(V/A)$ the reaction rate constant corrected by the ratio between the reactor volume and surface area of the catalyst. In addition, the Peclet (Pe) number is introduced to quantify the relative roles of advection and diffusive mass transport, with L being the characteristic length for advection [97],

$$\mathrm{Pe} = \frac{\text{diffusion time}}{\text{advection time}} = \frac{H^2/D}{L/u}.$$

To quantify the efficiency of the process, different concepts can be used. According to IUPAC, the correct statement of quantum yield would be the ratio between the number of reactive events and the number of photons absorbed, which are impractical to determine. A more direct parameter is defined as apparent quantum yield, which compares the rate of conversion of the target molecule with the rate of photons incident on the catalyst surface. However, other indexes of efficiency may be related to the energy consumption to degrade a given amount of pollutant. For instance, IUPAC defined the Electrical Energy per Order (E_{EO}) as the electrical energy required for the degradation of a contaminant by one order of magnitude, in a unit reactor volume of wastewater [97, 98] for both batch and continuous flow operation (P is the power consumption, t the time, V the reactor volume and F the volumetric flow rate),

$$E_{\text{EO-Batch}} = \frac{1000 \times P \times t}{V \times \log\left(\frac{c_0}{c_t}\right)} \quad \text{or} \quad E_{\text{EO-Continuous}} = \frac{P}{F \times \log\left(\frac{c_0}{c_t}\right)}.$$

Alternatively, a Reactor Characterisation (R_c) number has been introduced as a factor for selecting the best UV source from an economic point of view during dye degradation in a pilot scale photoreactor [10],

$$R_c = \frac{\text{input power}(W) \times \text{UV} - \text{C efficiency}(\%)}{\text{Reactor Volume}(L)}.$$

As said, homogeneous Fenton or Photo-Fenton approach can be used to abate pollutants from waste water [3, 99, 100]. The kinetic model of this reaction, applied to the oxidation of paracetamol, has been proposed including a form of LVRPA. The kinetic expression was expressed as.

$$\left[\boldsymbol{R}(x,t)\right] = \left[\boldsymbol{R}^T(x,t)\right] + \Phi \sum_{\lambda} e_{\lambda}^{a}(x,t)\,\tau(x,t),$$

where $\boldsymbol{R}(x,t)$ is the vector of formation/conversion rates of each product/reactant, $\boldsymbol{R}^T(x,t)$ is the contribution to the rate of thermal (dark) reactivity, Φ is the quantum yield averaged over the wavelength and the summation computes the LVRPA in polychromatic conditions. $\tau(x,t)$ is the term that completes the expression of the reaction rates, as detailed in [101].

This expression of the rate is therefore included in the mass balance of the reactor, setting $C = C^0$ at $t_0 = 0$ as initial condition and considering the volume averaged quantity indicated between $\langle\ \rangle$,

$$\frac{\mathrm{d}\boldsymbol{C}}{\mathrm{d}t} = \boldsymbol{R}^T(t) + \frac{V_{\mathrm{IRR}}}{V_{\mathrm{TOT}}}\Phi\sum_{\lambda} e_{\lambda}^{a}(x,t)\tau(x,t).$$

To compute the spatial distribution of the radiation, a line source radiation model was introduced, assuming a spherical and isotropic emission,

$$\mathrm{e}_{\lambda}^{a}(x,t) = \kappa_{\lambda}(x,t)\frac{P_{\lambda,\mathrm{s}}}{2\pi L_{\mathrm{L}}}\int_{\theta_1}^{\theta_2}\exp\left[-\frac{\kappa_{T,\lambda}(x,t)(r_i - r_{\mathrm{int}})}{\cos\theta}\right]\mathrm{d}\theta,$$

where $P_{\lambda,s}$ is the lamp emission power, $\kappa_{\lambda}(x,t)$ and $\kappa_{T,\lambda}(x,t)$ are the volumetric absorption coefficient of the reacting species and of the medium, respectively, r the radius and L_L the useful length of the lamp [101].

The overall quantum yield is a parameter that can allow the comparison of different catalysts and to assess the overall feasibility of the photocatalytic process. It may be calculated as the ratio between the pollutant degradation reaction rate and the LVRPA (both volume averaged) [9].

An interesting definition of energy and exergy balances has been proposed [34], though for the photoproduction of H_2 and not for the abatement of pollutants. At first, the irradiance of solar light was determined as follows: K_{λ} measures monochromatic intensity of the radiation and n represents the numbering of intervals of wavelengths selected,

$$j_{\mathrm{s}} = 4.329\times10^{-5}\pi\sum_{n}(K_{\lambda}\Delta\lambda)_n,$$

$$q_{\mathrm{e}} = \alpha_{\mathrm{s}}\varepsilon_{\mathrm{b}}\left(T^4 - T_0^4\right),$$

where q_e is the energy exchanged between the reactor and the surroundings, α_s is the absorptivity of the reactor surface and ε_b is the Boltzmann constant. The energy efficiency of the process has been calculated as

$$\eta = \frac{n_{\mathrm{H_2}}\mathrm{HHV}_{\mathrm{H_2}}}{\gamma\left[\alpha_{\mathrm{v}}j_{\mathrm{v}} + \alpha_{\mathrm{nv}}(j_{\mathrm{s}} - j_{\mathrm{v}})\right] + n_{\mathrm{H_2O}}h_{\mathrm{H_2O}} + 0.5V_{\mathrm{H_2O,in}}^2}$$

with γ being radiation weakening factor, α_v and α_{nv} the absorptivity of the catalyst in the visible region and the non-visible one, n the moles of each compound, HHV the higher heating value, V the volume and j_v the irradiance in the visible range. The results revealed the highest efficiency at the minimum visible light absorptivity and a significant energy and exergy loss when increasing the temperature, suggesting the need of an efficient heat exchange system.

The modelling of falling film reactors has been accomplished under turbulent flow regime considering a pilot scale photoreactor, 1.25 m × 1 m in size [102]. Different papers considered a laminar regime, but in order to keep the photocatalyst in suspension, turbulence was needed, with Re > 15,000. This induces waves to be statistically described since they determine the falling film thickness and, hence, the absorption of radiation by the suspension. The velocity profiles were then estimated by momentum balances and the LVRPA was solved according to a SFM approach.

9.4 Conclusions

The scale-up of photochemical or photocatalytic reactors is a required step to exploit advanced oxidation processes and similar remediation processes to treat water effluents. Besides searching for active materials, it is important to describe reactor geometry, flow patterns and solid properties in order to understand the interaction between the reactants, the catalyst and the third actor, i.e. radiation.

Different reactor layouts are available, yet conceptualised at a pilot scale, to maximise the exposure to solar light and to limit mass transfer and radiation distribution issues. Various modelling tools are also available to describe the reaction rate as dependent on reactants, catalyst and radiation distribution across the photoreactor.

Broader efforts are needed for radiation modelling and its coupling to reactor design. Specific attention should be put on microreactor devices and on those configurations, which reached a pilot scale development to allow validation and further improvement.

Tools which couple CFD with radiation modelling should be implemented to provide rigorous methods to scale-up photoreactors.

References

1. V.H. Nguyen, J.C.S. Wu, Recent developments in the design of photoreactors for solar energy conversion from water splitting and CO2 reduction. Appl. Catal. A Gen. **550**, 122–141 (2018). https://doi.org/10.1016/j.apcata.2017.11.002
2. D. Heggo, S. Ookawara, Multiphase photocatalytic microreactors. Chem. Eng. Sci. **169**, 67–77 (2017). https://doi.org/10.1016/j.ces.2017.01.019
3. F.S. Freyria, M. Compagnoni, N. Ditaranto, I. Rossetti, M. Piumetti, G. Ramis, B. Bonelli, Pure and Fe-doped mesoporous titania catalyse the oxidation of acid orange 7 by H2O2 under different illumination conditions: Fe doping improves photocatalytic activity under simulated solar light. Catalysts **7**, 213 (2017). https://doi.org/10.3390/catal7070213
4. S. Mozia, Photocatalytic membrane reactors (PMRs) in water and wastewater treatment. A review. Sep. Purif. Technol. **73**, 71–91 (2010). https://doi.org/10.1016/J.SEPPUR.2010.03.021
5. R. Molinari, C. Lavorato, P. Argurio, Recent progress of photocatalytic membrane reactors in water treatment and in synthesis of organic compounds. A review. Catal. Today **281**, 144–164 (2017). https://doi.org/10.1016/J.CATTOD.2016.06.047
6. R. Molinari, C. Lavorato, P. Argurio, K. Szymański, D. Darowna, S. Mozia, Overview of photocatalytic membrane reactors in organic synthesis, energy storage and environmental applications. Catalysts **9**, 239 (2019). https://doi.org/10.3390/catal9030239

7. R. Munter, Advanced oxidation processes—current status and prospects. Proc. Est. Acad. Sci. Chem. **50**, 59–80 (2001)
8. Y. Boyjoo, H. Sun, J. Liu, V.K. Pareek, S. Wang, A review on photocatalysis for air treatment: from catalyst development to reactor design. Chem. Eng. J. **310**, 537–559 (2017). https://doi.org/10.1016/j.cej.2016.06.090
9. Y. Boyjoo, M. Ang, V. Pareek, Some aspects of photocatalytic reactor modeling using computational fluid dynamics. Chem. Eng. Sci. **101**, 764–784 (2013). https://doi.org/10.1016/j.ces.2013.06.035
10. E. Bahadori, M. Rapf, A. Di Michele, I. Rossetti, Photochemical vs. photocatalytic azo-dye removal in a pilot free-surface reactor: is the catalyst effective? Sep. Purif. Technol. **237**, 116320 (2020). https://doi.org/10.1016/j.seppur.2019.116320
11. G. Ramis, E. Bahadori, I. Rossetti, Photoreactors design for hydrogen production. Chem. Eng. Trans. **74**, 481–486 (2019). https://doi.org/10.3303/CET1974081
12. M. Compagnoni, G. Ramis, F.S. Freyria, M. Armandi, B. Bonelli, I. Rossetti, Innovative photoreactors for unconventional photocatalytic processes: the photoreduction of CO2 and the photo-oxidation of ammonia. Rend Lincei **28**, 151–158 (2017). https://doi.org/10.1007/s12210-017-0617-z
13. F. Galli, M. Compagnoni, D. Vitali, C. Pirola, C.L. Bianchi, A. Villa, L. Prati, I. Rossetti, CO2 photoreduction at high pressure to both gas and liquid products over titanium dioxide. Appl. Catal. B Environ. **200**, 386–391 (2017). https://doi.org/10.1016/j.apcatb.2016.07.038
14. A. Olivo, E. Ghedini, M. Signoretto, M. Compagnoni, I. Rossetti, Liquid vs. gas phase CO2 photoreduction process: which is the effect of the reaction medium? Energies **10**, 1394 (2017). https://doi.org/10.3390/en10091394
15. I. Rossetti, A. Villa, M. Compagnoni, L. Prati, G. Ramis, C. Pirola, C.L. Bianchi, W. Wang, D. Wang, CO2 photoconversion to fuels under high pressure: effect of TiO2 phase and of unconventional reaction conditions. Cat. Sci. Technol. **5**, 4481–4487 (2015). https://doi.org/10.1039/C5CY00756A
16. E. Bahdori, A. Tripodi, A. Villa, C. Pirola, L. Prati, G. Ramis, N. Dimitratos, D. Wang, I. Rossetti, High pressure CO2 photoreduction using Au/TiO2: unravelling the effect of the co-catalyst and of the titania polymorph. Cat. Sci. Technol. **9**, 2253–2265 (2019)
17. E. Bahadori, A. Tripodi, A. Villa, C. Pirola, L. Prati, G. Ramis, I. Rossetti, High pressure photoreduction of CO2: effect of catalyst formulation, hole scavenger addition and operating conditions. Catalysts **8**, 430 (2018). https://doi.org/10.3390/catal8100430
18. I. Rossetti, A. Villa, C. Pirola, L. Prati, G. Ramis, A novel high-pressure photoreactor for CO2 photoconversion to fuels. RSC Adv. **4**, 28883–28885 (2014). https://doi.org/10.1039/c4ra03751k
19. C.S. Ong, W.J. Lau, P.S. Goh, B.C. Ng, A.F. Ismail, Investigation of submerged membrane photocatalytic reactor (sMPR) operating parameters during oily wastewater treatment process. Desalination **353**, 48–56 (2014). https://doi.org/10.1016/j.desal.2014.09.008
20. R.A. Damodar, S.J. You, S.H. Ou, Coupling of membrane separation with photocatalytic slurry reactor for advanced dye wastewater treatment. Sep. Purif. Technol. **76**, 64–71 (2010). https://doi.org/10.1016/j.seppur.2010.09.021
21. R.A. Damodar, S.J. You, G.W. Chiou, Investigation on the conditions mitigating membrane fouling caused by TiO2 deposition in a membrane photocatalytic reactor (MPR) used for dye wastewater treatment. J. Hazard. Mater. **203–204**, 348–356 (2012). https://doi.org/10.1016/j.jhazmat.2011.12.027
22. S.R. De Loos, J. Van Der Schaaf, R.M. Tiggelaar, T.A. Nijhuis, M.H.J.M. De Croon, J.C. Schouten, Gas-liquid dynamics at low Reynolds numbers in pillared rectangular micro channels. Microfluid. Nanofluid. **9**, 131–144 (2010). https://doi.org/10.1007/s10404-009-0525-3
23. A. Serizawa, Z. Feng, Z. Kawara, Two-phase flow in microchannels. Exp. Therm. Fluid Sci. **26**, 703–714 (2002). https://doi.org/10.1016/S0894-1777(02)00175-9

24. P.M.Y. Chung, M. Kawaji, The effect of channel diameter on adiabatic two-phase flow characteristics in microchannels. Int. J. Multiphase Flow **30**, 735–761 (2004). https://doi.org/10.1016/j.ijmultiphaseflow.2004.05.00
25. K.S. Elvira, R.C.R. Wootton, N.M. Reis, M.R. Mackley, A.J. DeMello, Through-wall mass transport as a modality for safe generation of singlet oxygen in continuous flows. ACS Sustain. Chem. Eng. **1**, 209–213 (2013). https://doi.org/10.1021/sc300093j
26. H. Eskandarloo, A. Badiei, M.A. Behnajady, G.M. Ziarani, UV-LEDs assisted preparation of silver deposited TiO2 catalyst bed inside microchannels as a high efficiency microphotoreactor for cleaning polluted water. Chem. Eng. J. **270**, 158–167 (2015). https://doi.org/10.1016/j.cej.2015.01.117
27. K.N. Knust, D. Hlushkou, R.K. Anand, U. Tallarek, R.M. Crooks, Electrochemically mediated seawater desalination. Angew. Chem. Int. Ed. **52**, 8107–8110 (2013). https://doi.org/10.1002/anie.201302577
28. W. Liao, N. Wang, T. Wang, J. Xu, X. Han, Z. Liu, X. Zhang, W. Yu, Biomimetic microchannels of planar reactors for optimized photocatalytic efficiency of water purification. Biomicrofluidics **10**, 014123 (2016). https://doi.org/10.1063/1.4942947
29. A. Castedo, E. Mendoza, I. Angurell, J. Llorca, Silicone microreactors for the photocatalytic generation of hydrogen. Catal. Today **273**, 106–111 (2016). https://doi.org/10.1016/j.cattod.2016.02.053
30. O. Ola, M.M. Maroto-Valer, Synthesis, characterization and visible light photocatalytic activity of metal based TiO2 monoliths for CO2 reduction. Chem. Eng. J. **283**, 1244–1253 (2016). https://doi.org/10.1016/j.cej.2015.07.090
31. Z. He, Y. Li, Q. Zhang, H. Wang, Capillary microchannel-based microreactors with highly durable ZnO/TiO2 nanorod arrays for rapid, high efficiency and continuous-flow photocatalysis. Appl. Catal. B Environ. **93**, 376–382 (2010). https://doi.org/10.1016/j.apcatb.2009.10.011
32. B.A. Marinho, R.O. Cristóvão, R. Djellabi, A. Caseiro, S.M. Miranda, J.M. Loureiro, R.A.R. Boaventura, M.M. Dias, J.C.B. Lopes, V.J.P. Vilar, Strategies to reduce mass and photons transfer limitations in heterogeneous photocatalytic processes: hexavalent chromium reduction studies. J. Environ. Manag. **217**, 555–564 (2018). https://doi.org/10.1016/j.jenvman.2018.04.003
33. S. Li, L. Yang, O. Ola, M. Maroto-Valer, X. Du, Y. Yang, Photocatalytic reduction of CO2 by CO co-feed combined with photocatalytic water splitting in a novel twin reactor. Energy Convers. Manag. **116**, 184–193 (2016). https://doi.org/10.1016/j.enconman.2016.03.001
34. E. Baniasadi, I. Dincer, G.F. Naterer, Radiative heat transfer and catalyst performance in a large-scale continuous flow photoreactor for hydrogen production. Chem. Eng. Sci. **84**, 638–645 (2012). https://doi.org/10.1016/j.ces.2012.09.013
35. L. Ghalamchi, M.H. Rasoulifard, Immobilization of Fe3O4/TiO2 nanocomposite thin layer on the glass tubes in a component parabolic collector for the treatment of DR23. Int. J. Environ. Sci. Technol. **16**, 7509–7522 (2019). https://doi.org/10.1007/s13762-018-2169-x
36. K.S. Ochoa-Gutiérrez, E. Tabares-Aguilar, M.Á. Mueses, F. Machuca-Martínez, G. Li Puma, A novel prototype offset multi tubular photoreactor (OMTP) for solar photocatalytic degradation of water contaminants. Chem. Eng. J. **341**, 628–638 (2018). https://doi.org/10.1016/j.cej.2018.02.068
37. I. Salmerón, K.V. Plakas, I. Sirés, I. Oller, M.I. Maldonado, A.J. Karabelas, S. Malato, Optimization of electrocatalytic H2O2 production at pilot plant scale for solar-assisted water treatment. Appl. Catal. B Environ. **242**, 327–336 (2019). https://doi.org/10.1016/j.apcatb.2018.09.045
38. A.M. Cahino, M.M.A. de Andrade, E.S. de Araújo, E.L. Silva, C. de Oliveira Cunha, E.M.R. Rocha, Degradation of tetracycline by solar photo-Fenton: optimization and application in pilot photoreactor. Environ. Qual. Manag. **28**, 101–106 (2018). https://doi.org/10.1002/tqem.21579
39. F. Cao, Q. Wei, H. Liu, N. Lu, L. Zhao, L. Guo, Development of the direct solar photocatalytic water splitting system for hydrogen production in Northwest China: design and

evaluation of photoreactor. Renew. Energy **121**, 153–163 (2018). https://doi.org/10.1016/j.renene.2018.01.016
40. I. Boiarkina, S. Norris, D.A. Patterson, Investigation into the effect of flow structure on the photocatalytic degradation of methylene blue and dehydroabietic acid in a spinning disc reactor. Chem. Eng. J. **222**, 159–171 (2013). https://doi.org/10.1016/j.cej.2013.02.025
41. I. Boiarkina, S. Norris, D.A. Patterson, The case for the photocatalytic spinning disc reactor as a process intensification technology: comparison to an annular reactor for the degradation of methylene blue. Chem. Eng. J. **225**, 752–765 (2013). https://doi.org/10.1016/j.cej.2013.03.125
42. M. Mirzaei, B. Dabir, M. Dadvar, M. Jafarikojour, Photocatalysis of phenol using a spinning disc photoreactor immobilized with TiO2 nanoparticles: hydrodynamic modeling and reactor optimization. Ind. Eng. Chem. Res. **56**, 1739–1749 (2017). https://doi.org/10.1021/acs.iecr.6b03204
43. M. Mirzaei, M. Jafarikojour, B. Dabir, M. Dadvar, Evaluation and modeling of a spinning disc photoreactor for degradation of phenol: impact of geometry modification. J. Photochem. Photobiol. A Chem. **346**, 206–214 (2017). https://doi.org/10.1016/j.jphotochem.2017.05.043
44. C. Boyle, N. Skillen, L. Stella, P.K.J. Robertson, Development and optimization of an immobilized photocatalytic system within a stacked frame photoreactor (SFPR) using light distribution and fluid mixing simulation coupled with experimental validation. Ind. Eng. Chem. Res. **58**, 2727–2740 (2019). https://doi.org/10.1021/acs.iecr.8b05709
45. M. Jafarikojour, B. Dabir, M. Sohrabi, S.J. Royaee, Application of a new immobilized impinging jet stream reactor for photocatalytic degradation of phenol: reactor evaluation and kinetic modelling. J. Photochem. Photobiol. A Chem. **364**, 613–624 (2018). https://doi.org/10.1016/j.jphotochem.2018.03.043
46. M. Jafarikojour, B. Dabir, M. Sohrabi, S.J. Royaee, Evaluation and optimization of a new design photocatalytic reactor using impinging jet stream on a TiO2 coated disc. Chem. Eng. Process. Process Intensif. **121**, 215–223 (2017). https://doi.org/10.1016/j.cep.2017.08.011
47. M. Jafarikojour, M. Sohrabi, S.J. Royaee, M. Rezaei, A new model for residence time distribution of impinging streams reactors using descending-sized stirred tanks in series. Chem. Eng. Res. Des. **109**, 86–96 (2016). https://doi.org/10.1016/j.cherd.2016.01.003
48. Y.K. Abdel-Maksoud, E. Imam, A.R. Ramadan, TiO2 water-bell photoreactor for wastewater treatment. Sol. Energy **170**, 323–335 (2018). https://doi.org/10.1016/j.solener.2018.05.053
49. A.A. Azzaz, A.A. Assadi, S. Jellali, A. Bouzaza, D. Wolbert, S. Rtimi, L. Bousselmi, Discoloration of simulated textile effluent in continuous photoreactor using immobilized titanium dioxide: effect of zinc and sodium chloride. J. Photochem. Photobiol. A Chem. **358**, 111–120 (2018). https://doi.org/10.1016/j.jphotochem.2018.01.032
50. O. Shvydkiv, C. Limburg, K. Nolan, M. Oelgemöller, Synthesis of juglone (5-hydroxy-1,4-naphthoquinone) in a falling film microreactor. J. Flow Chem. **2**, 52–55 (2012). https://doi.org/10.1556/jfchem.2012.00022
51. S. Baghbani Ghatar, S. Allahyari, N. Rahemi, M. Tasbihi, Response surface methodology optimization for photodegradation of methylene blue in a ZnO coated flat plate continuous photoreactor. Int. J. Chem. React. Eng. **16**, 1–14 (2018). https://doi.org/10.1515/ijcre-2017-0221
52. J.C. Espíndola, R.O. Cristóvão, S.R.F. Araújo, T. Neuparth, M.M. Santos, R. Montes, J.B. Quintana, R. Rodil, R.A.R. Boaventura, V.J.P. Vilar, An innovative photoreactor, FluHelik, to promote UVC/H2O2 photochemical reactions: tertiary treatment of an urban wastewater. Sci. Total Environ. **667**, 197–207 (2019). https://doi.org/10.1016/j.scitotenv.2019.02.335
53. J.C. Espíndola, R.O. Cristóvão, D.A. Mayer, R.A.R. Boaventura, M.M. Dias, J.C.B. Lopes, V.J.P. Vilar, Overcoming limitations in photochemical UVC/H2O2 systems using a mili-photoreactor (NETmix): oxytetracycline oxidation. Sci. Total Environ. **660**, 982–992 (2019). https://doi.org/10.1016/j.scitotenv.2019.01.012
54. J.C. Espíndola, R.O. Cristóvão, S.G.S. Santos, R.A.R. Boaventura, M.M. Dias, J.C.B. Lopes, V.J.P. Vilar, Intensification of heterogeneous TiO2 photocatalysis using the NETmix

mili-photoreactor under microscale illumination for oxytetracycline oxidation. Sci. Total Environ. **681**, 467–474 (2019). https://doi.org/10.1016/j.scitotenv.2019.05.066
55. S.G.S. Santos, L.O. Paulista, T.F.C.V. Silva, M.M. Dias, J.C.B. Lopes, R.A.R. Boaventura, V.J.P. Vilar, Intensifying heterogeneous TiO2 photocatalysis for bromate reduction using the NETmix photoreactor. Sci. Total Environ. **664**, 805–816 (2019). https://doi.org/10.1016/j.scitotenv.2019.02.045
56. V. Héquet, C. Raillard, O. Debono, F. Thévenet, N. Locoge, L. Le Coq, Photocatalytic oxidation of VOCs at ppb level using a closed-loop reactor: the mixture effect. Appl. Catal. B Environ. **226**, 473–486 (2018). https://doi.org/10.1016/j.apcatb.2017.12.041
57. X. Meng, N. Yun, Z. Zhang, Recent advances in computational photocatalysis: a review. Can. J. Chem. Eng. **97**, 1982–1998 (2019). https://doi.org/10.1002/cjce.23477
58. R. Acosta-Herazo, J. Monterroza-Romero, M.Á. Mueses, F. Machuca-Martínez, G. Li Puma, Coupling the six flux absorption-scattering model to the Henyey-Greenstein scattering phase function: evaluation and optimization of radiation absorption in solar heterogeneous photoreactors. Chem. Eng. J. **302**, 86–96 (2016). https://doi.org/10.1016/j.cej.2016.04.127
59. I. Rossetti, E. Bahadori, A. Tripodi, G. Ramis, Modelling of photoreactors for water treatment. Chem. Eng. Trans. **74**, 289–294 (2019). https://doi.org/10.3303/CET1974049
60. N. Padoin, C. Soares, An explicit correlation for optimal TiO2 film thickness in immobilized photocatalytic reaction systems. Chem. Eng. J. **310**, 381–388 (2017). https://doi.org/10.1016/j.cej.2016.06.013
61. M.T. Dhotre, B. Niceno, B.L. Smith, Large eddy simulation of a bubble column using dynamic sub-grid scale model. Chem. Eng. J. **136**, 337–348 (2008)
62. E. Olmos, C. Gentric, C. Vial, G. Wild, N. Midoux, Numerical simulation of multiphase flow in bubble column reactors. Influence of bubble coalescence and break-up. Chem. Eng. Sci. **56**, 6359–6365 (2001)
63. V.K. Pareek, S.J. Cox, M.P. Brungs, B. Young, A.A. Adesina, Computational fluid dynamic (CFD) simulation of a pilot-scale annular bubble column photo- catalytic reactor. Chem. Eng. Sci. **58**, 859–865 (2003)
64. J. Sanyal, S. Roy, M.P. Dudukovic, Numerical simulation of gas–liquid dynamics in cylindrical bubble column reactors. Chem. Eng. Sci. **54**, 5071–5083 (1999)
65. M.V. Tabib, S.A. Roy, J.B. Joshi, CFD simulation of bubble column—an analysis of interphase forces and turbulence models. Chem. Eng. J. **139**, 589–614 (2008)
66. Y. Boyjoo, M. Ang, V. Pareek, CFD simulation of a pilot scale slurry photocatalytic reactor and design of multiple-lamp reactors. Chem. Eng. Sci. **111**, 266–277 (2014). https://doi.org/10.1016/j.ces.2014.02.022
67. R.L. Romero, O.M. Alfano, A.E. Cassano, Radiation field in an annular, slurry photocatalytic reactor. 2. Model and experiments. Ind. Eng. Chem. Res. **42**, 2479–2488 (2003)
68. J. Wang, B. Deng, J. Gao, H. Cao, Numerical simulation of radiation distribution in a slurry reactor: the effect of distribution of catalyst particles. Chem. Eng. J. **357**, 169–179 (2019). https://doi.org/10.1016/j.cej.2018.09.126
69. O.I. Tokode, R. Prabhu, L.A. Lawton, P.K.J. Robertson, Effect of controlled periodic-based illumination on the photonic efficiency of photocatalytic degradation of methyl orange. J. Catal. **290**, 138–142 (2012)
70. O. Tokode, R. Prabhu, L.A. Lawton, P.K.J. Robertson, Mathematical modelling of quantum yield enhancements of methyl orange photooxidation in aqueous TiO2 suspensions under controlled periodic UV LED illumination. Appl. Catal. B Environ. **156–157**, 398–403 (2014)
71. O. Tokode, R. Prabhu, L.A. Lawton, P.K.J. Robertson, The effect of pH on the photonic efficiency of the destruction of methyl orange under controlled periodic illumination with UV-LED sources. Chem. Eng. J. **246**, 337–342 (2014)
72. C. Casado, R. Timmers, A. Sergejevs, C.T. Clarke, D.W.E. Allsopp, C.R. Bowen, R. van Grieken, J. Marugán, Design and validation of a LED-based high intensity photocatalytic reactor for quantifying activity measurements. Chem. Eng. J. **327**, 1043–1055 (2017). https://doi.org/10.1016/j.cej.2017.06.167

73. C. Casado, J. Marugán, R. Timmers, M. Muñoz, R. van Grieken, Comprehensive multiphysics modeling of photocatalytic processes by computational fluid dynamics based on intrinsic kinetic parameters determined in a differential photoreactor. Chem. Eng. J. **310**, 368–380 (2017). https://doi.org/10.1016/j.cej.2016.07.081
74. F.C. Moreira, E. Bocos, A.G.F. Faria, J.B.L. Pereira, C.P. Fonte, R.J. Santos, J.C.B. Lopes, M.M. Dias, M.A. Sanromán, M. Pazos, R.A.R. Boaventura, V.J.P. Vilar, Selecting the best piping arrangement for scaling-up an annular channel reactor: an experimental and computational fluid dynamics study. Sci. Total Environ. **667**, 821–832 (2019). https://doi.org/10.1016/j.scitotenv.2019.02.260
75. F. Denny, J. Scott, V. Pareek, G.D. Peng, R. Amal, Computational fluid dynamics modelling and optimal configuring of a channelled optical fibre photoreactor. Chem. Eng. Sci. **65**, 5029–5040 (2010). https://doi.org/10.1016/j.ces.2010.06.003
76. H. Einaga, J. Tokura, Y. Teraoka, K. Ito, Kinetic analysis of TiO2-catalyzed heterogeneous photocatalytic oxidation of ethylene using computational fluid dynamics. Chem. Eng. J. **263**, 325–335 (2015). https://doi.org/10.1016/j.cej.2014.11.017
77. J.S. Devia-Orjuela, L.A. Betancourt-Buitrago, F. Machuca-Martinez, CFD modeling of a UV-A LED baffled flat-plate photoreactor for environment applications: a mining wastewater case. Environ. Sci. Pollut. Res. **26**, 4510–4520 (2019). https://doi.org/10.1007/s11356-018-2431-2
78. J.E. Duran, F. Taghipour, M. Mohseni, Irradiance modeling in annular photoreactors using the finite-volume method. J. Photochem. Photobiol. A Chem. **215**, 81–89 (2010). https://doi.org/10.1016/j.jphotochem.2010.07.027
79. J. Kumar, A. Bansal, Photocatalytic degradation in annular reactor: modelization and optimization using computational fluid dynamics (CFD) and response surface methodology (RSM). J. Environ. Chem. Eng. **1**, 398–405 (2013). https://doi.org/10.1016/j.jece.2013.06.002
80. J. van Walsem, S.W. Verbruggen, B. Modde, S. Lenaerts, S. Denys, CFD investigation of a multi-tube photocatalytic reactor in non-steady-state conditions. Chem. Eng. J. **304**, 808–816 (2016). https://doi.org/10.1016/j.cej.2016.07.028
81. A.J. Guadarrama-Mendoza, H.I. Villafán-Vidales, P.J. Valadés-Pelayo, C.A. Arancibia-Bulnes, D. Riveros-Rosas, H. Romero-Paredes, Radiative analysis in a multichanneled monolith solar reactor coated with ZnFe2O4 thin film. Int. J. Therm. Sci. **132**, 275–284 (2018). https://doi.org/10.1016/j.ijthermalsci.2018.05.042
82. A. Brucato, A.E. Cassano, F. Grisafi, G. Montante, L. Rizzuti, G. Vella, Estimating radiant fields in flat heterogeneous photoreactors by the six flux model. AICHE J. **52**, 3882–3890 (2006)
83. A. Brucato, L. Rizzuti, Simplified modeling of radiant fields in heterogeneous photoreactors. 1. Case of zero reflectance. Ind. Eng. Chem. Res. **36**, 4740–4747 (1997). https://doi.org/10.1021/ie960259j
84. G. Li Puma, J.N. Khor, A. Brucato, Modeling of an annular photocatalytic reactor for water purification: oxidation of pesticides. Environ. Sci. Technol. **38**, 3737–3745 (2004). https://doi.org/10.1021/es0301020
85. A. Busciglio, O.M. Alfano, F. Scargiali, A. Brucato, A probabilistic approach to radiant field modeling in dense particulate systems. Chem. Eng. Sci. **142**, 79–88 (2016). https://doi.org/10.1016/j.ces.2015.11.025
86. G. Marotta, F. Scargiali, S. Lima, G. Caputo, F. Grisafi, A. Brucato, Vacuum air-lift bioreactor for microalgae production. Chem. Eng. Trans. **57**, 925–930 (2017). https://doi.org/10.3303/CET1757155
87. G. Marotta, J. Pruvost, F. Scargiali, G. Caputo, A. Brucato, Reflection-refraction effects on light distribution inside tubular photobioreactors. Can. J. Chem. Eng. **95**, 1646–1651 (2017). https://doi.org/10.1002/cjce.22811
88. I. Grcˇic, G.L. Puma, Photocatalytic degradation of water contaminants in multiple photoreactors and evaluation of reaction kinetic constants independent of photon absorption, irradiance, reactor geometry and hydrodynamics. Environ. Sci. Technol. **47**, 13702–13711 (2013)

89. P.J. Valades-Pelayo, J. Moreira, B. Serrano, H. de Lasa, Boundary conditions and phase functions in a Photo-CREC Water-II reactor radiation field. Chem. Eng. Sci. **107**, 123–136 (2014)
90. R. Acosta-Herazo, M.Á. Mueses, G.L. Puma, F. Machuca-Martínez, Impact of photocatalyst optical properties on the efficiency of solar photocatalytic reactors rationalized by the concepts of initial rate of photon absorption (IRPA) dimensionless boundary layer of photon absorption and apparent optical thickness. Chem. Eng. J. **356**, 839–849 (2019). https://doi.org/10.1016/j.cej.2018.09.085
91. D. Castilla-Caballero, F. Machuca-Martínez, C. Bustillo-Lecompte, J. Colina-Márquez, Photocatalytic degradation of commercial acetaminophen: evaluation, modeling, and scaling-up of photoreactors. Catalysts **8**, 179 (2018). https://doi.org/10.3390/catal8050179
92. F. Cao, H. Li, H. Chao, L. Zhao, L. Guo, Optimization of the concentration field in a suspended photocatalytic reactor. Energy **74**, 140–146 (2014). https://doi.org/10.1016/j.energy.2014.04.068
93. M.A. Mueses, F. Machuca-Martinez, G. Li Puma, Effective quantum yield and reaction rate model for evaluation of photocatalytic degradation of water contaminants in heterogeneous pilot-scale solar photoreactors. Chem. Eng. J. **215–216**, 937–947 (2013). https://doi.org/10.1016/j.cej.2012.11.076
94. B.M. da Costa Filho, A.L.P. Araujo, S.P. Padrão, R.A.R. Boaventura, M.M. Dias, J.C.B. Lopes, V.J.P. Vilar, Effect of catalyst coated surface, illumination mechanism and light source in heterogeneous TiO2 photocatalysis using a mili-photoreactor for n-decane oxidation at gas phase. Chem. Eng. J. **366**, 560–568 (2019). https://doi.org/10.1016/j.cej.2019.02.122
95. G. Charles, T. Roques-Carmes, N. Becheikh, L. Falk, J.M. Commenge, S. Corbel, Determination of kinetic constants of a photocatalytic reaction in micro-channel reactors in the presence of mass-transfer limitation and axial dispersion. J. Photochem. Photobiol. A Chem. **223**, 202–211 (2011). https://doi.org/10.1016/j.jphotochem.2011.08.019
96. O. Barlay Ergu, O.N. Sara, S. Yapici, M.E. Arzutug, Pressure drop and point mass transfer in a rectangular microchannel. Int. Commun. Heat Mass Transf. **36**, 618–623 (2009)
97. D.S. de Sá, L.E. Vasconcelos, J.R. de Souza, B.A. Marinkovic, T. Del Rosso, D. Fulvio, D. Maza, A. Massi, O. Pandoli, Intensification of photocatalytic degradation of organic dyes and phenol by scale-up and numbering-up of meso- and microfluidic TiO2 reactors for wastewater treatment. J. Photochem. Photobiol. A Chem. **364**, 59–75 (2018). https://doi.org/10.1016/j.jphotochem.2018.05.020
98. J.R. Bolton, K.G. Bircher, W. Tumas, C.A. Tolman, Figures of merit for the technical development and application of advanced oxidation technologies for both electric and solar driven systems. Pure Appl. Chem. **73**, 627–637 (2001)
99. M. Compagnoni, G. Ramis, F.S. Freyria, M. Armandi, B. Bonelli, I. Rossetti, Photocatalytic processes for the abatement of N-containing pollutants from waste water. Part 1: inorganic pollutants. J. Nanosci. Nanotechnol. **17**, 3632–3653 (2017)
100. F.S. Freyria, M. Armandi, M. Compagnoni, G. Ramis, I. Rossetti, B. Bonelli, Catalytic and photocatalytic processes for the abatement of N-containing pollutants from wastewater. Part 2: organic pollutants. J. Nanosci. Nanotechnol. **17**, 3654–3672 (2017). https://doi.org/10.1166/jnn.2017.14014
101. F. Audino, L. Conte, A. Schenone, M. Pérez-Moya, M. Graells, O. Alfano, A kinetic study for the fenton and photo-fenton paracetamol degradation in an annular photoreactor involving assessment of LVRPA. Environ. Sci. Pollut. Res. **26**, 4312–4323 (2019). https://doi.org/10.1007/s11356-018-3098-4
102. J. Colina-Marquez, D. Castilla-Caballero, F. Machuca-Martinez, Modeling of a falling-film photocatalytic reactor: fluid dynamics for turbulent regime. Appl. Math. Model. **40**, 4812–4821 (2016). https://doi.org/10.1016/j.apm.2015.12.007

Chapter 10
From Nano- to Macrostructured Carbon Catalysts for Water and Wastewater Treatment

João Restivo, Olívia Salomé G. P. Soares, and Manuel Fernando R. Pereira

10.1 Introduction

While catalysis is widely used in a variety of industrial applications, namely, fine chemical synthesis and oil refining, its application in water treatment is yet to be fully integrated into the current industrial landscape [1, 2]. Nevertheless, catalytic emission control technologies are extensively implemented for gas effluents, in particular, downstream from combustion processes. The most common application is the use of catalytic converters in automobiles for control of CO and NO_x, occasionally in combination with soot and catalytic regeneration of particle filters [3].

The lack of use of catalytic processes into water management is attributed mostly to the efficiency of the available technologies to provide sufficient treatment to comply with the applicable legislation [4]. The legislation is drawn taking into account what are the best available technologies (BTA) and is thus unlikely to significantly move forward to stricter requirements unless innovative cost-efficient technologies enter the market.

New findings in water quality have triggered interest in the development of efficient technologies that can address new rising concerns or facilitate the use or reuse of previously unavailable water sources. This includes the development of analytical techniques to detect trace organic pollutants [5–7]; the discovery of their public health and environmental impacts at very low (mg to µg per litre) concentrations [8]; the understanding of the toxicity associated with recalcitrant by-products of the degradation of emerging pollutants by conventional treatment solutions [7, 9]; and the interest in the recovery of contaminated water and recommendations for reuse of treated water where available technologies such as biological treatment may not be suitable for the intended final uses [10, 11]. The relevant authorities, faced with

J. Restivo · O. S. G. P. Soares (✉) · M. F. R. Pereira
Laboratory of Separation and Reaction Engineering—Laboratory of Catalysis and Materials (LSRE-LCM), Faculty of Engineering, University of Porto, Porto, Portugal
e-mail: salome.soares@fe.up.pt

M. Piumetti, S. Bensaid (eds.), *Nanostructured Catalysts for Environmental Applications*, https://doi.org/10.1007/978-3-030-58934-9_10

the advances here described, have recently shown the intention to implement stricter legislations towards the control of target emerging pollutants in waste and drinking water [12]. Thus, the development of efficient technologies for water and wastewater treatment to remove emerging pollutants, and for the reclaiming of contaminated ground and surface water sources, is an urgent requirement to ensure adequate water management, which in turn is becoming an even more critical concern due to the rising climate emergency [13].

Advanced non-catalytic technologies exist that can mitigate or solve these issues; however, they generally rely on the concentration of the contaminants onto a different phase—usually liquid or solid. The secondary waste streams still require further handling that represents undesirable energetic and environmental costs. Several solutions are commercially available and/or topics of on-going research. Membrane processes, from micro- to nanofiltration and reverse osmosis, are regularly used as tertiary water treatment, and their efficiency in the removal of emerging pollutants is widely reported, including for those found at trace levels [14, 15]. Similarly, ion exchange is also available for the removal of ionic contaminants [16]. The use of adsorbents is also widespread in conventional water and wastewater treatment plants, typically based on high specific surface area activated carbon [17]. All these solutions result in the concentration of the contaminants in a secondary waste stream, either solid when the contaminants are adsorbed or retained from the liquid phase or liquid in the case of reverse osmosis. The concentrated secondary streams require handling for regeneration or disposal with added environmental and financial costs [18]. Biological treatment is another alternative for the removal of emerging pollutants; while this technology is already used in the large majority of wastewater treatment plants for nitrogen and phosphorous control [19], its efficiency varies with the class of pollutants and generally results in by-products that still present toxicity [20]. Moreover, biological treatment requires further handling of the activated sludge, and its use for drinking water raises concerns regarding biological risks [13].

10.1.1 Catalytic Technologies for Water and Wastewater Treatment

Catalytic technologies for water and wastewater treatment have the potential to answer the issues raised above [21], either in isolation or in combination with other treatment solutions [2], presenting several potentially interesting advantages [2, 22, 23]. Two of these characteristics are here highlighted to illustrate the potential of catalytic water treatment. On the one hand, catalysts can be tailored for the selective (e.g. in complex matrices where competition may hinder reactions) or non-selective (e.g. where conventional methods are likely to target only some of the contaminants) degradation of target pollutants. On the other hand, catalytic systems can be designed towards the complete degradation or conversion of the target pollutants, therefore reducing the generation of secondary waste streams requiring further handling. Thus, catalytic technologies for water treatment have the potential to successfully eliminate emerging pollutants from water and wastewater without generation of concentrated secondary streams, leading to their complete mineralisation without the formation of toxic by-products. Such emerging pollutants include organic micropollutants, such as pharmaceuticals and personal care products (PPCP), pesticides, fertilisers, and detergents, and inorganic pollutants, such as nitrate, nitrite, bromate, and perchlorate. The successful implementation of catalytic water treatment technologies will likely involve their integration in urban and industrial treatment plants to help efficiently achieve water quality targets, rather than acting as a one-fit-for-all solution [2].

10.1.2 Carbon as a Catalyst

Carbon, specifically activated carbon, is extensively used as a traditional adsorbent in water treatment. Its use is very well established and firmly accepted as a methodology suitable for both waste and drinking water treatment [24]. Moreover, the capabilities of carbon as a catalyst are well-known; in particular, the appeal of carbon as a catalyst or catalyst supports resides on its easily tunable surface textural and chemical properties [25]. More recently, the structuring of carbon at the nanoscale has gained interest from researchers after the discovery and spread of materials such as graphene and carbon nanotubes [26]. In this chapter, it will be demonstrated how carbon-based catalysts can be tailored and structured at different scales (from nano to macro) towards the design of efficient systems to degrade both organic and inorganic pollutants in water.

Two main approaches to catalytic technologies for water and wastewater treatment are discussed in the chapter. This distinction is closely related to different preparation methods of the carbon-based catalysts. First, the use of carbon-based metal-free catalysts in the oxidation of organic pollutants will be described and discussed in detail. Afterwards, the description of the application of carbon-based supports for active metallic phases in the catalytic reduction of inorganic ionic contaminants is presented. The description of each approach is focused on the nano- to macroscale structuring of carbon materials (Fig. 10.1), with several examples provided to illustrate the behaviour of the materials and to explain how their textural and chemical properties influence their performance.

Nanostructured carbon can be defined as materials presenting observable structures (individual particles or agglomerates) in at least one external dimension in the 1–100 nm range (or at least a set percentage of the number size distribution, usually attributed as 50%) [27]. The approach to nanostructured carbon catalysts will focus on sp2 hybridised graphitic carbon nanomaterials, namely, 1D (carbon nanotubes and nanofibers) and 2D (graphene) materials (Fig. 10.2) [28].

Graphitic nanostructured carbon is understood to have properties of great interest for catalysis in general, which can be expanded to both catalytic wet air oxidation

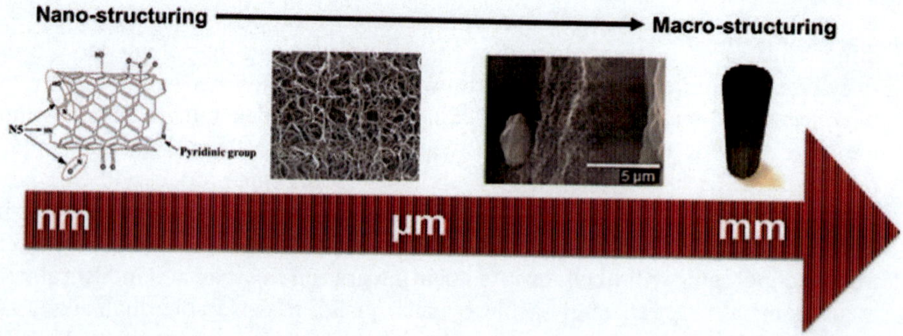

Fig. 10.1 Approaches to the modification of carbon materials in catalysis and their relation to structuring from the nano- to the macroscale

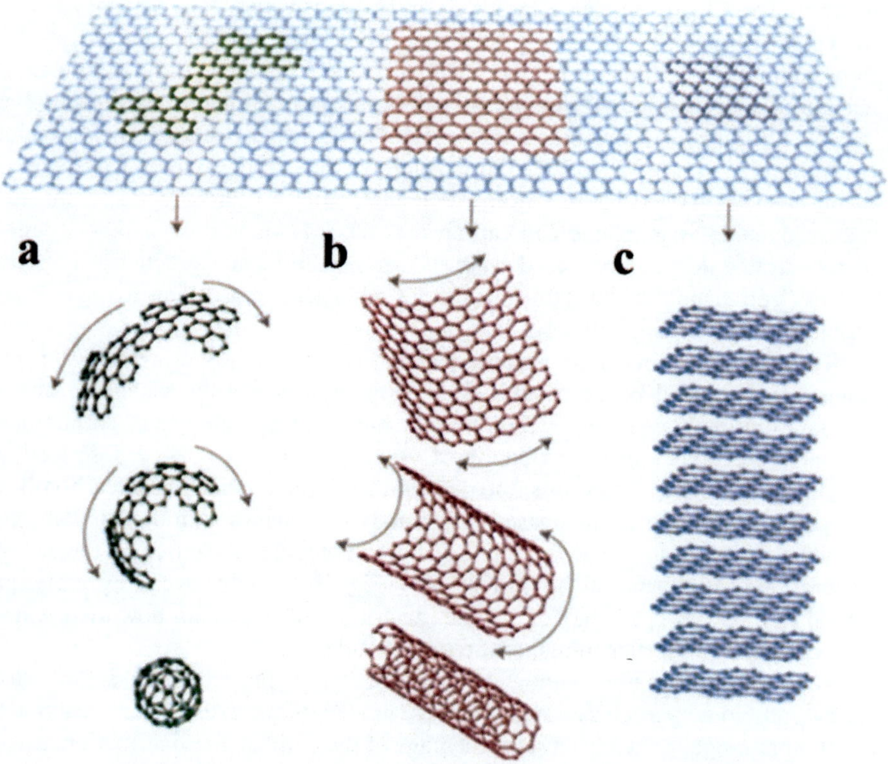

Fig. 10.2 Structural relation between typical (a) 0D, e.g. buckyballs, (b) 1D, e.g. carbon nanotubes, and (c) 2D, e.g. graphene sp2-hybridised carbon nanomaterials [29]

(CWAO) and catalytic ozonation (COz). These include the high density of electrons that can potentially be delocalised and the presence of defects and edges requiring saturation with heteroatoms, thus increasing the potential for doping [26].

10.1.3 Carbon Surface Nano- and Microengineering

The modification of the surface chemistry and texture of carbons are already a classic approach to explore its potential as an adsorbent, catalyst support, or metal-free catalyst [25]. Analogous procedures have been extended to carbon nanomaterials using similar methodologies.

The textural properties, and the nature and concentration of the surface functional groups of the carbon materials, can be tailored for specific applications by the adequate preparation methods. The presence of heteroatoms, such as oxygen, nitrogen, and sulphur, bound to the edges or defects of the graphene layers in the carbon material originates from several surface functional groups (Fig. 10.3). Such elements either are present in the starting material or become chemically bound to the structure during the preparation [25]. Oxygenated groups can be introduced onto the carbon material surface by oxidative treatments, either in the gas or liquid phase [30], and can be selectively removed by thermal treatments under inert atmosphere [31]. The introduction of heteroatoms on the carbon structure allows us to control the electronic properties by introducing electron acceptors or donors, which can enhance π-bonding, leading to improved stability and electron transfer rate, and consequently, improved performance and durability of the catalysts. The doping of the carbon materials can be tuned to the reactions of interest (Table 10.1) [32, 33].

Activated carbons are normally characterised by their very large surface area, ranging up to a few thousand square metres per gram, and their microporosity. Some reactions may be favoured by other types of porosity, especially in cases where the small size of the pores may hurt the access of reactants to the active sites. The nanostructure of carbon materials can form porosities in the mesoporous range, which can thus facilitate the access to larger molecules, while avoiding a complete sacrifice of the large surface areas, still reaching up to several hundred square metres per gram. The mesoporous texture is present for 1D materials, such as carbon nanotubes, and emerges from the formation of empty spaces or pores within entangled bundles (Fig. 10.4) [35].

10.1.4 Nanostructured Carbon Supported on Macrostructures

The handling of nanosized powders (and powder catalysts in general) and their application in water treatment have several inherent challenges, in particular, the requirements for separation of the solids from the resulting water [36]. One potential alternative is the immobilisation of the catalysts in macrostructures that mediate

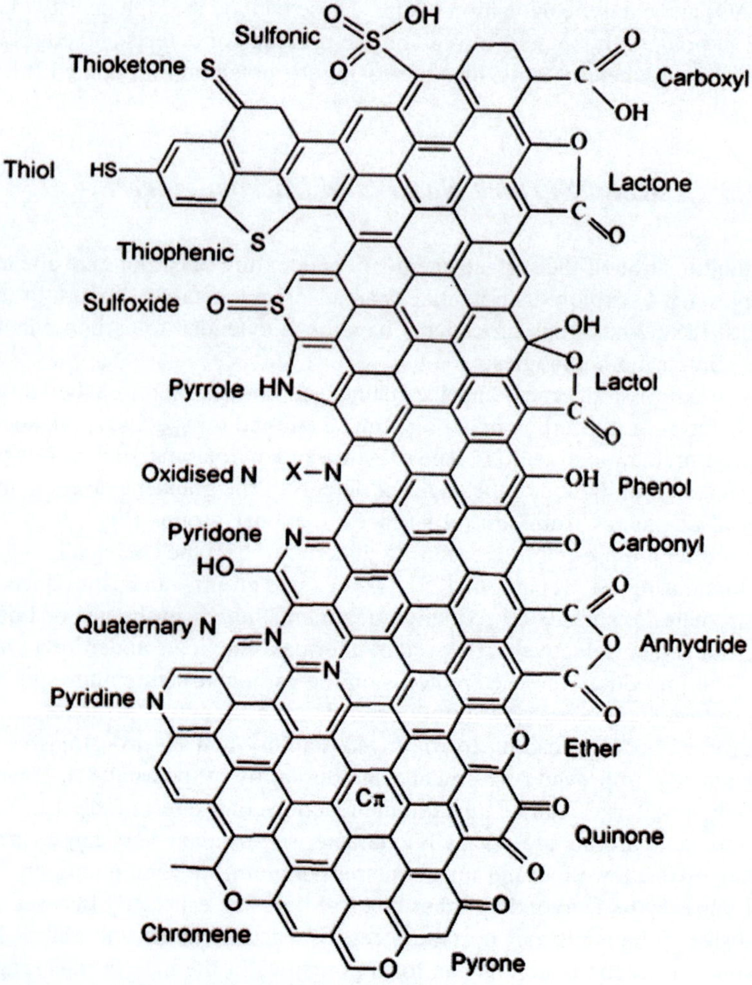

Fig. 10.3 Heteroatom functionalisation of MWCNT [34]

their contact with water and gases. Honeycomb monoliths, as a macrostructured support, have been proposed as a solution with several advantages over other configurations, including low pressure drop and the formation of favourable hydrodynamic regimes in multiphasic applications [37].

Monolithic catalysts are widely used in the treatment of gas effluents, in particular, in the automotive industry [1]. However, the preparation of nanostructured carbon layers presents challenges different from those generally faced in current industrial processes. Two approaches are reported in the literature: in situ formation of a nanostructured carbon layer or coating using traditional dip-coating techniques with pre-modified nanostructured carbons [38].

Table 10.1 Gas and liquid phase catalytic reactions catalysed by carbon and respective surface functionalities responsible for catalytic activity [34]

Reactions	Catalytic active components
Gas phase	
Oxidative dehydrogenation	Quinones
Dehydration of alcohols	Carboxylic acids
Dehydrogenation of alcohols	Lewis acid and basic sites
NO_x reduction (SCR with NH_3)	Acidic surface oxides (carboxyl, lactone) + basic sites (carbonyl or N5, N6)
NO oxidation	Basic surface
SO_2 oxidation	Basic sites; Pyridinic—N6
H_2S oxidation	Basic sites
Dehydrohalogenation	Pyridinic nitrogen sites
Liquid phase	
Hydrogen peroxide reactions	Basic sites
Catalytic ozonation	Basic sites
Catalytic wet air oxidation	Basic sites

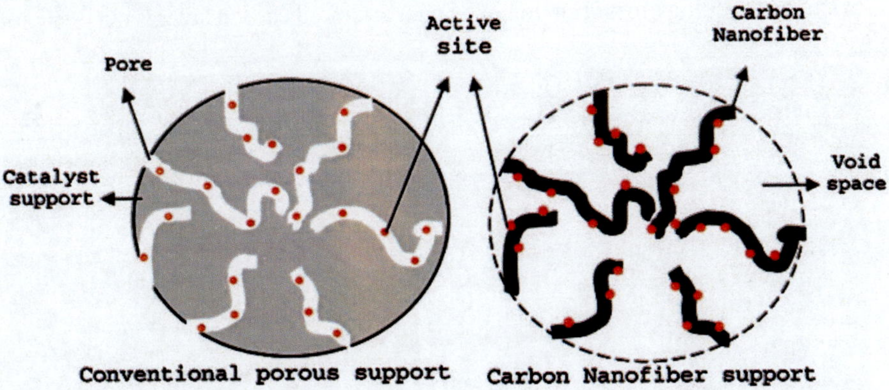

Fig. 10.4 Illustration of the nature of the 1D carbon nanomaterial mesoporosity by comparison with the microporosity of traditional porous supports [35]

10.1.4.1 In Situ Nanostructured Carbon Layer Formation

The in situ growth of nanostructured 1D carbon on a macrostructured support is an adaptation of the chemical vapour deposition (CVD) method for the growth of carbon nanotubes and nanofibers [39]. A supporting layer with adequate chemistry and porosity may be coated onto the support beforehand (when required), after which the metallic growth catalyst is dispersed on the support in the form of nanoparticles. The growth of the carbon nanofibers or tubes is then promoted by the decomposition of a carbon source at temperature [40]. The resulting nanostructured layer is often called a carbon nanotube/fibre forest, in particular when their alignment is promoted during growth [41], as it is formed by several individual tubes or fibres (Fig. 10.5).

The growth conditions can be modified to obtain carbon nanostructures with the desired characteristics, either in their morphology or chemical properties. The methods available to promote these modifications are similar to those used for the synthesis of carbon nanomaterials in powder form and are based on changing the growth catalyst (choice of metal, particle size and shape, and crystalline structure), the carbon source, and the growth parameters (temperature and time) [43]. Furthermore, the carbon nanostructures can be doped with heteroatoms by mixing a precursor with the carbon source, for example, ammonia for N-doped carbon [44]. However, this doping procedure is known to cause changes in the yield of carbon nanostructures, limiting the thickness of the layers, and to affect the structure of the carbon (from nanotubes to bamboo-like structures for example).

Post-growth modification of the carbon nanostructure layers can also be used to modify their surface chemistry and texture to an extent. Similar methods as those used to powder materials are applicable, with the required adaptations to accommodate the macrostructures. Nevertheless, the efficiency of these is highly dependent on the macrostructure and layer thickness and structure [45]. Additionally, the in situ methodology for formation of a nanostructured carbon layer is not compatible

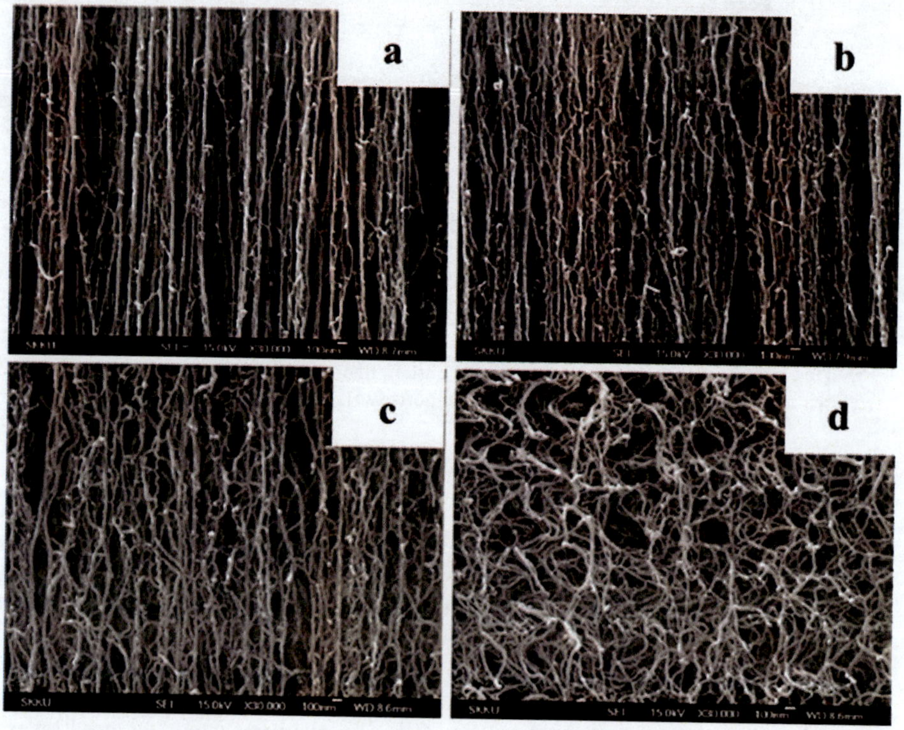

Fig. 10.5 Example of carbon nanotube layers grown in situ with different orders of alignment (from **a** to **d**) with changing growth catalyst particle size [42]

with mechanical methodologies for modification of the chemical and textural properties of carbon nanomaterials. The potential of these methods is further detailed in Sects. 10.2 and 10.3.

10.1.4.2 Coating with Pre-modified Nanostructured Carbon

The limitations to the modification of carbon nanomaterials grown in situ motivate the development of methods to create coatings with pre-modified materials. This approach has specific challenges related to the handling of the nanomaterials and the formation of the catalytic layer. On one hand, the dispersion of carbon nanomaterials in water, while preferable from a process point of view, is not readily achieved in most cases. Modification of the nanomaterials can improve their solubility and help form slurries that can be coated onto macrostructures. This can be achieved by chemical or physical surface functionalisation: chemical functionalisation refers to the covalent linkage onto the carbon scaffold, while physical functionalisation refers to non-covalent interactions with chemical functionalities (Fig. 10.6) [46]. The use of the former to enhance the catalytic performance of carbon materials is thoroughly described in Sect. 10.2.1. However, the latter (or a combination of both) is understood to be more successful for the coating of macrostructures with nanostructured carbon layers.

The use of polymers has the advantage of, besides resulting in stable dispersions, providing an organic polymeric framework upon coating that maintains the nanostructured layer integrity. This method is used effectively in preparing ceramic and steel macrostructures using a variety of polymeric binders such as glucose (Fig. 10.7) [47–50] and sucrose or dextrin [51, 52].

The use of polymeric wrapping of carbon nanotubes to create nanocomposite layers can create issues with the access of reactants to the catalytic surface. The carbonisation of the polymers can produce a separate carbon phase wrapping the carbon nanomaterials. The new carbon wrapping can be active on itself; in fact, this method is often used to create activated carbon coatings for catalytic applications [53]. Nevertheless, this type of carbon material is distinct from the actual nanomaterials in the coating. A compromise between stability and activity can be achieved

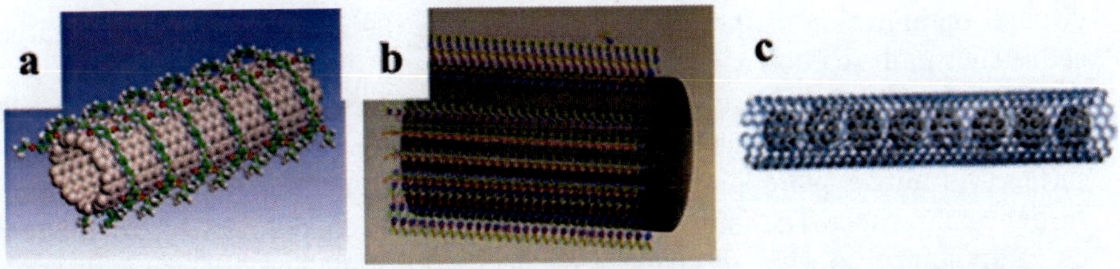

Fig. 10.6 Physical functionalisation of carbon nanotubes by (**a**) polymer wrapping, (**b**) surfactant adsorption, and (**c**) endohedral method [46]

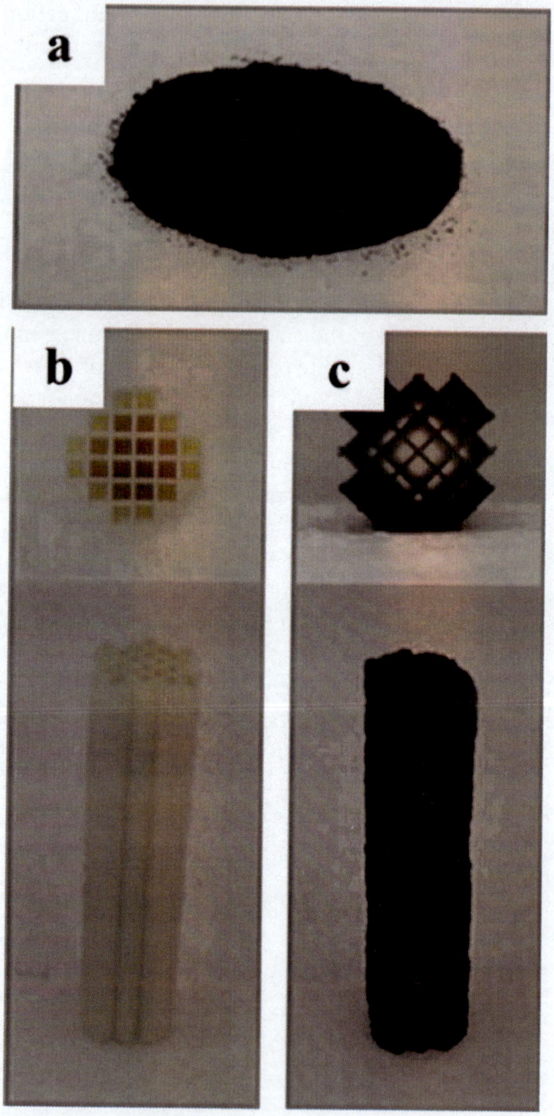

Fig. 10.7 Preparation of a macrostructured N-doped carbon nanotube catalyst from (**a**) N-doped carbon nanotubes in powder form and (**b**) cordierite bare support, resulting in (**c**) the carbon nanotube covered cordierite structure [47]

through optimisation of the process to maximise the availability of the catalytic active sites in the carbon nanomaterials.

Surfactants have also been successful in creating carbon nanotube dispersions to coat cordierite macrostructures [54]. Surfactants adsorb on the carbon nanomaterial surface, forming a hydrophilic barrier that hinders agglomeration and stabilises the suspension in water. The formation of a homogeneous coating was found to be significantly improved when mechanical means (ball milling and ultrasonication) was used to reduce the size of the original nanomaterial bundles, improving the

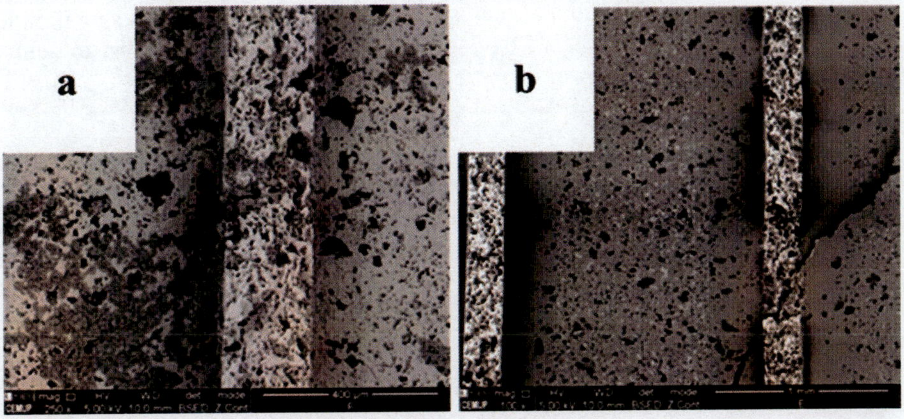

Fig. 10.8 Carbon nanotube coating (dark) on cordierite (light) top view: (**a**) as-received carbon nanotube and (**b**) ball-milled carbon nanotubes [54]

interaction between the carbon and the cordierite substrate (Fig. 10.8). Other methods available include covalent bonding of graphene to surface functionalities anchored on cordierite used to form catalytic nanostructured graphene layers, taking advantage of the 2D graphene structure [55, 56] and electrophoretic formation of a carbon nanotube layer on a conducting macrostructure [57].

10.2 Metal-Free Structured Catalysts for Water Treatment

The application of carbon (mostly activated carbon—AC) as a catalyst in advanced oxidation processes (AOPs) for water treatment is a relatively well-established research topic within the relevant community. In the past twenty years, several works have been published on the topic. Two highlights are the applications of AC in catalytic wet air oxidation (CWAO) [24] and in catalytic ozonation (COz) [58]. More recently, both processes have been expanded to include nano- and macroscale structured carbons. The discussion concerning the application of metal-free catalysts will mostly focus on these two processes, where some overlapping exists with the work previously carried out with AC.

While differing in targets and overall process design, CWAO and COz are fundamentally related in their reaction mechanism. Both achieve oxidation of organic contaminants by the action of highly reactive and non-selective radicals, formed upon the interaction of an oxidant with the carbon surface [58–61]. The main difference between these processes lies in the oxidant applied; CWAO uses oxygen, while COz uses ozone. CWAO requires high working pressures and/or temperatures and is generally applied in effluents with high organic loadings [62]. COz operates at room temperature and pressure, but requires in situ ozone generation, and usually targets effluents with recalcitrant pollutants at lower levels [63]. Moreover, the combination of ozonation and a catalyst looks to improve the process's total mineralisation rates, whereas wet air oxidation is combined with a catalyst to achieve efficient performance at lower operating temperatures and pressures.

10.2.1 Surface Chemistry and Catalytic Performance

The acid or basic character of multi-walled carbon nanotubes (MWCNTs) plays a major role in their performance as a catalyst in ozonation. The surface charge, as measured by calculation of the pH at the point of zero charge (pH_{pzc}), is linked directly to the type and amount of O-containing functionalities (Table 10.2). As per the seminal study detailing the relation between CO and CO_2 released during thermally programmed desorption (TPD) experiments and the nature of the surface functionalities on AC [31], a similar relationship is found for MWCNT. Nitric acid oxidation is then expected to introduce similar oxygen-containing groups (e.g. carboxyl, hydroxyl, and carbonyl groups) on the MWCNT surface [64], which are then selectively removed during thermal treatment [63].

The effect of the acid/basic character of MWCNT in its catalytic performance is made clear when tested in the ozonation of oxalic acid (OxAc). OxAc is a carboxylic acid, recalcitrant to molecular ozonation, and often found as a final product in the oxidation of many organic contaminants; moreover, a sufficiently concentrated oxalic acid solution hinders the spontaneous decomposition of dissolved ozone into hydroxyl radicals, thus serving as an ideal model to study the performance of catalysts in ozonation [58]. Figure 10.9 shows the improvement in specific surface area (SSA) normalised rate constants for ozone decomposition and oxalic acid degradation with pH_{pzc} increase [63]. More basic samples are increasingly efficient in the decomposition of ozone into non-selective highly active radicals, which in turn results in a faster degradation of oxalic acid. Heat treatment with hydrogen gas has been found to be even more efficient in emphasising the basic character of MWCNT, improving their performance in catalytic ozonation of oxalic acid beyond that of nitrogen gas heat-treated materials [65].

Table 10.2 Textural and chemical characterisation of modified MWCNT samples (nitric acid modified, with subsequent thermal treatment at 400, 600, and 900 °C and with O_2 and H_2O_2). [63]

Sample	CO_2 (μmol g^{-1})	CO (μmol g^{-1})	%m_O	pH ($_{pzc}$)	S_{BET} (m^2 g^{-1})
Original	25	478	0.84	7.0	331
HNO3	1514	2435	8.7	3.0	476
HNO3 400	561	2350	5.6	3.8	483
HNO3 600	125	1494	2.8	5.9	504
HNO3 900	15	308	0.54	7.3	529
O2	91	1339	2.5	5.2	508
H2O2	150	466	1.2	5.0	337
AC	63	579	1.1	8.5	909

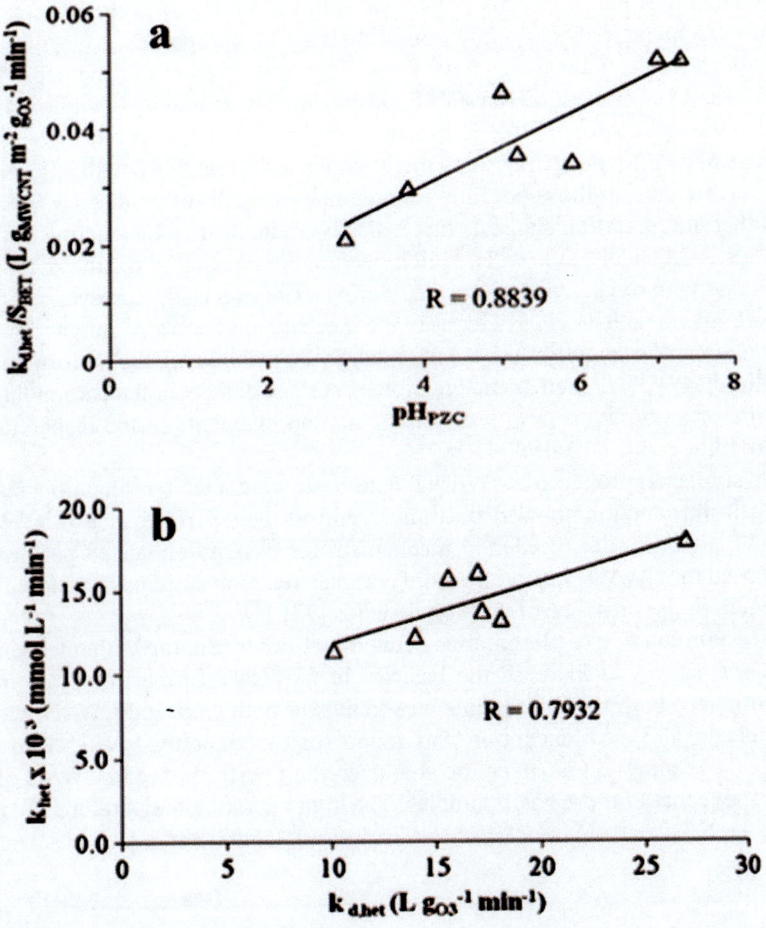

Fig. 10.9 Correlation of normalised rate constants for heterogeneous ozone decomposition (**a**) and oxalic acid degradation (**b**) with modified MWCNT surface charge [63]

The comparison of MWCNT and AC performances further highlights the potential of graphitic carbon materials to act as catalysts for COz. In the same conditions and with a much higher SSA than any MWCNT sample, AC was not able to completely degrade OxAc after 180 min; the unmodified MWCNT almost completely degrades OxAc in 60 min, with the most basic sample achieving complete mineralisation in 60 min. The improved performance of MWCNT is attributed to the higher amount of delocalised π-electrons on their surface [32, 33], highlighted by the behaviour of the more basic samples [58, 66], and the significant decrease of internal mass transfer resistances due to the mesoporous character of MWCNT vs. the microporous character of AC [67]. The re-utilisation of the basic catalysts showed a tendency of the catalytic performance towards what was seen for the most acidic samples. Ozone is known to oxidise the MWCNT surface by introducing oxygen functionalities, and thus, this behaviour is in agreement with the expected performance of the materials as a function of their surface properties.

Nevertheless, other works have found a different trend in ozone degradation vs. surface oxygen content on MWCNT. Despite the low reactivity of carboxylic groups with molecular ozone [68], it has been suggested that deprotonated carboxylic groups in neutral pH suspensions might enhance O_3 reactivity with adjacent aryl or alkene moieties, as those possibly found in the sidewalls of oxidised MWCNT [69]. Another interpretation suggests that hydroxyl radicals may be reacting with carbon centres in MWCNT sidewalls, which become less available for the reaction when occupied with oxygen groups formed during oxidation [70]. However, most of such reports are not able to clearly identify the mechanisms for ozone interaction with the oxygen surface groups, as other factors might be influencing the performance of the oxidised MWCNT, such as the creation of surface defects in the carbon lattice, formation of amorphous carbon materials, or improvements in the dispersion of the hydrophilic oxidised MWCNT [69].

A similar approach to MWCNT acid/basic character modification by surface oxygen introduction showed the same trend in their performance in CWAO [71] (Fig. 10.10). Similar to COz, a mechanism for basicity enhanced performance is proposed for CWAO [71], while more complex reaction systems occasionally report a benefit in the presence of carboxylic acids [72].

The introduction of N-containing functionalities by doping with nitrogen precursors was shown to increase the basicity of MWCNT further; thus, an improved performance with these materials was found for both COz and CWAO with OxAc and phenol [73]. An exception was found for S-containing MWCNT in CWAO; while these samples had an acidic character, their performance in CWAO of phenol was above that of more basic samples. The high operating temperature and pressure

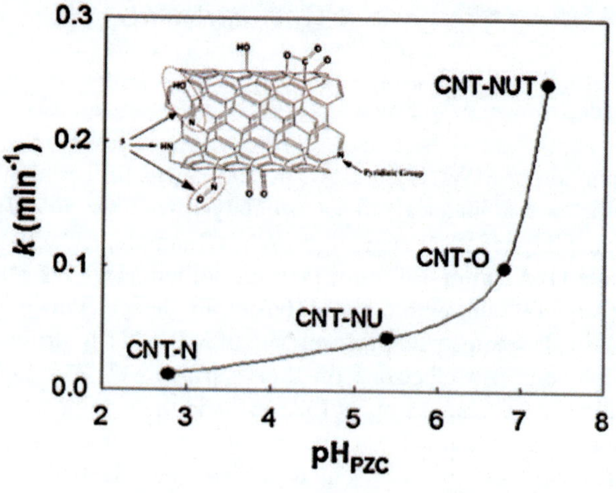

Fig. 10.10 Correlation of initial rate constants of oxalic acid degradation by CWAO with modified MWCNT surface charge [71]

of CWAO promote the formation of sulphonic radicals from surface S-containing functionalities. Sulphonic radicals react promptly with phenol, but are not able to further mineralise phenol intermediate oxidation products such as oxalic acid [73]. Thus, the S-containing MWCNT samples do not act as a true catalyst and the relationship between catalytic performance and basic character is maintained.

Solvent-free mechanical techniques (namely, ball milling) have also been used as an easy methodology to introduce large amounts of heteroatoms onto MWCNT [74]. While ball milling also affects the textural properties of the materials (discussed in next section), this method further highlights the effect of heteroatom doping in the MWCNT catalytic performance (Fig. 10.11), as seen in N-doped MWCNT in both CWAO and COZ [75]. The role of N-, P-, and B-containing surface groups is explored in the CWAO of OxAc beyond the acid/basic character of the materials. Mechanisms proposed for the enhanced performance of the MWCNT include electron donation by N-groups with unpaired electrons [47] and higher affinity to oxygen due to the effect of P and B atoms in the electroneutrality of the graphitic structure [74], enhancing the interaction with oxygen.

Graphene underwent a similar approach towards the tuning of its surface chemistry. Reduced graphene oxide (rGO) samples doped with nitrogen using melamine and urea as precursors, with an increased basic character when compared with the pristine rGO, were again found to have enhanced catalytic performance both in COz and CWAO [76, 77]. However, the most basic samples were not necessarily the most active, showing how other factors can affect the performance of N-doped samples (textural properties, nitrogen amount, and nature of nitrogen groups).

The remarkable performance of MWCNT and rGO in COz and CWAO prompted their use in the treatment of more complex pollutants. Numerous instances are found in the literature, from traditional pollutants like organic dyes [78, 79] or phenolic wastewaters [73, 80] to emerging organic pollutants [81–84]. Since conventional water and wastewater treatment solutions are generally efficient at degrading large organic molecules at sufficiently fast rates for their efficient removal from

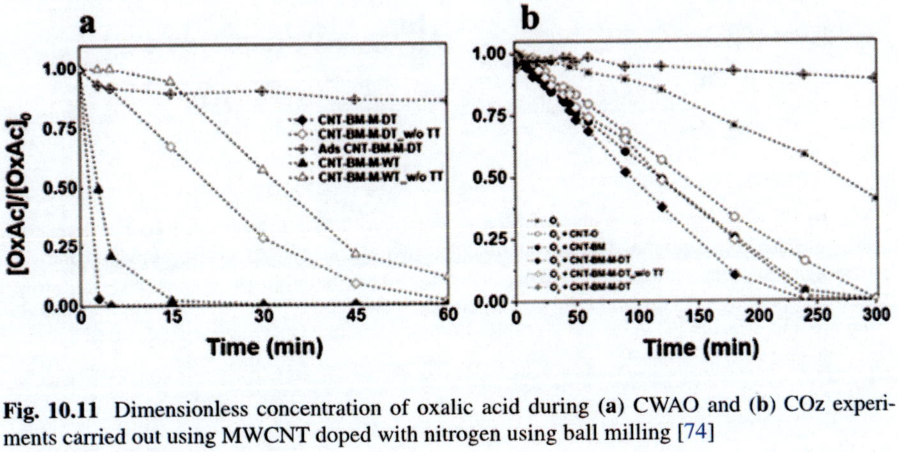

Fig. 10.11 Dimensionless concentration of oxalic acid during (**a**) CWAO and (**b**) COz experiments carried out using MWCNT doped with nitrogen using ball milling [74]

water and wastewater, the main concern of these studies lies on the total mineralisation of the contaminants. COz studies carried out with the pesticide atrazine (ATZ) have shown how molecular ozone readily reacts with ATZ, but leads to the accumulation of a series of recalcitrant intermediate products, often more toxic than the parent compound (Fig. 10.12) [85].

A similar behaviour is found for many other emerging micropollutants. Addition of a catalyst might even decrease the rate of degradation of the parent compound due to the decomposition of ozone and the unselective nature of the formed active radicals [82–84]. Nevertheless, these radicals are able to enhance the efficiency of the process, either by complete mineralisation of the contaminants into water and carbon dioxide or decomposition into less toxic by-products [82]. The complex nature of the involved reaction pathways means that it is not advisable to generalise on the mechanisms of their catalytic oxidation. In fact, it is found that besides leading to higher mineralisation degrees, the reaction pathways often change due to the tuning of the catalysts through surface modification [85]. Both COz and CWAO using several different variations of rGO and MWCNT have been tested towards the treatment of a large range of emerging organic pollutants [81, 85–97].

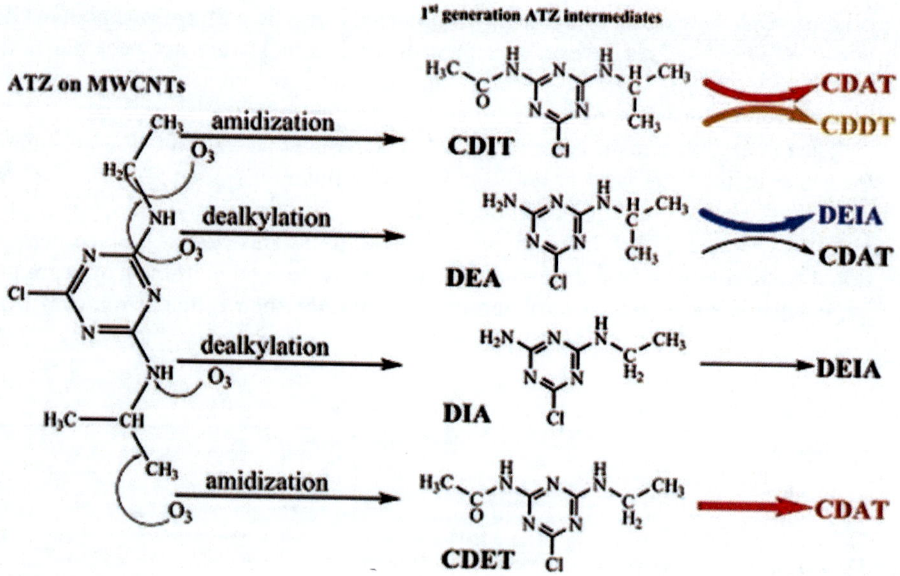

Fig. 10.12 Reaction of molecular ozone with atrazine (ATZ) and accumulation of first generation intermediates recalcitrant to direct ozonation (desethylatrazine (DEA), desisopropylatrazine (DIA), desethyldeisopropyl-atrazine (DEIA), 2-chloro-4-acetamido-6-isopropylamino-1,3,5-triazine (CDIT), 2-chloro-4-ethylamino-6-acetamido-1,3,5-triazine (CDET), 2-chloro-4-acetamido-6-amino-1,3,5-triazine (CDAT), and 2-chloro-4,6-diacetamido-1,3,5-triazine (CDDT)) [85]

10.2.2 Textural Properties and Catalytic Performance

The specific surface area and pore characteristics also play a key role in defining the catalytic potential of carbon materials. This effect has been widely studied for an extensive range of catalytic reactions, including CWAO and COz, using more traditional carbon materials such as activated carbon and carbon black [25, 59, 98].

A larger surface area allows for more existing active sites for reaction and thus can be expected to increase the catalytic performance of carbon materials. One of the main selling points of AC is, in fact, its potential to easily obtain high specific surface areas, up to 3000 $m^2\ g^{-1}$ [99]. However, the microporous nature of AC can restrict the access of large molecules to active reaction sites or create additional internal diffusion and mass transfer resistances [100].

While 1D carbon nanomaterials, such as MWCNT, typically present a lower surface area than AC, their porous structure consists mainly of mesopores [35, 67]. Treatments for modification of surface chemistry generally result in changes to the surface area and pore structure. It is not always straightforward to decouple the simultaneous effect on surface chemistry and texture of these modification methods. On the other hand, ball milling without any heteroatom precursor can produce changes in textural properties without significantly changing the surface chemistry [101]. A correlation between ball-milling parameters and catalytic performance of MWCNT in COz was found, which in turn translates into a correlation between MWCNT morphology and surface textural properties and performance (Fig. 10.13).

10.2.3 Macrostructured Catalysts

Macrostructured supports covered with carbon nanotubes or nanofibers have been used extensively as catalysts for the oxidation of organic pollutants. Most of these studies focus on continuous water treatment processes. Various examples can be found in the literature, including catalytic ozonation [102] and catalytic wet air oxidation [47]. The catalysts were found to translate the expected performance of the powders, including N-doped carbon nanofibers grown on cordierite macrostructured supports.

Ex situ prepared nanostructured carbon layers on macrostructured supports have also been successfully used in both CWAO (Fig. 10.14) [47] and COz of organic pollutants [54].

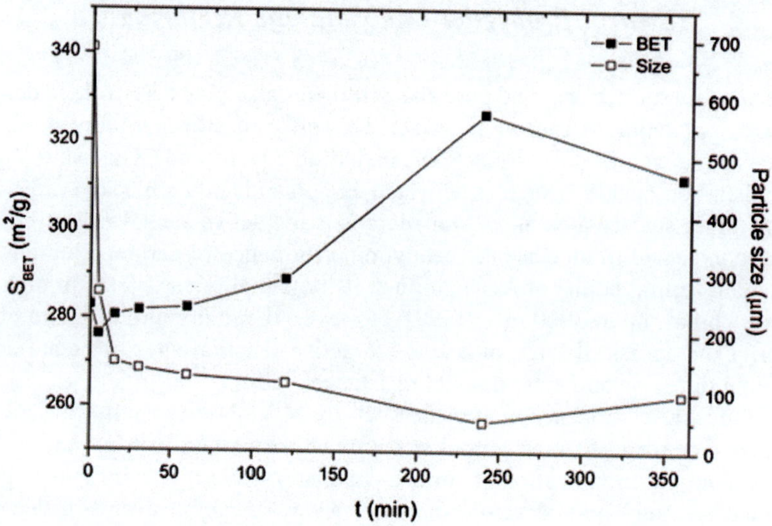

Fig. 10.13 Changes to specific surface area and particle size of MWCNT for increasing ball-milling time [101]

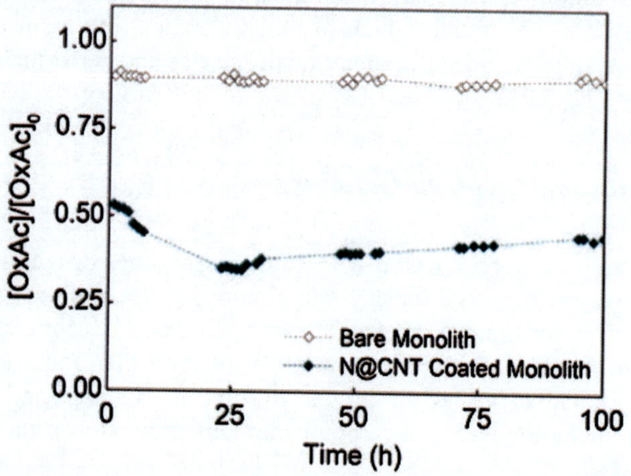

Fig. 10.14 Dimensionless oxalic acid concentration during continuous CWAO experiments carried out using ex situ prepared N-doped CNT coated onto a cordierite macrostructure [47]

10.2.4 Application in Real Conditions

The advances described here have mostly been achieved at the laboratory scale, using model compounds that allow the direct evaluation of the catalytic performance of the materials being tested. This is particularly relevant when targeting emerging organic micropollutants, as they are found with relevant impacts even at concentrations in the ppb or smaller ranges.

The use of carbon nanomaterials in a catalytic ozonation pilot plant (Fig. 10.15) has been demonstrated in the degradation of 5 selected organic micropollutants: atrazine, bezafibrate, erythromycin, metolachlor, and nonylphenol [22]. The plant was designed to compare the removal efficiency of the non-catalytic and the catalytic ozonation process. Teflon reactors holding carbon nanotube covered macrostructures were used to contact the catalyst with the effluent with or without pre-saturation with ozone. Each reactor held 6 macrostructured catalysts, each with 22 mm diameter and 60 mm length. The system showed that the catalytic process was efficient at removing the selected pollutants.

The influence of a real water matrix in the catalytic ozonation process over carbon nanotubes has also been investigated [82]. The influence of typical real water components was assessed, and the main inhibitors of the catalytic activity were identified. HCO_3^- is proposed to act as a radical scavenger in solution; SO_4^{2-} is also proposed to act as a radical scavenger, besides affecting the dissolution of ozone in water; Ca^{2+} was found to form precipitates when interacting with the organics in the effluent, resulting in blocking of the active sites available for the catalytic reaction.

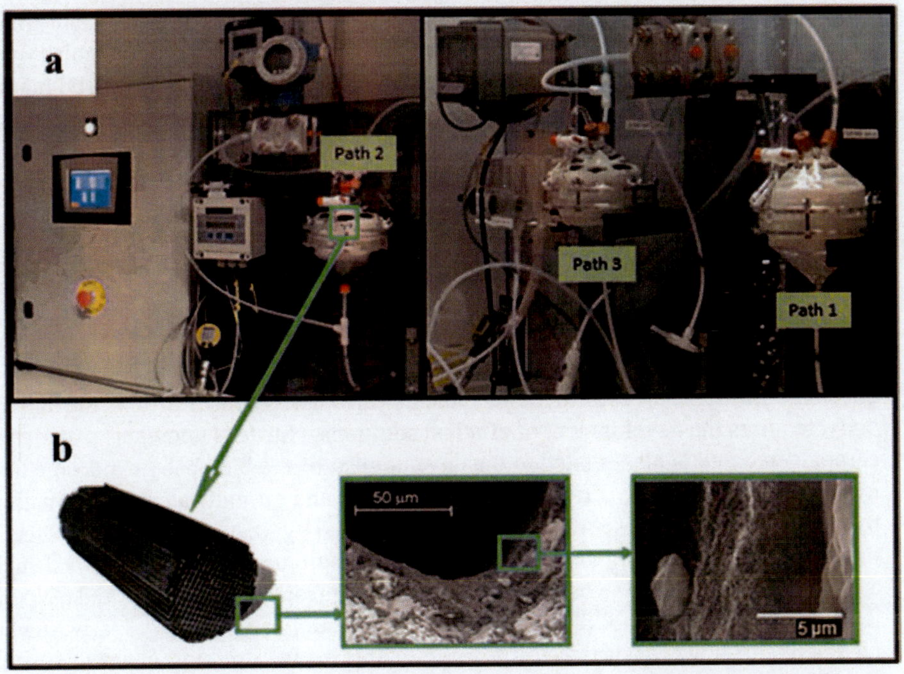

Fig. 10.15 Pilot-scale experimental set-up for the testing of carbon nanotube covered ceramic macrostructures in a catalytic ozonation process: (**a**) overview of set-up and teflon reactors holding 6 catalysts in parallel and (**b**) detailed view of each catalyst by SEM imaging of the nanostructured carbon layer [22]

Nevertheless, the catalytic process was still found to largely present good performances. Moreover, long-term testing of the catalysts has shown these to be stable after an initial period of deactivation due to ozone surface oxidation [82].

While CWAO has been tested at different scales using different catalysts, there are no reports on large-scale applications and/or with real effluents using carbon nanomaterials [103]. The EU funded FP7 project FREECATS culminated in the formulation of a business plan laying out the development of a continuous CWAO reactor using N-doped carbon nanomaterials as catalysts [104]. The stability of the carbon nanomaterials has been assessed in long-term experiments again, showing that the catalysts deposited on macrostructures maintain their activity under reaction conditions [47].

10.2.5 Outlooks

The use of carbon nanomaterials as metal-free catalysts for environmental applications has matured immensely in the past 20 years, showing great potential to tackle current urgent issues in water treatment. The research and development of nanomaterials with tailored properties have contributed largely for their application, but further research is required to overcome the constraints that limit their use in water treatment plants. The current research towards the adoption of the technology for practical applications focuses on the scaling-up of the process (from material manufacture to operation) and its integration in water and wastewater treatment plants to guarantee their performance and durability.

10.3 Carbon as a Catalyst Support in Water Treatment

Another opportunity for the development of catalytic technologies for water remediation exists for waters contaminated with inorganic pollutants. The contamination of drinking water sources with nitrate, which is metabolised into nitrite in the human body, requires the development of efficient solutions. Nitrite is known as a potential carcinogenic and is also related to the development of the blue baby syndrome and hypertension [105]. The diffuse sources of surface and groundwater contamination by nitrate, such as fertiliser run-off, cannot be tackled by treatment of effluents feeding into the natural water sources. Alternative technologies for nitrate removal such as ultrafiltration or reverse osmosis are effective but result in a concentrated effluent that requires further handling. Biological denitrification is widely used in wastewater treatment; however, it is yet not or rarely applied for drinking water due to its complexity and concerns regarding possible bacterial contamination, the presence of residual organics in treated water, and the possible increase in chlorine demand of purified water [106]. Bromate is also a reoccurring contaminant in both water and wastewater, which is a good candidate target for catalytic reduction. Besides being found in natural waters from diffuse sources such as sweater intrusion and, again, pesticide and fertiliser run-off, bromate is also formed during ozonation of bromide containing waters in drinking water plants. Bromate in drinking water is a public health concern due to its carcinogenic potential, and thus, efficient technologies for its removal are required. The available technologies are biological treatment, photocatalysis, electrochemical reduction, and catalytic reduction. The catalytic route presents advantages in terms of efficiency and rate of removal, without producing secondary waste streams with the accumulated remnants [107].

As is the case for metal-free catalytic applications, carbon is highly attractive as a support for an active catalytic phase. While carbon can be used to replace and avoid the use of often expensive metals for the reactions described in the previous section, other applications still require the use of an active metallic phase. The chemical reduction of inorganic pollutants generally requires a mono- or bimetallic catalyst able to activate the reduction agent (e.g. hydrogen or formic acid [108]) and to adsorb the pollutant to promote its reduction on the catalytic surface [23, 109]. The characteristics of carbon materials of interest for use as catalyst support largely overlap with those of metal-free catalysis, namely, their very high surface area suitable for metal particle dispersion, access to active sites, and potential for surface chemical and textural modification. The cost of activated carbon is also attractive when compared with other standard catalyst supports [25].

Nanostructured carbon materials have added advantages compared to traditional carbon supports, like activated carbon, due to their unique properties, while retaining the high surface area and surface textural and chemical tunability. Besides the modified pore structure detailed in Sect. 10.2, the electronic properties of the sp2 hybridised structure are particularly relevant to their role as catalyst support. The high electrical conductivity can promote specific metal-support interactions that directly affect the catalyst activity and selectivity [26].

The most common method for the preparation of nanostructured carbon supported metallic catalysts is through impregnation using the appropriated precursor salts for the desired metal. The improvement in activity of carbon nanotube supported catalysts has been demonstrated for monometallic catalysts supported on multi-walled carbon nanotubes for the catalytic reduction of bromate under hydrogen when compared with those using activated carbon as a support [110]. An improvement in the turn-over frequency (TOF, moles converted per available metallic surface area per unit time) of ~30 times was measured on average for Pd, Pt, Rh, and Ru monometallic catalysts. While the mesoporous structure of MWCNT is expected to contribute to the improvement in the activity, it has also been suggested that the intrinsic contribution of the support (i.e. discounting the contribution of the diffusion effects) is still higher for MWCNT when compared with other supports in the reduction of nitrite [111, 112]. It is suggested that the graphitic surface provides an additional site for anionic compound adsorption and reaction with spilled-over hydrogen formed on the metallic sites. This is in agreement with findings in other catalytic reactions over supported metallic catalysts, such as the hydrogenation of cinnamaldehyde [113]. The adsorption of spilled-over hydrogen on graphitic carbon has also been reported for non-catalytic cases [114, 115]. However, it has also been reported that addition of MWCNT to carbon nanofibers as a support for Pd catalyst improved the reduction of nitrite by promoting the formation of smaller metallic nanoparticles and thus improving the available metallic surface for reaction [116].

The hydrogenation of nitrate into nitrogen in water requires a bimetallic catalyst. Unlike bromate and nitrite, which can adsorb on the noble metal or graphitic planes, nitrate is preferentially adsorbed onto transition metals. However, transition metals cannot dissociate hydrogen, and thus, a combination of a noble metal and a transition metal is a necessary condition for the reduction of nitrate to nitrite, which is then further reduced on the noble metal (Fig. 10.16). This mechanism is well-understood with several reported examples using a variety of carbon and metal oxide supports [117–119]. The main challenge in the process is the selectivity towards ammonia, which is also a harmful pollutant. It is proposed that a combination of factors affects the selectivity to ammonia, ranging from the concentration of nitrite ions in solution to the spatial distribution of the metallic phases and including metal-support interaction effects [106].

Unlike with bromate and nitrite, the graphitic carbon in carbon nanotube supports cannot directly participate in the nitrate reduction reaction mechanism, as seen when monometallic catalysts are tested in the reaction . However, a very large impact on the catalyst activity and selectivity is still found. The effect varies with the metal pair chosen, the weight ratios between both, the calcination and reduction temperatures of the prepared catalysts, and the surface chemistry of the supports [121]. For the 1%Pd/1%Cu pair, the nitrate removal of the catalyst prepared with activated carbon was only matched with a surface-modified carbon nanotube support. On the other hand, the selectivity to nitrogen was improved with three of the eleven carbon nanotube supported catalysts tested. The same experiment carried out with the pair 1%Pt/0.3%Cu was very different, with five of eleven carbon nanotube supported catalysts outperforming activated carbon in nitrate reduction. This pair, on the other hand, resulted in much worse selectivity to nitrogen, independent of the support used. The kinetics of nitrate reduction over 1%Pd/1%Cu were studied in detail and suggested that the structured support was able to improve mass transfers

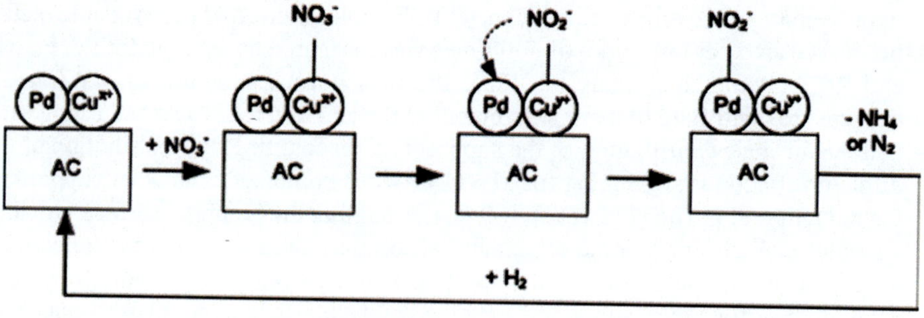

Fig. 10.16 Schematic reaction mechanism for nitrate catalytic reduction under hydrogen over bimetallic catalysts supported on carbon [120]

happening near the metallic catalyst surface, resulting in improved selectivity to nitrogen [122].

Another approach to the use of carbon nanotube as support for an active bimetallic phase is the formation of confined catalysts. Ag-cut carbon nanotubes were used to confine Pd-Cu alloyed particles on the internal and external surface of the nanotubes, with the former showing better nitrate conversion and selectivity to nitrogen [123]. It is suggested that the main driver for the improved activity was the electron deficient internal surface of the carbon nanotubes, leading to donation of electrons from the bimetallic catalysts, which in turn hinders nitrate adsorption on the metallic surface.

Graphene has also been used as a support for bimetallic catalysts for the reduction of nitrate using zero-valent iron as an electron source to form hydrogen in water and reduce nitrate, with similar results as with carbon nanotubes [124].

Similar techniques to those used in the development of metal-free catalysts for water and wastewater treatment can be used towards the modification of nanostructured carbon supports.

10.3.1 Support Surface Chemistry and Catalytic Performance

The effect of the surface chemistry of the nanocarbon supports on the catalytic performance of mono- and bimetallic catalysts has been studied for bromate and nitrate reduction.

Carbon nanotubes oxidised with different surface oxygen content were prepared and used as supports for Pd/Cu and Pt/Cu bimetallic catalysts at optimised ratios for nitrate reduction [121]. The carbon nanotubes were used as-received, after oxidation with 7 M HNO_3, and after thermal treatment under inert atmosphere at 400 °C post-oxidation. The activity of the oxidised carbon nanotubes as support for the Pd/Cu catalyst was found to be below that of the as-received supports, except for high calcination and reduction temperatures. In the as-received supports, the increase of the preparation temperature above 200 °C resulted in a decrease in activity due to the sintering of the metallic nanoparticles and alloy formation. The preparation of the catalysts on the oxidised support at temperatures below 200 °C could not form metallic particles from the precursor salts. In fact, thermal programmed reduction (TPR) experiments reveal that the oxidised support has a much less reducing character when compared with the as-received carbon nanotubes, thus requiring high temperatures to form a dispersion of metallic particles [125]. It is also suggested that the interaction of copper with surface oxygen can hinder the interaction with the palladium particles. A good compromise in the role of oxygen-containing surface functionalities is found for the catalyst prepared using oxidised carbon nanotubes subject to thermal treatment under inert gas at 400 °C prior to metal impregnation. The heat treatment at 400 °C removes only part of the acid oxygen-containing groups from the support surface. The oxygen still present on the surface promotes the anchoring and dispersion of the metals on the support surface [126],

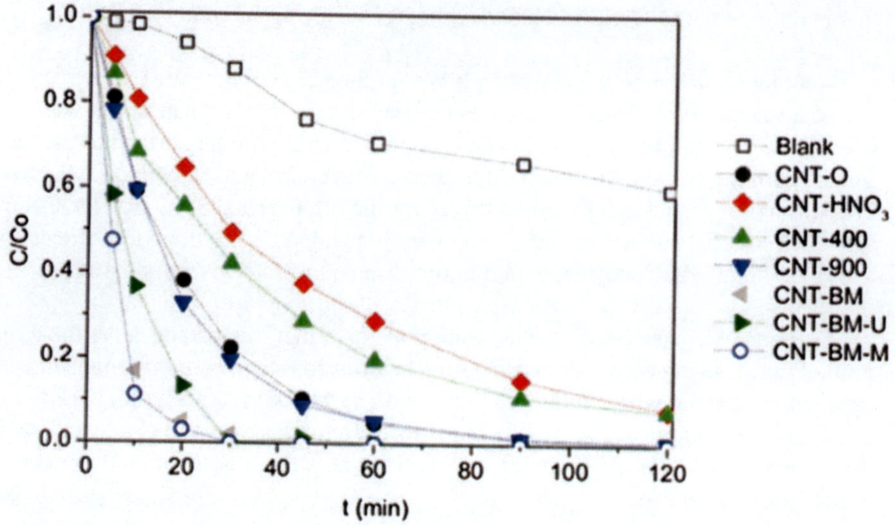

Fig. 10.17 Dimensionless concentration of bromate in catalytic reduction experiments over Pd mono-metallic catalysts supported on carbon nanotubes with different surface properties (as-received (CNT-O), nitric acid treated (CNT-HNO_3) and subsequent thermal treatments at 400 and 900 °C (CNT-400 and CNT-900), ball-milled (CNT-BM), and heteroatom doped with nitrogen precursors urea and melamine (CNT-BM-U and CNT-BM-M)) [127]

offering a compromise offset with the negative contribution of surface oxygen in the reduction of the metals or the interaction between palladium and copper, as seen for the support with the highest content of surface oxygen.

When similar supports were used in the preparation of monometallic Pd catalysts for bromate reduction (as received, HNO_3 oxidised, and subsequently thermally treated at 400 and 900 °C), the activity was directly related to the amount of surface-oxygen groups (Fig. 10.17) [127].

Basic supports were also prepared using a mechano-thermal approach by adding a nitrogen precursor during the ball milling of carbon nanotubes. The morphological and textural effect is discussed in detail in the next section; however, it is clear that the more basic supports shown an increased activity when compared with their counterparts with no nitrogen doping. The nitrogen heteroatoms increase the electronic density on the carbon surface and consequently its reducibility, acting then as promoters of the formation of metallic dispersions on the carbon surface.

10.3.2 *Support Textural Properties and Catalytic Performance*

The textural properties of carbon supports were also found to affect the performance of mono- and bimetallic catalysts supported on carbon nanotubes in the reduction of bromate [127]. Carbon nanotubes after ball milling were found to drastically improve the catalytic activity of both Pd (Fig. 10.14) and Pd/Cu catalysts. This effect was attributed to the increase in the surface area due to the opening of the carbon nanotubes. This effect is very prominent with bromate as reactions on the surface of the carbon support of the adsorbed bromate with spilled-over hydrogen from the metallic catalyst are possible, in addition to reactions exclusively on the metallic surface.

10.3.3 Macrostructured Nanocarbon Supports for Metallic Phases

Macrostructures containing carbon nanomaterials as support for active metallic phases have been developed and applied in catalytic reduction of inorganic contaminants in water.

The most common method for coating macrostructures with nanostructured carbon layers for catalytic reduction applications is still chemical vapour deposition (CVD). However, unlike in the case of metal-free catalysis, the impregnation of the nanostructured carbon materials with the active metallic phase is an additional challenge in the process. Incipient wetness impregnation, when applied to powder materials, offers excellent fine control of the uptake of the metallic precursor salts by the carbon supports, achieved by tuning the solution volumes used to the support pore structures. However, and since the impregnation step needs to be carried out following the CVD growth of the nanostructured carbon layers, it is not as straightforward to achieve a uniform contact of the carbon supports with the precursor solution using a limited optimised volume. Thus, most methods rely on impregnation from solution by excess [128, 129]. An alternative approach using ex situ catalyst preparation and subsequent coating is possible, but the post-deposition thermal treatments significantly affect the performance of the catalyst, leading, for example, to the sintering of metallic nanoparticles. Thus, some optimisation of those methods is still required to achieve an efficient methodology [52].

Nevertheless, despite the difficulties in optimising the preparation method, several different types of macrostructured supports have been successfully applied to the catalytic reduction of inorganic pollutants.

Three types of macrostructures have been used thus far in bromate and nitrate continuous removal, using different mono- and bimetallic catalysts prepared using a nanostructured carbon layer: activated carbon fibres [128, 130, 131], sintered metal filters (Fig. 10.18) [132], and honeycomb ceramic monoliths [110, 132].

Besides the effects found for the catalysts in powder form, additional effects are found for the different types of macrostructures used both in bromate and nitrate reduction. A comparison between activated carbon filters and sintered metal filters found that the latter promoted more favourable hydrodynamic conditions [128]. Honeycomb monolithic catalysts also promote formation of favourable

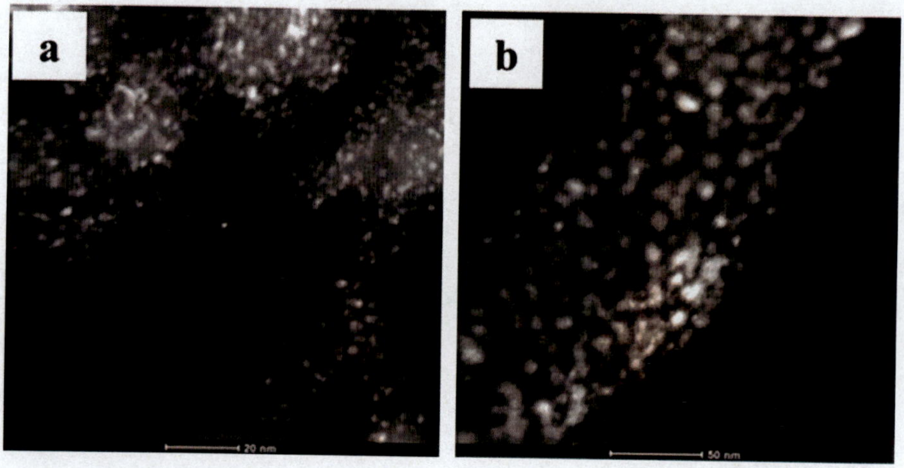

Fig. 10.18 STEM images of (**a**) Pd/Cu and (**b**) Pd/Sn catalysts supported on carbon nanofibers grown on sintered metal fibres [128]

hydrodynamic conditions (Taylor or slug flow), which resulted in very good bromate conversion rates [110].

10.3.4 Application in Real Conditions

Carbon nanofibers coated on activated carbon filters and sintered metal filters have been prepared with Pd mono-metallic and Pd/Sn bimetallic catalysts for continuous and simultaneous bromate and nitrate reduction [133]. A bubble-reactor was designed to contact an effluent sequentially with mono- and bimetallic catalyst, achieving conversions capable of bringing their concentrations below the legal limits when used in polluted industrial and natural waters (Fig. 10.19).

The behaviour of the catalysts has also been studied using natural waters. Nitrate reduction was found to be affected by high conductivities and hardness, due to precipitation of calcium salts and poisoning of Pd by sulphur [134]. In another study, using polluted groundwater contaminated with nitrate over a Pd/Cu catalyst on carbon nanotubes, it was reported that all water components besides sulphates actually had a negative impact in the performance of the catalysts due to competitive adsorption on the catalytic active sites [135]. The effect of water components is thus likely to be specific to the concentrations in which they are present and the specificities of the catalysts.

As of today, there are not reported large-scale applications of catalytic reduction for the removal of nitrate or bromate from water. One on-going project supported by the Portugal—University of Texas in Austin collaborative programme includes the development of a pilot scale plant for reduction of nitrate, bromate, and perchlorate.

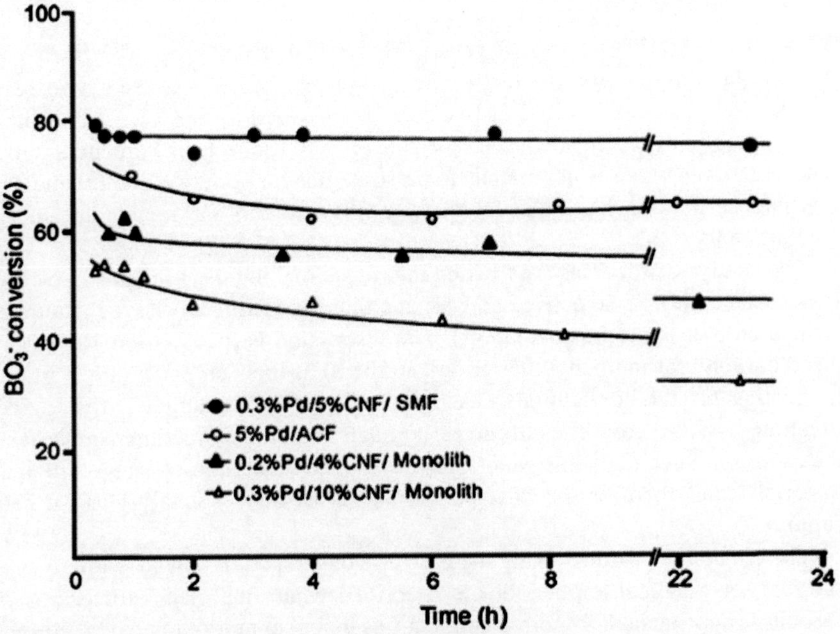

Fig. 10.19 Bromate conversion during long-term continuous reactions over carbon-supported Pd catalyst (carbon nano fibre (CNF), sintered metal filter (SMF), and activated carbon felt (ACF)) [133]

10.3.5 Outlooks

The use of nanostructured carbons as supports in the catalytic reduction of inorganic contaminants in water over mono- and bimetallic catalysts containing noble metals results in very promising increases in activity when compared with the same process using other traditional carbon supports.

The modification of carbon nanotubes by mechanical methods (e.g. ball milling) in particular is seen to greatly improve their performance in bromate reduction. Methods to make use of the mechanically modified carbon nanomaterials in macrostructured reactors need optimisation since thermal treatment affects the performance of the catalysts. On the other hand, the nitrate removal is particularly sensitive to the catalyst used in terms of selectivity to ammonia, suggesting that practical applications will require custom-designed solutions rather than a "one-size-fits-all" solution. Nevertheless, reactor design can also be used to tune the selectivity of the catalytic process towards nitrogen formation. As with metal-free environmental catalysts for water treatment, the gap between the laboratory experimental results and the application to a real case needs to be completed with further research into catalyst and reactor scaling-up and integration in water and wastewater treatment plants.

10.4 Conclusions

An introduction into the structuring of carbon from the nano- to the macroscale and its application in catalytic water treatment is provided to those seeking to have an overall picture of recent advances. The role of catalytic water treatment and its integration into the wider water management strategies for pollution control and mitigation is described, in particular for the removal of emerging organic and inorganic contaminants.

The catalytic processes are divided into those making use of metal-free catalysts (where the carbon is an active catalyst) and those requiring an active metallic phase (where carbon is used as a support). The discussion is focussed on 1D sp2 hybridised carbon (carbon nanotubes and nanofibers) as their properties are particularly interesting for the applications described here. Several examples are provided to illustrate how the properties of the carbon nanomaterials affect the performance of the catalysts. Succinctly, the general case is that basic and mesoporous carbon nanomaterials tend to offer the best performances as both a catalyst and a catalyst support.

The challenge of transferring the catalysts developed in powder form to systems suitable for practical applications is described where the macrostructuring of the carbons is approached. Reported applications in real or approximated conditions are used to provide insight into what can be expected from these technologies in the future. Pathways for future research are briefly suggested, focussing on the translation of the catalytic water treatment solutions to practical real-world applications.

Acknowledgements This work was financially supported by InTreat-POCI-01-0145-FEDER--31337-funded by FEDER funds through COMPETE2020-Programa Operacional Competitividade e Internacionalização (POCI), Base Funding—UIDB/50020/2020 of the Associate Laboratory LSRE-LCM—funded by national funds through FCT/MCTES (PIDDAC). OSGPS acknowledges FCT funding under the Scientific Employment Stimulus—Institutional Call CEECINST/00049/2018.

References

1. G. Centi, P. Ciambelli, S. Perathoner, P. Russo, Environmental catalysis: trends and outlook. Catal. Today **75**, 3–15 (2002). https://doi.org/10.1016/S0920-5861(02)00037-8
2. A.K. Choubey, Y. Shukla, A review on advancements and challenges of catalytic and non-catalytic waste water treatment processes. J. Water. Pollut. Purif. Res. **6**, 24–36 (2019). https://doi.org/10.37591/JOWPPR.V6I1.319
3. R.M. Heck, S. Gulati, R.J. Farrauto, The application of monoliths for gas phase catalytic reactions. Chem. Eng. J. **82**, 149–156 (2001). https://doi.org/10.1016/S1385-8947(00)00365-X
4. K. Pirkanniemi, M. Sillanpää, Heterogeneous water phase catalysis as an environmental application: a review. Chemosphere **48**, 1047–1060 (2002). https://doi.org/10.1016/S0045-6535(02)00168-6
5. A. Pal, K.Y.-H. Gin, A.Y.-C. Lin, M. Reinhard, Impacts of emerging organic contaminants on freshwater resources: review of recent occurrences, sources, fate and effects. Sci. Total Environ. **408**, 6062–6069 (2010). https://doi.org/10.1016/j.scitotenv.2010.09.026

6. M. Petrović, S. Gonzalez, D. Barceló, Analysis and removal of emerging contaminants in wastewater and drinking water. TrAC Trends Anal. Chem. **22**, 685–696 (2003). https://doi.org/10.1016/S0165-9936(03)01105-1
7. K. Noguera-Oviedo, D.S. Aga, Lessons learned from more than two decades of research on emerging contaminants in the environment. J. Hazard. Mater. **316**, 242–251 (2016)
8. N. Bolong, A.F. Ismail, M.R. Salim, T. Matsuura, A review of the effects of emerging contaminants in wastewater and options for their removal. Desalination **239**, 229–246 (2009). https://doi.org/10.1016/j.desal.2008.03.020
9. J. Restivo, E. Garcia-Bordejé, J.J.M. Órfão, M.F.R. Pereira, Carbon nanofibers doped with nitrogen for the continuous catalytic ozonation of organic pollutants. Chem. Eng. J. **293**, 102–111 (2016). https://doi.org/10.1016/j.cej.2016.02.055
10. A. Kapoor, T. Viraraghavan, Nitrate removal from drinking water. J. Environ. Eng. **123**, 371–380 (1997)
11. S. Mudgal, L. Van Long, N. Saïdi, L. Wisniewska, *Optimising Water Reuse in the EU: Public Consultation Analysis Report* (Publications Office of the European Union, Luxembourg, 2015)
12. M.O. Barbosa, N.F. Moreira, A.R. Ribeiro, M.F.R. Pereira, A.M.T. Silva, Occurrence and removal of organic micropollutants: an overview of the watch list of EU decision 2015/495. Water Res. **94**, 257–279 (2016). https://doi.org/10.1016/j.watres.2016.02.0472
13. J.C. Brown, R.D. Anderson, J.H. Min, L. Boulos, D. Prasifka, G.J.G. Juby, Fixed-bed biological treatment of perchlorate-contaminated: DRINKING WATER. J. Am. Water Works Assoc. **97**, 70–81 (2005). https://doi.org/10.1002/j.1551-8833.2005.tb07473.x
14. A. Egea-Corbacho, S. Gutiérrez Ruiz, J.M. Quiroga Alonso, Removal of emerging contaminants from wastewater using nanofiltration for its subsequent reuse: full–scale pilot plant. J. Clean. Prod. **214**, 514–523 (2019). https://doi.org/10.1016/j.jclepro.2018.12.297
15. J. Garcia-Ivars, L. Martella, M. Massella, C. Carbonell-Alcaina, M.I. Alcaina-Miranda, M.I. Iborra-Clar, Nanofiltration as tertiary treatment method for removing trace pharmaceutically active compounds in wastewater from wastewater treatment plants. Water Res. **125**, 360–373 (2017). https://doi.org/10.1016/j.watres.2017.08.070
16. A. Bergquist, M. Bertoch, G. Gildert, T.J. Strathmann, C.J. Werth, Catalytic denitrification in a trickle bed reactor: ion exchange waste brine treatment. J. Am. Water Works Assoc. **109** (2017)
17. M.F.R. Pereira, S.F. Soares, J.J.M. Órfão, J.L. Figueiredo, Adsorption of dyes on activated carbons: influence of surface chemical groups. Carbon N.Y. **41**, 811–821 (2003). https://doi.org/10.1016/S0008-6223(02)00406-2
18. S.J. Khan, D. Murchland, M. Rhodes, T.D. Waite, Management of concentrated waste streams from high-pressure membrane water treatment systems. Crit. Rev. Environ. Sci. Technol. **39**, 367–415 (2009). https://doi.org/10.1080/10643380701635904
19. F.L. Burton, G. Tchobanoglous, *Wastewater Engineering: Treatment, Disposal and Reuse* (Tata McGraw-Hill, New Delhi, 2000)
20. D.K. Kanaujiya, T. Paul, A. Sinharoy, K. Pakshirajan, Biological treatment processes for the removal of organic micropollutants from wastewater: a review. Curr. Pollut. Rep. **5**, 112–128 (2019)
21. D. Li, J. Qu, The progress of catalytic technologies in water purification: a review. J. Environ. Sci. **21**, 713–719 (2009). https://doi.org/10.1016/S1001-0742(08)62329-3
22. S. Derrouiche, D. Bourdin, P. Roche, B. Houssais, C. MacHinal, M. Coste, J. Restivo, J.J.M. Órfão, M.F.R. Pereira, Y. Marco, E. Garcia-Bordeje, Process design for wastewater treatment: catalytic ozonation of organic pollutants. Water Sci. Technol. **68**, 1377–1383 (2013)
23. J. Restivo, O.S.G.P. Soares, J.J.M. Órfão, M.F.R. Pereira, Metal assessment for the catalytic reduction of bromate in water under hydrogen. Chem. Eng. J. **263**, 119–126 (2015). https://doi.org/10.1016/j.cej.2014.11.052
24. N. Delgado, A. Capparelli, A. Navarro, D. Marino, Pharmaceutical emerging pollutants removal from water using powdered activated carbon: study of kinetics and adsorption equilibrium. J. Environ. Manag. **236**, 301–308 (2019). https://doi.org/10.1016/j.jenvman.2019.01.116

25. P. Serp, J.L. Figueiredo, *Carbon Materials for Catalysis* (John Wiley & Sons, Hoboken, NJ, 2008)
26. P. Serp, B. Machado, *Nanostructured Carbon Materials for Catalysis* (The Royal Society of Chemistry, London, 2015)
27. EU, European Union. Commission recommendation of 18 October 2011 on the definition of nanomaterial (2011/696/EU). Off. J. Eur. Union **54**, 38–40 (2011). https://doi.org/10.3000/19770677.L_2011.275.eng
28. C. Popov, *Nanostructured Carbon Materials* (Springer, Dordrecht, 2006), pp. 387–398
29. A.K. Geim, K.S. Novoselov, The rise of graphene. Nat. Mater. **6**, 183–191 (2007). https://doi.org/10.1038/nmat1849
30. J.L. Figueiredo, M.F.R. Pereira, The role of surface chemistry in catalysis with carbons. Catal. Today **150**, 2–7 (2010). https://doi.org/10.1016/j.cattod.2009.04.010
31. J. Figueiredo, M.F. Pereira, M.M. Freitas, J.J. Órfão, Modification of the surface chemistry of activated carbons. Carbon N.Y. **37**, 1379–1389 (1999). https://doi.org/10.1016/S0008-6223(98)00333-9
32. H.-P. Boehm, Catalytic properties of nitrogen-containing carbons, in *Carbon Materials for Catalysis*, (John Wiley & Sons, Hoboken, NJ, 2008), pp. 219–265
33. W.Y. Wong, W.R.W. Daud, A.B. Mohamad, A.A.H. Kadhum, K.S. Loh, E.H. Majlan, Recent progress in nitrogen-doped carbon and its composites as electrocatalysts for fuel cell applications. Int. J. Hydrog. Energy **38**, 9370–9386 (2013). https://doi.org/10.1016/j.ijhydene.2012.12.095
34. J.L. Figueiredo, M.F.R. Pereira, The role of surface chemistry in catalysis with carbons. Catal. Today **150**(1-2), 2–7 (2009)
35. J.K. Chinthaginjala, K. Seshan, L. Lefferts, Preparation and application of carbon-nanofiber based microstructured materials as catalyst supports. Ind. Eng. Chem. Res. **46**, 3968 (2007)
36. M.T. Kreutzer, F. Kapteijn, J.A. Moulijn, J.J. Heiszwolf, Multiphase monolith reactors: chemical reaction engineering of segmented flow in microchannels. Chem. Eng. Sci. **60**, 5895–5916 (2005). https://doi.org/10.1016/j.ces.2005.03.022
37. T. Boger, S. Roy, A.K. Heibel, O. Borchers, A monolith loop reactor as an attractive alternative to slurry reactors. Catal. Today **79–80**, 441–451 (2003)
38. M. Barletta, A. Gisario, G. Rubino, V. Tagliaferri, Electrostatic spray deposition (ESD) of 'self organizing' TiO2-epoxy powder paints: experimental analysis and numerical modeling. Surf. Coat. Technol. **201**, 3212–3228 (2006). https://doi.org/10.1016/j.surfcoat.2006.06.051
39. V. Meille, Review on methods to deposit catalysts on structured surfaces. Appl. Catal. A Gen. **315**, 1–17 (2006)
40. E. García-Bordejé, F. Kapteijn, J.A. Moulijn, Preparation and characterisation of carbon-coated monoliths for catalyst supports. Carbon N.Y. **40**, 1079–1088 (2002). https://doi.org/10.1016/S0008-6223(01)00252-4
41. G.D. Nessim, Properties, synthesis, and growth mechanisms of carbon nanotubes with special focus on thermal chemical vapor deposition. Nanoscale **2**, 1306–1323 (2010). https://doi.org/10.1039/b9nr00427k
42. S.P. Patole, P.S. Alegaonkar, H.-C. Shin, J.-B. Yoo, Alignment and wall control of ultra long carbon nanotubes in water assisted chemical vapour deposition. J. Phys. D. Appl. Phys. **41**, 155311 (2008). https://doi.org/10.1088/0022-3727/41/15/155311
43. A. Aqel, K.M.M.A. El-Nour, R.A.A. Ammar, A. Al-Warthan, Carbon nanotubes, science and technology part (I) structure, synthesis and characterisation. Arab. J. Chem. **5**, 1–23 (2012)
44. L. Roldán, S. Armenise, Y. Marco, E. García-Bordejé, Control of nitrogen insertion during the growth of nitrogen-containing carbon nanofibers on cordierite monolith walls. Phys. Chem. Chem. Phys. **14**, 3568–3575 (2012). https://doi.org/10.1039/c2cp23609e
45. S. Armenise, M. Nebra, E. García-Bordejé, A. Monzón, Functionalization of carbon nanofibers coated on cordierite monoliths by oxidative treatment, in *Studies in Surface Science and Catalysis*, ed. by E. Gaigneaux, M. Devillers, S. Hermans, P. A. Jacobs, J. Martens, P. Ruiz, (Elsevier, Amstrdam, 2010), pp. 483–486

46. P.-C. Ma, N.A. Siddiqui, G. Marom, J.-K. Kim, Dispersion and functionalization of carbon nanotubes for polymer-based nanocomposites: a review. Compos. Part A Appl. Sci. Manuf. **41**, 1345–1367 (2010). https://doi.org/10.1016/j.compositesa.2010.07.003
47. R.P. Rocha, D.F.M. Santos, O.S.M.P. Soares, A.M.T. Silva, M.F.R. Pereira, J.L. Figueiredo, Metal-free catalytic wet oxidation: from powder to structured catalyst using N-doped carbon nanotubes. Top. Catal. **61**, 1957–1966 (2018). https://doi.org/10.1007/s11244-018-1029-8
48. C. Pham-Huu, Carbon nanomaterials with controlled macroscopic shapes as new catalytic materials. Top. Catal. **40**, 49 (2006)
49. M.-J. Ledoux, C. Pham-Huu, Carbon nanostructures with macroscopic shaping for catalytic applications. Catal. Today **102–103**, 2–14 (2005). https://doi.org/10.1016/J.CATTOD.2005.02.036
50. Y. Liu, H. Ba, D.-L. Nguyen, O. Ersen, T. Romero, S. Zafeiratos, D. Begin, I. Janowska, C. Pham-Huu, Synthesis of porous carbon nanotubes foam composites with a high accessible surface area and tunable porosity. J. Mater. Chem. A **1**, 9508 (2013). https://doi.org/10.1039/c3ta10695k
51. F. Simescu-Lazar, V. Meille, S. Pallier, E. Chaînet, C. De Bellefon, Regeneration of deactivated catalysts coated on foam and monolith: example of Pd/C for nitrobenzene hydrogenation. Appl. Catal. A Gen. **453**, 28–33 (2013). https://doi.org/10.1016/j.apcata.2012.11.044
52. F. Simescu-Lazar, T. Chaieb, S. Pallier, L. Veyre, R. Philippe, V. Meille, Direct coating of carbon-supported catalysts on monoliths and foams—singular behaviour of Pd/MWCNT. Appl. Catal. A Gen. **508**, 45–51 (2015). https://doi.org/10.1016/J.APCATA.2015.09.042
53. T. Vergunst, F. Kapteijn, J.A. Moulijn, Carbon coating of ceramic monolithic substrates. Stud. Surf. Sci. Catal. **118**, 175–183 (1998). https://doi.org/10.1016/S0167-2991(98)80180-8
54. J. Restivo, C.A. Orge, A.S.G.G. Santos, O.S.G.P. Soares, M.F.R. Pereira, Nanostructured layers of mechanically processed multi-walled carbon nanotubes for catalytic applications. ASC Appl. Nano Mater. **3**(6), 5271–5284 (2020)
55. Y. Zhu, L. Yu, X. Wang, Y. Zhou, H. Ye, A novel monolithic Pd catalyst supported on cordierite with graphene coating. Catal. Commun. **40**, 98–102 (2013). https://doi.org/10.1016/J.CATCOM.2013.06.018
56. Y. Zhu, Y. Zhou, L. Yu, G. Liu, Y. Tian, H. Ye, A highly stable and active Pd catalyst on monolithic cordierite with graphene coating assisted by PDDA. RSC Adv. **4**, 9480 (2014). https://doi.org/10.1039/c3ra46316h
57. B.J.C. Thomas, A.R. Boccaccini, M.S.P. Shaffer, Multi-walled carbon nanotube coatings using electrophoretic deposition (EPD). J. Am. Ceram. Soc. **88**, 980–982 (2005). https://doi.org/10.1111/j.1551-2916.2005.00155.x
58. P.C.C. Faria, J.J.M. Órfão, M.F.R. Pereira, Activated carbon catalytic ozonation of oxamic and oxalic acids. Appl. Catal. B Environ. **79**, 237–243 (2008). https://doi.org/10.1016/j.apcatb.2007.10.021
59. M. Santiago, F. Stüber, A. Fortuny, A. Fabregat, J. Font, Modified activated carbons for catalytic wet air oxidation of phenol. Carbon N.Y. **43**, 2134–2145 (2005). https://doi.org/10.1016/j.carbon.2005.03.026
60. H.T. Gomes, B.F. Machado, A. Ribeiro, I. Moreira, M. Rosário, A.M.T. Silva, J.L. Figueiredo, J.L. Faria, Catalytic properties of carbon materials for wet oxidation of aniline. J. Hazard. Mater. **159**, 420–426 (2008). https://doi.org/10.1016/j.jhazmat.2008.02.070
61. F.J. Beltrán, *Ozone Reaction Kinetics for Water and Wastewater Systems* (Lewis Publishers, Boca Raton, FL, 2004)
62. M. Soria-Sánchez, A. Maroto-Valiente, J. Álvarez-Rodríguez, V. Muñoz-Andrés, I. Rodríguez-Ramos, A. Guerrero-Ruíz, Carbon nanostrutured materials as direct catalysts for phenol oxidation in aqueous phase. Appl. Catal. B Environ. **104**, 101–109 (2011). https://doi.org/10.1016/j.apcatb.2011.02.023
63. A.G. Gonçalves, J.L. Figueiredo, J.J.M. Órfão, M.F.R. Pereira, Influence of the surface chemistry of multi-walled carbon nanotubes on their activity as ozonation catalysts. Carbon N.Y. **48**, 4369–4381 (2010)

64. K.A. Wepasnick, B.A. Smith, K.E. Schrote, H.K. Wilson, S.R. Diegelmann, D.H. Fairbrother, Surface and structural characterization of multi-walled carbon nanotubes following different oxidative treatments. Carbon N.Y. **49**, 24–36 (2011). https://doi.org/10.1016/j.carbon.2010.08.034
65. Z.-Q. Liu, J. Ma, Y.-H. Cui, L. Zhao, B.-P. Zhang, Influence of different heat treatments on the surface properties and catalytic performance of carbon nanotube in ozonation. Appl. Catal. B Environ. **101**, 74–80 (2010). https://doi.org/10.1016/j.apcatb.2010.09.009
66. F.J. Beltrán, F.J. Rivas, L.A. Fernández, P.M. Álvarez, R. Montero-de-Espinosa, Kinetics of catalytic ozonation of oxalic acid in water with activated carbon. Ind. Eng. Chem. Res. **41**, 6510–6517 (2002). https://doi.org/10.1021/ie020311d
67. S.A.C. Carabineiro, T. Thavorn-Amornsri, M.F.R. Pereira, P. Serp, J.L. Figueiredo, Comparison between activated carbon, carbon xerogel and carbon nanotubes for the adsorption of the antibiotic ciprofloxacin. Catal. Today **186**, 29–34 (2012). https://doi.org/10.1016/j.cattod.2011.08.020
68. J. Hoigné, H. Bader, Rate constants of reactions of ozone with organic and inorganic compounds in water-I. Non-dissociating organic compounds. Water Res. **17**, 173–183 (1983). https://doi.org/10.1016/0043-1354(83)90098-2
69. R. Oulton, J.P. Haase, S. Kaalberg, C.T. Redmond, M.J. Nalbandian, D.M. Cwiertny, Hydroxyl radical formation during ozonation of multiwalled carbon nanotubes: performance optimization and demonstration of a reactive CNT filter. Environ. Sci. Technol. **49**, 3687–3697 (2015). https://doi.org/10.1021/es505430v
70. F. Morales-Lara, M.J. Pérez-Mendoza, D. Altmajer-Vaz, M. García-Román, M. Melguizo, F.J. López-Garzón, M. Domingo-García, Functionalization of multiwall carbon nanotubes by ozone at basic pH. Comparison with oxygen plasma and ozone in gas phase. J. Phys. Chem. C **117**, 11647–11655 (2013). https://doi.org/10.1021/jp4017097
71. R.P. Rocha, J.P.S. Sousa, A.M.T. Silva, M.F.R. Pereira, J.L. Figueiredo, Catalytic activity and stability of multiwalled carbon nanotubes in catalytic wet air oxidation of oxalic acid: the role of the basic nature induced by the surface chemistry. Appl. Catal. B Environ. **104**, 330–336 (2011). https://doi.org/10.1016/j.apcatb.2011.03.009
72. S. Yang, X. Li, W. Zhu, J. Wang, C. Descorme, Catalytic activity, stability and structure of multi-walled carbon nanotubes in the wet air oxidation of phenol. Carbon N.Y. **46**, 445–452 (2008). https://doi.org/10.1016/j.carbon.2007.12.006
73. J. Restivo, R.P. Rocha, A.M.T. Silva, J.J.M. Órfão, M.F.R. Pereira, J.L. Figueiredo, Catalytic performance of heteroatom-modified carbon nanotubes in advanced oxidation processes. Cuihua Xuebao/Chinese J. Catal. **35**, 896–905 (2014)
74. O.S.G.P. Soares, R.P. Rocha, J.J.M. Órfão, M.F.R. Pereira, J.L. Figueiredo, Mechanothermal approach for N-, S-, P-, and B-doping of carbon nanotubes: methodology and catalytic performance in wet air oxidation. C **5**, 30 (2019). https://doi.org/10.3390/c5020030
75. O.S.G.P. Soares, R.P. Rocha, A.G. Gonçalves, J.L. Figueiredo, J.J.M. Órfão, M.F.R. Pereira, Highly active N-doped carbon nanotubes prepared by an easy ball milling method for advanced oxidation processes. Appl. Catal. B Environ. **192**, 296–303 (2016). https://doi.org/10.1016/J.APCATB.2016.03.069
76. R.P. Rocha, A.G. Gonçalves, L.M. Pastrana-Martínez, B.C. Bordoni, O.S.G.P. Soares, J.J.M. Órfão, J.L. Faria, J.L. Figueiredo, A.M.T. Silva, M.F.R. Pereira, Nitrogen-doped graphene-based materials for advanced oxidation processes. Catal. Today **249**, 192–198 (2015). https://doi.org/10.1016/j.cattod.2014.10.046
77. M. Pedrosa, L.M. Pastrana-Martínez, M.F.R. Pereira, J.L. Faria, J.L. Figueiredo, A.M.T. Silva, N/S-doped graphene derivatives and TiO2 for catalytic ozonation and photocatalysis of water pollutants. Chem. Eng. J. **348**, 888–897 (2018). https://doi.org/10.1016/j.cej.2018.04.214
78. C. Tizaoui, H. Mohammad-Salim, J. Suhartono, Multiwalled carbon nanotubes for heterogeneous nanocatalytic ozonation. Ozone Sci. Eng. **37**, 269–278 (2015). https://doi.org/10.1080/01919512.2014.983455

79. S. Zhang, D. Wang, S. Zhang, X. Zhang, P. Fan, Ozonation and carbon-assisted ozonation of methylene blue as model compound: effect of solution pH. Procedia Environ. Sci. **18**, 493–502 (2013). https://doi.org/10.1016/j.proenv.2013.04.066
80. M. Farzadkia, Y. Dadban Shahamat, S. Nasseri, A.H. Mahvi, M. Gholami, A. Shahryari, Catalytic ozonation of phenolic wastewater: identification and toxicity of intermediates. J. Eng. **2014** (2014). https://doi.org/10.1155/2014/520929
81. J. Restivo, J.J.M. Órfão, S. Armenise, E. Garcia-Bordejé, M.F.R. Pereira, Catalytic ozonation of metolachlor under continuous operation using nanocarbon materials grown on a ceramic monolith. J. Hazard. Mater. **239–240**, 249–256 (2012)
82. J. Restivo, J.J.M. Órfão, M.F.R. Pereira, E. Garcia-Bordejé, P. Roche, D. Bourdin, B. Houssais, M. Coste, S. Derrouiche, Catalytic ozonation of organic micropollutants using carbon nanofibers supported on monoliths. Chem. Eng. J. **230**, 115–123 (2013)
83. A.G. Gonçalves, J.J.M. Órfão, M.F.R. Pereira, Ozonation of sulfamethoxazole promoted by MWCNT. Catal. Commun. **35**, 82–87 (2013). https://doi.org/10.1016/j.catcom.2013.02.012
84. A. Gonçalves, J.J.M. Órfão, M.F.R. Pereira, Ozonation of bezafibrate promoted by carbon materials. Appl. Catal. B Environ. **140–141**, 82–91 (2013). https://doi.org/10.1016/j.apcatb.2013.03.034
85. X. Fan, J. Restivo, J.J.M. Órfão, M.F.R. Pereira, A.A. Lapkin, The role of multiwalled carbon nanotubes (MWCNTs) in the catalytic ozonation of atrazine. Chem. Eng. J. **241**, 66–76 (2014)
86. J.-N. Liu, Z. Chen, Q.-Y. Wu, A. Li, H.-Y. Hu, C. Yang, Ozone/graphene oxide catalytic oxidation: a novel method to degrade emerging organic contaminant N, N-diethyl-m-toluamide (DEET). Sci. Rep. **6**, 31405 (2016). https://doi.org/10.1038/srep31405
87. Z. Cai, A.D. Dwivedi, W.-N. Lee, X. Zhao, W. Liu, M. Sillanpää, D. Zhao, C.-H. Huang, J. Fu, Application of nanotechnologies for removing pharmaceutically active compounds from water: development and future trends. Environ. Sci. Nano **5**, 27–47 (2018). https://doi.org/10.1039/C7EN00644F
88. Y. Wang, Y. Xie, H. Sun, J. Xiao, H. Cao, S. Wang, Efficient catalytic ozonation over reduced graphene oxide for *p*-hydroxylbenzoic acid (PHBA) destruction: active site and mechanism. ACS Appl. Mater. Interfaces **8**, 9710–9720 (2016). https://doi.org/10.1021/acsami.6b01175
89. Y. Ahn, H. Oh, Y. Yoon, W.K. Park, W.S. Yang, J.W. Kang, Effect of graphene oxidation degree on the catalytic activity of graphene for ozone catalysis. J. Environ. Chem. Eng. **5**, 3882–3894 (2017). https://doi.org/10.1016/j.jece.2017.07.038
90. Y. Wang, H. Cao, C. Chen, Y. Xie, H. Sun, X. Duan, S. Wang, Metal-free catalytic ozonation on surface-engineered graphene: microwave reduction and heteroatom doping. Chem. Eng. J. **355**, 118–129 (2019). https://doi.org/10.1016/j.cej.2018.08.134
91. L. Jothinathan, J. Hu, Kinetic evaluation of graphene oxide based heterogenous catalytic ozonation for the removal of ibuprofen. Water Res. **134**, 63–73 (2018). https://doi.org/10.1016/j.watres.2018.01.033
92. R. Yin, W. Guo, J. Du, X. Zhou, H. Zheng, Q. Wu, J. Chang, N. Ren, Heteroatoms doped graphene for catalytic ozonation of sulfamethoxazole by metal-free catalysis: performances and mechanisms. Chem. Eng. J. **317**, 632–639 (2017). https://doi.org/10.1016/j.cej.2017.01.038
93. G. Li, Y. Lu, C. Lu, M. Zhu, C. Zhai, Y. Du, P. Yang, Efficient catalytic ozonation of bisphenol-A over reduced graphene oxide modified sea urchin-like α-MnO2 architectures. J. Hazard. Mater. **294**, 201–208 (2015). https://doi.org/10.1016/j.jhazmat.2015.03.045
94. M. Sui, S. Xing, L. Sheng, S. Huang, H. Guo, Heterogeneous catalytic ozonation of ciprofloxacin in water with carbon nanotube supported manganese oxides as catalyst. J. Hazard. Mater. **227–228**, 227–236 (2012). https://doi.org/10.1016/j.jhazmat.2012.05.039
95. A.G. Gonçalves, J.J.M. Órfão, M.F.R. Pereira, Ozonation of erythromycin over carbon materials and ceria dispersed on carbon materials. Chem. Eng. J. **250**, 366–376 (2014). https://doi.org/10.1016/j.cej.2014.04.012

96. A.G. Gonçalves, J.J.M. Órfão, M.F.R. Pereira, Catalytic ozonation of sulphamethoxazole in the presence of carbon materials: catalytic performance and reaction pathways. J. Hazard. Mater. **239–240**, 167–174 (2012). https://doi.org/10.1016/j.jhazmat.2012.08.057
97. Z.Q. Liu, J. Ma, Y.-H. Cui, B.-P. Zhang, Effect of ozonation pretreatment on the surface properties and catalytic activity of multi-walled carbon nanotube. Appl Catal B **92**, 301–306 (2009)
98. M. Sanchez-Polo, U. von Gunten, J. Rivera-Utrilla, Efficiency of activated carbon to transform ozone into *OH radicals: influence of operational parameters. Water Res. **39**, 3189–3198 (2005). https://doi.org/10.1016/j.watres.2005.05.026
99. T. Otowa, R. Tanibata, M. Itoh, Production and adsorption characteristics of MAXSORB: high-surface-area active carbon. Gas Sep. Purif. **7**, 241–245 (1993). https://doi.org/10.1016/0950-4214(93)80024-Q
100. Z. Hu, M.P. Srinivasan, Mesoporous high-surface-area activated carbon. Microporous Mesoporous Mater. **43**, 267–275 (2001). https://doi.org/10.1016/S1387-1811(00)00355-3
101. O.S.G.P. Soares, A.G. Gonçalves, J.J. Delgado, J.J.M. Órfão, M.F.R. Pereira, Modification of carbon nanotubes by ball-milling to be used as ozonation catalysts. Catal. Today **249**, 199–203 (2015). https://doi.org/10.1016/J.CATTOD.2014.11.016
102. J. Restivo, J.J.M. Órfão, M.F.R. Pereira, E. Vanhaecke, M. Rönning, T. Iouranova, L. Kiwi-Minsker, S. Armenise, E. Garcia-Bordejé, Catalytic ozonation of oxalic acid using carbon nanofibres on macrostructured supports. Water Sci. Technol. **65**, 1854–1862 (2012). https://doi.org/10.2166/wst.2012.882
103. K.M. Sushma, A.K. Saroha, Performance of various catalysts on treatment of refractory pollutants in industrial wastewater by catalytic wet air oxidation: a review. J. Environ. Manag. **228**, 169–188 (2018)
104. FREECATS Final Report Summary—FREECATS (Doped carbon nanostructures as metal-free catalysts)|Report Summary|FREECATS|FP7|CORDIS|European Commission
105. C. Della Rocca, V. Belgiorno, S. Meriç, Overview of in-situ applicable nitrate removal processes. Desalination **204**, 46–62 (2007). https://doi.org/10.1016/j.desal.2006.04.023
106. A. Pintar, Catalytic processes for the purification of drinking water and industrial effluents. Catal. Today **77**(4), 451–465 (2003)
107. D.B. Thakur, R.M. Tiggelaar, Y. Weber, J.G.E. Gardeniers, L. Lefferts, K. Seshan, Ruthenium catalyst on carbon nanofiber support layers for use in silicon-based structured microreactors. Part II: catalytic reduction of bromate contaminants in aqueous phase. Appl. Catal. B Environ. **102**, 243–250 (2011). https://doi.org/10.1016/j.apcatb.2010.12.004
108. A. Garron, F. Epron, Use of formic acid as reducing agent for application in catalytic reduction of nitrate in water. Water Res. **39**, 3073–3081 (2005). https://doi.org/10.1016/j.watres.2005.05.012
109. A. Pintar, Catalytic hydrogenation of aqueous nitrate solutions in fixed-bed reactors. Catal. Today **53**, 35 (1999)
110. J. Restivo, O.S.G.P. Soares, J.J.M. Órfão, M.F.R. Pereira, Catalytic reduction of bromate over monometallic catalysts on different powder and structured supports. Chem. Eng. J. **309**, 197–205 (2017). https://doi.org/10.1016/j.cej.2016.10.025
111. J.K. Chinthaginjala, L. Lefferts, Support effect on selectivity of nitrite reduction in water. Appl. Catal. B Environ. **101**, 144–149 (2010). https://doi.org/10.1016/j.apcatb.2010.09.023
112. D. Shuai, J.K. Choe, J.R. Shapley, C.J. Werth, Enhanced activity and selectivity of carbon nanofiber supported Pd catalysts for nitrite reduction. Environ. Sci. Technol. **46**, 2847–2855 (2012). https://doi.org/10.1021/es203200d
113. M.L. Toebes, Support effects in hydrogenation of cinnamaldehyde over carbon nanofiber-supported platinum catalysts: kinetic modeling. Chem. Eng. Sci. **60**, 5682 (2005)
114. L. Chen, A.C. Cooper, G.P. Pez, H. Cheng, Mechanistic study on hydrogen spillover onto graphitic carbon materials. J. Phys. Chem. C **111**, 18995–19000 (2007). https://doi.org/10.1021/jp074920g

115. D.J. Browning, M.L. Gerrard, J.B. Lakeman, I.M. Mellor, R.J. Mortimer, M.C. Turpin, Studies into the storage of hydrogen in carbon nanofibers: proposal of a possible reaction mechanism. Nano Lett. **2**, 201–205 (2002). https://doi.org/10.1021/nl015576g
116. T. Ye, D.P. Durkin, M. Hu, X. Wang, N.A. Banek, M.J. Wagner, D. Shuai, Enhancement of nitrite reduction kinetics on electrospun Pd-carbon nanomaterial catalysts for water purification. ACS Appl. Mater. Interfaces **8**, 17739–17744 (2016). https://doi.org/10.1021/acsami.6b03635
117. F. Epron, F. Gauthard, C. Pinéda, J. Barbier, Catalytic reduction of nitrate and nitrite on Pt–Cu/Al2O3 catalysts in aqueous solution: role of the interaction between copper and platinum in the reaction. J. Catal. **198**, 309–318 (2001). https://doi.org/10.1006/jcat.2000.3138
118. F.A. Marchesini, L.B. Gutierrez, C.A. Querini, E.E. Miró, Pt,In and Pd,In catalysts for the hydrogenation of nitrates and nitrites in water. FTIR characterization and reaction studies. Chem. Eng. J. **159**, 203–211 (2010). https://doi.org/10.1016/j.cej.2010.02.056
119. N. Barrabés, A. Dafinov, F. Medina, J.E. Sueiras, Catalytic reduction of nitrates using Pt/CeO2 catalysts in a continuous reactor. Catal. Today **149**, 341–347 (2010). https://doi.org/10.1016/j.cattod.2009.05.029
120. O.S.G.P. Soares, J.J.M. Órfão, M.F.R. Pereira, Nitrate reduction with hydrogen in the presence of physical mixtures with mono and bimetallic catalysts and ions in solution. Appl. Catal. B Environ. **102**, 424–432 (2011)
121. O.S.G.P. Soares, J.J.M. Órfão, M.F.R. Pereira, Pd-Cu and Pt-Cu catalysts supported on carbon nanotubes for nitrate reduction in water. Ind. Eng. Chem. Res. **49**, 7183–7192 (2010)
122. O.S.G.P. Soares, X. Fan, J.J.M. Órfão, A.A. Lapkin, M.F.R. Pereira, Kinetic modeling of nitrate reduction catalyzed by Pd-Cu supported on carbon nanotubes. Ind. Eng. Chem. Res. **51**, 4854–4860 (2012)
123. S. Panić, Á. Kukovecz, G. Boskovic, Design of catalytic carbon nanotube-based reactor for water denitration—the impact of active metal confinement. Appl. Catal. B Environ. **225**, 207–217 (2018). https://doi.org/10.1016/j.apcatb.2017.11.078
124. Y. Yun, Z. Li, Y.-H. Chen, M. Saino, S. Cheng, L. Zheng, Reduction of nitrate in secondary effluent of wastewater treatment plants by Fe0 reductant and Pd–Cu/graphene catalyst. Water Air Soil Pollut. **227**, 111 (2016). https://doi.org/10.1007/s11270-016-2792-4
125. A. Sepulveda-Escribano, F. Coloma, F. Rodrıguez-Reinoso, Platinum catalysts supported on carbon blacks with different surface chemical properties. Appl. Catal. A Gen. **173**, 247–257 (1998)
126. X. Gu, W. Qi, X. Xu, Z. Sun, L. Zhang, W. Liu, X. Pan, D. Su, Covalently functionalized carbon nanotube supported Pd nanoparticles for catalytic reduction of 4-nitrophenol. Nanoscale **6**, 6609–6616 (2014). https://doi.org/10.1039/c4nr00826j
127. O.S.G.P. Soares, P.S.F. Ramalho, A. Fernandes, J.J.M. Órfão, M.F.R. Pereira, Catalytic bromate reduction in water: influence of carbon support. J. Environ. Chem. Eng. **7**, 103015 (2019). https://doi.org/10.1016/j.jece.2019.103015
128. T. Yuranova, C. Franch, A.E. Palomares, E. Garcia-Bordejé, L. Kiwi-Minsker, Structured fibrous carbon-based catalysts for continuous nitrate removal from natural water. Appl. Catal. B Environ. **123–124**, 221–228 (2012). https://doi.org/10.1016/j.apcatb.2012.04.007
129. E. Garcia-Bordeje, Vanadium supported on carbon coated honeycomb monoliths for the selective catalytic reduction of NO at low temperatures: influence of the oxidation pretreatment. Carbon N.Y. **44**, 407 (2006)
130. A.E. Palomares, C. Franch, T. Yuranova, L. Kiwi-Minsker, E. García-Bordeje, S. Derrouiche, The use of Pd catalysts on carbon-based structured materials for the catalytic hydrogenation of bromates in different types of water. Appl. Catal. B Environ. **146**, 186–191 (2014). https://doi.org/10.1016/j.apcatb.2013.02.056
131. P. Yaseneva, C.F. Marti, E. Palomares, X. Fan, T. Morgan, P.S. Perez, M. Ronning, F. Huang, T. Yuranova, L. Kiwi-Minsker, Efficient reduction of bromates using carbon nanofibre supported catalysts: experimental and a comparative life cycle assessment study. Chem. Eng. J. **248**, 230–241 (2014)

132. Y. Marco, E. García-Bordejé, C. Franch, A.E. Palomares, T. Yuranova, L. Kiwi-Minsker, Bromate catalytic reduction in continuous mode using metal catalysts supported on monoliths coated with carbon nanofibers. Chem. Eng. J. **230**, 605–611 (2013). https://doi.org/10.1016/J.CEJ.2013.06.040
133. T. Yuranova, L. Kiwi-Minsker, C. Franch, A.E. Palomares, S. Armenise, E. García-Bordejé, Springer, in *Industrial and Engineering Chemistry Research*, (American Chemical Society, Washington, DC, 2013), pp. 13930–13937
134. A.E. Palomares, C. Franch, A. Corma, Nitrates removal from polluted aquifers using (Sn or Cu)/Pd catalysts in a continuous reactor. Catal. Today **149**, 348–351 (2010). https://doi.org/10.1016/j.cattod.2009.05.013
135. O.S.G.P. Soares, J.J.M. Órfão, E. Gallegos-Suarez, E. Castillejos, I. Rodríguez-Ramos, M.F.R. Pereira, Nitrate reduction over a Pd-Cu/MWCNT catalyst: application to a polluted groundwater. Environ. Technol. **33**, 2353–2358 (2012)

Chapter 11
Rational Design of Graphene-based Sorbents for Water Purification

Asif Hussain, Muhammad Usman, Rana Zafar Abbas Manj, Fuqiang Liu, Dengxin Li, and Yanbiao Liu

11.1 Introduction

The textile industry, one of the major industries of China, plays a pivotal role in boosting the economic development [1, 2]. Meanwhile, it also causes serious environmental pollution due to discharge of excess dyes into water bodies. The textile industry has been listed by the Chinese government as the key industry that needs to be rectified and improved. However, the textile wastewater is usually characterized by complex compositions, large volume, poor biodegradability, and high chrominance which render it one of the hardest-to-treat forms of industrial wastewater [2, 3]. Therefore, it is highly desirable to develop advanced approaches to address this challenging issue [4]. Different processes have been developed and applied for the treatment of textile wastewater. For example, conventional biological processes demonstrated limited capability for dye degradation due to their poor biodegradability [5]. Chemical oxidation methods like photocatalytic oxidation [6] and ozonation [7] are usually energy intensive and expensive, while catalytic oxidation based on peroxymonosulfate activation may lead to the generation of insufficient

A. Hussain
College of Environmental Science and Engineering, Donghua University, Shanghai, China

Department of Environmental Science, Federal Urdu University of Arts, Science and Technology, Karachi, Pakistan

M. Usman
PEIE Research Chair for the Development of Industrial Estates and Free Zones, Center for Environmental Studies and Research, Sultan Qaboos University, Al Khoudh, Muscat, Oman

R. Z. A. Manj
College of Material Science and Engineering, Donghua University, Shanghai, China

F. Liu · D. Li · Y. Liu (✉)
College of Environmental Science and Engineering, Donghua University, Shanghai, China
e-mail: yanbiaoliu@dhu.edu.cn

M. Piumetti, S. Bensaid (eds.), *Nanostructured Catalysts for Environmental Applications*, https://doi.org/10.1007/978-3-030-58934-9_11

mineralization or byproducts, causing further secondary pollution [8]. Dye rejection by state-of-the-art membrane processes is feasible; however, the high turbidity and complex compositions of the dye wastewater can easily block membrane pores and decrease membrane efficacy [9]. Alternatively, adsorption has received significant interest from the scientific community to serve as a promising, affordable, and viable technology to remove dyes from water [10, 11].

Carbon-based materials have been extensively investigated due to their intriguing physicochemical properties [11, 12]. Among them, graphene compounds have been extensively studied for various environmental applications [9, 12, 13]. In this chapter, we summarized the recent advances on the rational design of graphene-based sorbents. Their applications in dye removal are introduced in detail in terms of sorption kinetics and thermodynamics. At last, existing challenges and future prospects are provided.

11.2 Overview of Graphene-based Sorbents

11.2.1 Graphene Oxide Adsorbents

Graphite oxide (GO) is an important derivative of graphene in highly oxidized form (i.e., containing oxo functionalities and hydrogen atoms) [15]. It is usually synthesized through chemical oxidation of pristine graphite followed by exfoliation [16]. The exact chemical structure and composition of GO are still debatable due to the complexity and variability of GO samples arising from different synthesis conditions and non-stoichiometric atomic compositions of CxH_yO_z [17]. Many chemical structural models have been proposed for GO (e.g., Hofmann model [18], Ruess model [19], Scholz–Boehm model [20], Nakajima–Matsuo model [21], Lerf–Klinowski model [22, 23], or Szabo model [24]). They show mostly a layered structure of graphite with diverse oxygen functionalities [14, 25]. Among these models, Lerf–Klinowski model is the most acceptable [26], in which the characteristic GO functional defect sites are decorated with carbonyl, epoxy, carboxyl, and hydroxyl groups [14, 26, 27]. The oxidation of graphite breaks the π–π conjugation of graphite flakes into nanoscale fragments of ordered-sp^2 graphitic domains (C=C) surrounded by disordered-sp^3 (C-C) oxidized domains [28]. The relative size of these domains depends on the oxidation degree of mother compounds [23]. The carbon-to-oxygen ratio (C/O) in GO usually varies from 1.5 to 2.5. The functional defect sites make GO highly hydrophilic to form a stable light brown suspension of single or multilayers GO sheets with an average dimension of 0.22 to ≥2.5 μm in water, depending on gradient differential centrifugation, GDC (Fig. 11.1) [14], and allow one to control the assembly behavior of macroscopic graphene [27].

The planar sp^2 aromatic domains and sp^3 hybridized edge sites make this nanomaterial amphiphilic or behave like a polymer for various covalent or noncovalent bonding interactions [29]. It should be noted that some dye molecules also contain

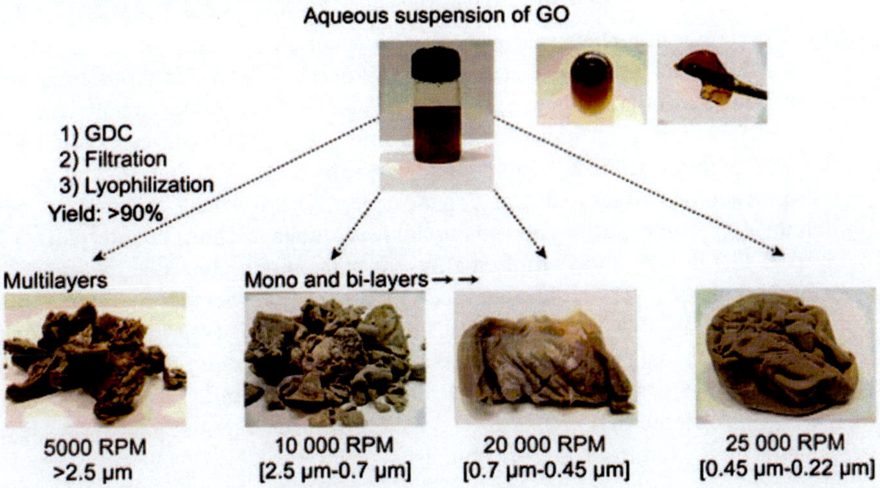

Fig. 11.1 Several fractions of GO aqueous suspensions after GDC and membrane filtration. Reproduced with permission [14]. Copyright 2020, Royal Society of Chemistry

aromatic structures, and exist usually in ionic form in water. GO is characterized by negatively charged surface over a wide pH range, which promotes an increased adsorption of the cationic dye methylene blue (MB) via electrostatic interaction [30]. GO has a larger and smoother surface than CNTs and activated carbon [31], which favors an additional contribution of π–π electron donor/acceptor interactions and causes an elevated adsorption capacity.

GO-based adsorbents can be prepared via Hummer's method with variable oxidation degree by using different oxidant doses (e.g., $KMnO_4$) and reaction conditions [32]. A quantitative measurement of GO oxidation degree has been performed using a combination of X-ray photon spectroscopy (XPS) and X-ray diffraction (XRD) to test for the MB removal. It was proposed that an improvement in oxidation degree can result in a strong affinity to MB due to alteration in the fundamental adsorption behavior of GO. For example, with an increase in the oxidation degree, the MB uptake capacity by GO can be increased exponentially [32]. It may change the isotherm adsorption behavior from Freundlich type to Langmuir type due to the generation of more active adsorption sites on the GO surface [32]. Furthermore, the relative proportion of these two functional regions in GO (e.g., oxidized GO-type, and non-oxidized graphite regions) was flexible to the oxidative conditions [32]. The binding features in the non-oxidative region of planar graphite tend to form parallel π–π stacking/hydrophobic interactions with MB molecules in water. In contrast, oxidized regions containing oxygen functionalities can bind these MB molecules vertically on the GO plane via electrostatic interactions or hydrogen bonding. It was proposed that the latter interaction mode usually exhibits stronger affinity toward cationic MB dye at higher oxidation degree. Furthermore, the adsorption of two anionic azo dyes (acid orange 8, AO8, and direct red 23, DR23) onto GO was also investigated [33]. The adsorption data fitted well with the Langmuir isotherm with a maximum monolayer adsorption capacity of 29 ± 0.2 and 15 ± 0.3 mg/g for AO8 and DR23, respectively. The adsorption process of AO8 was exothermic, while the adsorption of DR23 was endothermic in nature. Adsorption of both dyes onto GO mainly occurred through electrostatic interaction, but hydrogen bonding and π–π stacking interactions were also involved [33].

Several studies suggested that a nanoscale GO adsorbent possesses a better potential for water purification compared to other carbon counterparts [30]. However, GO usually shows limited affinity to the anionic dyes due to strong electrostatic repulsion [34]. Further engineering concerns associated with the applications of nanoscale GO in powder forms are expensive recovery and/or reuse, and negative impacts on living organisms that seriously restrict its feasibility at industrial scale [35]. Consequently, the true revolutionary potential lies in diverse pathways for the tailored manipulation of graphene surface chemistry and can only be rationalized via bottom-up design synthesis.

11.2.2 Graphene Adsorbents

Reduced GO (rGO) or graphene is a two-dimensional carbon nanomaterial comprising a single layer of sp^2 carbon atoms arranged in a hexagonal ring [36]. rGO has attracted significant worldwide attention due to its outstanding physical, chemical, electrical, and mechanical properties [37]. The key opportunity to employ graphene as an adsorbent material is due to its high surface area up to 2630 g/m^2 [38] and high hydrophobicity (i.e., highly ordered-sp^2 graphitic domains) [17]. rGO can be prepared through wet chemistry reduction using a GO precursor and appropriate reducing agents (e.g., hydrazine [34], sodium ascorbate [39], hypophosphorous acid [40], ascorbic acid [41], urea [35], or thiourea [38]). It appears that each method produces graphene with different physicochemical characteristics (i.e., functionalities or degree of reduction), thereby showing different performances.

To boost their sorption performance, three-dimensional (3D) rGO hydrogels were developed via chemical approaches [39]. It was proposed that graphene nanosheets can be self-organized into a 3D porous network through *van der Waals* and π–π interactions [39]. As-obtained 3D mesoporous material showed maximum adsorption capacities of 8 ± 0.2 mg/g for MB and 29 ± 0.5 mg/g for RhB based on a heterogeneous type Freundlich isotherm model. The strong π–π stacking and anion–cation interactions were proposed to dictate the adsorption process. Sorbent recycling experiments showed that the materials could easily be regenerated by washing with ethylene glycol [39]. Furthermore, 3D microstructures of rGO were explored, which were prepared by heating the GO suspension in a mixture of hypophosphorous acid (H_3PO_2) and iodine (I_2) at 80 °C for 12 h followed by freeze-drying [40]. The 3D-rGO macrostructures were examined to remove anionic acid red 1 (AC1) and cationic MB dyes. The surface charge of the 3D-rGO and the chemical structure of the dye molecules caused notable differences in the overall adsorption capacity, equilibrium time, and adsorption mechanism [40]. The equilibrium data were best fitted to a Langmuir and Freundlich model for MB and AC1, respectively. This chemically reduced 3D-rGO provides an ideal platform for a fast adsorption of MB with a capacity of 300 mg/g. The electrostatic and π–π interactions between the localized π-electrons of conjugated aromatic rings in rGO and dye molecules were involved in the MB adsorption. However, the as-prepared 3D-rGO sorbent was unsuitable for the uptake of negatively charged AC1 (adsorption capacity of 277 mg/g) due to dissociated SO_3^- groups in aqueous solution [40].

Most of these reported designs existed in powder form [32–34, 37, 39, 40], which tend to generate irreversible agglomerates due to strong π–π stacking and *van der Waals* interaction between the graphene sheets. This seriously decreases the surface area. Alternatively, functional composites of graphene were modified further with magnetic nanoparticles [42, 43] or by coating onto different macroscopic supports [44, 45]. However, magnetic adsorbent designs frequently showed limited adsorption capability, resulting from reduced surface area due to large aspect ratio of graphene. Loss of active adsorbent sites during cross-linking or coating of graphene on other material may also be responsible for the reduced adsorption performance. Thus, other innovative and advanced designs to enhance the functionality of graphene adsorbents should be given an imperative focus.

11.3 Graphene-based Macroscopic Sorbents

The synthesis of 3D macroscopic structures is of importance toward the practical applications of graphene-based sorbents. Until now, template-directed [46–49] and template-free (i.e., self-assembled) methods [50–57] are two main routes for the integration of graphene nanosheets into a 3D macroscopic network. These 3D designs not only promote the exposure of adsorption active sites and facile transmission of pollutant molecules [58], but also allow affordable recycling potential and further modifications. For example, a random arrangement of GO sheets into a 3D network can create a sufficient porosity or out-of-plane space in-between graphene sheets, which decreases the π–π stacking between graphene nanosheets [58] and increases the exposure of adsorptive active sites for organic molecules. The macroscopic dimensions with hierarchical porous network (i.e., micro-, meso-, and macropores) can promote facile diffusion of pollutant molecules and allow simple and affordable reuse [59–61]. Furthermore, such designs can also host various functional guests (e.g., metals oxide, polymers, carbon nanofibers, CNTs, etc.) via covalent or noncovalent interactions, which result in the synergy of hierarchical structures and multifunctional properties for desired applications [35, 62, 63]. To date, an increasing number of macroscopic designs of graphene have been reported toward environmental remediation applications [36].

11.3.1 Template-Directed Design of Graphene Adsorbents

The template-directed macroscopic design of graphene can be prepared by hard [46] or soft-template [47–49, 64] methods using GO solution. Recent advances in the template-directed macroscopic designs of graphene adsorbents are listed in Table 11.1. A GO sponge, synthesized by a centrifugal vacuum and evaporation method, showed good removal performance of cationic dyes (MB and methyl violate, MV) from water (Fig. 11.2) with a fast equilibrium time of 2 min and adsorption capacities of 397 and 467 mg/g for MB and MV, respectively [47]. An endothermic type chemical adsorption of MB and MV on GO sponge was proposed through strong π–π stacking and anion–cation interactions with activation energies of 50 ± 0.3 and 71 ± 0.2 kJ/mol, respectively. However, extra post-filtration and insufficient reusability, due to inevitable re-dissolution of nanoscale GO after dye treatment, limit its water phase application.

Further, GO foam with macroporous architecture was prepared by direct lyophilization of GO dispersions [48]. The GO foam was tested to remove cationic dyes such as RhB, MG, and acriflavine (AF) containing different functional groups (e.g., xanthene fluorine rhodamine, tri-aryl methane, and acridine, respectively), but with similar molecular size. The adsorption capacities of GO foam were in the order of 446 mg/g (RhB) > 321 mg/g (MG) > 228 mg/g (AF). The electrostatic interactions between the GO foam and the dye molecules were proposed based on the recorded shifts of G-band toward lower wavenumbers in the Raman spectra. Furthermore, GO aerogel foams were prepared by a unidirectional ice segregation induced self-assembly method, in which a cylindrical plastic tube containing water suspension of GO (4 mg/mL) is placed into a Styrofoam container to provide a uniaxial thermal

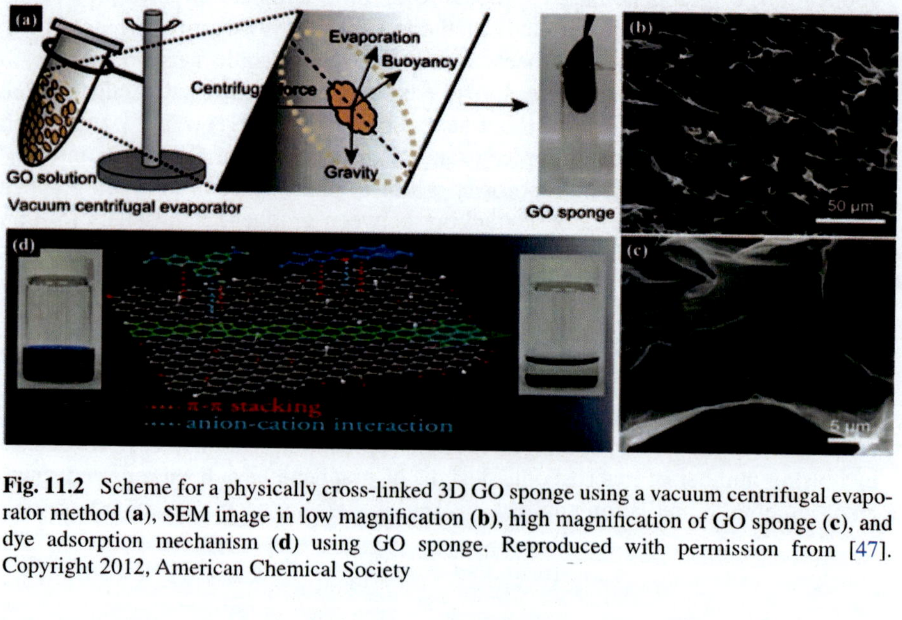

Fig. 11.2 Scheme for a physically cross-linked 3D GO sponge using a vacuum centrifugal evaporator method (**a**), SEM image in low magnification (**b**), high magnification of GO sponge (**c**), and dye adsorption mechanism (**d**) using GO sponge. Reproduced with permission from [47]. Copyright 2012, American Chemical Society

Table 11.1 Template-directed design of graphene sorbents to remove organic pollutants from water

Adsorbent	Synthesis protocol	Major additives	Pollutant	Q (mg/g)	References
GO foam	Centrifugal & vacuum evaporation	Additive free	MB	397	[47]
GO foam	Centrifugal & vacuum evaporation	Additive free	MV	467	[47]
GO foam	Ice template and freeze dry	Additive free	RhB	446	[48]
GO foam	Ice template and freeze dry	Additive free	MG	321	[48]
GO foam	Ice template and freeze dry	Additive free	AF	228	[48]
GO foam	Unidirectional ice segregation	Additive free	MB	416.67	[49]
rGO-LD sponge	Ice template and annealing	Laundry detergent	Diesel	–	[64]
rGO-SDS sponge	Ice template and annealing	SDS	Diesel	–	[64]
GO-PVA foams	Ice template & solvent exchange	PVA	MB	571.4	[65]
rGO-PVA foam	Ice template & solvent exchange	Ammonia / glucose	MB	81.0	[65]
GO-SA aerogel	Ice template & solvent exchange	Sodium alginate	MB	833.3	[25]
rGO-SA aerogel	Ice template & thermal reduction	Ammonia / glucose	MB	192.3	[25]
GO-agar aerogel	Ice template and freeze dry	Biopolymer agar	MB	476.8	[66]
GO-ECNFs aerogel	Ice template and freeze dry	ECNF	CV	>800	[67]
GO-ECNFs aerogel	Ice template and freeze dry	ESCNFs	MG	>800	[67]
GO-ESCNFs aerogel	Ice template and freeze dry	ESCNFs	RhB	>800	[67]
GO-PAA aerogels	Ice template and vacuum dry	PAA as cross-linker	Oil	120 g/g	[68]
GO-cellulose aerogel	Bidirectional freeze dry	Cellulose and saline	Oil	197 times	[69]
rGO-CNTs monolith	CVD method	CNTs	Oil	–	[70]
3D-rGO-Fe_3O_4 aerogel	Polymer template & calcination	PMMA and Fe_3O_4	MB	170.1	[46]
rGO-Cys monolith	3D printing	L-cysteine	IC	1005.7	[71]
rGO-Cys monolith	3D printing	L-cysteine	NR	1301.8	[71]

gradient using liquid N_2 [49, 72]. Next, the frozen suspension is freeze-dried to sublimate ice crystals and to obtain a GO aerogel, which is used to remove MB from water. The GO aerogel showed aligned porosity in the micrometer range, which allows the creation of an increased number of adsorption sites. A maximum adsorption capacity of 417 ± 0.5 mg/g was achieved based on the Langmuir isotherm model, while the adsorption process was suggested to be spontaneous and endothermic in nature.

Graphene sponge with a hierarchical porous network was developed using laundry detergent bubbles (B-LN) as a soft template by freezing rapidly in liquid N_2 and a subsequent annealing at 300 °C in Ar environment [64]. The formation mechanism involved the self-assembly of GO sheets copying the morphologies of original liquid-filled detergent bubbles. The structure can be tuned by changing the freezing media, adjusting the stirring rate, or adding functional additives. However, pure GO foams or aerogels prepared by a hard ice or a soft soap template method generally show weak and delicate 3D networks, which complicate their handling and limit their commercial implementation. Alternatively, various kinds of chemical additives (e.g., polyvinyl alcohol (PVA), sodium alginate (SA), agar, cellulose, chitosan, or CNTs) were employed to design mechanically stable GO foams via the template-directed methods.

For example, a macroporous composite foam of GO-PVA was synthesized through a process involving aqueous precursor solution freezing, solvent exchange (in ethanol), and drying [65]. The presence of frozen ice crystals created a porous structure of randomly oriented GO sheets surrounded by PVA. The yellow GO-PVA foam was reduced further by heating at 95 °C for 1 h with a tannery mixture of water/ammonia/glucose to obtain a black-colored rGO-PVA foam. GO-PVA and rGO-PVA monolithic foams were tested for the removal of soybean oil, ethanol, or the cationic MB dye from water. GO-PVA showed higher MB adsorption capacity of 571 ± 0.5 mg/g than rGO-PVA (81.0 mg/g) according to the Langmuir isotherm model. The GO-PVA foam followed electrostatic type adsorption interaction for the MB dye, while the rGO-PVA showed a π–π stacking mechanism with adsorbed MB molecules.

A biopolymer agar containing plenty of oxygen functional groups was used as reinforcing agent to prepare a GO-agar aerogel (GO-A) by mixing GO with agar suspension to obtain a hydrogel followed by freeze-drying [66]. The GO-A aerogel had an adsorption capacity of 578 mg/g for cationic MB dye. The adsorption kinetics follows the pseudo-second-order model. The adsorption mechanism was attributed to the electrostatic interactions between MB and the GO-A aerogel. Besides, the small sheets of electrospun cellulose nanofibers (CNFs) were utilized to fabricate a GO/nanofiber (G-CN) aerogel by an ice-templating freeze-dry technique to prevent over-stacking of GO sheets and to enhance connectivity of the cell walls [67]. The G-CN aerogel showed higher and selective adsorption capacity of >800 mg/g toward cationic dyes such as crystal violet (CV), methylene green, MB, or RhB in water solutions. However, the G-CN aerogel showed an insufficient adsorption performance for the removal of the anionic indigo carmine (IC) dye.

Three-dimensional printing is another emerging technique to design macroscopic graphene adsorbents toward dye removal applications. GO suspension is mixed with chemical additives such as a polymer and silica powder to obtain a highly viscous 3D printing ink [73]. Finally, the printing ink flows through a nozzle under shear force and rapid pseudoplastic for dilatant recovery after deposition. For example, Xiao et al. [71] fabricated a macroscopic rGO architecture by 3D printing technique using L-cysteine as reducing agent (rGO-Cys). The 3D adsorbent was tested to remove anionic, nonionic, and cationic dyes with conjugated aromatic structure. The rGO-Cys exhibited a high adsorption capacity of 1005 ± 1 and 1301 ± 1 mg/g for the anionic indigo carmine (IC) or the cationic dye neutral red (NR), respectively. The rGO-Cys also showed excellent adsorption performance even for multiple dye contaminations (>3500 mg/g). The π–π stacking adsorption mechanism was proposed for the high removal efficiency of rGO-Cys.

To summarize, GO sponge or graphene adsorbent foams prepared by template-directed protocols showed a high adsorption capacity to remove organic pollutants from water. However, a microporous network of GO foams with weak connections between the GO nanosheets causes instability and release of nanosheets during pollutant removal applications. While the CVD protocols are usually associated with a complicated method, high production cost and have limited potential for an application at large scale. Using soft or hard templating generally requires high-temperature calcination or complex etching processes.

11.3.2 Template-Free Design of Graphene Sorbents via Sol–Gel Route

The template-free design of macroscopic graphenes is usually achieved by self-concentration of physically or chemically modified GO sheets at the liquid–solid interface (Table 11.2). These designs typically employ chemical additives as cross-linkers (e.g., acid, metal ions, small, and macromolecules) to tune the gelation properties and mechanical stability of graphene hydrogels through alteration in attractive or repulsive forces between the GO in aqueous solution (Fig. 11.3c, d) [82]. Interactive mechanisms to obtain 3D graphene hydrogel can be divided into two basic categories: weak noncovalent bonding and strong covalent bonding [36]. For example, noncovalent interactions include hydrogen bond and π–π stacking between GO and hydroxyl or oxygen-rich polymers such as PVA, polyethylene oxide (PEO), polyvinyl pyrrolidone (PVP), or hydroxyl-propyl-cellulose (HPC) [83]. Ion-induced mechanism using divalent and trivalent metal ions (Ca^{+2}, Mg^{+2}, Cu^{+2}, Pb^{+2}, Cr^{+3}, Fe^{+3}) was also proposed through coordination of metal ions and oxygen functionalities of GO (e.g., hydroxyl and carboxyl groups) [84]. Furthermore, covalent bonding mechanisms include the interactions between hydroxyl and carboxyl groups of GO and nitrogen containing organic molecules (e.g., melamine, polydimethyldiallylammonium chloride (PDDA), polyethyleneimine (PHI), cetyltrimethyl

Table 11.2 Template-free designs of graphene sorbents to remove organic pollutants from water

Adsorbent	Synthesis protocol	Major additives	Pollutant	Q (mg/g)	References
Graphene hydrogel	Sol–gel at 90 °C, 5 min	DNA	Safranin	960	[74]
GO-chitosan foam	Sol–gel and vacuum dried	Chitosan	RB5	1.0	[75]
rGO-MWCNTs aerogel	Sol–gel 12 h, critical dried	MWCNTs and Vc	MB	191	[41]
Graphene-GA aerogel	Sol–gel at 95 °C, 8 h	GA	Calcein	1226	[50]
Graphene-GA aerogel	Sol–gel at 95 °C, 8 h	GA	MO	115	[50]
GO-PEI foam	Sol–gel and freeze dry	PEI	Amaranth	>800	[51]
GO-PDA hydrogel	Sol–gel at 60 °C, 6 h	PDA	RhB	207.06	[52]
GO-PDA hydrogel	Sol–gel at 60 °C, 6 h	PDA	Nitrophenol	324.89	[52]
GO-chitosan hydrogel	Sol–gel and freeze dry	Chitosan	MB and EY	>300	[53]
GO/G/PVA aerogel	Sol–gel for 12 h	PVA & glutaraldehyde	Oil	118	[54]
N and S doped rGO	Sol–gel at 90 °C, 3 h	NH_3H_2O & glutathione	MG	738.1	[62]
GO-gelatin hydrogel	Sol–gel at 95 °C	Gelatin	RhB	280	[26]
Graphene-PVA aerogel	Sol–gel and freeze dry	PVA & LC	Ionic dyes	–	[59]
PEI graphene aerogel	Sol–gel at 90 °C, 24 h	PEI	Amaranth	2043	[60]
PEI graphene aerogel	Sol–gel at 90 °C, 24 h	PEI	MO	331	[61]
Graphene-CA aerogel	Sol–gel at 90 °C	Cysteamine	MB	207.8	[15]
Graphene foam	Hydrothermal 180 °C, 12 h	Additive free	MB	215.35	[55]
Graphene foam	Hydrothermal 180 °C, 12 h	Additive free	CV	462	[56]
Spongy graphene	Hydrothermal 180 °C, 24 h	$NH_3{\cdot}H_2O$/NaOH	Oil	–	[76]
Graphene sponge	Hydrothermal 180 °C, 5 h	Thiourea	RhB	72.5	[38]
Graphene monolith	Hydrothermal 180 °C, 5 h	Urea	MB	50.76	[35]
Graphene monolith	Hydrothermal 95 °C, 5 h	Divalent ions & PVA	Oil	–	[77]

(continued)

Table 11.2 (continued)

Adsorbent	Synthesis protocol	Major additives	Pollutant	Q (mg/g)	References
Graphene monolith	Hydrothermal treatment	Sodium ascorbate	Bisphenol	204	[63]
rGO/CNTs aerogel	Solvothermal 200 °C, 6 h	CNTs and ethanol	MB	81.97	[78]
rGO/CNTs aerogel	Hydrothermal 120 °C, 12 h	CNT, PVP, EDTA	Oil	270	[79]
Graphene/α--FOOH	Hydrothermal 95 °C, 5 h	Ferrous ions	Oil	–	[80]
rGO-MOF monolith	Hydrothermal 180 °C, 24 h	ZIF-67 and $NH_3{\cdot}H_2O$	CV	1714.2	[56]
rGO-CNFs monolith	Hydrothermal 180 °C, 5 h	TEMPO-CNFs & urea	MB	227.27	[81]

ammonium bromide (CTAB), tetramethylammonium chloride (TMAC) [82]) or biomolecules, DNA [58], and hemoglobin [85].

Physically cross-linked 3D hydrogels of GO also demonstrate through an additive-free ultra-sonication technique to convert aqueous dispersions of GO into 3D hydrogels (Fig. 11.3a, b) [16]. In this process, sonication fractures the GO nanosheets into freshly exposed smaller fragment sheet edges, which do not possess the carboxylic functional groups. Those groups are known to stabilize the GO sheets in solution (i.e., reduced repulsion force between GO sheets). It was proposed that this nonchemical induced change in surface chemistry of GO leads to a gelation of a cross-linked 3D hydrogel [16]. As-prepared hydrogels exhibit exceptionally low critical gelation concentrations ranging from ~0.05 to ~0.125 mg/mL and a porous network to host biomolecules as a carrier. However, mechanical properties and the highly hydrophilic nature of a GO hydrogel may cause serious recontamination of purified water after a pollutant removal application. Shi et al. [58] prepared graphene hydrogel by heating GO suspension and DNA at 90 °C for 5 min and tested it for the first time to remove a cationic organic dye, safranin, from water (Fig. 11.3e–g). The authors suggested that the GO/DNA hydrogel can withstand various harsh environments (e.g., acidic, alkaline, salty, or organic solvents) due to multiple non-covalent and electrostatic interactions. The hydrogel exhibited an adsorption capacity of 960 mg/g for safranin dye due to strong electrostatic attraction between the cationic dye and the negatively charged GO/DNA. Furthermore, a large surface area and porous structure were proposed to facilitate the diffusion and to increase adsorption affinity of dye molecules on the hydrogel [58].

A composite foam of chitosan and GO (GO-CS) with 3D-mesostructure was fabricated by heating a sol-mixture at 150 °C under Ar atmosphere [75]. The GO-CS foam was examined for the removal of 1 mg/L reactive black 5 (RB5) from water. The interactions of electrostatic, hydrogen bonding and van der Waals interactions between the rGO-CS and RB5 molecules were proposed to be responsible for the efficient removal of RB5 (97.5%). In another study, a hybrid aerogel of MWCNTs

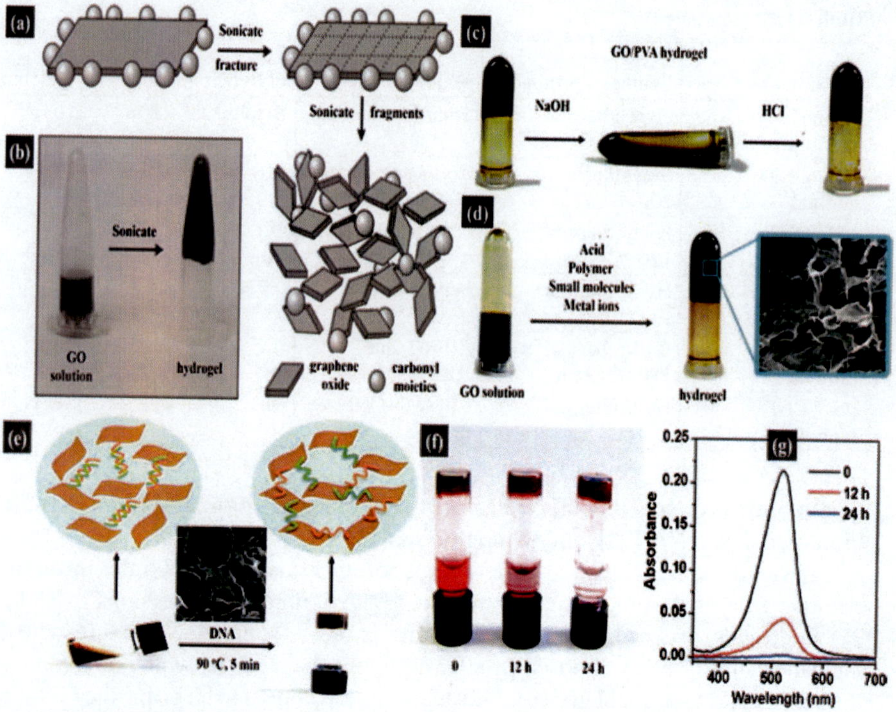

Fig. 11.3 Schematic (**a**) and optical images (**b**) of sonication assisted physically cross-linked GO hydrogel. Reproduced with permission from [16]. Copyright 2012, Elsevier. Sol–gel transition routes of chemically cross-linked graphene hydrogel using PVA (**c**). Acid, polymers, small molecules, and metal ions (**d**). Reproduced with permission from [82]. Copyright 2011, American Chemical Society. Formation of rapid GO hydrogel using single strain DNA, and inset SEM image showing 3D morphology (**e**), time resolved digital (**f**) and UV spectra of dye safranin removal (**g**) using GO/DNA hydrogel. Reproduced with permission from [58]. Copyright 2010, American Chemical Society

and graphene was prepared via a sol–gel protocol using vitamin C (Vc) as a reducing agent [41]. The mechanically robust hydrogel was obtained after gelation for 12 h followed by treatment with supercritical CO_2. The graphene-MWCNTs aerogel possessed a surface area of 435 m^2/g and a pore volume of 2.58 cm^3/g. It demonstrated a decent sorption capacity of 191 mg/g MB, 180.8 mg/g fuchsine, 150.2 mg/g RhB, and 35.8 mg/g acid fuchsine. The electrostatic interactions between oxygen functionalities of the hybrid aerogel and basic dyes was responsible for such excellent performance, while a removal of an acidic dye was associated to π–π interactions. However, the use of supercritical CO_2 drying limits the upscaling of this design to some extent. The use of different natural phenolic acids (e.g., mono-, di-, and tri-hydroxy gallic acids, GA) as reducing agents to synthesize monolithic graphene hydrogel was also demonstrated by a sol–gel method [50].

Polyethylenimine (PEI) is another effective cross-linker to prepare a self-assembled GO-PEI 3D foam by sol–gel and freeze-drying protocols. The as-obtained

GO-PEI foam showed hierarchical morphology and large specific surface area (476 m^2/g) [51]. Two acidic dyes (amaranth and orange G) and one basic dye (RhB) were tested for adsorption performance. Owing to its pore-rich and amine-rich structure, the GO-PEI foam showed excellent adsorption performance for the amaranth dye (800 mg/g). In addition, polydopamine (PDA) could be used to prepare a PDA-modified graphene hydrogel (PDA-GH) via heating of a PDA and GO (2.0 mg/mL) mixture [52]. The spontaneous polymerization of dopamine was proposed for the formation of graphene hydrogel structures. The PDA-GH hydrogel was tested to remove RhB dye from water. The PDA-GH showed a maximum adsorption capacity of 207.06 mg/g for RhB, higher than that of the physically cross-linked graphene hydrogel. The chemical structure of RhB with cationic atoms (N^+) was found favorable for their adsorption on PDA-GH through electrostatic interactions. Macroscopic hydrogels of GO-CS could also be prepared using chitosan (CS) [53], PVA [54], or cysteamine (CA) [15] as cross-linker.

Generally speaking, an application of graphene in hydrogel forms usually requires complex handling processes due to high water absorption, density, and poor mechanical properties. Furthermore, large pore dimensions and complicated recollections using post-filtration in water flow-through systems call for a more rational design with robust stability and tunable porosity for water purification applications.

11.3.3 Template-Free Design of Graphene Sorbents via a Hydrothermal Pathway

A monolithic network produced by a hydrothermal reduction of GO is another alternative to create cross-linking sites between graphene sheets (Table 11.2). Xu et al. [74] developed a bottom-up synthesis to fabricate a physically cross-linked rGO hydrogel by a one-step hydrothermal method. The formation of a monolithic graphene was driven by the π–π stacking interactions between graphene sheets. It was proposed that these physical cross-linking sites were largely dependent on the GO concentration and reaction time (Fig. 11.4). For example, a GO concentration of 2 mg/mL can produce a cylindrical 3D monolith type as compared to 0.5 mg/mL GO suspension due to abundant cross-linking sites. Physically cross-linked graphene foams (GFs) can also be prepared using different lateral dimensions of GO sheets by a hydrothermal treatment at 180 °C for 12 h with a subsequent freeze-drying [55]. The pore dimensions of GFs using different sheet sizes of GO were ranked as: small sheet < medium sheet < large sheet. Consequently, the as-fabricated GFs showed an MB adsorption performance in the order of 215.35 mg/g (small pore size, S-GF) > 175.10 mg/g (medium pore size, M-GF) > 90.70 mg/g (large pore size, L-GF). The MB molecules strongly adhered to the GF walls via electrostatic attraction and π–π interaction.

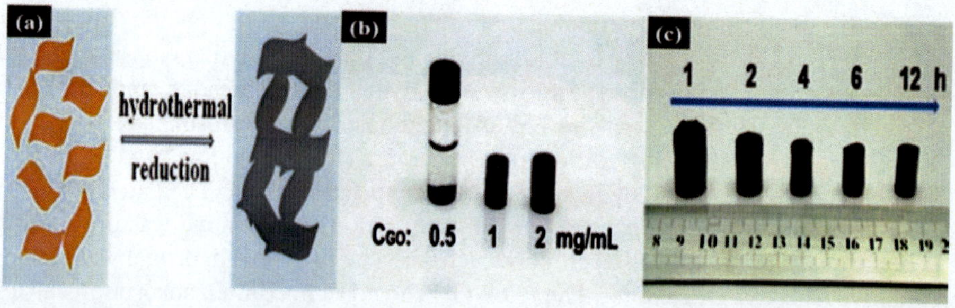

Fig. 11.4 Schematic illustration of a self-assembled graphene hydrogel (SGH) by an additive-free hydrothermal reduction (**a**), optical images of SGH of different concentrations at 180 °C for 12 h (**b**) and the products prepared with 2 mg/mL GO at 180 °C for different heating times. Reproduced with permission from [74]. Copyright 2010, American Chemical Society

Generally, the additive-free macroscopic adsorbent usually leads to poor structural strength due to loss of hydrogen bonding (i.e., physically cross-linked sites) [76] and expansion of ice crystals upon direct heat or freeze-drying [35, 38]. Alternatively, various chemical additives have been tested (e.g., ammonia solution [76], heteroatoms [35, 38], metal ions [77, 80], CNTs [41, 78, 79], polymers [86], or metal organic frameworks (MOF) [9, 56]) as cross-linkers or dopants to improve the mechanical and surface properties of 3D graphene adsorbents. For example, Yang et al. [56] prepared a monolithic rGO hydrogel by hydrothermal treatment of a GO suspension with ammonia solution ($NH_3 \cdot H_2O$, 25–28 w/w %). The removal of hydrophilic groups (–OH, –COOH, etc.) from the GO surface and a π–π stacking mechanism were proposed for the formation of the rGO hydrogel. This hydrogel showed an adsorption capacity of 462 ± 0.2 and 119 ± 0.3 mg/g for CV and MO, respectively.

Other chemical reduction routes for macroscopic rGO include the use of $NaHSO_3$, Na_2S, or hydroquinone during the hydrothermal process [56]. However, the use of expensive or toxic reducing agents not only increases the cost, but also requires prolonged purification time (i.e., dialysis) to remove excess reducing agent and absorbed water during freeze-drying. Alternatively, the use of affordable and nontoxic additives (e.g., urea [35] and thiourea [38], divalent ions [77], vitamin C (Vc) or sodium ascorbate (Vc Na) [63]) as cross-linkers also yields macroscopic graphene. For example, Zhao et al. [38] used thiourea as cross-linker and reducing agent to design graphene sponge (GS). The GSs showed a tunable porous structure and surface properties with enhanced mechanical strength (140 kPa at a strain of 82%). The GS was applied to remove MB, RhB, and MO dye pollutants from water with specific sorption capacities of 184, 72.5, and 11.5 mg/g, respectively. The variability in the adsorption capacity was attributed to surface charge concentration (i.e., electrostatic interaction) and specific surface area of graphene sponges [38]. Regeneration of GS using different alcohols (methanol or ethanol) showed poor desorption performance for MB, but was suitable for desorption of RhB and MO dye. In addition, the use of divalent ions (Ca^{2+}, Ni^{2+}, or Co^{2+}) as cross-linker and/or reducing agent allowed to fabricate a self-standing graphene monoliths by hydrothermal treatment [77]. In this case, the 3D monolith was further immersed into a PVA solution before freeze-drying in order to strengthen the mechanical properties of the monolithic graphene aerogel.

Generally speaking, the mechanical strength and dye removal performance of monolithic graphene preparations prepared through a hydrothermal method are not satisfying due to large pore dimensions in a few hundred to micrometer range sizes. Furthermore, macroscopic 3D graphene designs usually produce reduced GO of hydrophobic nature. The reduction of oxygen-containing functions from GO sheets will result in a decreased mass transfer of water-soluble dyes and an insufficient adsorption capacity. Given that, more efforts have been made to enhance the surface and mechanical properties of existing monolithic graphene adsorbents using chemical dopants.

The chemical doping using other carbonaceous nanomaterial such as single- or multi-walled CNTs is an effective approach to assemble 3D graphene composite adsorbents with superior mechanical, pore morphologies and chemical properties [41]. The hydroxyl groups of oxidized CNTs can act as hydrogen-bonding acceptor associated with oxygen functionalities of GO [78]. Furthermore, CNTs doping also enlarged the interlayer space between GO sheets [41, 79]. For instance, Ai and Jiang [78] fabricated a hybrid monolith of graphene-CNTs (G-CNTs) by a solvothermal method using a mixture of ethanol and water at 200 °C for 6 h. The reduction of oxygen-containing functional groups and self-induced π–π interactions was proposed for the random stacking interactions between graphene and CNTs to form self-assembled monoliths. The G-CNTs monolith showed an adsorption capacity of 82 ± 0.3 mg/g for the cationic MB dye than CNTs (46.2 mg/g) [87], exfoliated GO (17.3 mg/g) [34], and granular activated carbon (21.5 mg/g) [88]. The adsorption kinetics was in accordance with a pseudo-second-order kinetic model, and a Freundlich adsorption isotherm was applicable to the adsorption process.

Further efforts involve decoration with hydrophilic and affordable chemical additives such as metal organic frameworks (MOF) [9, 56], polymers (PVA) [86], or TEMPO-oxidized cellulose nanofibers (CNFs) [81] to improve the mechanical and adsorptive performance of 3D graphene adsorbents. For example, an rGO/PVA composite aerogel was fabricated by a hydrothermal and template technique [86]. The rGO/PVA aerogel had a good adsorption performance toward a wide range of dyes including cationic (i.e., RB, CV, MG, and MB), anionic (i.e., EY, AF, MO, and AB), and nonionic dyes (OR) as well as aromatic drugs (i.e., PV and TC). The π–π interaction was proposed to be the major adsorption mechanism. The removal of those compounds by the rGO-PVA aerogel was independent of dyes charge properties.

Meanwhile, the graphene materials may also serve as support to host other functional materials, which may endow the sorbents additional oxidative capabilities. For example, the loading of nanoscale TiO_2 onto macroscopic rGO has been proposed to serve as dual-functional materials for simultaneous adsorption and photocatalytic oxidation of organic pollutants from solution [13, 89]. Such design has proven to be effective for the degradation of various dyes (e.g., MB, MO, RhB, Congo red (CR), safranin, eosin Y (EY), and Azure B) from water [89]. It is worth mentioning that the potential release of metal ions in the course of the reaction may be of concern from environmental perspective. This needs to be taken into consideration to avoid secondary pollution. Moreover, if combining with other nanoscale photocatalysts, the potential coverage of the incident light by the graphene sheets seems inevitable. This may deteriorate the overall photocatalytic oxidation performance. The integration of graphene with other advanced oxidation processes (AOPs) may be promising, since the dye molecules adsorb the incident light significantly. The integration of other processes like Fenton-like reactions or persulfate-based oxidation processes would avoid the shortcoming caused by light adsorption. These aspects should be considered, if designing other advanced dual-functional systems.

11.4 Outlook and Summary

Currently, several companies are investing in the diverse applications of graphene such as energy storage or wastewater treatment. Owing to the progress in mass production technology, the costs for graphene and derivatives have decreased significantly over the past decade. Though numerous studies suggested graphene-based materials as promising adsorbent materials, their environmental applications still face several challenges that need to be resolved further.

Firstly, the physicochemical properties of functional materials are dictated by their microstructure. GO, for example, has a highly hydrophilic nature and its nanoscale dimension make it inconvenient for practical applications. Certain GO synthesis protocols still include toxic solvents, which need to be minimized. Furthermore, most of the reports were conducted only under laboratory conditions or at bench scale. Their applications were rather limited mainly to the removal of one or a few dyes from synthetic wastewater. Such reported superior performance may fail under real conditions for the treatment of industrial dyeing wastewater with complex compositions. Moreover, the presence of large amounts of salts may significantly deteriorate or impact the overall efficiency.

Secondly, the structural integrity of a macroscopic 3D hydrogel or aerogel prepared by sol–gel route is still insufficient for flow-through systems. Harsh operating conditions (e.g., high temperature, calcination, or chemical etching), expensive chemical additives, or complex formation strategies have been observed in several cases. Development of mechanically robust 3D graphene aerogels, fine-tuning their surface functional groups and compositions, and exploring their sorption capacity and kinetics deserve further investigations. Further systematic studies on the stability, fate, and life cycle analysis of graphene-based materials cannot be overlooked and should constitute the subject of future work. In addition, mechanic and theoretical approaches will provide new insight and understanding on the rational design of advanced graphene-based sorbents for practical applications. This chapter may help to improve the knowledge of a next-generation water remediation technology based on graphene materials and may provide the *know-how* of the rational design of functional sorbents for environmental applications.

Acknowledgments This work was supported by the Iron and Steel Joint Research Fund of National Natural Science Foundation-China BaoWu Steel Group Co., Ltd. (No. U1660107), and Specialized research fund for the Doctoral Program of Senior Education in China (Grant No. 20130075110006).

References

1. F.L. Chen, J.X. Zhu, Y.D. Yang, L.L. Wang, Assessing environmental impact of textile production with water alkalization footprint. Sci. Total Environ. **719**, 137522–137524 (2020)
2. X. Gang, Q. Wang, Y.J. Qian, P. Gao, Y.M. Su, Z.H. Liu, H. Chen, X. Li, J.B. Chen, Simultaneous removal of aniline, antimony and chromium by ZVI coupled with H_2O_2: implication for textile wastewater treatment. J. Hazard. Mater. **368**, 840–848 (2019)
3. G. Asgari, A. Shabanloo, M. Salari, F. Eslami, Sonophotocatalytic treatment of AB113 dye and real textile wastewater using ZnO/persulfate: modeling by response surface methodology and artificial neural network. Environ. Res. **184**, 109367–109384 (2020)
4. C. Li, H.Y. Ma, S. Venkateswaran, B.S. Hsiao, Highly efficient and sustainable carboxylated cellulose filters for removal of cationic dyes/heavy metals ions. Chem. Eng. J. **389**, 123458–123470 (2020)
5. J.M.S. Oliveira, M. Silva, C.G. Issa, J.J. Corbi, M. Damianovic, E. Foresti, Intermittent aeration strategy for azo dye biodegradation: a suitable alternative to conventional biological treatments? J. Hazard. Mater. **385**, 121558–121565 (2020)
6. F. Teng, Z.L. Liu, A. Zhang, M. Li, Photocatalytic performances of Ag_3PO_4 polypods for degradation of dye pollutant under natural indoor weak light irradiation. Environ. Sci. Technol. **49**, 9489–9494 (2015)
7. K. El Hassani, D. Kalnina, M. Turks, B.H. Beakou, A. Anouar, Enhanced degradation of an azo dye by catalytic ozonation over Ni-containing layered double hydroxide nanocatalyst. Sep. Purif. Technol. **210**, 764–774 (2019)
8. A. Hussain, Y. Liu, T. Bin-aftab, D. Li, W. Sand, Engineering reusable sponge of cobalt heterostructures for highly efficient organic pollutants degradation via peroxymonosulfate activation. Chem. Nano. Mat. **5**, 547–557 (2019)
9. W. Cheng, X.L. Lu, M. Kaneda, W. Zhang, R. Bernstein, J. Ma, M. Elimelech, Graphene oxide-functionalized membranes: the importance of nanosheet surface exposure for biofouling resistance. Environ. Sci. Technol. **54**, 517–526 (2020)
10. A.A. Alqadami, M. Naushad, Z.A. Alothman, T. Ahamad, Adsorptive performance of MOF nanocomposite for methylene blue and malachite green dyes: kinetics, isotherm and mechanism. J. Environ. Manag. **233**, 29–36 (2018)
11. H.Q. Zhao, Y. Cheng, H.L. Lv, G.B. Ji, Y.W. Du, A novel hierarchically porous magnetic carbon derived from biomass for strong lightweight microwave absorption. Carbon **142**, 245–253 (2019)
12. J. Ma, Y.R. Sun, M.Z. Zhang, M.X. Yang, X. Gong, F. Yu, J. Zheng, Comparative study of graphene hydrogels and aerogels reveals the important role of buried water in pollutant adsorption. Environ. Sci. Technol. **51**, 12283–12292 (2017)
13. Y. Zhang, W.Q. Cui, W.J. An, L. Liu, Y.H. Liang, Y.F. Zhu, Combination of photoelectrocatalysis and adsorption for removal of bisphenol A over TiO_2-graphene hydrogel with 3D network structure. Appl. Catal. B Environ. **221**, 36–46 (2018)
14. P.P. Brisebois, M. Siaj, Harvesting graphene oxide-years 1859 to 2019: a review of its structure, synthesis, properties and exfoliation. J. Mater. Chem. C **8**, 1517–1547 (2020)

15. C. Chen, X.Y. Zhu, B.L. Chen, Covalently cross-linked graphene oxide aerogel with stable structure for high-efficiency water purification. Chem. Eng. J. **354**, 896–904 (2018)
16. O.C. Compton, Z. An, K.W. Putz, B.J. Hong, B.G. Hauser, L.C. Brinson, S.T. Nguyen, Additive-free hydrogelation of graphene oxide by ultrasonication. Carbon **50**, 3399–3406 (2012)
17. D.R. Dreyer, S. Park, C.W. Bielawski, R.S. Ruoff, The chemistry of graphene oxide. Chem. Soc. Rev. **39**, 228–240 (2010)
18. U. Hofmann, R. Holst, Über die Säurenatur und die methylierung von graphitoxyd. Ber. Dtsch. Chem. Ges. **72**, 754–771 (1939)
19. G. Ruess, Über das graphitoxyhydroxyd (graphitoxyd). Monatsh. Chem. Verw. Teile. Anderer. Wiss. **76**, 381–417 (1947)
20. W. Scholz, H.P. Boehm, Untersuchungen am Graphitoxid VI. Betrachtungen zur struktur des graphitoxids. Z. Anorg. Allg. Chem. **369**, 327–340 (1969)
21. T. Nakajima, A. Mabuchi, R.A. Hagiwara, New structure model of graphite oxide. Carbon **26**, 357–361 (1988)
22. A. Lerf, H. He, T. Riedl, M. Forster, J. Klinowski, ^{13}C and ^{1}H MAS NMR studies of graphite oxide and its chemically modified derivatives. Solid State Ionics **101-103**, 857–862 (1997)
23. A. Lerf, H. He, M. Forster, J. Klinowski, Structure of graphite oxide revisited. J. Phys. Chem. B **102**, 4477–4482 (1998)
24. T. Szabó, Evolution of surface functional groups in a series of progressively oxidized graphite oxides. Chem. Mater. **18**, 2740–2749 (2006)
25. T.T. Ma, P.R. Chang, P.W. Zheng, F. Zhao, X.F. Ma, Fabrication of ultra-light graphene-based gels and their adsorption of methylene blue. Chem. Eng. J. **240**, 595–600 (2014)
26. C.Y. Liu, H.Y. Liu, A.R. Xu, K.Y. Tang, Y. Huang, C. Lu, In situ reduced and assembled three-dimensional graphene aerogel for efficient dye removal. J. Alloys Compd. **714**, 522–529 (2017)
27. C. Li, G. Shi, Three-dimensional graphene architectures. Nanoscale **4**, 5549–5563 (2012)
28. D. Krishnan, F. Kim, J.Y. Luo, R. Cruz-Silva, L.J. Cote, H.D. Jang, J.X. Huang, Energetic graphene oxide: challenges and opportunities. Nano Today **7**, 137–152 (2012)
29. J.J. Shao, W. Lv, Q.H. Yang, Self-assembly of graphene oxide at interfaces. Adv. Mater. **26**, 5586–5612 (2014)
30. S.T. Yang, S. Chen, Y. Chang, A. Cao, Y. Liu, H. Wang, Removal of methylene blue from aqueous solution by graphene oxide. J. Colloid Interface Sci. **359**, 24–29 (2011)
31. Y. Li, Q. Du, T. Liu, X. Peng, J. Wang, J. Sun, Comparative study of methylene blue dye adsorption onto activated carbon, graphene oxide, and carbon nanotubes. Chem. Eng. Res. Des. **91**, 361–368 (2013)
32. H. Yan, X. Tao, Z. Yang, K. Li, H. Yang, A.M. Li, R.S. Cheng, Effects of the oxidation degree of graphene oxide on the adsorption of methylene blue. J. Hazard. Mater. **268**, 191–198 (2014)
33. W. Konicki, M. Aleksandrzak, D. Moszynski, E. Mijowska, Adsorption of anionic azo-dyes from aqueous solutions onto graphene oxide: equilibrium, kinetic and thermodynamic studies. J. Colloid Interface Sci. **496**, 188–200 (2017)
34. G.K. Ramesha, A.V. Kumara, H.B. Muralidhara, S. Sampath, Graphene and graphene oxide as effective adsorbents toward anionic and cationic dyes. J. Colloid Interface Sci. **361**, 270–277 (2011)
35. A. Hussain, T.B. Aftab, G. Qihao, J. Li, D. Li, The adsorptive potential of graphene monolith for organic pollutant treatment: kinetics, isotherms, and thermodynamics of interactions. Desalin. Water Treat. **139**, 156–165 (2019)
36. H.P. Cong, J.F. Chen, S.H. Yu, Graphene-based macroscopic assemblies and architectures: an emerging material system. Chem. Soc. Rev. **43**, 7295–7325 (2014)
37. K.S. Novoselov, A.K. Geim, S.V. Morozov, D. Jiang, Y. Zhang, S.V. Dubonos, I.V. Grigorieva, A.A. Firsov, Electric field effect in atomically thin carbon films. Science **306**, 666–669 (2004)
38. J.P. Zhao, W.C. Ren, H.M. Cheng, Graphene sponge for efficient and repeatable adsorption and desorption of water contaminations. J. Mater. Chem. **22**, 20197–20202 (2012)

39. J.N. Tiwari, K. Mahesh, N.H. Le, K.C. Kemp, R. Timilsina, R.N. Tiwari, K.S. Kim, Reduced graphene oxide-based hydrogels for the efficient capture of dye pollutants from aqueous solutions. Carbon **56**, 173–182 (2013)
40. H. Kim, S.O. Kang, S. Park, H.S. Park, Adsorption isotherms and kinetics of cationic and anionic dyes on three-dimensional reduced graphene oxide macrostructure. J. Ind. Eng. Chem. **21**, 1191–1196 (2015)
41. Z.Y. Sui, Q.H. Meng, X.T. Zhang, R. Ma, B. Cao, Green synthesis of carbon nanotube-graphene hybrid aerogels and their use as versatile agents for water purification. J. Mater. Chem. **22**, 8767–8771 (2012)
42. P. Wang, M. Cao, C. Wang, Y. Ao, J. Qian, Kinetics and thermodynamics of adsorption of methylene blue by a magnetic graphene-carbon nanotube composite. Appl. Surf. Sci. **290**, 116–124 (2014)
43. Y. Yao, S. Miao, S. Liu, L.P. Ma, H. Sun, S. andWang, Synthesis, characterization, and adsorption properties of magnetic Fe_3O_4@graphene nanocomposite. Chem. Eng. J. **184**, 326–332 (2012)
44. W. Gao, M. Majumder, L.B. Alemany, T.N. Narayanan, M.A. Ibarra, B.K. Pradhan, Engineered graphite oxide materials for application in water purification. ACS Appl. Mater. Interfaces **3**, 1821–1826 (2011)
45. M. Zhang, B. Gao, Y. Yao, Y. Xue, M. Inyang, Synthesis, characterization, and environmental implications of graphene-coated biochar. Sci. Total Environ. **435–436**, 567–572 (2012)
46. Y.R. Liu, R. Hu, Z.Q. Zhang, A facile colloidal crystal templating method to produce three-dimensional hierarchical porous graphene-Fe_3O_4 nanocomposite for the removal of dyes from aqueous solution. J. Porous. Mater. **26**, 271–280 (2019)
47. F. Liu, S. Chung, G. Oh, T.S. Seo, Three-dimensional graphene oxide nanostructure for fast and efficient water-soluble dye removal. ACS Appl. Mater. Interfaces **4**, 922–927 (2012)
48. S. Jayanthi, N.K. Eswar, S.A. Singh, K. Chatterjee, G. Madras, A.K. Sood, Macroporous three-dimensional graphene oxide foams for dye adsorption and antibacterial applications. RSC Adv. **6**, 1231–1242 (2016)
49. S. Zamani, N.S. Tabrizi, Removal of methylene blue from water by graphene oxide aerogel: thermodynamic, kinetic, and equilibrium modeling. Res. Chem. Intermed. **41**, 7945–7963 (2015)
50. J.L. Wang, Z.X. Shi, J.C. Fan, Y. Ge, J. Yin, G.X. Hu, Self-assembly of graphene into three-dimensional structures promoted by natural phenolic acids. J. Mater. Chem. **22**, 22459–22466 (2012)
51. Z.Y. Sui, Y. Cui, J.H. Zhu, B.H. Han, Preparation of three-dimensional graphene oxide-polyethylenimine porous materials as dye and gas adsorbents. ACS Appl. Mater. Interfaces **5**, 9172–9179 (2013)
52. H.C. Gao, Y.M. Sun, J.J. Zhou, R. Xu, H.W. Duan, Mussel-Inspired synthesis of polydopamine-functionalized graphene hydrogel as reusable adsorbents for water purification. ACS Appl. Mater. Interfaces **5**, 425–432 (2013)
53. Y.Q. Chen, L.B. Chen, H. Bai, L. Li, Graphene oxide-chitosan composite hydrogels as broad-spectrum adsorbents for water purification. J. Mater. Chem. A **1**, 1992–2001 (2013)
54. S.B. Ye, Y. Liu, J.C. Feng, Low-density, mechanical compressible, water-induced self-recoverable graphene aerogels for water treatment. ACS Appl. Mater. Interfaces **9**, 22456–22464 (2017)
55. W. Deng, Q.L. Fang, X.F. Zhou, H.L. Cao, Z.P. Liu, Hydrothermal self-assembly of graphene foams with controllable pore size. RSC Adv. **6**, 20843–20849 (2016)
56. Q.X. Yang, R. Lu, S.S. Ren, C.T. Chen, Z.J. Chen, X.Y. Yang, Three dimensional reduced graphene oxide/ZIF-67 aerogel: effective removal cationic and anionic dyes from water. Chem. Eng. J. **348**, 202–211 (2018)
57. W.F. Chen, L.F. Yan, In situ self-assembly of mild chemical reduction graphene for three-dimensional architectures. Nanoscale **3**, 3132–3137 (2011)

58. Y.X. Xu, Q.O. Wu, Y.Q. Sun, H. Bai, G.Q. Shi, Three-dimensional self-assembly of graphene oxide and DNA into multifunctional hydrogels. ACS Nano **4**, 7358–7362 (2010)
59. J.L. Xiao, W.Y. Lv, Z. Xie, Y.H. Song, Q. Zheng, L-cysteine-reduced graphene oxide/poly(vinyl alcohol) ultralight aerogel as a broad-spectrum adsorbent for anionic and cationic dyes. J. Mater. Sci. **52**, 5807–5821 (2017)
60. D. Shu, F. Feng, H.L. Han, Z.F. Ma, Prominent adsorption performance of amino-functionalized ultra-light graphene aerogel for methyl orange and amaranth. Chem. Eng. J. **324**, 1–9 (2017)
61. Q. Zhao, X.Y. Zhu, B.L. Chen, Stable graphene oxide/poly(ethyleneimine) 3D aerogel with tunable surface charge for high performance selective removal of ionic dyes from water. Chem. Eng. J. **334**, 1119–1127 (2018)
62. Y.C. Shi, A.J. Wang, X.L. Wu, J.R. Chen, J.J. Feng, Green-assembly of three-dimensional porous graphene hydrogels for efficient removal of organic dyes. J. Colloid Interface Sci. **484**, 254–262 (2016)
63. Z. Fang, Y. Hu, W. Zhang, X. Ruan, Shell-free three-dimensional graphene-based monoliths for the aqueous adsorption of organic pollutants. Chem. Eng. J. **316**, 24–32 (2017)
64. R.J. Zhang, Y.C. Cao, P.X. Li, X.B. Zang, P.Z. Sun, K.L. Wang, M.L. Zhong, J.Q. Wei, D.H. Wu, F.Y. Kang, H.W. Zhu, Three-dimensional porous graphene sponges assembled with the combination of surfactant and freeze-drying. Nano Res. **7**, 1477–1487 (2014)
65. N. Wang, P.R. Chang, P.W. Zheng, X.F. Ma, Graphene-poly(vinyl alcohol) composites: fabrication, adsorption and electrochemical properties. Appl. Surf. Sci. **314**, 815–821 (2014)
66. L. Chen, Y.H. Li, Q.J. Du, Z.H. Wang, Y.Z. Xia, E. Yedinak, J. Lou, L.J. Ci, High performance agar/graphene oxide composite aerogel for methylene blue removal. Carbohydr. Polym. **155**, 345–353 (2017)
67. J.L. Xiao, W.Y. Lv, Y.H. Song, Q. Zheng, Graphene/nanofiber aerogels: performance regulation towards multiple applications in dye adsorption and oil/water separation. Chem. Eng. J. **338**, 202–210 (2018)
68. H. Ha, K. Shanmuganathan, C.J. Ellison, Mechanically stable thermally cross linked poly(acrylic acid)/reduced graphene oxide aerogels. ACS Appl. Mater. Interfaces **7**, 6220–6229 (2015)
69. H.Y. Mi, X. Jing, A.L. Politowicz, E. Chen, H.X. Huang, L.S. Turng, Highly compressible ultra-light anisotropic cellulose/graphene aerogel fabricated by bidirectional freeze drying for selective oil absorption. Carbon **132**, 199–209 (2018)
70. X.C. Dong, J. Chen, Y.W. Ma, J. Wang, M.B. Chan-Park, X.M. Liu, L.H. Wang, W. Huang, P. Chen, Superhydrophobic and superoleophilic hybrid foam of graphene and carbon nanotube for selective removal of oils or organic solvents from the surface of water. Chem. Commun. **48**, 10660–10662 (2012)
71. J.L. Xiao, W.Y. Lv, Z. Xie, Y.Q. Tan, Y.H. Song, Q. Zheng, Environmentally friendly reduced graphene oxide as a broad-spectrum adsorbent for anionic and cationic dyes via pi-pi interactions. J. Mater. Chem. A **4**, 12126–12135 (2016)
72. J.L. Vickery, A.J. Patil, S. Mann, Fabrication of graphene-polymer nanocomposites with higher-order three-dimensional architectures. Adv. Mater. **21**, 2180–2184 (2009)
73. A.E. Jakus, E.B. Secor, A.L. Rutz, S.W. Jordan, M.C. Hersam, R.N. Shah, Three-dimensional printing of high-content graphene scaffolds for electronic and biomedical applications. ACS Nano **9**, 4636–4648 (2015)
74. Y.X. Xu, K.X. Sheng, C. Li, G.Q. Shi, Self-assembled graphene hydrogel via a one-step hydrothermal process. ACS Nano **4**, 4324–4330 (2010)
75. J.S. Cheng, J. Du, W.J. Zhu, Facile synthesis of three-dimensional chitosan-graphene mesostructures for reactive black 5 removal. Carbohydr. Polym. **88**, 61–67 (2012)
76. H. Bi, X. Xie, K. Yin, Y. Zhou, S. Wan, L. He, F. Xu, F. Banhart, L. Sun, R.S. Ruoff, Spongy graphene as a highly efficient and recyclable sorbent for oils and organic solvents. Adv. Funct. Mater. **220**, 4401–4401 (2012)

77. X. Jiang, Y.W. Ma, J.J. Li, Q.L. Fan, W. Huang, Self-assembly of reduced graphene oxide into three-dimensional architecture by divalent ion linkage. J. Phys. Chem. C **114**, 22462–22465 (2010)
78. L.H. Ai, J. Jiang, Removal of methylene blue from aqueous solution with self-assembled cylindrical graphene-carbon nanotube hybrid. Chem. Eng. J. **192**, 156–163 (2012)
79. W. Wan, R. Zhang, W. Li, H. Liu, Y. Lin, L. Li, Y. Zhou, Graphene-carbon nanotube aerogel as an ultra-light, compressible and recyclable highly efficient absorbent for oil and dyes. Environ. Sci. Nano **3**, 107–113 (2016)
80. H.P. Cong, X.C. Ren, P. Wang, S.H. Yu, Macroscopic multifunctional graphene-based hydrogels and aerogels by a metal ion induced self-assembly process. ACS Nano **6**, 2693–2703 (2012)
81. A. Hussain, J. Li, J. Wang, F. Xue, Y. Chen, T.B. Aftab, D. Li, Hybrid monolith of graphene/TEMPO-oxidized cellulose nanofiber as mechanically robust, highly functional, and recyclable adsorbent of methylene blue dye. J. Nanomater. **2018**, 1–12 (2018)
82. H. Bai, C. Li, X.L. Wang, G.Q. Shi, On the gelation of graphene oxide. J. Phys. Chem. C **115**, 5545–5551 (2011)
83. C.C. Huang, H. Bai, C. Li, G.Q. Shi, A graphene oxide/hemoglobin composite hydrogel for enzymatic catalysis in organic solvents. Chem. Commun. **47**, 4962–4964 (2011)
84. H. Bai, C. Li, X.L. Wang, G.Q. Shi, A pH-sensitive graphene oxide composite hydrogel. Chem. Commun. **46**, 2376–2378 (2010)
85. H. Bai, C. Li, G.Q. Shi, Functional composite materials based on chemically converted graphene. Adv. Mater. **23**, 1089–1115 (2011)
86. J.L. Xiao, J.F. Zhang, W.Y. Lv, Y.H. Song, Q. Zheng, Multifunctional graphene/poly(vinyl alcohol) aerogels: in situ hydrothermal preparation and applications in broad-spectrum adsorption for dyes and oils. Carbon **123**, 354–363 (2017)
87. Y.J. Yao, F.F. Xu, M. Chen, Z.X. Xu, Z.W. Zhu, Adsorption behavior of methylene blue on carbon nanotubes. Biores. Technol. **101**, 3040–3046 (2010)
88. J. Yener, T. Kopac, G. Dogu, T. Dogu, Dynamic analysis of sorption of Methylene Blue dye on granular and powdered activated carbon. Chem. Eng. J. **144**, 400–406 (2008)
89. Y. Zhao, Y. Zhang, A. Liu, Z. Wei, S. Liu, Construction of three-dimensional hemin-functionalized graphene hydrogel with high mechanical stability and adsorption capacity for enhancing photodegradation of methylene blue. ACS Appl. Mater. Interfaces **9**, 4006–4014 (2017)

Chapter 12
Carbon Nanomaterials for Air and Water Remediation

Sergio Morales-Torres, Luisa M. Pastrana-Martínez, and Francisco José Maldonado-Hódar

12.1 Generalities of Carbon Materials: Fitting Porosity and Surface Chemistry

The control of the pollution level is currently a social challenge from multiple points of view, including aspects such as the preservation of the environment, the human health, or the quality of life (QOL), but pollution also causes an enormous economic impact associated to the materials degradation, increasing medical/pharmaceutical costs, and loss of workers' productivity by disease or progressively deeper damages associated to the global climate change (e.g., drought, flooding, fires, etc.) [1–6]. Pollution is directly related to human, and thus, the appropriate treatment of liquid or gaseous effluents from domestic, industrial, mining, or agricultural/animal breeding activities is essential for the control of atmospheric and water resources quality [7–10]. Because the nature and concentration of pollutants present in the effluents are directly related with the developed activity, specific in situ treatments could increase the thereof efficacy, or at least, fulfill with the legislation, before the effluent being casted into the public network. Legislation is progressively more restricted in this sense, but it needs the cooperation of all the population sectors. In Europe, the Commission Implementing Decision (EU) 2018/1147 establishes best available techniques (BAT) for wastewater treatment, under Directive 2010/75/EU of 24 November 2010 on industrial emissions (integrated pollution prevention and control).

The original version of this chapter was revised: This chapter was previously published with error in reference. This has now been corrected. The correction to this chapter is available at https://doi.org/10.1007/978-3-030-58934-9_15

S. Morales-Torres · L. M. Pastrana-Martínez (✉) · F. J. Maldonado-Hódar
Carbon Materials Research Group, Department of Inorganic Chemistry, Faculty of Sciences, University of Granada, Granada, Spain
e-mail: lpastrana@ugr.es

M. Piumetti, S. Bensaid (eds.), *Nanostructured Catalysts for Environmental Applications*, https://doi.org/10.1007/978-3-030-58934-9_12

Typical air pollutant can be classified into various groups: particulate matter (PM10 and PM2.5), ozone (O_3), volatile organic compounds (VOCs), and gaseous oxides (NO_x, CO_x, and SO_2). The large part of air pollution is related with the combustion of fossil fuels for electricity production and transport, and thus, as an example, the use of biofuels and biomass has been proposed as an alternative to maintain the carbon cycle and reduce the carbon footprint [11]. Indoor air pollution is produced by paints, solvents, smoke, etc. [12]. Similarly, water effluents contain solids in suspension (turbidity) and soluble compounds, including hazardous metals (e.g., Hg^{+2}, Cd^{+2}, and Pb^{+2}) or fertilizers (e.g., mainly nitrates and phosphates), together with a vast list of organic pollutants (e.g., herbicides, fungicides, drugs, personal care products, hormones, endocrine disruptors, etc.). Special attention should be paid in this case to the emerging and not biodegradable contaminants, which are not removed in the conventional wastewater treatment plants (WWTPs), and thereby, are accumulated in the water supply reservoirs and recirculated until the tap water [13, 14].

In general, the treatment of effluents is based on the performance of adsorbents and catalysts. Both approaches are useful to remove efficiently pollutants, but in the case of the adsorption processes the contaminants are only transferred from the effluent to the solid adsorbent, requiring a posterior treatment and regeneration of the saturated adsorbents. On contrast, the aim of catalytic processes is the transformation of pollutants into nontoxic species. Evidently, both adsorption and catalysis are surface processes. Thus, porous materials with the appropriate surface physicochemical properties are progressively developed in order to favor the performance of adsorbents and catalysts.

Among the porous solids, carbon materials present several advantages such as a low cost and flexibility of production, because they can be prepared from several cheap raw materials and in many different formats (e.g., pellets, grains, powder, cloths, monoliths, films, foams, etc.). Carbons are also stable in acidic or basic media and at very high temperature (in non-oxidant conditions). Nevertheless, their main advantage is related with the ability to tailor their physicochemical properties, both porosity and surface chemistry, for specific requirements using available activation and/or functionalization methodologies.

Physical (e.g., partial gasification in CO_2, air, or steam flow) or chemical activation (e.g., thermal treatments with chemicals such as, KOH, $ZnCl_2$, or H_3PO_4) processes are usually used to develop the porosity and surface area of carbon materials [15–17]. The precursor nature, activating agent, temperature, atmosphere, and soak time determine the final porosity and performance of materials. This fact allows fitting the diffusion of pollutants into the porosity (controlled by macro and mesopores) and the adsorption capacity (determined mainly by the microporosity).

Furthermore, specific interactions with pollutants can be induced by generating specific chemical groups on the carbon surface. Typically, oxygenated surface groups are generated by oxidation (Fig. 12.1), but nitrogen, sulfur, phosphorus, boron, halogens, coordination complex, etc. can be also anchored on the surface by different treatments [18–24]. Carbons can also act as catalysts in many different reactions, but they are mainly used as supports of metallic active phases [18, 25, 26]. The developed porosity and adjusted chemical properties of carbon supports allow to obtain highly dispersive active phase, achieving small nanoparticles. When carbon materials are

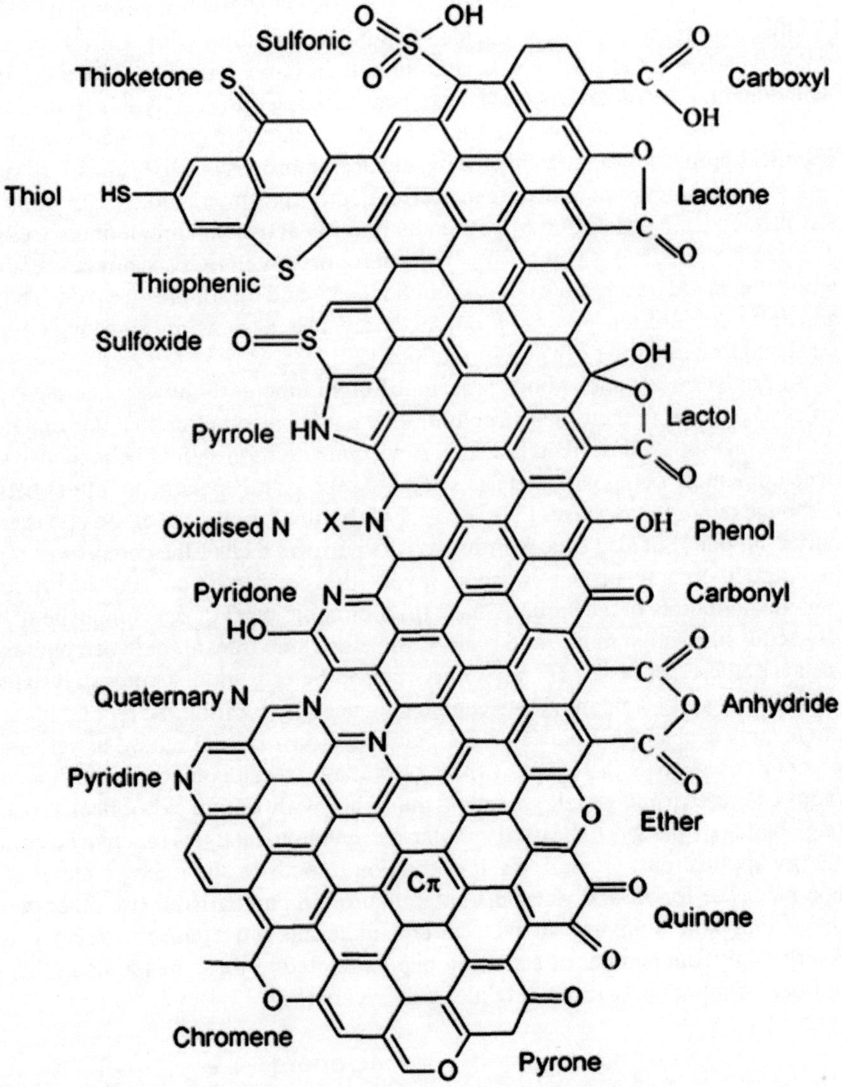

Fig. 12.1 Schematic representation of oxygen, nitrogen, and sulfur-containing functional groups on the carbon surface. (Reprinted with permission from Ref. [22]. Copyright 2018, Elsevier)

used as valuable noble metals, an additional advantage is related with the ability to recover the valuable noble metal used as the active phase once the catalyst has been deactivated, by simply burning the carbon support, in contrast with other complicated procedures required when classical inorganic supports such as alumina (Al_2O_3), silica (SiO_2), zeolites, or metal-organic frameworks (MOFs) are used.

Besides the classical and cheap activated carbons (ACs), there is a wide range of carbon nanomaterials, developed using more specific and advanced synthesis procedures. The nanostructure of these carbons changes from unidimensional (1D) for carbon nanofibers (CNFs) and nanotubes (CNTs) to two-dimensional (2D) for graphene or graphitic carbon nitride (g-C_3N_4) and tridimensional (3D) for foams, ordered carbons, and carbon gels, although other nanostructures of carbon, such as nanodiamonds or fullerenes (0D), have been also used for catalytic applications.

The composition and morphology of carbon materials determine their physicochemical properties (e.g., conductivity, electronic properties, hybridization, surface groups, aromaticity, porosity), and consequently, the interactions with pollutants. For instance, the adsorption of pollutants into 1D activated carbon fibers (ACFs) is compared with 3D ACs (Fig. 12.2). The microporosity in ACFs is directly available, while it is mainly accessible only through macro- and mesopores for ACs. Thus, the adsorption/desorption processes on ACFs are faster, showing high mass transfer rates and low internal diffusion limitations [27].

Although these nanocarbons present different inherently physicochemical properties associated to their nanostructure, they can be also tailored and/or functionalized, as previously described [28, 29]. A special attention should be paid also to the development of composite materials for the sake of the synergistic effect between carbon and inorganic phases [21, 28, 30, 31]. In this context, the carbon phase is not only a support, but also clearly improves the performance of the composites regarding pure inorganic phases. Carbon favors the concentration and adsorption of organic pollutants in the carbon phase (hydrophobic part) of the composite through a specific adsorption mode, and mainly, the electronic transfer between phases. For instance, these processes are very important in heterogenous photocatalysis, since the efficiency of semiconductors can be enhanced by limiting the recombination of photoexcited electron–hole (e^- / h^+) pairs, carbon material acting as an electron acceptor, as showed in Fig. 12.3 [32]. Nowadays, these composites have an enormous interest in the development of materials with optimized optical properties (e.g., bandgap energy) to be used as solar-driven photocatalysts for environment and energy applications. Overall, the introduction intends to show the potential of carbon materials for air and water decontamination in comparison with other conventional inorganic supports. In fact, carbon materials are a current research topic. Accordingly, the number of scientific papers related to these materials in this field has been continuously increasing along the years (Fig. 12.4).

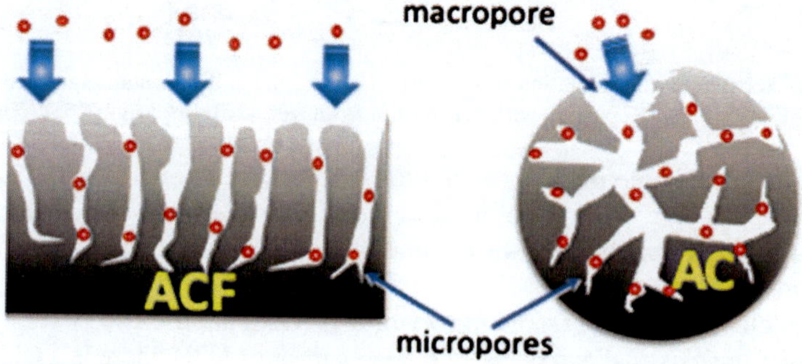

Fig. 12.2 Scheme of the adsorption process of pollutants on ACFs and ACs

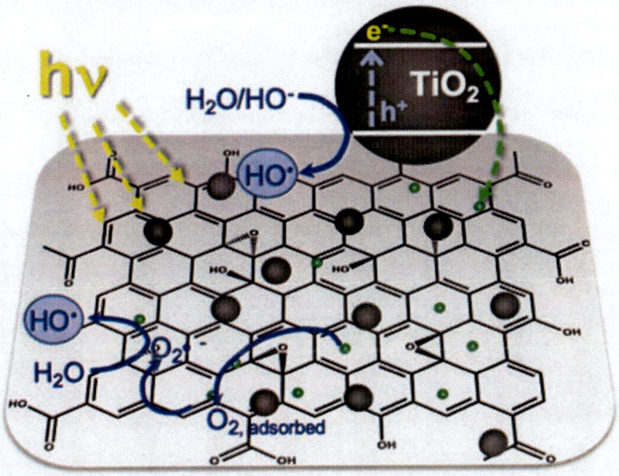

Fig. 12.3 Electron transfer mechanism through a graphene-TiO_2 composite under irradiation. (Adapted with permission from Ref. [32]. Copyright 2010, Springer)

Based on the exposed, the aim of this book chapter is to provide an overview on the properties of carbon materials and their application for air and water treatment and shows the interdisciplinary character of a field of interest for many scientists working in chemistry, environmental engineering, material science, nanotechnology, and so on.

12.2 Carbon Nanomaterials for Air Purification

As previously commented in Sect. 12.1, the purification of gaseous effluents is carried out by adsorption (recuperative techniques) or catalytic processes (destructive techniques). In this section, the performance of carbon nanomaterials in the removal of pollutants from air streams is reviewed.

12.2.1 Removal of Volatile Organic Compounds (VOCs)

The concept of VOCs is given to all of chemical organic compounds with low boiling point (e.g., alkane, alkene, aromatics, ketones, etc.). Nevertheless, World Health Organization (WHO) or U.S. Environmental Protection Agency (EPA) considers VOCs to those compounds with a wide range of boiling point (between 50–400 °C), semivolatile organic compounds, and even, the particulate organic matters (POMs) present in the atmosphere [33, 34]. VOCs are dangerous for the health because they are toxic producing different types of diseases and mutagenic and carcinogenic effects. From an environmental point of view, they are involved in atmospheric

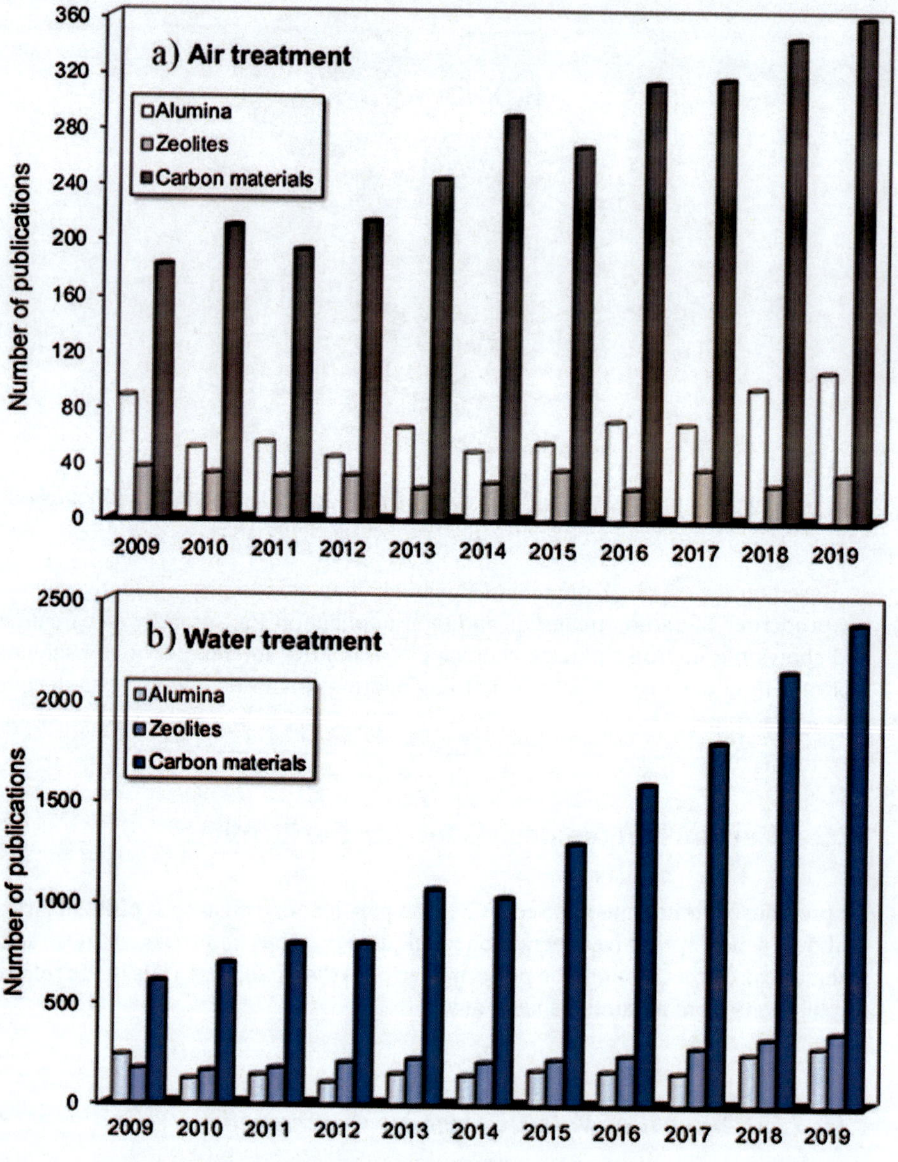

Fig. 12.4 Number of publication entries referring to keywords: "air treatment" or "water treatment" and "carbon materials" (included "graphene" and "nanotubes") from 2009 to 2019 (search on Scopus, April 2020). A comparison with other inorganic materials, "alumina" and "zeolites," is given

photochemical reactions that generate radicals responsible for the stratospheric ozone depletion and formation of the ozone pollutant at the tropospheric level. In addition, VOCs have a powerful greenhouse effect contributing to the global warming.

The anthropogenic VOC emissions are mainly associated to the production, storage, and use of fossil fuels, and each industrial activity (e.g., pharmaceutical, agro-food, textile, electronic, etc.) generates specific VOC emissions. Biogenic VOCs are also emitted even at larger levels than those produced by human activities [35] and this problem is progressively increasing because of the methane emissions from the permafrost associated with the decomposition of the methane hydrates [36] caused by the global warming. It is noteworthy that CH_4 traps up to 100 times more heat in the atmosphere than CO_2 within a period of 5 year or 72 times more within a period of 20 years. Nevertheless, these are low values compared with the global warming potential (WGP) of most VOCs. Among them, CF_4 presents a lifetime in the atmosphere of 50,000 years and the WGP value of 1200 times greater than CO_2 for a period of 500 years.

VOCs are very valuable materials in many industrial processes, and thus, different technologies based on adsorption, absorption, condensation, or membrane separation have been developed to simultaneously remove and recover them from the streams. Among these technologies, adsorption is probably the most extended technology proving a good effectiveness, simplicity, and low energy consumption, as well as low cost. Different types of porous materials can be used as VOC adsorbents, such as carbon materials, zeolites, MOFs, clays, or silica gels, although carbon materials possess the ability to tailor their shape, porosity, and surface chemistry, as previously commented.

ACs are cheap materials produced from a large variety of raw precursors and are considered as a versatile adsorbent with high VOC adsorption capacity based on a developed porosity and elevated surface area. In spite of the fact that the preparation and application of ACs as adsorbents for the remediation of soil, water, and air are widely studied, an important interest is still devoted. Recently, Yang et al. [37] prepared ACs from wood, coals, and coconut shells; ACs obtained from wood present the maximal adsorption capacity of toluene (184 mg g^{-1}) at 200 ppm in N_2, due to its largest surface area and total pore volume. The adsorption capacity is also influenced by the interactions between adsorbent-pollutants. In this context, Yu et al. [38] pointed out that the oxidation of ACs with nitric acid favored the acetone adsorption up to 433.9 mg g^{-1} at 500 ppm in N_2 and 27 °C, and this effect was related to the interaction between carboxylic acid groups and acetone. Oxygenated functional groups (OFG) can act as active sites with a strong adsorption affinity on the polar acetone molecules. Analogously, Vivo et al. [39] employed tailored ACs for the adsorption of vapors from blended gasoline. ACs were developed from olive stone by carbonization and chemical activation, and subsequent oxidation with ammonium persulfate to anchor OFG on AC surface. Then, these groups were selectively removed by thermal treatments. The adsorption of n-octane and ethanol was studied under dynamic conditions in dry air, obtaining the corresponding breakthrough curves, which provided information about the mass transfer zone in each adsorption bed. While n-octane adsorption is mainly determined by the porosity of ACs, ethanol adsorption is related to the surface chemistry, and in particular, is enhanced by the presence of carboxylic acid groups. The importance of this type of OFG is even more marked when wet air flow is used [40], because it determines the strength of interactions with the adsorbents. The competitive adsorption rate for n-octane was significantly reduced when water or ethanol was present, while adsorption capacity was almost equivalent, more specifically when more oxidized samples were tested. In addition, it was demonstrated that polar (ethanol) and non-polar VOCs (n-octane) were adsorbed in different adsorption sites under competitive adsorption experiments.

As previously described, different types of carbon nanomaterials are continuously being developed and optimized for specific applications. For instance, ACFs are expensive 1D materials due to the high cost of raw materials (e.g., viscose, polyacrylonitrile) and synthesis procedure through electrospinning and thermal treatment. The structure of ACFs, mainly their characteristic direct microporosity, allows fast adsorption kinetics and high mass transfer rates. They contain low oxygen content and the adsorption of nonpolar VOCs is favored. Liu et al. [41] reported a high performance in the capture of isobutane at different relative humidities. Doping with oxygen-containing species (metal oxides) allows the adsorption of polar species. The impregnation of ACFs with basic oxides (i.e., La_2O_3) increased a 20% acetaldehyde adsorption compared to the neat ACFs [42].

CNTs are other 1D carbon nanomaterials used in the VOC adsorption. The main difference with regard to ACFs is not only the presence or absence of the central hollow, but also the different chemical bonds between carbons. In the CNT case, carbons are linked between them by sp^2 hybrids, in a similar configuration than in graphene or graphite. CNTs can be synthetized by different methods such as laser ablation, arc discharge, or chemical vapor deposition (CVD). They can be arranged in single-walled carbon nanotubes (SWCNTs) with only a graphene sheet rolled up around the central hole, or multi-walled carbon nanotubes (MWCNTs), where at least two graphene sheets are rolled. CNTs present good mechanical properties, hydrophobicity and conductivity, and tend to form aggregates (bundles), which reduce the surface area but generate important mesopore volumes that can be used as pollutant sorption sites. In this context, the adsorption process can take place at (1) the external, and/or (2) internal surface of CNTs, (3) the interstitial pathways between individual nanotube sheet, and (4) the external groove sites [43] (Fig. 12.5).

In general, the spaces between graphene layers in CNTs are too small, and VOCs are preferably adsorbed on their external or internal surface. This configuration of the CNT surface favors the strong adsorption of aromatic rings. In this context, adsorption capacity of aromatic compounds on CNTs is negatively correlated with their melting, and oxygen-containing surface groups and surface area ratios of mesopores to total pores of CNTs, but positively correlated with total surface area of CNTs [45]. The adsorption of nonpolar VOCs in CNTs implies only physisorption processes, while polar VOCs like acetone imply chemical interactions [46]. Analogously, while physisorption is predominant in the phenol adsorption, chemisorption is enhanced by the presence of chlorine in 2-chlorophenol [47] . Chemical

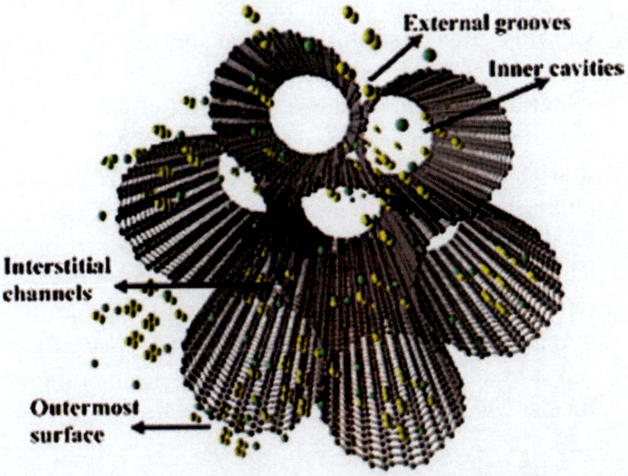

Fig. 12.5 Four adsorption sites of pollutants in SWCNTs. (Reprinted with permission from Ref. [44]. Copyright 2017, Elsevier)

interactions with polar VOCs are favored by morphological defects and amorphous carbon present in CNTs. The CNT functionalization also improves the polar VOC adsorption. Thus, carboxylation and nitration treatments led to the generation of polar functional groups on the CNT surface, ethanol adsorption being increased by 300% [48]. The application of CNTs in adsorption processes from gas streams is limited by the tendency of CNTs to bundling, which hinders the permeation and diffusion of gases. In fact, the use of CNTs for VOC adsorption is not common; they can be supported on SiO_2 to overcome the bundling limitation [47].

Graphene is a 2D carbon nanomaterial, in which layers present also the sp^2 hybridization between carbons in hexagonal arrangement, providing theoretical exceptional high surface area, conductivity, mechanical strength, and possibilities of functionalization. The theoretical specific surface area of completely exfoliated graphene is assumed to be 2620 $m^2\ g^{-1}$. However, the surface area of graphene calculated by several experimental techniques is rather less. Graphene can be produced by different methods, such as nanotube slicing, CVD, and by exfoliation of graphite, among others. Of course, the method used in the fabrication provides different properties and/or defects, which can be or not desirable for a specific application.

Pristine graphene prefers to interact with hydrophobic pollutants, while other graphene derivatives such as graphene oxide (GO), which is obtained by Hummer's method, strongly interact with polar molecules due to the large amount of oxygenated surface groups, in particular hydroxyl/epoxide and carboxyl acid groups. GO can be progressively reduced by thermal and/or chemical treatments to yield reduced graphene oxide (rGO), which possesses still some oxygenated surface groups and structural defects, and thereby, should not be confused with pristine graphene. The adsorption of aromatic compounds is also favored in rGO as consequence of its hydrophobicity and the π-π interactions. However, Yu et al. [49] reported that the surface areas for rGO and GO were 293 and 236 $m^2\ g^{-1}$ and the benzene adsorption capacities are 276.4 and 216.2 mg g^{-1}, respectively. Thus, the highest benzene adsorption capacity was only related with the largest surface area of rGO. Thermal regeneration of spent samples did not influence significantly the toluene adsorption capacity after four consecutive adsorption-desorption cycles.

An interesting approach is to combine graphene derivatives with other carbon materials to yield composites with high adsorption performance. Graphene-based materials, as previously described for CNTs, undergo also significant agglomeration restricting their performance. Thus, Wu et al. [50] prepared composites with amino-functionalized graphene and CNTs (Fig. 12.6) in order to favor the exposure of the graphene layers to the functionalization and increase the formaldehyde adsorption. The combination of GO with MOFs (e.g., Cu-BTC and ZIF-8) was also reported to improve the removal capacity of toluene [51]. The breakthrough of graphene-based materials in the development of filters for air treatments is still hindered by (1) the current high cost of production; (2) toxicity due to the small size sheets, which can be inhaled and/or absorbed in the body with influence on human health; and (3) possible adverse effects on the environment [52].

The dispersion of the adsorbent is clearly avoided using monolithic or particulate materials with appropriate size of grain or pellet. In this sense, the sol-gel technology allows to obtain carbon materials with fitted characteristics and macrostructure. For instance, carbon gels can be prepared as coatings, powders, monoliths, or grains [53]. This carbon material is typically prepared from mixture of resorcinol and formaldehyde, and during the polymerization step, nearly spherical particles, known as primary particles, are obtained forming a sol and overlapping between them along the gelation, leaving between them significant macro-mesopore volumes. Microporosity is formed during carbonization by the release of pyrolysis gases and can be fitted by activation methods [54]. Therefore, the final porosity of the materials depends on the experimental conditions in each step of synthesis. These materials prove to be excellent VOC adsorbent, with a fast adsorption-desorption rate allowing the complete regeneration of spent samples [55]. In addition, the porosity of carbon gels can be tailored to achieve an excellent performance as molecular sieves for the concentration, separation, and purification of VOCs [16]. Carbon gels, namely xerogels, with appropriate porosity, mechanical resistance, and adsorptive performance have been prepared from different precursors (even wastes), avoiding the use of specific drying procedures (supercritical drying) and long synthesis procedures, thus saving cost of materials, facilities, or energy, allowing an easier scalability of the process to the industrial level and cheaper carbon gels [54, 56–58].

In general, carbon materials can be also combined with other inorganic adsorbents (e.g., MOFs, silicas, alumina, zeolites, titania, etc.) to obtain carbon-metal

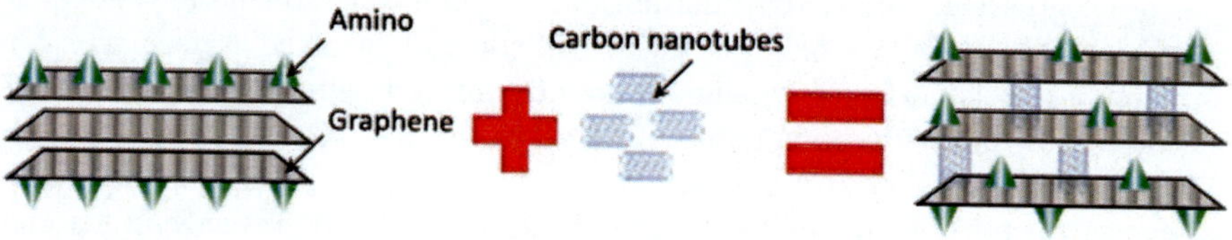

Fig. 12.6 Illustration on composites prepared with amino-functionalized graphene and CNTs for enhanced formaldehyde removal. (Adapted from Ref. [50])

oxide composites in order to satisfy specific requirements [59, 60]. The integration of MOFs into carbon materials enhances the composite performance regarding the pure MOF phase in terms of electrical conductivity or thermal stability, but also new functionalities and types of porosity are formed [61]. Carbon phases work as fillers, building blocks, templates, or supports in the composites and facilitate MOF crystallization, influencing on the coordination chemistry, reacting with metal-containing nodes, increasing the dispersive forces, and providing extra adsorptive sites.

Metal/metal oxide doping is used to enhance the selective adsorption of porous adsorbents by interaction between the reduced metal particles and some specific VOC molecules. The principle is to impregnate porous materials with a high valence metal compound. For instance, the adsorption of formaldehyde on ACs improved from 59.3 mg g^{-1} to 119.3 mg g^{-1} after impregnation with $AgNO_3$ and reduction in N_2 atmosphere, in spite of the fact that the surface area decreased by the pore blockage with Ag-nanoparticles [62].

12.2.2 Removal of CO_2, SO_x, and NO_x

It is well-known that the strong effect on the climatic change is produced by the progressive increase of the CO_2 levels in the atmosphere. In spite of the small WGP of CO_2 regarding other atmospheric compounds, its concentration is higher inducing the major effects. In this context, many adsorbents were developed for CO_2 capture. Different types of carbon materials with large volume of ultramicropores (i.e., $d_{pore} < 0.7$ nm) were doped with different heteroatoms (e.g., N, O, S, P), metals (typically alkaline-metals), or amine groups [63–65] for this propose.

The release of CO_2, SO_x, and NO_x is provoking not only the greenhouse effect, but also other atmospheric processes, such as the formation of acid rain or the strongly reactive photochemical smog. Izquierdo et al. [66] prepared cheap materials by carbonization and/or activation of lignite to be used in a combined two-step process, where first the SO_2 was removed and retained in the carbon material as H_2SO_4, and then, NO is reduced by ammonia in a second step. The efficiency of the first process determines the second one, the NO conversion being independent of the SO_2 concentration below a SO_2 concentration of 200 ppm, but at higher concentrations the previously adsorbed H_2SO_4 significantly reduced the concentration of NH_3 forming $(NH_4)_2SO_4$ and avoiding the NO reduction. The performance of cokes used strongly depends on the chemical characteristics and distribution of surface groups. Semicokes have a better performance than activated cokes due to the large amount of basic and lactone groups [67], as well can be regenerated at low temperatures.

Carbon materials can be also modified by impregnation with metal oxides to improve the performance of SO_2 and NO_x removal. The performance of the metallic phase is also dependent on the surface chemistry of the supports. Thus, the incorporation of oxygenated surface groups by previous oxidation of ACs with nitric acid before impregnation with CuO induced to a larger dispersion of the metallic phase and to the highest capacity of NO adsorption. This oxidation treatment also favored the catalytic reduction of NO, because the NO-adsorption may be the rate limiting step [68]. Bashkova et al. [69] studied the effect of doping with Ag-nanoparticles, a wood-based AC on the retention of NO_x at room temperature. The capacity to retain NO_2 and NO increases with the Ag-content until the plateau is reached, then the formation of large particles of Ag_2O occurred. Therefore, the carbon surface is also active for the NO_2 retention.

The synthesis method and the thermal treatments determine the interactions of the metal phases with the carbon supports, which can induce different oxidation states of the metallic phase [70], as observed in carbon xerogels-vanadium oxide composites. Similarly, the performance of metal-doped carbon aerogels in the reduction of NO depends on the reducibility of the active phase [71]. Although the carbon phase is active for the NO reduction, as denoted by using an undoped carbon aerogel, the conversion increased and reaction started at lower temperature in the presence of transition metals, Cu-doped carbon aerogel achieving higher NO conversion and higher N_2 selectivity than the Co-doped catalysts. The metals participate on both chemisorption/dissociation of NO and the transfer of species to the carbon surface, undergoing consecutive oxidation–reduction cycles by NO and carbon surface species, respectively. Below 150 °C, only NO-adsorption occurs, but experiments of temperature-programmed surface reaction (TPSR) carried out in absence of additional reductant agent showed that NO decay occurs together with the evolution of $CO + CO_2$ and N_2 by reaction with the carbon surface and increasing temperature.

The interactions of SO_2 and NO_x with the carbon surfaces have been studied by different techniques, but mainly by FTIR and temperature-programmed desorption (TPD). In this context, several works were published by Zawadzki et al. [72]. The formation of different intermediate species, including C– NO_2, C–ONO, C–NCO, and anhydride structures, were pointed out by the formation of specific vibration bands, which thereafter evolve as carbon oxides ($CO + CO_2$). Thus, the carbon phase is not only a catalyst of the NO_x reaction, but also takes part in the reaction. They also showed that the SO_2 retards the carbon combustion by oxygen, while the oxidation of SO_2 to sulfate is favored by the presence of $NO + O_2$.

12.2.3 *Catalytic Combustion of Volatile Organic Compounds (VOCs)*

The catalytic oxidation of VOCs to CO_2 is an interesting alternative for the removal of VOCs from air streams, namely when the recovery of these compounds is not economically suitable. The use of active combustion catalysts is mandatory due to the low concentration of VOCs in the treated effluents and the need to decrease the temperature of conventional incinerators, saving energy and avoiding the formation of intermediate organic compounds, which can be more dangerous than the original

VOCs treated, dioxins or NO_x [73, 74]. The advances in VOC combustion were recently reviewed [74], and probably because carbon materials are evidently combustible, the scientific community paid low attention to the use of carbon materials in this kind of reactions, despite that, as commented, they clearly improve the performance of analogous catalysts prepared on other inorganic supports. Carbon materials, in particular those with a large mesoporosity, such as xerogels, CNTs, CNFs, among others, allow the achievement of large dispersions of the phase active when it is supported on these supports. In fact, supported catalysts are usually more active when they present large metal dispersions throughout the carbon support, as commented beyond. Furthermore, the carbon surface possesses a reducing nature and provides anchoring sites for the metal, stabilizing and avoiding the negative effects on the metal sintering, as well as preserving its oxidation state during the reaction. The surface chemistry of carbon supports can also play a role on the interaction with the active phase, and consequently, the performance in reaction. Overall, the combination of hydrophobic supports and noble metals is considered an excellent option to develop combustion catalysts [75] with improved activity, able to work selectively at low temperatures, saving energy, avoiding the formation of toxic by-products, and evidently, preserving the integrity of carbon-based catalysts.

Metal oxides are usually used as combustion catalysts [76], but they present lower activity than noble metals, and for that, Pt and Pd-catalysts are the most studied metals. The selection of the support is a crucial key taking into account that water chemisorption from the humidity of the air feed and/or generated as a product of the combustion can block effectively the active sites of the catalysts. The benzene, toluene, and xylene (BTX) combustion were studied using Pd and Pt-catalysts supported on cordierite monoliths coated with carbon nanofibers or Al_2O_3 (ACNF or A supports, respectively, Fig. 12.7) [77]. The conversion values obtained with Pt-catalysts were always better than those corresponding to analogous Pd-catalysts and the catalysts supported on CNF-coated monoliths were the most active, independent of the metal catalyst (Pd or Pt) or the type of the tested aromatic compound (BTX). Catalysts were active at low temperature obtaining in all cases the total conversion of BTX selectively to CO_2.

The influence of the porosity and surface chemistry of carbon supports on the activity of combustion catalysts was also previously studied [45, 46]. Pérez-Cadenas

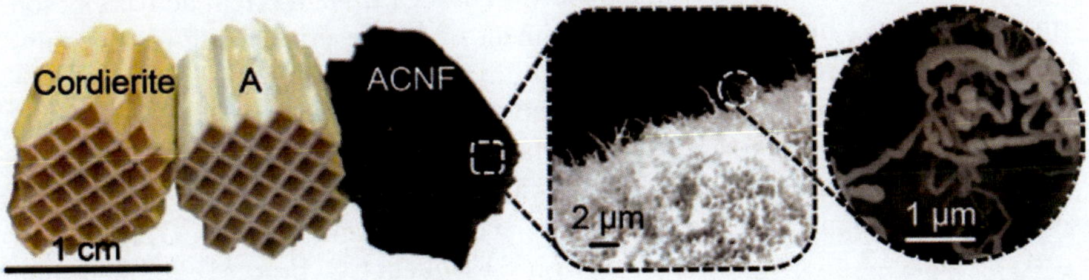

Fig. 12.7 Images of both γ-Al_2O_3 and CNF coatings on a cordierite monolith. Details of CNFs were also analyzed by scattering electron microscopy (SEM). (Reprinted with permission from Ref. [75]. Copyright 2015, MDPI)

et al. [78] studied the influence of the Pd distribution on different porosity range over the xylene combustion, using three different carbon-coated or carbon/ceramic monoliths. The importance of macro and mainly mesopores favoring the contact of pollutant molecules with the Pd-active phase was pointed out. Analogously, Morales-Torres et al. [23] prepared a series of Pt-catalysts supported on carbon aerogels with monomodal porosity ranging from narrow mesopores to large macropores. The influence of surface chemistry was also analyzed after oxidizing these supports with hydrogen peroxide (H_2O_2) or ammonium persulfate ($(NH_4)_2S_2O_8$). The benzene combustion using this Pt-catalysts took place at low temperatures (between 150–180 °C) and selectively to CO_2, the activity being controlled mainly by the Pt-particle size and in concordance with the sensitive character of the combustion reaction. The activity was favored using mesoporous materials regarding the macroporous ones. After oxidation, carboxylic acid groups acted as Pt-anchoring sites, but they favored the concentration and sintering of the Pt-nanoparticles during thermal pretreatments due to their low thermal stability, especially in the case of macroporous supports with the lowest external surface area. Nevertheless, the beneficial effect of the oxidation treatments on carbon supports was pointed using Co_3O_4/CNTs for the combustion of toluene [79]. Thus, the surface defect structures of CNT increased the proportion of the adsorbed oxygen species to the surface lattice oxygen ones, obtaining the complete conversion of toluene at lower temperature and increasing the selectivity to CO_2 with higher COOH amounts of CNTs. In cited work, the highest performance of the carbon supports was also reported, Co_3O_4/CNT catalyst being much better than when supported on zeolites or silica (i.e., Co_3O_4/Beta, Co_3O_4/ZSM-5, or Co_3O_4/SBA-15), even with similar performance to materials impregnated with noble metals (i.e., Pd/Beta and Pd/SBA-15).

The influence of Pt-particle size supported on carbon aerogels over the xylene combustion was studied by varying the pretreatment conditions in He or H_2 atmosphere at different temperatures [80]. In general, H_2-pretreatment produces larger Pt-nanoparticles than the He ones at the same temperature, because in last case the reduction of the precursor salts gives rise to a gradual decomposition along two intermediates. The Pt-particle size ranged between 2.3 and 9.5 nm and the activity of the catalysts progressively increased in this trend. The lowest xylene conversion was obtained for catalysts with smaller Pt particle sizes due to the stronger Pt–O bonds formed during xylene combustion.

Alternatively to the classical combustion of VOCs via thermal activation, today there is a strong interest in the development of this reaction using photocatalysts, especially when solar light can be used directly. Different strategies of synthesis and/or functionalization have been already proposed. For instance, Liu et al. [81] obtained C-N/TiO_2 nanocomposites, where N-doped mesoporous carbon (C-N)-encapsulated anatase-rutile phase junction was first obtained by hydrothermal synthesis, then carbonized in Ar atmosphere, and finally, annealed in CO_2 for different periods (Fig. 12.8). Nanocomposites with subsequent CO_2 process exhibited significantly improved photocatalytic activity and mineralization efficiencies in the styrene degradation using visible light irradiation.

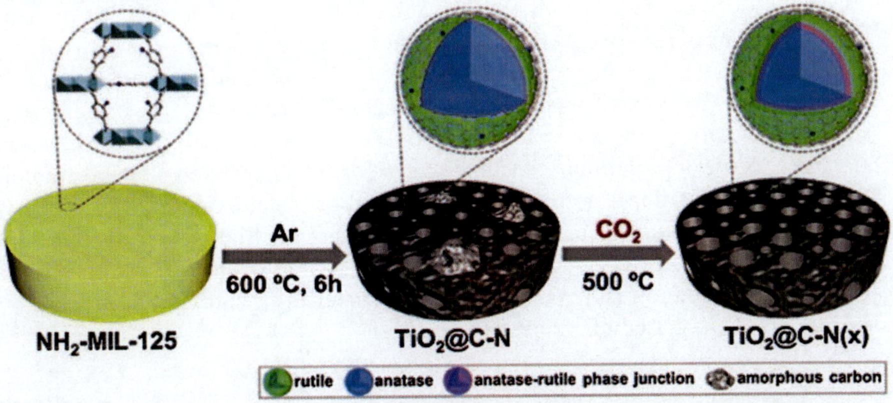

Fig. 12.8 Illustration of the preparation strategy of C-N/TiO_2 nanocomposites. (Reprinted with permission from Ref. [81]. Copyright 2019, Elsevier)

As final remark, there is also a strong interest in the application of carbon nanohybrids structures for sensing, because carbon nanostructures, in particular CNTs, graphene [82], and carbon aerogels [83], are extremely sensitive towards tiny changes in the surrounding gas atmosphere; in other words, their conductance (or resistance) varies greatly with the adsorption–desorption of VOC or NO_x gas molecules [82]. The activated carbon aerogel/polybutadiene composite displayed better response to toluene and n-hexane than that obtained for the graphite/polybutadiene composite [83].

12.3 Carbon Nanomaterials in Water Remediation and Disinfection

Carbon nanomaterials can be used also as adsorbents for pollutants in aqueous solutions due to their inherent properties, namely, high surface area, sorption capacity, and selectivity; however, the removal and/or detoxification of organic pollutants can be achieved with the application of advanced oxidation processes (AOPs). The use of these technologies has shown a great advantage in the treatment of water and wastewater treatment over the last four decades. AOPs are based on the production of highly reactive oxygen species, mainly hydroxyl radicals ($^{\bullet}OH$) using oxidants like O_2, H_2O_2, O_3, ultraviolet (UV), or visible light. Fenton and photo-Fenton processes, ozonation, heterogeneous photocatalysis with semiconductors, and electrochemical techniques, among others, can not only oxidize and mineralize almost each organic compound, but also damage pathogens, including bacteria and viruses [84].

Carbon nanomaterials have been extensively explored over the years for water remediation and disinfection using different AOPs, such as ozonation [85, 86], photocatalysis [28, 87–89], Fenton process, and catalytic wet air or peroxide oxidation (CWAO or CWPO, respectively), among others. Carbon materials can be used as (1) supports for anchoring metal complexes, (2) for the preparation of supported metal catalysts, and (3) as catalysts on their own (metal-free catalysts). These applications depend typically on the adequate tuning of the carbon surface properties by functionalization or doping. In this section, the use of catalysts based on nanostructured carbon materials, alone or combined with metal or metal oxides, will be reviewed for water remediation using AOPs. More emphasis will be put on photocatalytic process, while some examples using other AOPs such as ozonation, catalytic wet air or peroxide oxidation (CWAO or CWPO, respectively), and Fenton process over mostly graphene-based materials will be also addressed.

12.3.1 Heterogeneous Photocatalysis

The high mechanical strength, the chemical, electrical, and field emission properties of carbon materials, as well as the possibility to control these properties make these materials good candidates for the preparation of composites for photocatalytic applications. In the field of heterogeneous photocatalysis, carbon materials, namely, fullerenes, CNTs, graphene-based materials, nanofibers, nanodiamonds (NDs), and carbon gels [30, 90], among others, have been successfully combined with metal oxides for decontamination of water containing organic pollutants [91]. Semiconductors such as TiO_2, ZnO, CdS, ZnS, Fe_2O_3, WO_3, and Bi_2WO_4 [92, 93], among others, are some of the most studied photocatalysts so far. Different strategies have been developed, such as the addition of electron donors [94, 95], noble metal loading [96, 97], metal ion or anion doping [98, 99], and dye sensitization [100], to enhance the photocatalytic activity of semiconductor materials.

The presence of carbon nanomaterials has demonstrated to improve the photoefficiency of the semiconductors in environmental applications under UV and visible light conditions due to the synergistic effect between both metal oxide and carbon phases. This effect is normally related (1) to the decrease of the bandgap energy of the composite catalysts, (2) to an enhancement of the adsorption, (3) and the inhibition of electron-hole recombination.

In this section, the effect of different types of carbon materials (i.e., graphene-based materials, carbon nanotubes, nanodiamonds and fullerenes, among other) and metal oxide materials (TiO_2 or ZnO) as well as the effect of surface groups on the photocatalytic performance will be described in the heterogeneous photocatalysis process under UV-Vis irradiation.

The synthesis of graphene oxide-TiO_2 composites has been reported using sol-gel method followed by thermal reduction at different temperatures [87]. The composite materials were evaluated as photocatalysts for the degradation of a pharmaceutical compound and an azo-dye under both UV/Vis and visible light irradiation. The photocatalytic efficiency of the composites was related with the content of GO and calcination temperature. The highest degradation was achieved with the composite containing 3.3% wt. of GO and treated at 200 °C (GO-T) under UV-Vis and visible light irradiation, exceeding that of the benchmark P25 photocatalyst. In another work [28], GO and reduced graphene oxide (rGO) materials were synthetized in order to evaluate the effect of the amounts of functionalized groups on the photocatalytic performance of TiO_2-based composites under UV/Vis and visible irradiation. Among other characterization techniques, the presence of rGO and GO in the composite materials was identified by the characteristic D and G bands, whose intensity was appreciably enhanced over those of TiO_2 at 785 nm (Fig. 12.9a). Regarding the photoactivity of the prepared composites, it was proved that the presence of oxygen functionalities (such as epoxy and hydroxyl groups) in the GO sheets can be responsible for the optimal assembly and interfacial coupling between GO sheets and TiO_2 nanoparticles during the preparation of the composite, resulting in the most efficient photocatalyst under near UV/Vis and visible light (Fig. 12.9b). Thus, the composites prepared with rGO exhibited lower photocatalytic activities in comparison with the composite containing GO (GO-T) as observed in Fig. 12.9b. Graphene-based composites have also proven to be excellent photocatalysts under solar light and visible light illumination for different emerging contaminants, such as microcystin-LR and off-odor compounds (Geosmin and 2-methylisoborneol) [101], microcystin-LA (MC-LA) [88], 17-beta-estradiol [102], endocrine disruptor [103], trinitrophenol [104], and pesticides (i.e., diuron, atrazine, alachlor, and isoproturon) [105] (Fig. 12.10b). The photocatalytic mechanism has been also proposed (Fig. 12.10a) [105]. It has been reported that during the reaction over the GO-TiO_2 composite, hydroxyl radicals ($HO^{\bullet}$) seem to be the major active species during the photocatalytic reaction under UV radiation, while other reactive species, such as radical superoxide anion ($O_2^{\bullet-}$) and singlet oxygen ($^1O_2$), are involved in the photocatalytic mechanism under visible light illumination.

On the other hand, heteroatom doping of graphene is another strategy that can change the properties of carbon material, such as electric conductivity (bandgap), thermal stability or chemical reactivity, improving the catalytic activity, and elimi-

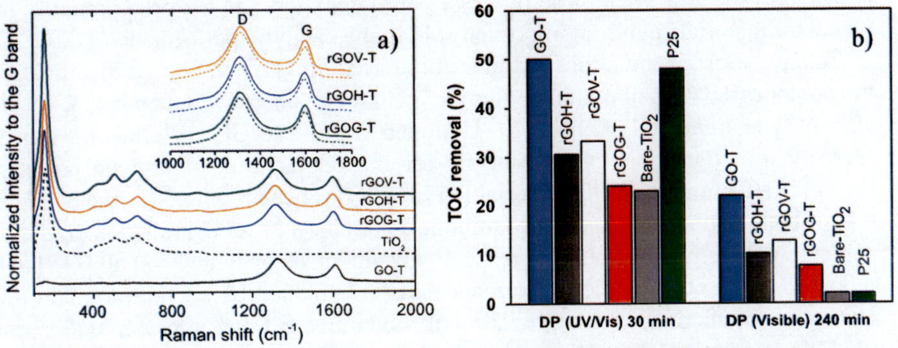

Fig. 12.9 (**a**) Raman spectra of rGO-T, GO-T, and bare-TiO_2 samples at 785 nm; (**b**) Total organic carbon (TOC) removal for the prepared catalysts and P25 under near-UV/Vis and visible light irradiation. (Adapted with permission from Ref. [28]. Copyright 2014, Elsevier)

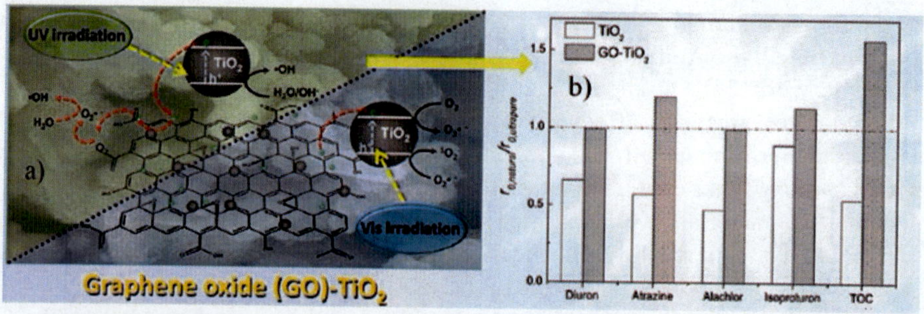

Fig. 12.10 (**a**) Schematic diagram of proposed reaction pathway over the GO-TiO_2 composite under UV and visible light irradiation; (**b**) Photodegradation of four EU/EPA priority pollutant pesticides by GO-TiO_2 composite. (Adapted with permission from Ref. [105]. Copyright 2017, Elsevier)

nating the need of metals in catalytic processes [106]. The presence of heteroatoms in graphene and its derivatives provides an interesting approach to develop effective catalysts for AOPs [107–109], including photocatalysis [109]. Pedrosa et al. [110] synthetized GO using the Hummers and Brodie methods and N and S-doped rGO materials by thermal treatment of GO using ammonia gas and benzyl disulfide, respectively. These metal-free doped rGO materials were tested in catalytic ozonation of oxalic acid. Moreover, TiO_2-based composites were prepared by the liquid phase deposition method using these materials. The performance of the composites was evaluated for degradation of oxalic acid (by ozonation) and for the degradation of DP (by photocatalysis). It is noteworthy that the same effects of reducing or doping GO on the performance of catalytic ozonation and photocatalysis were observed when studying composites from both the Hummers and Brodie methods, but worst performances were achieved with those prepared by the Brodie's method, in agreement with the lower oxygen (27.3 at.%) and, as consequence, nitrogen (3.3 at.%) contents of the respective precursors. Thus, the advantage of using these composite materials as catalysts was demonstrated for catalytic ozonation and heterogeneous photocatalysis, and when these processes are integrated, the oxygen content in the precursor materials plays an important role in the catalytic performance.

To date, most of the published literature based on graphene-based materials for the photodegradation of organic pollutant includes its union with metals [28, 87, 99, 109–113] or nonmetals [114–116]. Undoped metal-free GO materials have been scarcely investigated in photocatalysis for the degradation of organic pollutant [117–119]. Commercial GO material [117, 119] has been reported for the degradation of both bisphenol A and hydroquinone achieving 11% of degradation after 24 h and complete removal in 30 min, respectively, under solar light. Oh et al. [118] also reported the photocatalytic performance of GO (in this case synthetized by the Hummers' method) for the degradation of rhodamine B (19% after 2 h) under visible light irradiation. In another study, Pedrosa et al. [120] reported the photocatalytic degradation of phenol over GO materials prepared by the Hummers' (GO-H) and Brodie's (GO-B) methods (Fig. 12.11a). The results showed a higher photocata-

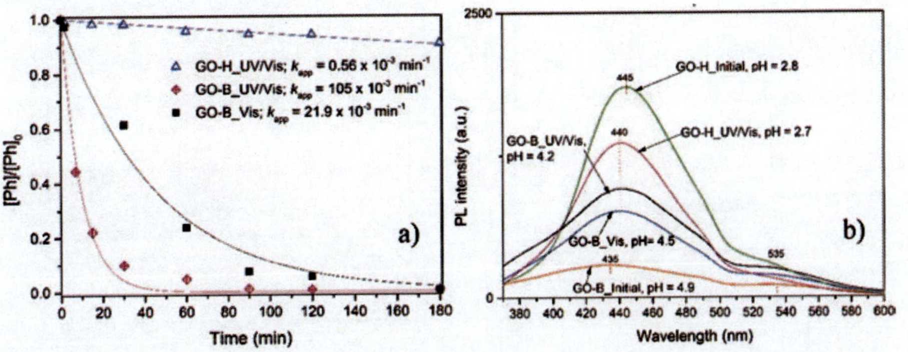

Fig. 12.11 (**a**) Normalized concentration of phenol under near-UV/Vis irradiation using GO-H and GO-B; (**b**) photoluminescence (PL) measurements of phenol solution before (initial) and after UV/Vis or Vis irradiations. (Adapted with permission from Ref. [120]. Copyright 2020, Elsevier)

lytic performance for GO-B due to the lower photoluminescence intensity (Fig. 12.11b) in liquid dispersion as observed in Fig. 12.11b.

Other nanocarbon-TiO_2 composites were also prepared using CNTs [121–123], NDs [91, 124], and fullerenes (C_{60}) [125] instead of GO. Silva et al. [121] reported the oxidation of carbon nanotubes (CNTs) with nitric acid (CNT_{ox}) with the subsequent generation of large amounts of oxygenated functional groups, namely carboxylic acid and phenol groups on the CNTs. It was found that the oxygen functionalities facilitate the anchoring of TiO_2 and CNTs with the generation of C–O–Ti bonds. The resulting composite leads to an increase in the photocatalytic efficiency towards the degradation of methylene blue under UV-Vis irradiation.

Regarding the NDs, it is noteworthy that the oxidation of NDs to ND_{ox} (430 °C in air) produces not only oxygen functionalities on the surface (as shown in TPD results, Fig. 12.12a, b), but also the purification of the nanodiamond powders [126, 127]. ND_{ox}-TiO_2 composite prepared with 15 wt.% of functionalized ND (ND_{ox}T-15) showed lower specific surface area in comparison with ND_{ox} (Fig. 12.12c). The higher photocatalytic efficiency of this composite was attributed to the presence of carboxylic anhydride, lactone and phenolic groups [121, 128] and to the increased purity of the nanodiamond particles after the oxidation treatment.

A comparative study regarding the nature of carbon materials (i.e., GO, CNT or fullerenes, C_{60}) as well as the carbon content (4 or 12 wt.%) on TiO_2-based composites has been also reported for the photodegradation of water pollutants under UV-Vis irradiation [122]. All the carbon-TiO_2 composites showed higher efficiency for the degradation of the contaminants in comparison with bare TiO_2, although the higher photocatalytic performance was obtained with the composite containing GO. These results were related with the larger amount of oxygen surface functionalities in the GO sheets. In fact, the oxidation method of GO (Hummers method) was more efficient than the oxidation of CNT or C_{60} (i.e., oxidation with persulfate).

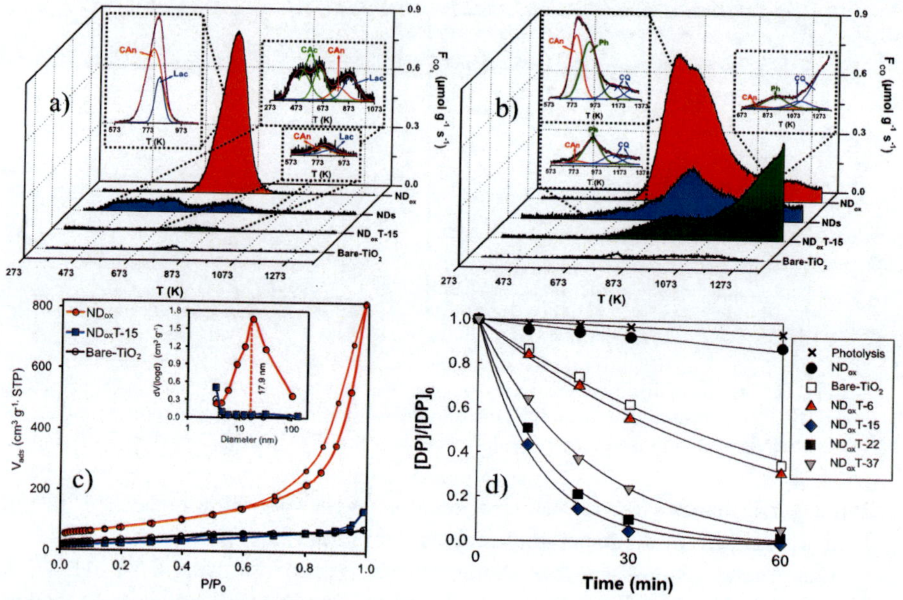

Fig. 12.12 (**a**) and (**b**) TPD profiles; (**c**) N_2 adsorption isotherms and pore size distribution (inset); (**d**) Concentration of DP over ND composites under near-UV/Vis light irradiation. (Adapted with permission from Ref. [91]. Copyright 2013, WILEY-VCH Verlag GmbH & Co)

ZnO is a very interesting semiconductor due to its adaptable morphology (such as nanofibers, nanorods, and flowers), low cost, and high photoefficiency. However, its bandgap around 3.3 eV makes it a necessity to improve its activity under visible light illumination and carbon materials have been also reported for this objective. Sampaio et al. [88] reported the synthesis of ZnO by chemical vapor deposition (CVD) and ZnO-based composites containing different carbon materials such as few-layers graphene (FLG), carbon fibers (CNFs), C_{60}, NDs, CNTs, and N-doped CNTs. The photocatalytic activity of the prepared composites was tested for the degradation of a model compound (i.e., phenol) under simulated solar light. The results showed that, among all the photocatalysts tested, the composite containing N-doped CNTs (i.e., N-CNT/ZnO) exhibited the highest performance (Fig. 12.13) due to the presence of electron-rich nitrogen-containing groups on the CNT surface.

12.3.2 Catalytic Wet (Air or Peroxide) Oxidation, Ozonation, and Fenton-Based Processes

Wet oxidation is a well-known AOP for the treatment of highly polluted industrial wastewaters and sludge [129–131]. Catalytic wet air oxidation (CWAO) uses oxygen or air to achieve pollutant oxidation by the means of active oxygen species, such

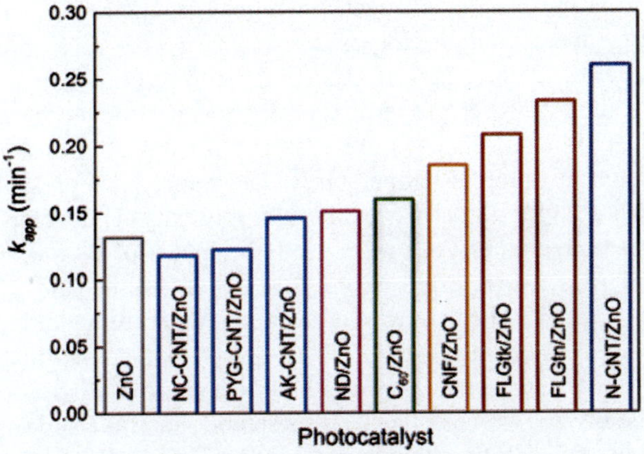

Fig. 12.13 Apparent first order rate constants (k_{app}) for the photocatalytic reactions using ZnO and carbon/ZnO materials. (Adapted with permission from Ref. [88]. Copyright 2015, Elsevier)

as hydroxyl radicals at high reaction temperatures (130–300 °C) and elevated pressures (20–200 bar). The addition of catalyst (homo- or heterogeneous) decreases the operation conditions (low temperatures and pressures), enhances the reaction rate, and reduces the reaction time and operating cost [131]. Catalytic wet peroxide oxidation (CWPO) is a liquid phase oxidation which uses hydrogen peroxide (H_2O_2) as oxygen source and mild or low operating conditions, namely, temperatures in the range of 25–100 °C and pressures of normal to 60 bar. Carbon or carbon-modified catalysts are normally used to generate hydroxyl radicals from the H_2O_2 decomposition [132]. In both processes, hydroxyl and hydroperoxyl radicals oxidize the organic compounds into CO_2 and H_2O, or alternatively, into easily biodegradable by-products [133].

Catalytic ozonation is another AOP for contaminants' removal involving the oxidation of ozone (O_3) in aqueous solution. This process produced hydroxyl radicals that can initiate the unselective oxidation of contaminants. In the presence of irradiation, H_2O_2, reduced metal ions, or heterogeneous catalyst, the radicals yield can be significantly improved [134].

Fenton-Based process is based on the Fenton reagent that is a mixture of Fe^{2+} (catalyst) and H_2O_2 (oxidizing agent) in acidic conditions (pH = 2–4). The mechanism involves the oxidation of Fe^{2+} by H_2O_2 with generation of hydroxyl radicals [135, 136].

It has been reported that the presence of oxygen and nitrogen functionalities on carbon materials is related with catalytic active sites in these processes, the basic sites being the most effective for environmental catalysis [137]. Nitrogen-doped carbon materials, including graphene-based materials, carbon xerogels, CNTs, and carbon blacks, have been tested in CWAO [133, 138–141] and catalytic ozonation [133, 142, 143]. Rocha et al. [139] reported the synthesis of N-doped rGO materials for the degradation of oxalic acid by CWAO and phenol by ozonation. The results showed that the presence of nitrogen functional groups remarkably influences the catalytic performance in both processes.

On the other hand, the classical ACs continue to be attractive as catalysts or supports due to their textural properties, easily tuned surface chemistry, and being cheap. In this context, three ACs were prepared from olive stones using KOH as activating agent at different temperatures. These materials present similar surface chemistry, but a larger porosity (mainly microporosity) is developed by increasing the activation temperature. ACs were used as support of Pt and both materials, Pt catalysts and ACs, were tested in the CWAO of aniline and TNP, respectively (Fig. 12.14) [17, 144]. The increased activation temperature yields a porosity development in ACs, and consequently, Pt catalysts that are highly dispersed were obtained. These changes in the porous structure and the Pt-dispersion influenced the activity in the CWAO process. Thus, all prepared materials showed higher conversion of aniline and TNP than the un-catalyzed process (WAO). For TNP, the WAO degradation is negligible, while a total mineralization is obtained with the use of ACs. The TNP removal and oxidation is related with the porous properties of AC. Thus, highly microporous ACs adsorb the TNP and the oxidation slowly takes places in the wider micropores. On the contrary, in ACs with macroporous and very few micropores, TNP degradation occurs more rapidly and on the external surface (Fig. 12.14b). Therefore, degradation mechanism was demonstrated to be always heterogeneous, but takes place in different ranges of porosity [17]. The use of Pt/ACs catalysts corroborates that the CWAO reaction is sensitive to the structure, and an optimum value for the dispersion of Pt was determined [144]. Analysis of the catalysts after reaction indicated some loss of microporosity, accompanied by an increase of the oxygenated surface groups and changes in the Pt oxidation state. In addition, no Pt leaching was detected. The performance of the catalysts was evaluated during three consecutive cycles without any reactivation treatment. The activity decreased after the first cycle, which is due to the saturation of the adsorption capac-

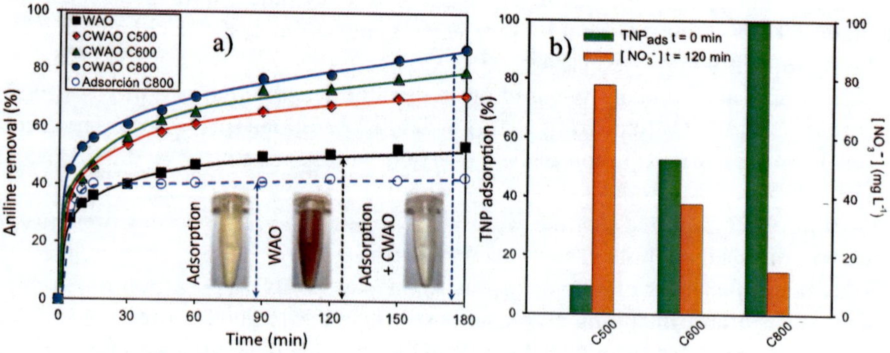

Fig. 12.14 (**a**) Aniline removal obtained with ACs by adsorption, WAO, and CWAO processes; (**b**) TNP removal and formation of nitrates during CWAO of TNP. (Adapted with permission from Ref. [17, 144]. Copyright 2010 and 2011, Elsevier)

ity; however, the catalytic performance was constant during the second and third cycles, showing the high stability of the tested catalysts.

CWPO involves the transference of electrons from the active sites of a catalyst to H_2O_2 molecules, with the generation of hydroxyl radicals ($HO^•$) and hydroxide ions (HO^-) [145–147]. Carbon materials such as carbon xerogels, CNTs, and graphene-based materials have been reported as metal-free catalysts for the CWPO of organic pollutants in aqueous solution [148–152]. Ribeiro et al. [152] reported the CWPO of 4-nitrophenol (4-NP) solutions with high concentration (5 g L^{-1}) at 50 °C, pH = 3, and $[H_2O_2]_0$ = 17.8 g L^{-1} over rGO materials obtained from the chemical reduction of GO using vitamin C (rGOV), glucose (rGOG), and hydrazine (rGOH). The obtained results showed the highest performance for the rGOV sample, achieving a 4-NP degradation of 65% and TOC removal of 23%, although after three reuse cycles the most stable catalyst was the rGOH sample, concluding that the rGO materials are very effective for the CWPO at mild conditions.

Martín-Martínez et al. [29] prepared four magnetic CNT samples, namely undoped, completely N-doped, and two selectively N-doped, by CVD. The materials were tested in the CWPO of highly concentrated 4-NP solutions. N doping was identified to influence considerably the CWPO performance of the materials. In particular, undoped CNTs, with a moderate hydrophobicity, favor the controllable and efficient decomposition of H_2O_2 into highly reactive hydroxyl radicals ($HO^•$), thus showing high catalytic activity for 4-NP degradation. On the other hand, the completely N-doped catalyst, fully hydrophilic, favors a quick decomposition of H_2O_2 into nonreactive O_2 and H_2O species. The selectively N-doped amphiphilic catalysts, i.e., hybrid structures containing undoped sections followed by N-doped ones (with high affinity for polar molecules), provided intermediate results, namely, a higher N content favored H_2O_2 decomposition towards nonreactive H_2O and O_2 species, while a lower N content resulted in the formation of $HO^•$, increasing 4-NP mineralization.

Fenton-based process is also an efficient method for wastewater treatment and groundwater remediation because of its high performance and easy operation [135, 136]. However, it should be pointed out that the homogeneous Fenton process has significant disadvantages, namely, the concentration of Fe ions leached in solution is well above the European Union directives and the removal/treatment of the sludge-containing iron ions at the end of the wastewater treatment is expensive and needs large amount of chemicals and human resources. To overcome the disadvantages of the homogeneous Fenton or Fenton-like processes, the immobilization of the catalyst on carbon supports (ACs, carbon aerogels) has been studied in order to avoid the catalyst-recovering step, also allowing the use of continuous reactors for the treatment of effluents [153–155]. The results showed a high efficiency of the catalysts for the degradation of nonbiodegradable industrial dyes with high TOC removal and low deactivation by Fe-leaching, allowing the reuse of the samples.

One version of this kind of process is the electro-Fenton (EF). This process is based on the generation of $HO^•$ radicals through the in situ production of H_2O_2 at the cathode during the oxygen reduction reaction, ORR (following a two-electron pathway), and the addition of Fe^{2+} ions [136]. So far, ACs such as carbon sponge, graph-

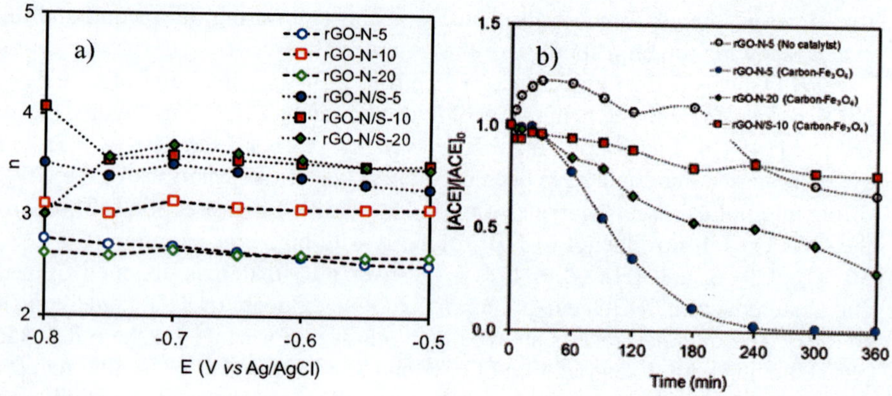

Fig. 12.15 (**a**) Variation of n with E vs Ag/AgCl for heteroatom graphene-based materials; (**b**) Normalized ACE concentration as function of time for electro-Fenton experiments at potential of −0.55 V. (Adapted with permission from Ref. [160]. Copyright 2019, Elsevier)

ite felt, CNTs, and carbon aerogel have been widely developed as the cathodes for H_2O_2 production [156–158]. Moreover, the use of iron supported on AC has been previously reported as catalyst in EF reaction for the degradation of several pollutants, such as phenol [157] or methyl blue [159]. Recently, Fernández-Saéz et al. [160] have reported the synthesis of N-and N/S-doped graphene aerogels and magnetic carbon-iron oxide particles for the degradation of a pharmaceutical compound (acetaminophen) in aqueous solution using the EF process. They observed different catalytic performances for the prepared materials in the ORR. The materials containing N/S heteroatoms exhibited better performance for ORR with the generation of H_2O, leading to an ideal mechanism of nearly four electrons. On the contrary, the materials with the N functionalities showed the generation of H_2O_2 in the ORR process following a two-electron pathway (Fig. 12.15a). Thus, the N-doped graphene aerogels were used for the degradation of acetaminophen by EF (Fig. 12.15b). The results showed a complete degradation of the pollutant (100%) with a mineralization of 45% and low iron leaching (0.24 mg L^{-1}).

12.3.3 Advanced Oxidation Processes in Water Disinfection

AOP and in particular solar photocatalysis (i.e., heterogeneous photocatalysis and photo-Fenton) have been widely used for the decontamination of organic pollutant and as efficient disinfection method (inactivating emerging pathogens and microorganisms) for irrigation and for drinking water [161].

Different carbon nanomaterials including graphene derivatives, CNTs, carbon dots, and graphitic carbon nitride (g-C_3N_4) have developed as effective catalysts on their own or combined with metal oxides to prepare carbon-based composites for water disinfection [162]. Fernandez-Ibañez et al. [163] reported the synthesis of rGO-TiO_2 composites for the photocatalytic inactivation of *E. coli* and *Fusarium solani* in water under real sun. They observed the formation of singlet oxygen under visible light illumination, evidencing a higher inactivation of *E. coli* under solar illumination. Graphene-TiO_2 semiconductor has been also tested on the removal of micropollutants and disinfection of a secondary effluent of an UWWTP (conventional urban wastewater treatment plants) by solar-H_2O_2, heterogeneous photocatalysis (with and/or without the addition of H_2O_2), and the photo-Fenton process in a pilot-scale photoreactor [164]. Organic pollutants such as sulfamethoxazole, carbamazepine, and diclofenac, biological contaminants (human pathogen indicators), as well as the whole bacterial community were characterized. Heterogeneous photocatalysis using TiO_2-P25 and assisted with H_2O_2 (P25/H_2O_2) was the most efficient process on the degradation of the chemical organic micropollutants, attaining levels below the limits of quantification in less than 4 h of treatment.

Zeng et al. [165] showed a high photocatalytic performance of carbon dots and TiO_2 on rGO towards the inactivation of *E. coli* under simulated solar light. The higher photocatalytic performance was attributed to the presence of carbon dots that promoted surface oxygen reduction by the production of reactive oxygen species. In another work, TiO_2/rGO/WO_3 showed enhanced photocatalytic activity for the inactivation of *E. coli* comparatively to that of TiO_2/WO_3, indicating the role of rGO for improved charge separation in the photocatalytic process [166]. Granular activated carbon (GAC) has been also used as a cathode to generate H_2O_2 for water disinfection (inhibiting *E. coli*) through the electro-Fenton process [167]. The results showed an efficient inactivation of the pathogen with high H_2O_2 yield and excellent reusability proving that GAC is a promising material for large-scale water disinfection. Thus, there is evidence that AOPs and carbon nanomaterials can be used for an efficient water disinfection treatment.

12.4 Conclusions and Future Perspectives

The pollution of air, soil, and water is progressively increasing and regulations and laws are getting stricter year by year. The occurrence of contaminants of emerging concern in effluents is claiming for the need to improve the conventional treatment technologies and the research of novel processes. Nanotechnology opens a gate to develop advanced materials with high efficiency in oxidation technologies. In this context, carbon (nano)-materials possess unique properties, such as high surface area, inertness, stability in acidic and basic media, and mainly, the ability of their surface chemistry, porosity, and shape being tailored. The aforementioned properties of carbon (nano)-materials are responsible to take into consideration as excellent candidates for water and air treatment in comparison with other common inorganic materials.

Carbon materials are typically employed as adsorbents in the removal of pollutants from air (e.g., VOCs, CO_2, NO_x, and SO_x) and water (e.g., dyes, pharmaceuticals, herbicides, etc.). The type of porosity and surface chemistry influence markedly the interactions with the pollutant molecules. For instance, the hydrophobic nature of carbons allows a large affinity for nonpolar pollutants, but the incorporation of heteroatoms (e.g. N, O, S, P) or nitrogen-, sulfur, and oxygen-containing functionalities on the carbon surface leads to more hydrophilic materials, increasing the interactions with polar pollutants. So, there is a large potential to modify not only the porosity and surface area of these materials, but also the chemical properties and thus the performance as adsorbent. Even so, a special attention should be given to the potential of carbon materials, as self-catalysts and catalyst supports, in specific catalytic processes for air and decontamination and disinfection of water. In particular, nanomaterials based on CNTs or graphene derivatives have demonstrated a good performance, stability, and selectivity in different AOPs, such CWAO, CWPO, ozonation, photocatalysis, and Fenton-based processes, to effectively remove pollutants by themselves or combined with metal oxides.

Acknowledgments This work was financially supported by the Spanish Project ref. RTI2018-099224-B-I00 from ERDF/Ministry of Science, Innovation and Universities—State Research Agency. LMPM (RYC-2016-19347) and SMT (RYC-2019-026634-I) acknowledge the Spanish Ministry of Economy and Competitiveness (MINECO) and the European Social Found for Ramón y Cajal research contracts. “Unidad de Excelencia Química Aplicada a Biomedicina y Medioambiente” of the University of Granada (UEQ - UGR) is gratefully acknowledged for the technical assistance.

References

1. M.-Y. Tai, C.-C. Chao, S.-W. Hu, Pollution, health and economic growth. North Am. J. Econom. Finance **32**, 155–161 (2015)
2. P. Das, R. Horton, Pollution, health, and the planet: time for decisive action. Lancet **391**(10119), 407–408 (2018)
3. J. Huang, X. Pan, X. Guo, G. Li, Impacts of air pollution wave on years of life lost: a crucial way to communicate the health risks of air pollution to the public. Environ. Int. **113**, 42–49 (2018)
4. J. Christodoulakis, C.A. Varotsos, A.P. Cracknell, G.A. Kouremadas, The deterioration of materials as a result of air pollution as derived from satellite and ground based observations. Atmos. Environ. **185**, 91–99 (2018)
5. M.R. Miller, Oxidative stress and the cardiovascular effects of air pollution. Free Radic. Biol. Med. (2020)
6. A. Oudin, Short review: air pollution, noise and lack of greenness as risk factors for Alzheimer’s disease-epidemiologic and experimental evidence. Neurochem. Int. **134**, 104646 (2020)
7. J.P. Vareda, A.J.M. Valente, L. Durães, Assessment of heavy metal pollution from anthropogenic activities and remediation strategies: a review. J. Environ. Manag. **246**, 101–118 (2019)
8. A.E.V. Evans, J. Mateo-Sagasta, M. Qadir, E. Boelee, A. Ippolito, Agricultural water pollution: key knowledge gaps and research needs. Curr. Opin. Environ. Sustain. **36**, 20–27 (2019)
9. A. Müller, H. Österlund, J. Marsalek, M. Viklander, The pollution conveyed by urban runoff: a review of sources. Sci. Total Environ. **709**, 136125 (2020)
10. D. Cheng, H.H. Ngo, W. Guo, S.W. Chang, D.D. Nguyen, Y. Liu, Q. Wei, D. Wei, A critical review on antibiotics and hormones in swine wastewater: water pollution problems and control approaches. J. Hazard. Mater. **387**, 121682 (2020)

11. M.F. Othman, A. Adam, G. Najafi, R. Mamat, Green fuel as alternative fuel for diesel engine: a review. Renew. Sust. Energ. Rev. **80**, 694–709 (2017)
12. C. Liu, X. Miao, J. Li, Outdoor formaldehyde matters and substantially impacts indoor formaldehyde concentrations. Build. Environ. **158**, 145–150 (2019)
13. M.-K. Kim, K.-D. Zoh, Occurrence and removals of micropollutants in water environment. Environ. Eng. Res. **21**(4), 319–332 (2016)
14. R. Álvarez-Ruiz, Y. Picó, Analysis of emerging and related pollutants in aquatic biota. Trends Environ. Anal. Chem. **25**, e00082 (2020)
15. M.J. Prauchner, F. Rodríguez-Reinoso, Chemical versus physical activation of coconut shell: a comparative study. Microporous Mesoporous Mater. **152**, 163–171 (2012)
16. J.F. Vivo-Vilches, F. Carrasco-Marín, A.F. Pérez-Cadenas, F.J. Maldonado-Hódar, Fitting the porosity of carbon xerogel by CO_2 activation to improve the TMP/n-octane separation. Microporous Mesoporous Mater. **209**, 10–17 (2015)
17. S. Morales-Torres, A.M.T. Silva, A.F. Pérez-Cadenas, J.L. Faria, F.J. Maldonado-Hódar, J.L. Figueiredo, F. Carrasco-Marín, Wet air oxidation of trinitrophenol with activated carbon catalysts: effect of textural properties on the mechanism of degradation. Appl. Catal. B Environ. **100**(1–2), 310–317 (2010)
18. D.S. Su, S. Perathoner, G. Centi, Catalysis on nano-carbon materials: Going where to? Catal. Today **186**(1), 1–6 (2012)
19. J.F. Vivo-Vilches, E. Bailón-García, A.F. Pérez-Cadenas, F. Carrasco-Marín, F.J. Maldonado-Hódar, Tailoring the surface chemistry and porosity of activated carbons: evidence of reorganization and mobility of oxygenated surface groups. Carbon **68**, 520–530 (2014)
20. A. Elmouwahidi, J. Castelo-Quibén, J.F. Vivo-Vilches, A.F. Pérez-Cadenas, F.J. Maldonado-Hódar, F. Carrasco-Marín, Activated carbons from agricultural waste solvothermally doped with sulphur as electrodes for supercapacitors. Chem. Eng. J. **334**, 1835–1841 (2018)
21. H. Hamad, J. Castelo-Quibén, S. Morales-Torres, F. Carrasco-Marín, F.A. Pérez-Cadenas, J.F. Maldonado-Hódar, On the interactions and synergism between phases of carbon–phosphorus–titanium composites synthetized from cellulose for the removal of the orange-G dye. Materials **11**(9), 1766 (2018)
22. J.L. Figueiredo, Nanostructured porous carbons for electrochemical energy conversion and storage. Surf. Coat. Technol. **350**, 307–312 (2018)
23. S. Morales-Torres, F.J. Maldonado-Hódar, A.F. Pérez-Cadenas, F. Carrasco-Marín, Design of low-temperature Pt-carbon combustion catalysts for VOC's treatments. J. Hazard. Mater. **183**(1–3), 814–822 (2010)
24. S. Morales-Torres, T.L.S. Silva, L.M. Pastrana-Martinez, A.T.S.C. Brandao, J.L. Figueiredo, A.M.T. Silva, Modification of the surface chemistry of single- and multi-walled carbon nanotubes by HNO_3 and H_2SO_4 hydrothermal oxidation for application in direct contact membrane distillation. Phys. Chem. Chem. Phys. **16**(24), 12237–12250 (2014)
25. V. Calvino-Casilda, A.J. López-Peinado, C.J. Durán-Valle, R.M. Martín-Aranda, Last decade of research on activated carbons as catalytic support in chemical processes. Catal. Rev. **52**(3), 325–380 (2010)
26. F.J. Maldonado-Hódar, Advances in the development of nanostructured catalysts based on carbon gels. Catal. Today **218-219**, 43–50 (2013)
27. G.B. Baur, I. Yuranov, A. Renken, L. Kiwi-Minsker, Activated carbon fibers for efficient VOC removal from diluted streams: the role of surface morphology. Adsorption **21**(6), 479–488 (2015)
28. L.M. Pastrana-Martínez, S. Morales-Torres, V. Likodimos, P. Falaras, J.L. Figueiredo, J.L. Faria, A.M.T. Silva, Role of oxygen functionalities on the synthesis of photocatalytically active graphene–TiO_2 composites. Appl. Catal. B Environ. **158-159**, 329–340 (2014)
29. M. Martin-Martinez, R.S. Ribeiro, B.F. Machado, P. Serp, S. Morales-Torres, A.M.T. Silva, J.L. Figueiredo, J.L. Faria, H.T. Gomes, Role of nitrogen doping on the performance of carbon nanotube catalysts: a catalytic wet peroxide oxidation application. ChemCatChem **8**(12), 2068–2078 (2016)

30. E. Bailón-García, A. Elmouwahidi, F. Carrasco-Marín, A.F. Pérez-Cadenas, F.J. Maldonado-Hódar, Development of Carbon-ZrO_2 composites with high performance as visible-light photocatalysts. Appl. Catal. B Environ. **217**, 540–550 (2017)
31. Y.N. Liang, W.-D. Oh, Y. Li, X. Hu, Nanocarbons as platforms for developing novel catalytic composites: overview and prospects. Appl. Catal. A Gen. **562**, 94–105 (2018)
32. S. Morales-Torres, L.M. Pastrana-Martínez, J.L. Figueiredo, J.L. Faria, A.M.T. Silva, Design of graphene-based TiO_2 photocatalysts-a review. Environ. Sci. Pollut. Res. **19**(9), 3676–3687 (2012)
33. World Health Organization Occupational; Environmental Health Team, *Guidelines for Air Quality* (World Health Organization, Geneva, 2000)
34. United States Environmental Protection Agency. Technical Overview of Volatile Organic Compounds. https://www.epa.gov/indoor-air-quality-iaq/technical-overview-volatile-organic-compounds. Accessed 20 April 2020
35. X. Zhang, B. Gao, A.E. Creamer, C. Cao, Y. Li, Adsorption of VOCs onto engineered carbon materials: a review. J. Hazard. Mater. **338**, 102–123 (2017)
36. E.A. Kolchanova, T.P. Lyubimova, Interface instability of methane hydrate deposits of variable permeability under permafrost conditions. Int. J. Heat Mass Transf. **98**, 329–340 (2016)
37. X. Yang, H. Yi, X. Tang, S. Zhao, Z. Yang, Y. Ma, T. Feng, X. Cui, Behaviors and kinetics of toluene adsorption-desorption on activated carbons with varying pore structure. J. Environ. Sci. **67**, 104–114 (2018)
38. X. Yu, S. Liu, G. Lin, X. Zhu, S. Zhang, R. Qu, C. Zheng, X. Gao, Insight into the significant roles of microstructures and functional groups on carbonaceous surfaces for acetone adsorption. RSC Adv. **8**(38), 21541–21550 (2018)
39. J.F. Vivo-Vilches, E. Bailón-García, A.F. Pérez-Cadenas, F. Carrasco-Marín, F.J. Maldonado-Hódar, Tailoring activated carbons for the development of specific adsorbents of gasoline vapors. J. Hazard. Mater. **263**, 533–540 (2013)
40. J.F. Vivo-Vilches, A.F. Pérez-Cadenas, F. Carrasco-Marín, F.J. Maldonado-Hódar, About the control of VOC's emissions from blended fuels by developing specific adsorbents using agricultural residues. J. Environ. Chem. Eng. **3**(4)., Part A, 2662–2669 (2015)
41. Y. Liu, K. Mallouk, H. Emamipour, M.J. Rood, X. Liu, Z. Yan, Isobutane adsorption with carrier gas recirculation at different relative humidities using activated carbon fiber cloth and electrothermal regeneration. Chem. Eng. J. **360**, 1011–1019 (2019)
42. G.B. Baur, I. Yuranov, L. Kiwi-Minsker, Activated carbon fibers modified by metal oxide as effective structured adsorbents for acetaldehyde. Catal. Today **249**, 252–258 (2015)
43. A.M. Awad, R. Jalab, A. Benamor, M.S. Nasser, M.M. Ba-Abbad, M. El-Naas, A.W. Mohammad, Adsorption of organic pollutants by nanomaterial-based adsorbents: an overview. J. Mol. Liq. **301**, 112335 (2020)
44. G. Ersan, O.G. Apul, F. Perreault, T. Karanfil, Adsorption of organic contaminants by graphene nanosheets: a review. Water Res. **126**, 385–398 (2017)
45. W. Wu, K. Yang, W. Chen, W. Wang, J. Zhang, D. Lin, B. Xing, Correlation and prediction of adsorption capacity and affinity of aromatic compounds on carbon nanotubes. Water Res. **88**, 492–501 (2016)
46. Y.-H. Shih, M.-S. Li, Adsorption of selected volatile organic vapors on multiwall carbon nanotubes. J. Hazard. Mater. **154**(1), 21–28 (2008)
47. S. Tulaphol, S. Bunsan, E. Kanchanatip, H.Y. Miao, N. Grisdanurak, W. Den, Influence of chlorine substitution on adsorption of gaseous chlorinated phenolics on multi-walled carbon nanotubes embedded in SiO_2. Int. J. Environ. Sci. Technol. **13**(6), 1465–1474 (2016)
48. C.M. Hussain, C. Saridara, S. Mitra, Modifying the sorption properties of multi-walled carbon nanotubesvia covalent functionalization. Analyst **134**(9), 1928–1933 (2009)
49. L. Yu, L. Wang, W. Xu, L. Chen, M. Fu, J. Wu, D. Ye, Adsorption of VOCs on reduced graphene oxide. J. Environ. Sci. **67**, 171–178 (2018)

50. L. Wu, Z. Qin, L. Zhang, T. Meng, F. Yu, J. Ma, CNT-enhanced amino-functionalized graphene aerogel adsorbent for highly efficient removal of formaldehyde. New J. Chem. **41**(7), 2527–2533 (2017)
51. F. Chu, Y. Zheng, B. Wen, L. Zhou, J. Yan, Y. Chen, Adsorption of toluene with water on zeolitic imidazolate framework-8/graphene oxide hybrid nanocomposites in a humid atmosphere. RSC Adv. **8**(5), 2426–2432 (2018)
52. V. Kumar, Y.-S. Lee, J.-W. Shin, K.-H. Kim, D. Kukkar, Y. Fai Tsang, Potential applications of graphene-based nanomaterials as adsorbent for removal of volatile organic compounds. Environ. Int. **135**, 105356 (2020)
53. S. Morales-Torres, F.J. Maldonado-Hódar, A.F. Pérez-Cadenas, F. Carrasco-Marín, Structural characterization of carbon xerogels: from film to monolith. Microporous Mesoporous Mater. **153**, 24–29 (2012)
54. E. Gallegos-Suárez, A.F. Pérez-Cadenas, F.J. Maldonado-Hódar, F. Carrasco-Marín, On the micro- and mesoporosity of carbon aerogels and xerogels. The role of the drying conditions during the synthesis processes. Chem. Eng. J. **181-182**, 851–855 (2012)
55. F.J. Maldonado-Hódar, C. Moreno-Castilla, F. Carrasco-Marín, A.F. Pérez-Cadenas, Reversible toluene adsorption on monolithic carbon aerogels. J. Hazard. Mater. **148**(3), 548–552 (2007)
56. H. Jirglová, A.F. Pérez-Cadenas, F.J. Maldonado-Hódar, Synthesis and properties of phloroglucinol–phenol–formaldehyde carbon aerogels and xerogels. Langmuir **25**(4), 2461–2466 (2009)
57. D. Espinosa-Iglesias, C. Valverde-Sarmiento, A.F. Pérez-Cadenas, M.I. Bautista-Toledo, F.J. Maldonado-Hódar, F. Carrasco-Marín, Mesoporous carbon-xerogels films obtained by microwave assisted carbonization. Mater. Lett. **141**, 135–137 (2015)
58. S. Han, Q. Sun, H. Zheng, J. Li, C. Jin, Green and facile fabrication of carbon aerogels from cellulose-based waste newspaper for solving organic pollution. Carbohydr. Polym. **136**, 95–100 (2016)
59. C. Moreno-Castilla, F.J. Maldonado-Hódar, F. Carrasco-Marín, E. Rodríguez-Castellón, Surface characteristics of titania/carbon composite aerogels. Langmuir **18**(6), 2295–2299 (2002)
60. L. Zhu, L. Meng, J. Shi, J. Li, X. Zhang, M. Feng, Metal-organic frameworks/carbon-based materials for environmental remediation: a state-of-the-art mini-review. J. Environ. Manag. **232**, 964–977 (2019)
61. X.-W. Liu, T.-J. Sun, J.-L. Hu, S.-D. Wang, Composites of metal–organic frameworks and carbon-based materials: preparations, functionalities and applications. J. Mater. Chem. A **4**(10), 3584–3616 (2016)
62. W.D.P. Rengga, A. Chafidz, M. Sudibandriyo, M. Nasikin, A.E. Abasaeed, Silver nanoparticles deposited on bamboo-based activated carbon for removal of formaldehyde. J. Environ. Chem. Eng. **5**(2), 1657–1665 (2017)
63. Y. Wang, M. Wang, Z. Wang, S. Wang, J. Fu, Tunable-quaternary (N, S, O, P)-doped porous carbon microspheres with ultramicropores for CO_2 capture. Appl. Surf. Sci. **507**, 145130 (2020)
64. S. Cao, H. Zhao, D. Hu, J.-A. Wang, M. Li, Z. Zhou, Q. Shen, N. Sun, W. Wei, Preparation of potassium intercalated carbons by in-situ activation and speciation for CO_2 capture from flue gas. J. CO2 Utiliz. **35**, 59–66 (2020)
65. J. Gao, Y. Hoshino, G. Inoue, Honeycomb-carbon-fiber-supported amine-containing nanogel particles for CO_2 capture using a rotating column TVSA. Chem. Eng. J. **383**, 123123 (2020)
66. M.T. Izquierdo, B. Rubio, C. Mayoral, J.M. Andrés, Low cost coal-based carbons for combined SO_2 and NO removal from exhaust gas. Fuel **82**(2), 147–151 (2003)
67. S. Ding, Y. Li, T. Zhu, Y. Guo, Regeneration performance and carbon consumption of semi-coke and activated coke for SO_2 and NO removal. J. Environ. Sci. **34**, 37–43 (2015)

68. Y. Xue, Y. Guo, Z. Zhang, Y. Guo, Y. Wang, G. Lu, The role of surface properties of activated carbon in the catalytic reduction of NO by carbon. Appl. Surf. Sci. **255**(5, Part 2), 2591–2595 (2008)
69. S. Bashkova, D. Deoki, T.J. Bandosz, Effect of silver nanoparticles deposited on micro/mesoporous activated carbons on retention of NO_x at room temperature. J. Colloid Interface Sci. **354**(1), 331–340 (2011)
70. E. Bailón-García, A. Elmouwahidi, F. Ribeiro, C. Henriques, A.F. Pérez-Cadenas, F. Carrasco-Marín, F.J. Maldonado-Hódar, Reduction of NO with new vanadium-carbon xerogel composites. Effect of the oxidation state of vanadium species. Carbon **156**, 194–204 (2020)
71. R.A. Catalão, F.J. Maldonado-Hódar, A. Fernandes, C. Henriques, M.F. Ribeiro, Reduction of NO with metal-doped carbon aerogels. Appl. Catal. B Environ. **88**(1), 135–141 (2009)
72. J. Zawadzki, M. Wiśniewski, An infrared study of the behavior of SO_2 and NO_x over carbon and carbon-supported catalysts. Catal. Today **119**(1), 213–218 (2007)
73. H. Huang, Y. Xu, Q. Feng, D.Y.C. Leung, Low temperature catalytic oxidation of volatile organic compounds: a review. Cat. Sci. Technol. **5**(5), 2649–2669 (2015)
74. C. He, J. Cheng, X. Zhang, M. Douthwaite, S. Pattisson, Z. Hao, Recent advances in the catalytic oxidation of volatile organic compounds: a review based on pollutant sorts and sources. Chem. Rev. **119**(7), 4471–4568 (2019)
75. S. Morales-Torres, F. Carrasco-Marín, A.F. Pérez-Cadenas, F.J. Maldonado-Hódar, Coupling noble metals and carbon supports in the development of combustion catalysts for the abatement of BTX compounds in air streams. Catalysts **5**(2), 774–799 (2015)
76. T. García, B. Solsona, S.H. Taylor, Naphthalene total oxidation over metal oxide catalysts. Appl. Catal. B Environ. **66**(1), 92–99 (2006)
77. S. Morales-Torres, A.F. Pérez-Cadenas, F. Kapteijn, F. Carrasco-Marín, F.J. Maldonado-Hódar, J.A. Moulijn, Palladium and platinum catalysts supported on carbon nanofiber coated monoliths for low-temperature combustion of BTX. Appl. Catal. B Environ. **89**(3), 411–419 (2009)
78. A.F. Pérez-Cadenas, S. Morales-Torres, F. Kapteijn, F.J. Maldonado-Hódar, F. Carrasco-Marín, C. Moreno-Castilla, J.A. Moulijn, Carbon-based monolithic supports for palladium catalysts: the role of the porosity in the gas-phase total combustion of m-xylene. Appl. Catal. B Environ. **77**(3), 272–277 (2008)
79. S. Jiang, S. Song, Enhancing the performance of Co_3O_4/CNTs for the catalytic combustion of toluene by tuning the surface structures of CNTs. Appl. Catal. B Environ. **140-141**, 1–8 (2013)
80. M.N. Padilla-Serrano, F.J. Maldonado-Hódar, C. Moreno-Castilla, Influence of Pt particle size on catalytic combustion of xylenes on carbon aerogel-supported Pt catalysts. Appl. Catal. B Environ. **61**(3), 253–258 (2005)
81. H. Liu, Y. Ma, J. Chen, M. Wen, G. Li, T. An, Highly efficient visible-light-driven photocatalytic degradation of VOCs by CO_2-assisted synthesized mesoporous carbon confined mixed-phase TiO_2 nanocomposites derived from MOFs. Appl. Catal. B Environ. **250**, 337–346 (2019)
82. N. Iqbal, A. Afzal, N. Cioffi, L. Sabbatini, L. Torsi, NO_x sensing one- and two-dimensional carbon nanostructures and nanohybrids: Progress and perspectives. Sensors Actuators B Chem. **181**, 9–21 (2013)
83. U. Thubsuang, D. Sukanan, S. Sahasithiwat, S. Wongkasemjit, T. Chaisuwan, Highly sensitive room temperature organic vapor sensor based on polybenzoxazine-derived carbon aerogel thin film composite. Mater. Sci. Eng. B **200**, 67–77 (2015)
84. O. Tsydenova, V. Batoev, A. Batoeva, Solar-enhanced advanced oxidation processes for water treatment: simultaneous removal of pathogens and chemical pollutants. Int. J. Environ. Res. Public Health **12**(8), 9542 (2015)

85. A. Sanchez-Polo, J. Rivera-Utrilla, Effect of the ozone-carbon reaction on the catalytic activity of activated carbon during the degradation of 1,3,6-naphthalenetrisulphonic acid with ozone. Carbon **41**(2), 303–307 (2003)
86. A. Gonçalves, J.J.M. Órfão, M.F.R. Pereira, Ozonation of bezafibrate promoted by carbon materials. Appl. Catal. B Environ. **140-141**, 82–91 (2013)
87. L.M. Pastrana-Martínez, S. Morales-Torres, V. Likodimos, J.L. Figueiredo, J.L. Faria, P. Falaras, A.M.T. Silva, Advanced nanostructured photocatalysts based on reduced graphene oxide-TiO_2 composites for degradation of diphenhydramine pharmaceutical and methyl orange dye. Appl. Catal. B Environ. **123-124**, 241–256 (2012)
88. M.J. Sampaio, R.R. Bacsa, A. Benyounes, R. Axet, P. Serp, C.G. Silva, A.M.T. Silva, J.L. Faria, Synergistic effect between carbon nanomaterials and ZnO for photocatalytic water decontamination. J. Catal. **331**, 172–180 (2015)
89. S. Morales-Torres, L.M. Pastrana-Martínez, J.L. Figueiredo, J.L. Faria, A.M.T. Silva, Graphene oxide-P25 photocatalysts for degradation of diphenhydramine pharmaceutical and methyl orange dye. Appl. Surf. Sci. **275**, 361–368 (2013)
90. E. Bailón-García, A. Elmouwahidi, M.A. Álvarez, F. Carrasco-Marín, A.F. Pérez-Cadenas, F.J. Maldonado-Hódar, New carbon xerogel-TiO_2 composites with high performance as visible-light photocatalysts for dye mineralization. Appl. Catal. B Environ. **201**, 29–40 (2017)
91. L.M. Pastrana-Martínez, S. Morales-Torres, S.A.C. Carabineiro, J.G. Buijnsters, J.L. Faria, J.L. Figueiredo, A.M.T. Silva, Nanodiamond–TiO_2 composites for heterogeneous photocatalysis. ChemPlusChem **78**(8), 801–807 (2013)
92. S. Gupta, M. Tripathi, An overview of commonly used semiconductor nanoparticles in photocatalysis. High Energy Chem. **46**(1), 1–9 (2012)
93. Q. Xiang, J. Yu, M. Jaroniec, Graphene-based semiconductor photocatalysts. Chem. Soc. Rev. **41**(2), 782–796 (2012)
94. Y. Li, G. Lu, S. Li, Photocatalytic production of hydrogen in single component and mixture systems of electron donors and monitoring adsorption of donors by in situ infrared spectroscopy. Chemosphere **52**(5), 843–850 (2003)
95. K. Maeda, M. Higashi, D. Lu, R. Abe, K. Domen, Efficient nonsacrificial water splitting through two-step photoexcitation by visible light using a modified oxynitride as a hydrogen evolution photocatalyst. J. Am. Chem. Soc. **132**(16), 5858–5868 (2010)
96. J. Yu, L. Qi, M. Jaroniec, Hydrogen production by photocatalytic water splitting over Pt/TiO_2 nanosheets with exposed (001) facets. J. Phys. Chem. C **114**(30), 13118–13125 (2010)
97. M. Murdoch, G.I.N. Waterhouse, M.A. Nadeem, J.B. Metson, M.A. Keane, R.F. Howe, J. Llorca, H. Idriss, The effect of gold loading and particle size on photocatalytic hydrogen production from ethanol over Au/TiO_2 nanoparticles. Nat. Chem. **3**(6), 489–492 (2011)
98. G. Liu, P. Niu, C. Sun, S.C. Smith, Z. Chen, G.Q. Lu, H.-M. Cheng, Unique electronic structure induced high photoreactivity of sulfur-doped graphitic C_3N_4. J. Am. Chem. Soc. **132**(33), 11642–11648 (2010)
99. A.S. Weber, A.M. Grady, R.T. Koodali, Lanthanide modified semiconductor photocatalysts. Cat. Sci. Technol. **2**(4), 683–693 (2012)
100. W. Kim, T. Tachikawa, T. Majima, W. Choi, Photocatalysis of dye-sensitized TiO_2 nanoparticles with thin overcoat of Al_2O_3: enhanced activity for H_2 production and dechlorination of CCl_4. J. Phys. Chem. C **113**(24), 10603–10609 (2009)
101. T. Fotiou, T.M. Triantis, T. Kaloudis, L.M. Pastrana-Martínez, V. Likodimos, P. Falaras, A.M.T. Silva, A. Hiskia, Photocatalytic degradation of microcystin-LR and off-odor compounds in water under UV-A and solar light with a nanostructured photocatalyst based on reduced graphene oxide–TiO_2 composite. Identification of intermediate products. Ind. Eng. Chem. Res. **52**(39), 13991–14000 (2013)
102. V.M. Mboula, V. Héquet, Y. Andrès, Y. Gru, R. Colin, J.M. Doña-Rodríguez, L.M. Pastrana-Martínez, A.M.T. Silva, M. Leleu, A.J. Tindall, S. Mateos, P. Falaras, Photocatalytic degra-

dation of estradiol under simulated solar light and assessment of estrogenic activity. Appl. Catal. B Environ. **162**(0), 437–444 (2015)
103. V. Maroga Mboula, V. Héquet, Y. Andrès, L.M. Pastrana-Martínez, J.M. Doña-Rodríguez, A.M.T. Silva, P. Falaras, Photocatalytic degradation of endocrine disruptor compounds under simulated solar light. Water Res. **47**(12), 3997–4005 (2013)
104. A. Katsoni, H.T. Gomes, L.M. Pastrana-Martínez, J.L. Faria, J.L. Figueiredo, D. Mantzavinos, A.M.T. Silva, Degradation of trinitrophenol by sequential catalytic wet air oxidation and solar TiO_2 photocatalysis. Chem. Eng. J. **172**(2–3), 634–640 (2011)
105. M. Cruz, C. Gomez, C.J. Duran-Valle, L.M. Pastrana-Martínez, J.L. Faria, A.M.T. Silva, M. Faraldos, A. Bahamonde, Bare TiO_2 and graphene oxide TiO_2 photocatalysts on the degradation of selected pesticides and influence of the water matrix. Appl. Surf. Sci. **416**, 1013–1021 (2017)
106. H. Liu et al., Chemical doping of graphene. J. Mater. Chem. **21**, 3335–3345 (2011)
107. X.-K. Kong, C.-L. Chen, Q.-W. Chen, Doped graphene for metal-free catalysis. Chem. Soc. Rev. **43**(8), 2841–2857 (2014)
108. X. Wang, G. Sun, P. Routh, D.H. Kim, W. Huang, P. Chen, Heteroatom-doped graphene materials: syntheses, properties and applications. Chem. Soc. Rev. **43**(20), 7067–7098 (2014)
109. L.K. Putri, W.J. Ong, W.S. Chang, S.P. Chai, Heteroatom doped graphene in photocatalysis: a review. Appl. Surf. Sci. **358**, 2–14 (2015)
110. M. Pedrosa, L.M. Pastrana-Martínez, M.F.R. Pereira, J.L. Faria, J.L. Figueiredo, A.M.T. Silva, N/S-doped graphene derivatives and TiO_2 for catalytic ozonation and photocatalysis of water pollutants. Chem. Eng. J. **348**, 888–897 (2018)
111. T.-D. Nguyen-Phan, V.H. Pham, E.W. Shin, H.-D. Pham, S. Kim, J.S. Chung, E.J. Kim, S.H. Hur, The role of graphene oxide content on the adsorption-enhanced photocatalysis of titanium dioxide/graphene oxide composites. Chem. Eng. J. **170**(1), 226–232 (2011)
112. L.M. Pastrana-Martínez, S. Morales-Torres, A.G. Kontos, N.G. Moustakas, J.L. Faria, J.M. Doña-Rodríguez, P. Falaras, A.M.T. Silva, TiO_2, surface modified TiO_2 and graphene oxide-TiO_2 photocatalysts for degradation of water pollutants under near-UV/Vis and visible light. Chem. Eng. J. **224**, 17–23 (2013)
113. L.K. Putri, L.-L. Tan, W.-J. Ong, W.S. Chang, S.-P. Chai, Graphene oxide: exploiting its unique properties toward visible-light-driven photocatalysis. Appl. Mater. Today **4**, 9–16 (2016)
114. M. Xing, W. Fang, X. Yang, B. Tian, J. Zhang, Highly-dispersed boron-doped graphene nanoribbons with enhanced conductibility and photocatalysis. Chem. Commun. **50**(50), 6637–6640 (2014)
115. Z.R. Tang, Y. Zhang, N. Zhang, Y.J. Xu, New insight into the enhanced visible light photocatalytic activity over boron-doped reduced graphene oxide. Nanoscale **7**(16), 7030–7034 (2015)
116. W. Peng, X. Li, Synthesis of a sulfur-graphene composite as an enhanced metal-free photocatalyst. Nano Res. **6**(4), 286–292 (2013)
117. A.S. Adeleye, X. Wang, F. Wang, R. Hao, W. Song, Y. Li, Photoreactivity of graphene oxide in aqueous system: reactive oxygen species formation and bisphenol A degradation. Chemosphere **195**, 344–350 (2018)
118. J. Oh, Y.H. Chang, Y.H. Kim, S. Park, Thickness-dependent photocatalytic performance of graphite oxide for degrading organic pollutants under visible light. Phys. Chem. Chem. Phys. **18**(16), 10882–10886 (2016)
119. H. Zhang, Y. Niu, W. Hu, Nitrogen/sulfur-doping of graphene with cysteine as a heteroatom source for oxygen reduction electrocatalysis. J. Colloid Interface Sci. **505**, 32–37 (2017)
120. M. Pedrosa, E.S. Da Silva, L.M. Pastrana-Martínez, G. Drazic, P. Falaras, J.L. Faria, J.L. Figueiredo, A.M.T. Silva, Hummers' and Brodie's graphene oxides as photocatalysts for phenol degradation. J. Colloid Interface Sci. **567**, 243–255 (2020)

121. C.G. Silva, J.L. Faria, Photocatalytic oxidation of benzene derivatives in aqueous suspensions: synergic effect induced by the introduction of carbon nanotubes in a TiO_2 matrix. Appl. Catal. B Environ. **101**(1–2), 81–89 (2010)
122. L.M. Pastrana-Martínez, S. Morales-Torres, S.K. Papageorgiou, F.K. Katsaros, G.E. Romanos, J.L. Figueiredo, J.L. Faria, P. Falaras, A.M.T. Silva, Photocatalytic behaviour of nanocarbon-TiO_2 composites and immobilization into hollow fibres. Appl. Catal. B Environ. **142-143**, 101–111 (2013)
123. M.J. Sampaio, C.G. Silva, A.M.T. Silva, L.M. Pastrana-Martínez, C. Han, S. Morales-Torres, J.L. Figueiredo, D.D. Dionysiou, J.L. Faria, Carbon-based TiO_2 materials for the degradation of Microcystin-LA. Appl. Catal. B Environ. **170-171**, 74–82 (2015)
124. M.J. Sampaio, L.M. Pastrana-Martinez, A.M.T. Silva, J.G. Buijnsters, C. Han, C.G. Silva, S.A.C. Carabineiro, D.D. Dionysiou, J.L. Faria, Nanodiamond-TiO_2 composites for photocatalytic degradation of microcystin-LA in aqueous solutions under simulated solar light. RSC Adv. **5**(72), 58363–58370 (2015)
125. L.M. Pastrana-Martínez, S. Morales-Torres, S.K. Papageorgiou, F.K. Katsaros, G.E. Romanos, J.L. Figueiredo, J.L. Faria, P. Falaras, A.M.T. Silva, Photocatalytic behaviour of nanocarbon–TiO_2 composites and immobilization into hollow fibres. Appl. Catal. B **142-143**, 101–111 (2013)
126. A.M. Schrand, S.A.C. Hens, O.A. Shenderova, Nanodiamond particles: properties and perspectives for bioapplications. Crit. Rev. Solid State Mater. Sci. **34**(1–2), 18–74 (2009)
127. V.N. Mochalin, O. Shenderova, D. Ho, Y. Gogotsi, The properties and applications of nanodiamonds. Nat. Nanotechol. **7**(1), 11–23 (2012)
128. R.R.N. Marques, M.J. Sampaio, P.M. Carrapiço, C.G. Silva, S. Morales-Torres, G. Dražić, J.L. Faria, A.M.T. Silva, Photocatalytic degradation of caffeine: developing solutions for emerging pollutants. Catal. Today **209**, 108–115 (2013)
129. K. Ikehata, M.G. El-Din, Aqueous pesticide degradation by ozonation and ozone-based advanced oxidation processes: a review (part II). Ozone Sci. Eng. **27**(3), 173–202 (2005)
130. K. Ikehata, M.G. El-Din, Aqueous pesticide degradation by ozonation and ozone-based advanced oxidation processes: a review (part I). Ozone Sci. Eng. **27**(2), 83–114 (2005)
131. J. Levec, A. Pintar, Catalytic wet-air oxidation processes: a review. Catal. Today **124**(3–4), 172–184 (2007)
132. R.S. Ribeiro, A.M.T. Silva, J.L. Figueiredo, J.L. Faria, H.T. Gomes, Catalytic wet peroxide oxidation: a route towards the application of hybrid magnetic carbon nanocomposites for the degradation of organic pollutants. A review. Appl. Catal. B Environ. **187**, 428–460 (2016)
133. R.P. Rocha, J. Restivo, J.P.S. Sousa, J.J.M. Órfão, M.F.R. Pereira, J.L. Figueiredo, Nitrogen-doped carbon xerogels as catalysts for advanced oxidation processes. Catal. Today **241**(0), 73–79 (2015)
134. J. Rivera-Utrilla, M. Sánchez-Polo, Ozonation of 1,3,6-naphthalenetrisulphonic acid catalysed by activated carbon in aqueous phase. Appl. Catal. B Environ. **39**(4), 319–329 (2002)
135. BRILLAS, Enrique; MUR, Eva; CASADO, Juan, Iron (II) Catalysis of the Mineralization of Aniline Using a Carbon-PTFE O 2-Fed Cathode. Journal of the Electrochemical Society. **143**(3), L49 (1996)
136. E. Brillas, I. Sirés, M.A. Oturan, Electro-fenton process and related electrochemical technologies based on Fenton's reaction chemistry. Chem. Rev. **109**(12), 6570–6631 (2009)
137. J.L. Figueiredo, M.F.R. Pereira, The role of surface chemistry in catalysis with carbons. Catal. Today **150**(1–2), 2–7 (2010)
138. H. Chen, G. Yang, Y. Feng, C. Shi, S. Xu, W. Cao, X. Zhang, Biodegradability enhancement of coking wastewater by catalytic wet air oxidation using aminated activated carbon as catalyst. Chem. Eng. J. **198–199**(0), 45–51 (2012)
139. R.P. Rocha, A.G. Gonçalves, L.M. Pastrana-Martínez, B.C. Bordoni, O.S.G.P. Soares, J.J.M. Órfão, J.L. Faria, J.L. Figueiredo, A.M.T. Silva, M.F.R. Pereira, Nitrogen-doped graphene-based materials for advanced oxidation processes. Catal. Today **249**, 192–198 (2015)

140. J.L. Diaz de Tuesta, A. Quintanilla, J.A. Casas, S. Morales-Torres, J.L. Faria, A.M.T. Silva, H.T. Gomes, The pH effect on the kinetics of 4-nitrophenol removal by CWPO with doped carbon black catalysts. Catal. Today (2019)
141. M. Martin-Martinez, B.F. Machado, P. Serp, S. Morales-Torres, A.M.T. Silva, J.L. Figueiredo, J.L. Faria, H.T. Gomes, Carbon nanotubes as catalysts for wet peroxide oxidation: the effect of surface chemistry. Catal. Today (2019)
142. L. Xing, Y. Xie, H. Cao, D. Minakata, Y. Zhang, J.C. Crittenden, Activated carbon-enhanced ozonation of oxalate attributed to HO oxidation in bulk solution and surface oxidation: effects of the type and number of basic sites. Chem. Eng. J. **245**(0), 71–79 (2014)
143. H. Cao, L. Xing, G. Wu, Y. Xie, S. Shi, Y. Zhang, D. Minakata, J.C. Crittenden, Promoting effect of nitration modification on activated carbon in the catalytic ozonation of oxalic acid. Appl. Catal. B Environ. **146**(0), 169–176 (2014)
144. S. Morales-Torres, A.M.T. Silva, F.J. Maldonado-Hódar, B.F. Machado, A.F. Pérez-Cadenas, J.L. Faria, J.L. Figueiredo, F. Carrasco-Marín, Pt-catalysts supported on activated carbons for catalytic wet air oxidation of aniline: activity and stability. Appl. Catal. B Environ. **105**(1–2), 86–94 (2011)
145. M. Kimura, I. Miyamoto, Discovery of the activated-carbon radical AC+ and the novel oxidation-reactions comprising the AC/AC+ cycle as a catalyst in an aqueous-solution. Bull. Chem. Soc. Jpn. **67**(9), 2357–2360 (1994)
146. P.R. Gogate, A.B. Pandit, A review of imperative technologies for wastewater treatment I: oxidation technologies at ambient conditions. Adv. Environ. Res. **8**(3–4), 501–551 (2004)
147. R.S. Ribeiro, A.M.T. Silva, J.L. Figueiredo, J.L. Faria, H.T. Gomes, The influence of structure and surface chemistry of carbon materials on the decomposition of hydrogen peroxide. Carbon **62**(0), 97–108 (2013)
148. F. Lücking, H. Köser, M. Jank, A. Ritter, Iron powder, graphite and activated carbon as catalysts for the oxidation of 4-chlorophenol with hydrogen peroxide in aqueous solution. Water Res. **32**(9), 2607–2614 (1998)
149. A. Dhaouadi, N. Adhoum, Heterogeneous catalytic wet peroxide oxidation of paraquat in the presence of modified activated carbon. Appl. Catal. B Environ. **97**(1–2), 227–235 (2010)
150. H.T. Gomes, S.M. Miranda, M.J. Sampaio, A.M.T. Silva, J.L. Faria, Activated carbons treated with sulphuric acid: catalysts for catalytic wet peroxide oxidation. Catal. Today **151**(1–2), 153–158 (2010)
151. C.M. Domínguez, P. Ocón, A. Quintanilla, J.A. Casas, J.J. Rodriguez, Graphite and carbon black materials as catalysts for wet peroxide oxidation. Appl. Catal. B Environ. **144**(0), 599–606 (2014)
152. R.S. Ribeiro, A.M.T. Silva, L.M. Pastrana-Martínez, J.L. Figueiredo, J.L. Faria, H.T. Gomes, Graphene-based materials for the catalytic wet peroxide oxidation of highly concentrated 4-nitrophenol solutions. Catal. Today **249**, 204–212 (2015)
153. J.H. Ramirez, F.J. Maldonado-Hódar, A.F. Pérez-Cadenas, C. Moreno-Castilla, C.A. Costa, L.M. Madeira, Azo-dye Orange II degradation by heterogeneous Fenton-like reaction using carbon-Fe catalysts. Appl. Catal. B Environ. **75**(3), 312–323 (2007)
154. F. Duarte, F.J. Maldonado-Hódar, A.F. Pérez-Cadenas, L.M. Madeira, Fenton-like degradation of azo-dye Orange II catalyzed by transition metals on carbon aerogels. Appl. Catal. B Environ. **85**(3), 139–147 (2009)
155. I. Mesquita, L.C. Matos, F. Duarte, F.J. Maldonado-Hódar, A. Mendes, L.M. Madeira, Treatment of azo dye-containing wastewater by a Fenton-like process in a continuous packed-bed reactor filled with activated carbon. J. Hazard. Mater. **237-238**, 30–37 (2012)
156. C.S.D. Rodrigues, O.S.G.P. Soares, M.T. Pinho, M.F.R. Pereira, L.M. Madeira, p-Nitrophenol degradation by heterogeneous Fenton's oxidation over activated carbon-based catalysts. Appl. Catal. B Environ. **219**, 109–122 (2017)
157. A.I. Zárate-Guzmán, L.V. González-Gutiérrez, L.A. Godínez, A. Medel-Reyes, F. Carrasco-Marín, L.A. Romero-Cano, Towards understanding of heterogeneous Fenton reaction using

carbon-Fe catalysts coupled to in-situ H_2O_2 electro-generation as clean technology for wastewater treatment. Chemosphere **224**, 698–706 (2019)

158. W. Zhou, L. Rajic, L. Chen, K. Kou, Y. Ding, X. Meng, Y. Wang, B. Mulaw, J. Gao, Y. Qin, A.N. Alshawabkeh, Activated carbon as effective cathode material in iron-free Electro-Fenton process: integrated H_2O_2 electrogeneration, activation, and pollutants adsorption. Electrochim. Acta **296**, 317–326 (2019)
159. W. Chen, X. Yang, J. Huang, Y. Zhu, Y. Zhou, Y. Yao, C. Li, Iron oxide containing graphene/carbon nanotube based carbon aerogel as an efficient E-Fenton cathode for the degradation of methyl blue. Electrochim. Acta **200**, 75–83 (2016)
160. N. Fernández-Sáez, D.E. Villela-Martinez, F. Carrasco-Marín, A.F. Pérez-Cadenas, L.M. Pastrana-Martínez, Heteroatom-doped graphene aerogels and carbon-magnetite catalysts for the heterogeneous electro-Fenton degradation of acetaminophen in aqueous solution. J. Catal. **378**, 68–79 (2019)
161. S. Malato, P. Fernández-Ibáñez, M.I. Maldonado, J. Blanco, W. Gernjak, Decontamination and disinfection of water by solar photocatalysis: recent overview and trends. Catal. Today **147**(1), 1–59 (2009)
162. L. Wang, Z. Yuan, H.E. Karahan, Y. Wang, X. Sui, F. Liu, Y. Chen, Nanocarbon materials in water disinfection: state-of-the-art and future directions. Nanoscale **11**(20), 9819–9839 (2019)
163. P. Fernández-Ibáñez, M.I. Polo-López, S. Malato, S. Wadhwa, J.W.J. Hamilton, P.S.M. Dunlop, R. D'Sa, E. Magee, K. O'Shea, D.D. Dionysiou, J.A. Byrne, Solar photocatalytic disinfection of water using titanium dioxide graphene composites. Chem. Eng. J. **261**, 36–44 (2015)
164. N.F.F. Moreira, C. Narciso-da-Rocha, M.I. Polo-López, L.M. Pastrana-Martínez, J.L. Faria, C.M. Manaia, P. Fernández-Ibáñez, O.C. Nunes, A.M.T. Silva, Solar treatment (H_2O_2, TiO_2-P25 and GO-TiO_2 photocatalysis, photo-Fenton) of organic micropollutants, human pathogen indicators, antibiotic resistant bacteria and related genes in urban wastewater. Water Res. **135**, 195–206 (2018)
165. X. Zeng, Z. Wang, N. Meng, D.T. McCarthy, A. Deletic, J.-h. Pan, X. Zhang, Highly dispersed TiO_2 nanocrystals and carbon dots on reduced graphene oxide: Ternary nanocomposites for accelerated photocatalytic water disinfection. Appl. Catal. B Environ. **202**, 33–41 (2017)
166. X. Zeng, Z. Wang, G. Wang, T.R. Gengenbach, D.T. McCarthy, A. Deletic, J. Yu, X. Zhang, Highly dispersed TiO_2 nanocrystals and WO_3 nanorods on reduced graphene oxide: Z-scheme photocatalysis system for accelerated photocatalytic water disinfection. Appl. Catal. B Environ. **218**, 163–173 (2017)
167. L. Chen, A. Pinto, A.N. Alshawabkeh, Activated carbon as a cathode for water disinfection through the electro-fenton process. Catalysts **9**(7), 601 (2019)

Chapter 13
Natural Zeolites as Sustainable Materials for Environmental Processes

Melodj Dosa, Marco Piumetti, Elahe Davarpanah, Giulia Moncaglieri, Samir Bensaid, and Debora Fino

13.1 Introduction

Zeolites are tectosilicates with three-dimensional aluminosilicate structure containing water molecules and metals in their structural framework [1, 2]. This particular framework can create different cavities interconnected to each other according to the chemical composition of the zeolite [2]. Zeolites with pore sizes lower than 2 nm are classified as microporous materials, while zeolites with larger pores (in the range between 2 and 50 nm) are known as mesoporous systems [3].

Zeolites are known as "molecular sieves" because of their high selectivity toward the molecules. In addition, these porous materials may interact with several compounds thanks to the presence of electrostatic forces into the cavities. Indeed, the presence of intra- and extra-framework cations can be possible for compensating the negative charges of $[AlO_4]^-$ groups [4–6]. Such unique physico-chemical properties render zeolites as sustainable materials for several environmental applications, including the treatment of air, water, and soil [7–13]. In fact, during the last decades, natural zeolites have received interest in academia and industry because of their potential applications [14, 15].

Among the natural zeolites, clinoptilolite is the most used. This natural zeolite is structurally characterized by two parallel channels with sizes 0.3 × 0.76 and 0.33 × 0.46 nm. The latter are connected with another one (0.26 × 0.47 nm) [16]. The general clinoptilolite unit cell formula is $(NaKCa)_4(Al_6Si_{30}O_{72})\cdot 24H_2O$, although the chemical composition also depends on the geographic area of mining districts [17–20]. Thanks to these physico-chemical properties, the clinoptilolite seems to be a suitable material for several environmental applications [12–15, 21–33].

M. Dosa · M. Piumetti (✉) · E. Davarpanah · G. Moncaglieri · S. Bensaid · D. Fino
Department of Applied Science and Technology, Corso Duca degli Abruzzi, Turin, Italy
e-mail: marco.piumetti@polito.it

M. Piumetti, S. Bensaid (eds.), *Nanostructured Catalysts for Environmental Applications*, https://doi.org/10.1007/978-3-030-58934-9_13

The actual technologies for CO_2 capturing are often expensive [34–36] and the clinoptilolite may represent a sustainable alternative for specific applications, as the cost can be lower than 200 \$ tons^{-1} [37]. Several studies have investigated the CO_2 adsorption capacity of clinoptilolite, confirming that CO_2 molecules are mainly adsorbed by the material through Van der Waals forces. Furthermore, modifications of the material may increase the effectiveness in capture and selectivity [40]. In fact, the presence of extra framework cations in the solid promotes the adsorption capacity [26, 38–42]. It has been demonstrated that the CO_2 adsorption capacity of clinoptilolite follows these orders: $Cs^+ > Rb^+ > K^+ > Na^+ > Li^+$ (for alkaline cations) and $Ba^{2+} > Sr^{2+} > Ca^{2+} > Mg^{2+}$ (for alkali cations).

On the other hand, the clinoptilolite can be an efficient material for the removal of organic molecules, like dyes, in the wastewater [13, 25, 28, 43–45]. These substances are the wastes of several industries (i.e., textile industry) that use dyes for their products [46, 47]. During the last years, several de-colorization processes have been proposed, since the dyes are resistant to aerobic digestion and chemically stable [25, 45, 46, 48–54]. Among these techniques, adsorption and degradation represent the best methods because of the efficient removal of dyes, versatility, and low costs [45, 48]. Thus, during the last decades, several adsorbents have been studies, including carbons, zeolites, clays, and so on [25]. On the other hand, the degradation of dyes can be operated by Fenton-like processes, in which the catalyst (e.g. Fe species in either homogeneous or heterogeneous systems) is able to decompose the dye in the presence of H_2O_2 and ascorbic acid [55–57]. For heterogeneous Fenton-type systems, micro- and mesoporous materials can be used as supports [55–57] and then clinoptilolite appears a suitable candidate.

In this chapter, the adsorption properties of clinoptilolite and Na-containing clinoptilolite are investigated for considering the possible use of this material in CO_2 separation processes and wastewater treatments. In the latter case, a Fe-containing clinoptilolite sample is prepared and tested for the catalytic abatement of the azo dye Acid Orange 7 (AO7) via a Fenton-like mechanism.

13.2 Experimental Part

13.2.1 Cation Exchange Preparation

The Zeolado company (Greece) provided the clinoptilolite used for this study. Clinoptilolite was washed with ultrapure water to remove impurities. After this step, the clinoptilolite was placed into drying oven and heated overnight at 60 °C. The washed clinoptilolite, herein labeled as "Clin," was used for the adsorption tests as well as for the preparation of Na- and Fe-containing clinoptilolite (Na/Clin and Fe/Clin, respectively) by the ion exchange method [13, 57, 58]. Briefly, 3 g of Clin were added to a solution containing proper amounts of Fe_2Cl or NaCl (Sigma-Aldrich) and stirred at 50 °C for 2 h. Then, the solution was centrifuged, washed, and dried overnight at 60 °C. Finally, the dried powder was calcined at 550 °C for 4 h.

The Na/Clin sample is used as adsorbent material for CO_2 capture and Fe/Clin is used as catalyst for the degradation of Acid Orange 7 (AO7).

13.2.2 *Characterization Techniques*

The powder X-ray diffraction patterns were collected on a X'Pert Philips PW3040 diffractometer using Cu Ka radiation (2θ range = 5–70, step = 0.013θ, time per step = 0.2 s). The diffraction peaks were indexed according to the Powder Data File database (PDF-21999, International Centre of Diffraction Data).

The N_2 physisorption at −196 °C, performed using a Micromeritics Tristar II 3020 (v1.03, Micromeritics Instrument Corp.), was used to evaluate the total pore volume (V_p), the micropore volume (V_{micro}), and the specific surface area (SSA, calculated using the Brunauer–Emmett–Teller method) on samples previously outgassed at 200 °C for 4 h. The V_p was estimated by the Barrett-Joyner Halenda (BJH) method, during the desorption phase. The V_{micro} was evaluated through the *t*-plot method.

The morphology of the samples was examined using the field emission scanning electron microscopy (FESEM Zeiss MERLIN, Gemini-II column). Elemental analysis was carried out via energy dispersive X-ray (EDX) analysis (AZTec, Oxford Instruments).

13.2.3 *Adsorption and Catalytic Activity Tests*

13.2.3.1 CO_2 Adsorption Tests

The adsorbent material was placed in a quartz U-shaped reactor with a sorbent bed (height = 3 cm, diameter = 1 cm). The reactor was put inside a furnace and connected to mass flow controllers that regulate the inlet gas mixture. The outlet was connected to an analyzer in order to analyze the outflow composition. The temperature was controlled by K-type thermocouple positioned above the catalytic bed of the adsorbent material.

The chemical composition of the inlet gas (40 mL min^{-1}) was 10 vol% CO_2 in N_2. Before each CO_2 adsorption test, a pretreatment at 100 °C for 1 h was performed to eliminate atmospheric contaminants, pollutants, and other molecules adsorbed on the sample surface. Then, the furnace was cooled down to reach the operative temperature. For each adsorption test, the adsorbed amount of CO_2 can be calculated using the Eqs. (13.1)–(13.3):

$$CO_{2_{adsorbed}} = \int_{t_0+\Delta t}^{t_f} \left(1 - \frac{C_{CO_{2out}}}{C_0}\right) d(Q_{out} t) \tag{13.1}$$

$$Q_{out} = Q_{in} \frac{1 - y_{CO_2}}{1 - y_{CO_2} \frac{C_{CO_{2out}}}{C_0}} \tag{13.2}$$

$$n_{CO_2(t)} = \frac{y_{CO_2} PV}{RT} \tag{13.3}$$

Where:

- $C_{CO_{2out}}$ is the volumetric concentration of the CO_2 measured by the gas analyzer.
- $Q_{out}\left(\frac{m^3}{s}\right)$ is the outlet volumetric flow rate of the reactor.
- y_{CO_2} is the molar fraction of the CO_2 (inlet gas).
- C_0 is the initial concentration of CO_2 (inlet gas).
- Δt (s) is the time delay for the gas analyzer.

Moreover, the adsorption loading capacities for a proper amount (weight) of sample was calculated by the Eq. (13.4):

$$\alpha_{zeolite} = \frac{n_{CO_{2adsorbed}}}{m_{sorbent}} \tag{13.4}$$

13.2.3.2 Catalytic Activity Tests

The AO7 was chosen as a probe molecule for azo dyes in wastewater that can be decomposed via Fenton-type reactions. Briefly, 1 g L^{-1} of powder (Fe/Clin), proper amounts of H_2O_2 and ascorbic acid (namely "AA") were put inside a AO7 solution (350 mg L^{-1}, pH = 6.8) in order to investigate the Fenton-like mechanism. H_2O_2 and AA were added so that have several possible combinations.

A test was performed with equal amounts of Fe^{2+}, H_2O_2 and AA (2.67 mM, stoichiometric balanced), namely 1:1:1. Other possible combinations were performed in order to have the excess of 10, 20 or 30 times with respect to Fe^{2+}, namely 1:10:10, 1:20:20 or 1:30:30.

During a typical test, the Fenton reagents (powder Fe-containing catalyst, H_2O_2 and AA) were added to a AO7 solution and mixed in dark condition for 150 min (r.t.). Then, aliquots of the suspension were collected every 10 min. The latter were centrifuged and then the liquid was analyzed by a UV-Vis spectrophotometer (Cary 5000 UV-Vis-NIR spectrophotometer, Varian instrument). The corresponding AO7 concentration was evaluated by the intensity of the peak at 485 nm, corresponding to the hydrazone form (Scheme 13.1). The hydrazone form is stable in the solid phase. However, when AO7 is found in solution, it goes to a tautomerism (azo form) by an intramolecular proton allocation [55–57]. The absorbance peaks at 230, 256, and 310 are due to the presence of aromatic ring into AO7 structure. On the other hand, the peak at 485 nm is due to the π-π* transition. This peak is related to the lone pair on N atoms and the conjugated system extending over the two aromatic moieties and encompassing the N–N group of the hydrazone form.

The signal at 403 nm has a similar nature, involving the N–N group of the azo form. The AO7 conversion (%) was calculated using the following equuation (13.5):

$$\text{AO7 conversion}\left(\%\right) = \frac{C_0 - C_t}{C_0} \tag{13.5}$$

where C_0 is the initial AO7 concentration (350 mg L^{-1}) and C_t is the concentration at a specific reaction time.

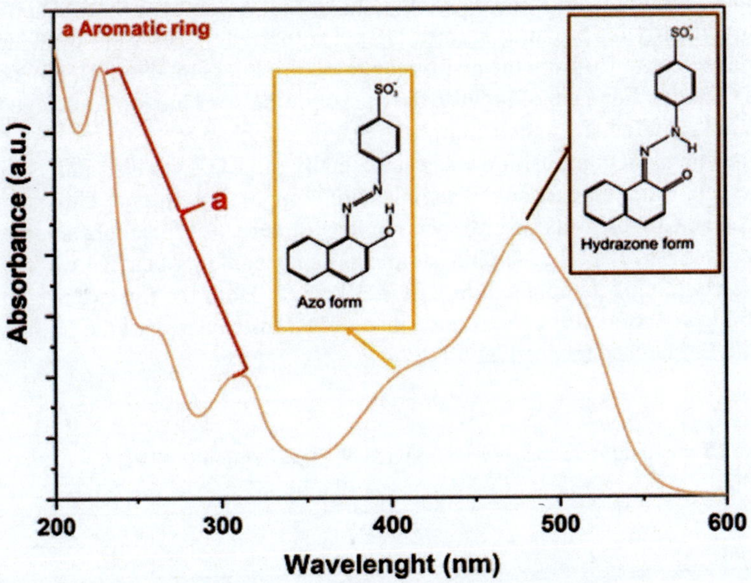

Scheme 13.1 AO7 spectrum along with the corresponding molecular structures

13.3 Results and Discussion

13.3.1 N_2 Physisorption at −196 °C and XRD

In Table 13.1 the textural results are reported as derived from the N_2 physisorption at −196 °C. As a whole, all the samples exhibit similar surfaces areas (32–37 $m^2\ g^{-1}$) and total pore volume (0.12–0.14 $cm^3\ g^{-1}$). On the other hand, a lower micropore volume can be achieved with the Fe- and Na-containing samples.

In Fig. 13.1 the X-ray diffractogram for the Clin sample is reported, along with the reference code of the clinoptilolite in PDF2 database (00-039-1383).

Additional peaks related to the presence of Mordenite are present in the sample whereas there is no signal corresponding to α-Quartz phase (references codes 00-022-1339 and 01-085-1054 respectively) [20, 59, 60]. The X-ray diffractograms of the ion exchanged samples are not reported in this work for sake of brevity.

13.3.2 FESEM and EDX Analysis

In Fig. 13.2 are reported both the FESEM and TEM images of the Clin sample. It appears that this material is characterized by self-assembled flake-shape particles, in agreement with previous studies [19]. The particle sizes were evaluated by the ImageJ software [61] and the results suggest a thickness of about 35 nm and a length of ca. 120 nm. For sake of brevity, the micrographics for the ion exchanged samples were not reported since their morphologies do not change.

The chemical composition was evaluated by the EDX analysis and is reported in Table 13.2. The ion exchanged samples are compared with pure Clin sample. As a whole, the Clin exhibits several metal in its structure, as typical observed with natural zeolites [18, 20]. The Na/Clin sample has a Na content of ca. 2.4 wt.%. Similarly, the Fe-containing sample exhibits Fe ~ 3.1 wt.%. Both the ion exchanged samples exhibit lower contents of Ca, K, etc., due to the ion-exchange of the framework with Na and Fe species, respectively.

Table 13.1 Textural results as derived from the N_2 physisorption at −196 °C

Sample	SSA ($m^2\ g^{-1}$)[a]	V_p ($cm^3\ g^{-1}$)[b]	V_{micro} ($cm^3\ g^{-1}$)[c]
Clin	37	0.14	0.005
Na/Clin	32	0.14	0.004
Fe/Clin	33	0.12	0.003

[a]Specific surface area evaluated by the BET method
[b]Pore volume evaluated by the BJH method during the desorption phase
[c]Micropore volume evaluated by the t-plot method

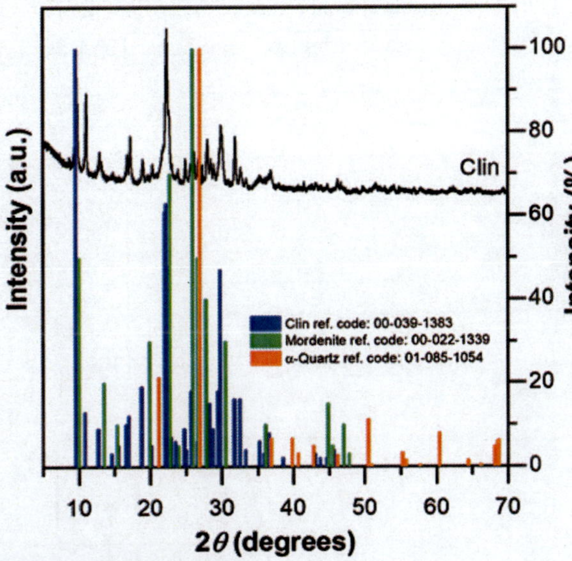

Fig. 13.1 X-ray diffractogram of the clinoptilolite (Clin sample)

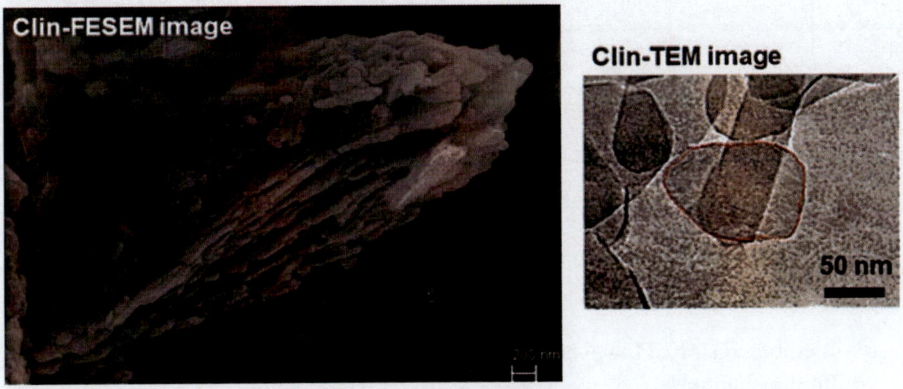

Fig. 13.2 FESEM (left) and TEM (right) images of the clinoptilolite (Clin sample)

13.3.3 Adsorption and Catalytic Activity Tests

13.3.3.1 CO_2 Adsorption Tests

The results of the CO_2 adsorption tests with Clin and Na/Clin, performed at 20 °C, are reported in Fig. 13.3. By comparing their relevant breakthrough curves, it appears that the Na/Clin exhibits higher adsorption performances. Their corresponding adsorption breakthrough times are 15 and 10 min. For Na/Clin and Clin, respectively. The higher adsorption uptake of the Na-exchanged sample can be explained by its higher charge density and electrostatic strength [41]. The Na^+ species are localized in the middle of the largest channels in Clin (Channel A with

Table 13.2 Chemical composition of the samples as evaluated by the EDX analysis

Elements	Clin weight (%)	Na/Clin weight (%)	Fe/Clin weight (%)
O	41.7	53.2	55.0
Mg	0.5	0.6	0.7
Na	–	2.4	–
Al	7.8	5.9	6.2
Si	39.9	35.4	31.2
K	5.5	0.7	1.8
Ca	2.8	0.5	2.1
Fe	1.7	1.3	3.1
Tot.	100.0	100.0	100.0

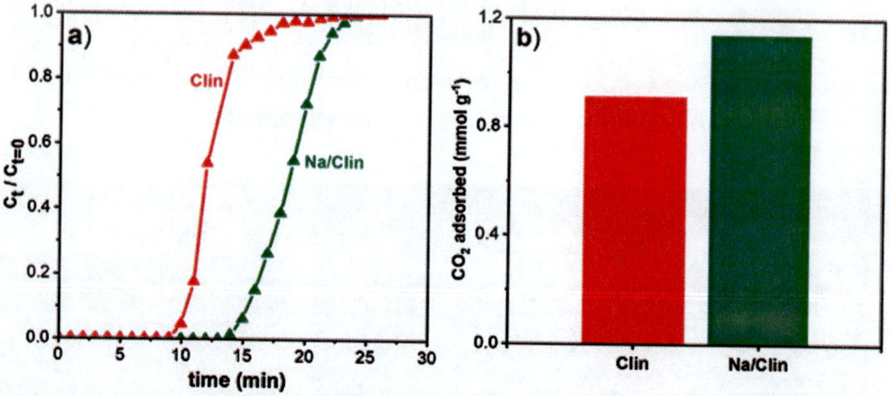

Fig. 13.3 (**a**) CO_2 adsorption breakthrough curves over the time for Clin and Na/Clin; (**b**) CO_2 adsorption capacity (mmol g^{-1}) of the studied samples

ten-membered ring). However, the latter species do not limit mass-transfer diffusion within the channels.

In Fig. 13.3b are reported the CO_2 adsorption capacities for the studied samples. The following increasing adsorption trend can be drawn: Clin < Na/Clin. This tendency can be explained by the different physico-chemical properties of the samples (textural properties, chemical composition, adsorption sites distribution, electrostatic force, etc.). In addition, the mass-transfer phenomena play a crucial role for the CO_2 adsorption in porous material [62].

Similarly, the CO_2 adsorption capacities at 65 °C were evaluated to study the adsorption capacity of the materials at higher temperature. The results are summarized in Fig. 13.4.

As a whole, lower adsorption capacities were observed at 65 °C compared to 20 °C, due to the exothermicity of the adsorption phenomenon. Noteworthy, better adsorption capacities at 65 °C were observed for the Clin sample rather than the Na-Clin, thus suggesting that pure clinoptilolite can be effective in this temperature range.

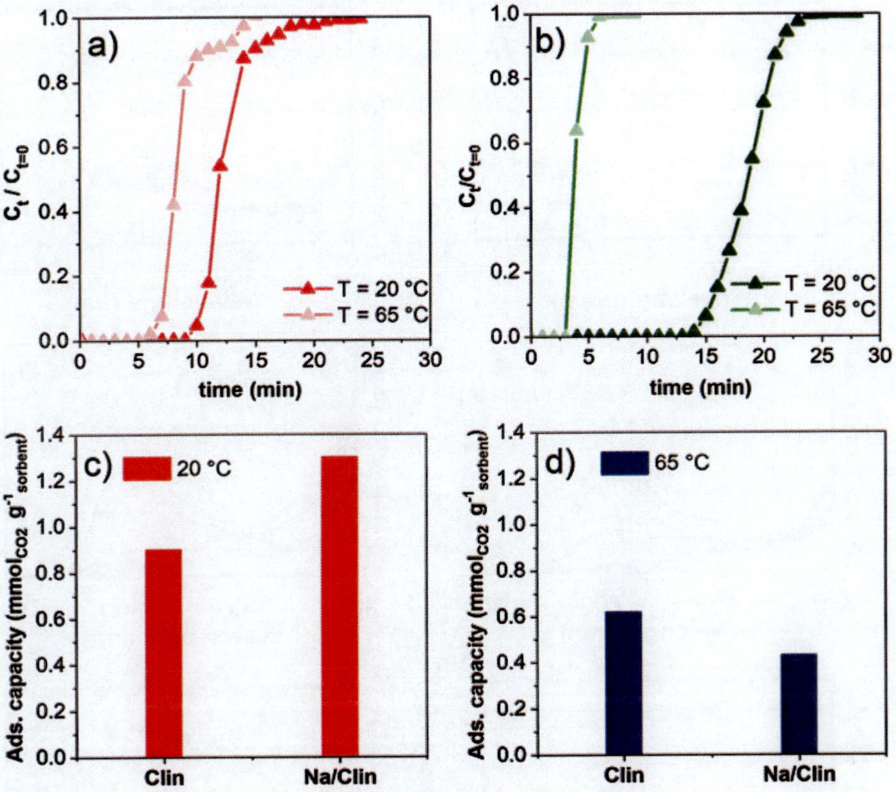

Fig. 13.4 Adsorption breakthrough curves evaluated at different temperatures (20 and 65 °C) over (**a**) Clin and (**b**) Na/Clin. (**c**) Comparison of the adsorption capacities evaluated at 20 °C and (**d**) at 65 °C

13.3.3.2 AO7 Abatement Via Fenton-Like Reaction

In Fig. 13.5 the results for the AO7 degradation are reported. The UV-Vis spectra as a function of time are reported in Fig. 13.5. As a whole, no reaction occurs with the only presence of Fe/Clin in the AO7 solution (Section A). However, when H_2O_2 and AA are added to the AO7 solution (containing Fe/Clin), the degradation process takes place as confirmed with the degreasing intensity of the signal at 486 nm. The AO7 degradation occurs via a Fenton-like mechanism (Eqs. (13.6)–(13.10)):

$$Fe^{2+} + H_2O_2 \rightarrow Fe^{3+} + OH^- + HO\cdot \quad (13.6)$$

$$Fe^{3+} + H_2O_2 \rightarrow Fe^{2+} + H^+ + HO_2\cdot \quad (13.7)$$

$$HO\cdot + H_2O_2 \rightarrow HO_2\cdot + H_2O \quad (13.8)$$

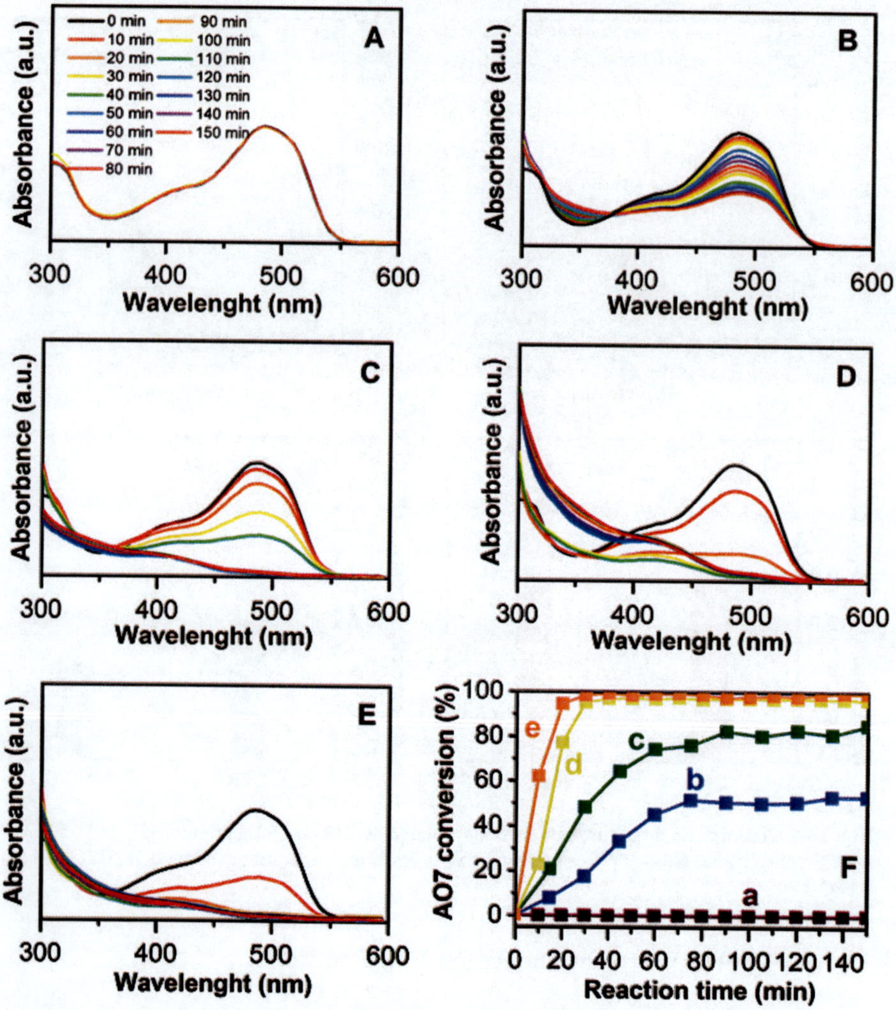

Fig. 13.5 Absorbance peaks over time (from 0 to 150 min) of: (**a**) Fe/Clin, (**b**) Fe/Clin:H_2O_2:AA (1:1:1), (**c**) Fe/Clin:H_2O_2:AA (1:10:10), (**d**) Fe/Clin:H_2O_2:AA (1:20:20) and (**e**) Fe/Clin:H_2O_2:AA (1:30:30). (**f**) AO7 conversion (%) over time for *a*-Fe/Clin, *b*-Fe/Clin:H_2O_2:AA (1:1:1), *c*-Fe/Clin:H_2O_2:AA (1:10:10), *d*-Fe/Clin:H_2O_2:AA (1:20:20) and *e*-Fe/Clin:H_2O_2:AA (1:30:30)

$$HO_2\cdot + H_2O_2 \rightarrow HO\cdot + H_2O + O_2 \tag{13.9}$$

$$AO7 + HO\cdot \rightarrow \text{degradation products} \tag{13.10}$$

The presence of both reagents (H_2O_2 and AA), along with Fe^{2+}/Fe^{3+} species, initializes the Fenton-type reaction. Indeed, the presence of H_2O_2 and AA contributes to the formation of ·OH radicals [13, 56–58].

Since both H_2O_2 and AA are key reagents for this Fenton-type reaction, it was analyzed how the AO7 conversion can be modified when the two reactants are in large excess compared to the Fe content. The first test is reported in Fig. 13.5c: H_2O_2 and AA are ten times in excess of the Fe content. Comparing with the previous case study in equimolar conditions, the adsorption spectra over the time decrease significantly and reach almost the total abatement of AO7. However, when the excess is increased again with the second and the third tests (1:20:20 and 1:30:30) the decreasing of the peak at 486 nm is much more evident.

These outcomes are resumed and elaborated in Fig. 13.5f. The absorbance peaks are converted to a specific AO7 concentration. Then, the AO7 concentration is used to elaborate the AO7 conversion by Eq. (13.2) (*see* Sect. 13.2.3): each absorbance peak represents a point of each curve.

In Fig. 13.5f it is possible to appreciate and clearly observe the effect on AO7 abatement by the amount of H_2O_2 and AA. With high excess of both reagents (1:20:20 curve d and 1:30:30 curve e) it is possible to obtain the 100% of abatement after 30 min. Moreover, the case 1:10:10 (curve c) shows interesting results for higher reaction time (almost 80% of abatement at 150 min). However, with equal molar amount of reagent respect to Fe, 1:1:1 curve b, the conversion does not reach the complete conversion (only about 50% at 150 min). Finally, as expected by the corresponding absorbance peaks in Fig. 13.5a, the non-catalytic reaction is 0% of AO7 abatement. Depending on the operating conditions, it is possible to vary the reagent amounts to perform the Fenton-like reaction: less reaction time is required if the test is performed with higher excess of reagents. On the other hand, it is possible to work with less amount of H_2O_2 and AA but with longer time to render the degradation process possible.

13.4 Conclusions

In this chapter, the application of clinoptilolite was investigated for capturing carbon dioxide (CO_2) emitted from industrial processes at moderate outlet temperatures and for wastewater remediation.

For CO_2 capture and storage, the clinoptilolite was used as adsorbent solid and it was tested at 20 and 65 °C. The results suggest that the clinoptilolite can be a sustainable material for capturing CO_2, because of the interesting adsorption capacity at moderate temperature.

On the other hand, the clinoptilolite was studied as a catalyst for the dye abatement in wastewater remediation. The AO7 was considered as a dye model molecule and the Fenton-like reaction was investigated using a Fe-containing clinoptilolite sample. The results demonstrated that it is possible to reach the 100% of AO7 abatement after 30 min if the test is performed with higher excess of H_2O_2 and AA.

Thus, both the clinoptilolite and the ion-exchanged clinoptilolite can be used as possible alternatives for environmental applications.

Acknowledgments The authors thank the Zeolado Company (Greece) that provided the clinoptilolite for this study.

References

1. A.F. Cronstedt, Observation and description of an unknown kind of rock to be named zeolites. Kongl. Vetenskaps Acad. Handl. Stock. **17**, 120–123 (1756)
2. R.W. Broach, D.-Y. Jan, D.A. Lesch, S. Kulprathipanja, E. Roland, P. Kleinschmit, Zeolites, in *Ullmann's Encyclopedia of Industrial Chemistry*, (Wiley-VCH Verlag GmbH & Co. KGaA, Weinheim, 2012). https://doi.org/10.1002/14356007.a28_475.pub2
3. M. Sprynskyy, R. Golembiewski, G. Trykowski, B. Buszewski, Heterogeneity and hierarchy of clinoptilolite porosity. J. Phys. Chem. Solids **71**, 1269–1277 (2010). https://doi.org/10.1016/j.jpcs.2010.05.006
4. Y. Li, J.N. Armor, Selective catalytic reduction of NOx with methane over metal exchange zeolites. Appl. Catal. B Environ. **2**, 239–256 (1993). https://doi.org/10.1016/0926-3373(93)80051-E
5. L. Ćurković, Š. Cerjan-Stefanović, T. Filipan, Metal ion exchange by natural and modified zeolites. Water Res. **31**, 1379–1382 (1997). https://doi.org/10.1016/S0043-1354(96)00411-3
6. M. Iwamoto, H. Yahiro, Y. Mine, S. Kagawa, Excessively copper ion-exchanged ZSM-5 zeolites as highly active catalysts for direct decomposition of nitrogen monoxide. Chem. Lett. **18**, 213–216 (1989). https://doi.org/10.1246/cl.1989.213
7. T. Bein, Host-guest interactions in zeolites and periodic mesoporous materials. Stud. Surf. Sci. Catal. **168**, 611-XIX (2007). https://doi.org/10.1016/S0167-2991(07)80806-8
8. K. Tanabe, Industrial application of solid acid–base catalysts. Appl. Catal. A Gen. **181**, 399–434 (1999). https://doi.org/10.1016/S0926-860X(98)00397-4
9. P.B. Venuto, Organic catalysis over zeolites: a perspective on reaction paths within micropores. Micropor. Mater. **2**, 297–411 (1994). https://doi.org/10.1016/0927-6513(94)00002-6
10. K. Sakaguchi, M. Matsui, F. Mizukami, Applications of zeolite inorganic composites in biotechnology: current state and perspectives. Appl. Microbiol. Biotechnol. **67**, 306–311 (2005). https://doi.org/10.1007/s00253-004-1782-4
11. B. Yilmaz, A. Sacco, J. Deng, Electrical transport through monatomic titania chains. Appl. Phys. Lett. **90**, 152101 (2007). https://doi.org/10.1063/1.2720742
12. M. Dosa, M. Piumetti, S. Bensaid, N. Russo, O. Baglieri, F. Miglietta, D. Fino, Properties of the clinoptilolite: characterization and adsorption tests with methylene blue. J. Adv. Catal. Sci. Technol. **5**, 1–10 (2018). https://www.cosmosscholars.com/phms/index.php/jacst/article/view/867/588
13. M. Dosa, M. Piumetti, C. Galletti, N. Russo, D. Fino, S. Bensaid, G. Mancini, F.S. Freyria, G. Saracco, A novel Fe-containing clinoptilolite for wastewater remediation: degradation of azo-dyes acid orange 7 by H2O2 and ascorbic acid. Desalin. Water Treat. **159**, 121–129 (2019). https://doi.org/10.5004/dwt.2019.24424
14. E. Chmielewská-Horváthová, J. Lesný, Study of sorption equilibria in the systems: water solutions of inorganic ions—clinoptilolite. J. Radioanal. Nucl. Chem. Lett. **201**, 293–301 (1995). https://doi.org/10.1007/BF02164048
15. A. Aquino, E. Bonamente, C. Buratti, F. Cotana, B. Castellani, V. Paolini, F. Petracchini, Carbon dioxide removal with tuff: experimental measurement of adsorption properties and breakthrough modeling using CFD approach. Energy Procedia **101**, 392–399 (2016). https://doi.org/10.1016/j.egypro.2016.11.050
16. D.A. Kennedy, F.H. Tezel, Cation exchange modification of clinoptilolite – screening analysis for potential equilibrium and kinetic adsorption separations involving methane, nitrogen, and carbon dioxide. Micropor. Mesopor. Mater. **262**, 235–250 (2018). https://doi.org/10.1016/j.micromeso.2017.11.054

17. F. Mumpton, Clinoptilolite redefined. Am. Mineral. **45**, 351–369 (1960)., http://ci.nii.ac.jp/naid/10011941740/. Accessed 27 Sept 2017
18. G. Tsitsishvili, T. Andronikashvili, G. Kirov, L. Filizova, *Natural Zeolites* (Ellis Horwood, Chichester, 1992). https://doi.org/10.1163/_q3_SIM_00374
19. F.A. Mumpton, W.C. Ormsby, Morphology of zeolites in sedimentary rocks by scanning electron microscopy. Clay Clay Miner. **24**, 1–23 (1976). https://doi.org/10.1346/CCMN.1976.0240101
20. F.A. Mumpton, La roca magica: uses of natural zeolites in agriculture and industry. Proc. Natl. Acad. Sci. U. S. A. **96**, 3463–3470 (1999). https://doi.org/10.1073/pnas.96.7.3463
21. A.E. Osmanlioglu, Treatment of radioactive liquid waste by sorption on natural zeolite in Turkey. J. Hazard. Mater. **137**, 332–335 (2006). https://doi.org/10.1016/j.jhazmat.2006.02.013
22. E. Erdem, N. Karapinar, R. Donat, The removal of heavy metal cations by natural zeolites. J. Colloid Interface Sci. **280**, 309–314 (2004). https://doi.org/10.1016/j.jcis.2004.08.028
23. A. Arefi Pour, S. Sharifnia, R. NeishaboriSalehi, M. Ghodrati, Performance evaluation of clinoptilolite and 13X zeolites in CO2 separation from CO2/CH4 mixture. J. Nat. Gas Sci. Eng. **26**, 1246–1253 (2015). https://doi.org/10.1016/j.jngse.2015.08.033
24. G. Aguilar-armenta, M.E. Patiño-iglesias, R.L. Centro, Adsorption kinetic behaviour of pure CO2, N2 and CH4 in natural clinoptilolite at different temperatures. Adsorpt. Sci. Technol. **21**, 81–91 (2003). https://doi.org/10.1260/026361703606699831
25. M. Qiu, C. Qian, J. Xu, J. Wu, G. Wang, Studies on the adsorption of dyes into clinoptilolite. Desalination **243**, 286–292 (2009). https://doi.org/10.1016/j.desal.2008.04.029
26. D.S. Karousos, A.A. Sapalidis, E.P. Kouvelos, G.E. Romanos, N.K. Kanellopoulos, A study on natural clinoptilolite for CO_2/N_2 gas separation. Sep. Sci. Technol. **51**, 83–95 (2016). https://doi.org/10.1080/01496395.2015.1085880
27. C. Galletti, M. Dosa, N. Russo, D. Fino, Zn2+ and Cd2+ removal from wastewater using clinoptilolite as adsorbent. Environ. Sci. Pollut. Res. (2020). https://doi.org/10.1007/s11356-020-08483-z
28. S. Wang, Y. Peng, Natural zeolites as effective adsorbents in water and wastewater treatment. Chem. Eng. J. **156**, 11–24 (2010). https://doi.org/10.1016/j.cej.2009.10.029
29. N. Gargiulo, F. Pepe, D. Caputo, CO2 adsorption by functionalized nanoporous materials: a review. J. Nanosci. Nanotechnol. **14**, 1811–1822 (2014). https://doi.org/10.1166/jnn.2014.8893
30. M.S. Kamal, S.A. Razzak, M.M. Hossain, Catalytic oxidation of volatile organic compounds (VOCs)—a review. Atmos. Environ. **140**, 117–134 (2016). https://doi.org/10.1016/j.atmosenv.2016.05.031
31. E. Muzenda, J. Kabuba, F.N.M. Mollagee, Kinetics study of ammonia removal from synthetic waste water. Int. J. Chem. Mol. Nucl. Mater. Metall. Eng. 5(7), 524–527 (2011). http://waset.org/publications/1075/kinetics-study-of-ammonia-removal-from-synthetic-waste-water
32. T.E. Rufford, S. Smart, G.C.Y. Watson, B.F. Graham, J. Boxall, J.C. Diniz da Costa, E.F. May, The removal of CO2 and N2 from natural gas: a review of conventional and emerging process technologies. J. Pet. Sci. Eng. **94–95**, 123–154 (2012). https://doi.org/10.1016/j.petrol.2012.06.016
33. S.U. Rege, R.T. Yang, M.A. Buzanowski, Sorbents for air prepurification in air separation. Chem. Eng. Sci. **55**, 4827–4838 (2000). https://doi.org/10.1016/S0009-2509(00)00122-6
34. M. Khraisheh, S. Mukherjee, A. Kumar, F. Al Momani, G. Walker, M.J. Zaworotko, An overview on trace CO2 removal by advanced physisorbent materials. J. Environ. Manage. **255**, 109874 (2020). https://doi.org/10.1016/j.jenvman.2019.109874
35. D. Saha, Z. Bao, F. Jia, S. Deng, Adsorption of CO2, CH4, N2O, and N2 on MOF-5, MOF-177, and zeolite 5A. Environ. Sci. Technol. **44**, 1820–1826 (2010). https://doi.org/10.1021/es9032309
36. S. Sjostrom, H. Krutka, Evaluation of solid sorbents as a retrofit technology for CO2 capture. Fuel **89**, 1298–1306 (2010). https://doi.org/10.1016/j.fuel.2009.11.019
37. V.J. Inglezakis, A.A. Zorpas, *Handbook of Natural Zeolites* (Bentham Science Publishers, Sharjah, 2012). https://doi.org/10.2174/97816080526151120101

38. D.A. Kennedy, M. Mujčin, C. Abou-Zeid, F.H. Tezel, Cation exchange modification of clinoptilolite–thermodynamic effects on adsorption separations of carbon dioxide, methane, and nitrogen. Micropor. Mesopor. Mater. **274**, 327–341 (2019). https://doi.org/10.1016/j.micromeso.2018.08.035
39. G. Aguilar-Armenta, G. Hernandez-Ramirez, E. Flores-Loyola, A. Ugarte-Castaneda, R. Silva-Gonzalez, C. Tabares-Munoz, A. Jimenez-Lopez, E. Rodriguez-Castellon, Adsorption kinetics of CO2, O2, N2, and CH4 in cation-exchanged clinoptilolite. J. Phys. Chem. B **105**, 1313–1319 (2001). https://doi.org/10.1021/jp9934331
40. R.V. Siriwardane, M.S. Shen, E.P. Fisher, Adsorption of CO2, N2, and O2 on natural zeolites. Energy Fuels **17**, 571–576 (2003). https://doi.org/10.1021/ef0201351
41. M.W. Ackley, R.F. Giese, R.T. Yang, Clinoptilolite: untapped potential for kinetics gas separations. Zeolites **12**, 780–788 (1992). https://doi.org/10.1016/0144-2449(92)90050-Y
42. R.T. Pabalan, F.P. Bertetti, Cation-exchange properties of natural zeolites. Rev. Mineral. Geochem. **45**, 453–517 (2001). https://doi.org/10.2138/rmg.2001.45.14
43. D. Karadag, Modeling the mechanism, equilibrium and kinetics for the adsorption of Acid Orange 8 onto surfactant-modified clinoptilolite: the application of nonlinear regression analysis. Dyes Pigments **74**, 659–664 (2007). https://doi.org/10.1016/j.dyepig.2006.04.009
44. B. Armağan, M. Turan, O. Özdemir, M.S. Çelik, Color removal of reactive dyes from water by clinoptilolite. J. Environ. Sci. Heal. Part A Toxic Hazard. Subst. Environ. Eng. **39**, 1251–1261 (2004). https://doi.org/10.1081/ESE-120030329
45. T. Sismanoglu, Y. Kismir, S. Karakus, Single and binary adsorption of reactive dyes from aqueous solutions onto clinoptilolite. J. Hazard. Mater. **184**, 164–169 (2010). https://doi.org/10.1016/j.jhazmat.2010.08.019
46. T. Robinson, G. McMullan, R. Marchant, P. Nigam, Remediation of dyes in textile effluent: a critical review on current treatment technologies with a proposed alternative. Bioresour. Technol. **77**, 247–255 (2001). https://doi.org/10.1016/S0960-8524(00)00080-8
47. Z. Wang, M. Xue, K. Huang, Z. Liu, Textile dyeing wastewater treatment, in *Advances in Treating Textile Effluent*, ed. by P. Hauser, (InTech, Shanghai, 2011). https://doi.org/10.5772/22670
48. Q. Sun, L. Yang, The adsorption of basic dyes from aqueous solution on modified peat-resin particle. Water Res. **37**, 1535–1544 (2003). https://doi.org/10.1016/S0043-1354(02)00520-1
49. J. Chen, L. Zhu, Catalytic degradation of Orange II by UV-Fenton with hydroxyl-Fe-pillared bentonite in water. Chemosphere **65**, 1249–1255 (2006). https://doi.org/10.1016/j.chemosphere.2006.04.016
50. F. Fu, Y. Xiong, B. Xie, R. Chen, Adsorption of Acid Red 73 on copper dithiocarbamate precipitate-type solid wastes. Chemosphere **66**, 1–7 (2007). https://doi.org/10.1016/j.chemosphere.2006.05.054
51. V.K. Garg, M. Amita, R. Kumar, R. Gupta, Basic dye (methylene blue) removal from simulated wastewater by adsorption using Indian Rosewood sawdust: a timber industry waste. Dyes Pigments **63**, 243–250 (2004). https://doi.org/10.1016/j.dyepig.2004.03.005
52. I.M. Banat, P. Nigam, D. Singh, R. Marchant, Microbial decolorization of textile-dye-containing effluents: a review. Bioresour. Technol. **58**, 217–227 (1996). https://doi.org/10.1016/S0960-8524(96)00113-7
53. C.I. Pearce, J.R. Lloyd, J.T. Guthrie, The removal of colour from textile wastewater using whole bacterial cells: a review. Dyes Pigments **58**, 179–196 (2003). https://doi.org/10.1016/S0143-7208(03)00064-0
54. G. McMullan, C. Meehan, A. Conneely, N. Kirby, T. Robinson, P. Nigam, I.M. Banat, R. Marchant, W.F. Smyth, Microbial decolourisation and degradation of textile dyes. Appl. Microbiol. Biotechnol. **56**, 81–87 (2001). https://doi.org/10.1007/s002530000587
55. M. Piumetti, F. Freyria, M. Armandi, F. Geobaldo, E. Garrone, B. Bonelli, Degradation of Acid Orange 7 by transition metals containing mesoporous titania in the presence of H2O2 and ascorbic acid. Catal. Struct. React. **33**, 10–11 (2013)

56. M. Piumetti, F.S. Freyria, M. Armandi, G. Saracco, E. Garrone, G.E. Gonzalez, B. Bonelli, Catalytic degradation of Acid Orange 7 by H2O2 as promoted by either bare or V-loaded titania under UV light, in dark conditions, and after incubating the catalysts in ascorbic acid. Catal. Struct. React. **1**, 183–191 (2015). https://doi.org/10.1080/2055074X.2015.1105618
57. F. Freyria, M. Compagnoni, N. Ditaranto, I. Rossetti, M. Piumetti, G. Ramis, B. Bonelli, Pure and Fe-doped mesoporous titania catalyse the oxidation of Acid Orange 7 by H2O2 under different illumination conditions: Fe doping improves photocatalytic activity under simulated solar light. Catalysts **7**, 213 (2017). https://doi.org/10.3390/catal7070213
58. M. Piumetti, F.S. Freyria, M. Armandi, F. Geobaldo, E. Garrone, B. Bonelli, Fe- and V-doped mesoporous titania prepared by direct synthesis: characterization and role in the oxidation of AO7 by H2O2 in the dark. Catal. Today **227**, 71–79 (2014). https://doi.org/10.1016/j.cattod.2013.11.013
59. A. Arcoya, J.A. González, N. Travieso, X.L. Seoane, Physicochemical and catalytic properties of a modified natural clinoptilolite. Clay Miner. **29**, 123–131 (1994). https://doi.org/10.1180/claymin.1994.029.1.14
60. L. Mihaly Cozmuta, A. Mihaly Cozmuta, A. Peter, C. Nicula, E. Bakatula Nsimba, H. Tutu, The influence of pH on the adsorption of lead by Na-clinoptilolite: kinetic and equilibrium studies. Water SA **38**, 269–278 (2012). https://doi.org/10.4314/wsa.v38i2.13
61. M.D. Abràmoff, P.J. Magalhães, S.J. Ram, Image processing with imageJ. Biophoton. Int. **11**, 36–41 (2004). https://doi.org/10.1201/9781420005615.ax4
62. S. Brandani, E. Mangano, H. Ahn, D. Friedrich, X. Hu, Diffusion mechanism of CO2 in 13X zeolite beads. Adsorption **20**, 121–135 (2013). https://doi.org/10.1007/s10450-013-9554-z

Chapter 14
Enzyme Biocatalysis and Sustainability

Carminna Ottone, Oscar Romero, Paulina Urrutia, Claudia Bernal, Andrés Illanes, and Lorena Wilson

14.1 Introduction

Enzymes are protein catalysts bearing an active site where specific amino acid residues are capable to recognize a substrate and catalyze its chemical conversion into product. Enzyme catalyzed reactions go from the simple hydrolysis of a substrate into smaller molecules to reactions of synthesis of highly valuable products [1, 2]. In some cases, enzymes are made by more than one polypeptide chain that interacts with each other by electrostatic forces or covalent bonds, forming an active quaternary structure [3].

The most significant property of enzymes is their high specificity toward substrates since they can recognize and act upon a molecule from a pool of similar compounds [2, 4]. In addition, enzymes work well under mild conditions of temperature and pH [1]. Despite these interesting properties, enzymes are vulnerable to environmental conditions, leading to loss of activity in time. To tackle this problem, enzyme immobilization by attachment to a solid support has proven to be a good alternative to stabilize their three-dimensional structure and preserve their catalytic properties [5]. Thus, the term "biocatalyst" refers to a catalyst having enzymatic activity, and frequently it is used to denote an immobilized enzyme. Enzyme immobilization has not only the advantage of producing robust and stable biocatalysts, but also the benefits of biocatalyst reuse for several cycles in batch operation or prolonged continuous use, and the easy removal of the catalyst from the reaction medium delivering a catalyst-free product stream [4, 6, 7].

C. Ottone (✉) · O. Romero · P. Urrutia · A. Illanes · L. Wilson
Escuela de Ingeniería Bioquímica, Pontificia Universidad Católica de Valparaíso, Valparaiso, Chile
e-mail: carminna.ottone@pucv.cl

C. Bernal
Tecnología Enzimática para Bioprocesos, Departamento de Ingeniería de Alimentos, Universidad de La Serena, La Serena, Chile

M. Piumetti, S. Bensaid (eds.), *Nanostructured Catalysts for Environmental Applications*, https://doi.org/10.1007/978-3-030-58934-9_14

With regard to molecular biology, focus has been on the development of novel enzymes by means of directed evolution [8, 9] or genetic engineering techniques [10], that allow to enhance the stability and catalytic performance of the enzymes by altering the sequence of the corresponding structural gene. Recently, nanozymes [11, 12] have been also proposed as promising catalysts making use of bioinformatics tools, where small peptides are synthesized aiming to preserve only the catalytic site of the protein. Independent of the chosen strategy to increase stability, it will need to be coupled with an adequate support material and the correct selection of the immobilization method. In addition, even if a highly active biocatalyst is obtained, the conditions of the process must be taken into consideration in order to obtain the desired product at an affordable economic cost.

Enzyme biocatalysis has become a mature technology for some industrial processes, including the synthesis of fructose syrup, the production of antibiotic precursors and bulk chemicals, like acrylamide, among others [7]. Enzymes have been extensively used as additives in a wide variety of food products, in detergents and cleansing products, in textiles, in winemaking, and in fruit juice processing [13, 14]. With regard to sustainability and the environmental field, enzymes have attracted much attention due to their biodegradability, which makes them a sound option to replace chemical processes [15, 16]. In particular, great progress has been experienced in using biocatalysts for the remediation of polluted areas [17–20] and for the upgrading of agro-industrial residues [21–23], which is in line with a circular economy approach.

This chapter summarizes the fundamentals of heterogeneous enzyme catalysis with immobilized enzymes and their use in environmental applications and sustainable development. The most suitable support materials for the environmental applications of immobilized enzymes are enlisted, considering the most common immobilization methods. The use of enzyme biocatalysts in bioremediation, in biofuel production, and in the valorization of waste streams is reviewed.

14.2 Enzyme Immobilization Techniques

The immobilization of enzymes is one of the most powerful tools to address the main problems of the industrial application of enzymes, namely, the lack of long term-stability and the difficulty to recover and reuse them. A benefit of the immobilization of enzymes is the enhanced stability under storage and operational conditions [4, 24]. These advantages are in compliance with technical and economic requirements of most chemical processes, where a continuous use of the catalysts for a long period of time is often necessary.

14.2.1 Methods of Enzyme Immobilization

Nowadays, a great diversity of immobilization methods for enzymes are available. Despite the basic methods of enzyme immobilization can be classified in few categories, several variations based on the original methods have been developed. The fundamental methods of enzyme immobilization can be classified into a few categories, however, several variations have been developed from them.

There are no general guidelines for choosing a specific immobilization method. A proper selection of the immobilization method should take into consideration the properties of the enzyme and the intended application of the biocatalyst. A summary of enzyme immobilization techniques and their advantages and disadvantages is shown in Table 14.1.

14.2.1.1 Chemical Interaction

This category includes methods in which the enzyme molecules are bound to an inert carrier, by a covalent or non-covalent linkage, and also the methods where the enzyme molecules are covalently linked among themselves.

In covalent immobilization, the enzyme is linked to the support by a covalent bond between amino acid residues of the enzyme and functional groups on the support surface. Most frequent amino acid residues for enzyme immobilization are

Table 14.1 Main advantages and disadvantages of the immobilization methods

Immobilization method	Advantages	Disadvantages
Chemical interaction		
a. Carrier bound		
Covalent immobilization	Strong interaction High stabilization	Significant loss of activity Enzyme structure distortion Complex protocols
Non-covalent immobilization	Reversibility of enzyme-support binding (recoverable support) Inexpensive and simple High expression of activity	Weak enzyme-support interaction (enzyme leakage) Low stabilization
b. Carrier free	High specific activities Simple, easy, and cheap	Mass transfer limitation Conformational changes Small and variable particle size. Poor mechanical properties
Physical containment		
a. Entrapment	Inexpensive and simple Minimal enzyme structure modification	Low stabilization Enzyme leakagel Mass transfer limitations
b. Membrane retention		Requirement of expensive equipment Low stabilization Membranes fouling

lysine, aspartic acid, glutamic acid, tyrosine, and tryptophan (Fig. 14.1a). Regarding the support, several functionalities can be introduced on its surface, with amine, carboxylic, epoxide, and aldehyde groups being the most used. A scheme of the most applied immobilization chemistry is shown in Fig. 14.1b.

Among the functional groups mentioned, aldehydes stand out for being able to generate several points of attachment between the enzyme and the support, promoting a strong stabilization of the immobilized enzyme [25]. This methodology has been widely used with a large number of enzymes, obtaining high stabilization factors [26]. However, supports functionalized with amines and carboxylic acids are the most widespread carriers for enzyme immobilization. The covalent linkage is formed by the addition of a crosslinking agent, usually carbodiimide, which activates the carboxylic acid to form an amide bond [27].

In the case of non-covalent immobilization, the enzyme is bound to the support by relatively weak and reversible interactions, like hydrophobic interactions, electrostatic interactions, Van der Waals forces, and hydrogen bonding. The multiple interactions formed between enzyme and support have a cooperative effect that promotes the stabilization of the enzyme. Non-covalent immobilization methodologies can be divided into: ionic [28], hydrophobic [29–31], and affinity adsorption [32, 33].

In carrier-free systems, the enzyme is immobilized in its own protein structure, without the need of an inert support. The enzyme is first insolubilized and then crosslinked using bifunctional reagents, like glutaraldehyde. Insolubilization can be done either by protein crystallization or precipitation under non-denaturing condition. The former are termed crosslinked enzyme crystals (CLECs) and require the enzyme protein to be in a pure state [34]; the latter are termed crosslinked enzyme aggregates (CLEAs) and can be produced by non-denaturing protein precipitation even from crude protein mixtures [35, 36].

Fig. 14.1 Schemes of the most frequently used amino acid (AA) residues, (**a**), and functionalization groups of the support, (**b**), in covalent immobilization

14.2.1.2 Physical Containment

It corresponds to those methods in which the enzymes are retained in a confined space by physical means, and includes entrapment within polymer matrices and retention by permeable membranes.

In enzyme entrapment system, enzymes are embedded into a polymer matrix formed by chemical or physical means, like crosslinking or gelation. Usually the polymeric matrix is formed along with enzyme immobilization, so that the enzyme should be compatible with the matrix precursor (the corresponding monomer) and withstand the conditions required for polymerization. Polymeric matrices are quite flexible and can adopt different shapes, such as beads, films, fibers, and foams [37].

In membrane retention system, the enzyme is retained in a semipermeable membrane that allows the free passage of substrates and products [38].

14.2.2 *Materials for Enzyme Support*

Different materials have been used for the immobilization of enzymes, including a large variety of organic, inorganic, and hybrid materials [39]. The type of material used as support plays a crucial role having a strong effect on the performance of the immobilized enzyme. Table 14.2 shows some of the most frequently used materials for enzyme immobilization, including the commercial names of some. Although different companies supply these materials, few companies are manufacturing specially designed materials for the immobilization of enzymes and other biomolecules. Among the metal oxide materials, siliceous materials are the most reported. However, porous alumina [40] and titania nanoparticles have been also proposed as support materials for enzyme immobilization [41, 42].

Table 14.2 Classification of the most frequent materials used in enzyme immobilization, with the commercial name in parenthesis

Type of material	Examples
a. Organics	
Natural polymers	Agarose (Sepharose), chitosan (Chitopearl), alginate, dextran (Shephadex), cellulose (Cellufine)
Synthetic polymers	Polyacrylic acid (ChiralVision), poly(methacrylate) (ReliZyme), polystyrene—polyvinylbenzene (Purolite)
b. Inorganic	Silica, controlled pore glass (PureBiotech), zeolites, graphene, silico aluminates ceramics, titania, alumina

14.2.2.1 Organic Materials

The most traditional and frequently used polymeric matrices for enzyme encapsulation are: agarose, alginate, polyacrylamide, chitosan, and polyvinyl alcohol (PVA) [43]. Agarose beads activated with different functional groups have been extensively used for enzyme immobilization because they allow a high protein load. However, the use of agarose beads is limited to small-scale operations because of their small size and poor mechanical stability. Polymers, like alginate, chitosan, and PVA, have been extensity used for cell encapsulation with excellent results. However, the same encapsulation technique is not suitable for enzyme immobilization because of enzyme leakage, which has to be prevented by combining it with other immobilization technique. Enzyme leakage can be reduced by using a high concentration of polymer leading to small pores, but this will magnify mass transfer limitations. Other strategies to retain the enzymes molecules are the physical adsorption or chemical attachment of the enzyme to the polymer matrix by using functionalized polymers or by the addition of a crosslinking agent [44].

14.2.2.2 Siliceous Materials for Enzyme Immobilization

Siliceous materials are silica nets made of siloxane and Si-OH groups that exist as 3D polymers, whose units are regular SiO_4 tetrahedral structures with their vertices shared through Si-0-Si bonds [45]. Furthermore, these materials may contain some metallic oxides [40] and organic groups [46] providing suitable characteristics for specific enzyme immobilization processes. The synthesis of siliceous materials depends on the final requirements (e.g. morphology, porosity and chemical surface), the precursor type, and the enzyme to be immobilized, but usually the process involves hydrolysis and polycondensation reactions of siloxane groups [47]. Traditionally, a partially condensed silica source is used as precursor. The most used are tetraethyl orthosilicate (TEOS) and tetramethyl orthosilicate (TMOS), with which the hydrolysis and polycondensation reactions can be easily modulated. However, completely hydrolyzed precursor are also used, such as sodium silicate [48] and silicic acid [49].

Porosity is an important characteristic of siliceous materials that can be modulated, being possible to obtain hierarchical materials. Mesoporous and macroporous siliceous-based particles are most studied materials for enzyme immobilization. The pore size diameter ranges between 2 and 50 nm for mesoporous materials, and it is bigger than 50 nm for macroporous materials. The size of the pore is fundamental for achieving an effective enzyme immobilization: it needs to be higher than the enzyme molecule average size and provide enough space for a proper catalytic process in terms of substrates and products diffusion rates. In this regard, pore diameter, pore volume, and surface area have a strong effect on biocatalyst performance [50].

The other important characteristic is the chemical surface of the siliceous material, which can be modified with many functional groups, with the octyl, glyoxyl, epoxide, amino, and sulfonate groups being the most used for enzyme immobilization [51]. Recently, there is the tendency of using supports with double chemical functionality, where covalent bond is formed with one functional group and non-covalent interaction is produced with the other. This heterofunctional strategy allows a proper enzyme orientation during the immobilization process, resulting in good biocatalyst performance [52]. Activation with glyoxyl or amino groups, combined with post-derivatization with glutaraldehyde, are the preferred options for covalent enzyme attachment to siliceous supports [53]. For non-covalent bonding the selection depends more on the enzyme to be immobilized. The immobilization of lipases is performed on silica activated with aliphatic groups, like octyl; for other enzymes, non-covalent immobilization is done using several functional groups and there is no specific trend, so that there is no general protocol to follow [10].

Conducting enzyme immobilization along with the formation of the siliceous materials is another way to obtain a heterogeneous biocatalyst. In this case, the enzyme is added during the sol-gel formation, which leads to siliceous precursors that interact with proteins forming complexes that after an ageing stage can produce an active and stable biocatalyst [54]. This strategy has many advantages, highlighting the possibility to have a one-step process of immobilization.

The applications of silica-based biocatalysts are quite diverse and are approached by multidisciplinary areas, including enzyme biocatalysis and materials science. There are still important challenges to be solved, including the reduction of diffusional restrictions that limit the enzymatic potential, the pore adaptation for big enzymes, and/or for enzyme acting on high molecular weight substrates, like proteases, polymer hydrolases, and synthetases. Moreover, finding a standard methodology that can be used for any enzyme to yield to an active and stable biocatalyst is a major challenge for future research.

14.2.2.3 Hybrid Matrices

Silica-organic hybrid matrices have gained importance as enzyme supports, since this kind of materials bring the best of two worlds together: the high surface area and stability of silica and the high enzyme-support compatibility of organic materials [55]. Several silica-hybrid carriers have been used for enzyme immobilization, by constructing silica-chitosan [56], silica-cellulose [57], silica-alginate [58], and silica-lignin composites [59]. In all cases, the catalytic performance of immobilized enzymes on hybrid silica materials is better than with the silica counterpart biocatalysts.

14.2.3 Coimmobilization of Enzymes

Enzyme coimmobilization is a recent feature in enzyme biocatalysis that consists in the immobilization of more than one enzyme in a single support particle, allowing the development of cascade reactions. In principle, coimmobilization allows a more efficient catalysis approaching the conditions of a metabolic route inside a cell. Having several enzymes in close proximity may increase the efficiency of cascade reactions, reducing product or substrate inhibition and mass transfer limitations [60, 61].

The proportion of each enzyme in the so-called combi catalyst is quite important for the proper balancing of the corresponding reaction rates [62]. Therefore, the proportion of the enzymes offered to the support should be determined considering their respective kinetic parameters [63].

Some recent illustrative examples are: the coimmobilization of dehydrogenases for the conversion of CO_2 into methanol; the coimobilization of glycerol dehydrogenase, NADH oxidase and catalase [64]; the coimmobilization of glucose oxidase and horseradish peroxidase [65]; the coimmobilization of pyruvate kinase and lactic dehydrogenase [66]; and the coimmobilization of glucose dehydrogenase and malate dehydrogenase [67]. Many multi-enzyme processes involve coenzyme-requiring enzymes and therefore coenzyme regeneration is necessary; in such cases it is possible to coimmobilize cofactors and enzymes in the same support [68].

CLEA technology is well suited for enzyme coimmobilization. In this case, the resulting catalyst is termed combi-CLEA [69–71]. In this case, two or more enzymes are precipitated and crosslinked forming a combined catalytic particle. Combi-CLEAs have the same advantages and constraints of CLEAs: they have very high specific activities and are easy to prepare, while their mechanical properties may not be robust enough and particle size is difficult to control. In the case of combi-CLEAs, an additional problem is the non-uniform distribution of the enzymes within the catalyst particle.

14.2.4 Assessment of the Immobilization Process

Enzyme immobilization is a multivariable process, in which several factors influence the final result. In the last decades substantial efforts and progress have been made in understanding the immobilization process, but until now their molecular and physicochemical bases have not been fully elucidated [5].

Main parameters for evaluating the immobilization process are:

- Immobilization Yield (Y_A), which is the percentage of the contacted enzyme that is expressed in the biocatalyst, being calculated by a simple balance of activity. The determination of the residual activity in the supernatant allows knowing how much enzyme has been immobilized. Y_A reflects the deactivation of the enzyme due to the immobilization conditions (pH, temperature), conformational changes induced by the support, and the reduction in the activity by mass transfer limitations.
- Protein Immobilization Yield (Y_P), which is the percentage of the contacted protein that is immobilized in the biocatalyst. The most frequently used protein determination methods are not suitable for insoluble protein, so that the

immobilized protein is determined by the difference between the contacted protein and the residual protein in the supernatant.

- Specific activity (a_{sp}), which represents the enzyme activity per unit of mass of the resulting biocatalyst, being a very relevant parameter of enzyme immobilization.

14.3 Applications of Enzyme Biocatalysis for Sustainability

The most efficient systems for the reintegration of contaminant compounds to the natural cycles are biological systems, hence the efforts to develop different biotechnologies aiming to set up a sustainable development. In particular, enzymatic biocatalysis proposes several solutions to achieve this goal.

Enzyme biocatalysts can be used in different applications within a sustainability approach. Such applications can be classified into three main groups: bioremediation, biorefinery, and biofuels. Figure 14.2 shows a scheme of the main enzymes involved in each process.

In the field of green chemistry, new applications are being reported in the literature about the use of immobilized enzymes for the synthesis of different chemical compounds [72–74]. However, this type of application is beyond the scope of this chapter and will not be reviewed here.

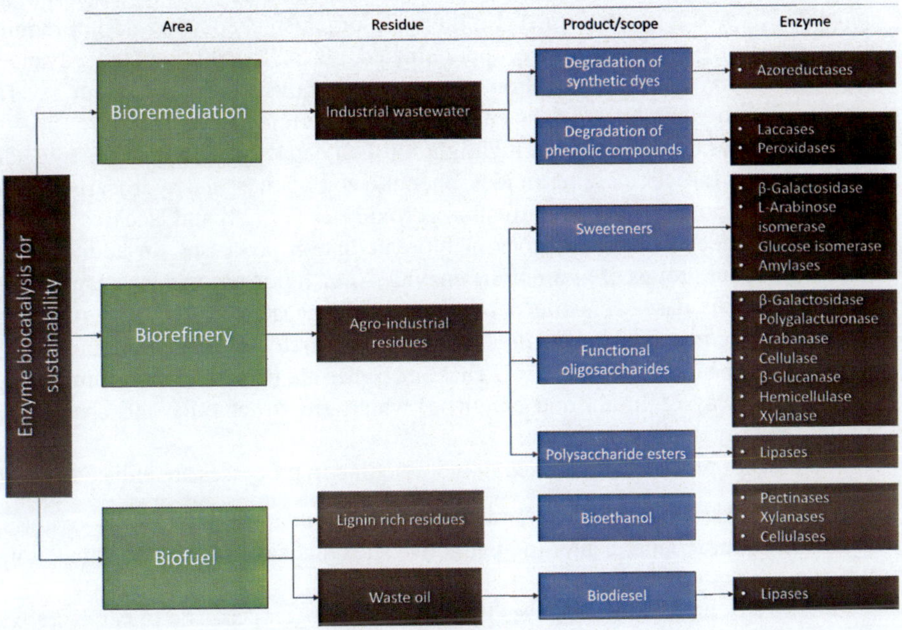

Fig. 14.2 Scheme of the main applications of enzymatic biocatalysis for sustainability

Different examples of the use of immobilized enzymes for bioremediation, biofuel production, and revalorization of carbohydrate-rich residues will be summarized.

14.3.1 Enzymatic Bioremediation

Bioremediation is a technique that goes hand in hand with the environmentally responsible industrial growth. Depending on the degree of pollution, type, and concentration of the contaminant, bioremediation can be carried out with plants, microorganisms, or enzymes. The type and number of water and soil pollutants that are degraded by enzymes increase along with research and innovation in the enzyme biocatalysis field.

The discovery of an enzyme that is able to hydrolyze polyethylene terephthalate, better known as PET, has been recently reported [75]. This finding has a paramount importance for our current lifestyle, where huge amounts of plastics are being disposed into the environment, with PET being one of the most abundant.

The use of enzymes is gaining relevance for the degradation of micropollutants that go through the microbiological waste treatments. Several hazardous endocrine disrupting chemicals (EDCs) coming from human activities, including fertilizers, dyes, and pharmaceutical bioactive components, are released into natural environments.

The addition of the enzymes into contaminated places can be designed according to the requirements of the place to be remediated, and care must be taken that no undesired byproducts are formed. Unlike microorganisms, enzymes do not require the presence of nutrients in the contaminated place. In addition, the advances achieved in enzyme immobilization allow the biocatalysts to be reused for several cycles, reducing the impact of the enzyme cost on bioremediation.

Extracellular oxidoreductases highlight for their capacity to degrade organic pollutants such as insecticides, herbicides, phenolic compounds, and hydrocarbons into less toxic compounds [76]. In particular, peroxidases [77, 78] and laccases [76, 79] are the two most studied enzymes in bioremediation processes. In addition, the study of the four groups of ligninolytic enzymes (i.e., lignin peroxidase, manganese-dependent peroxidase, versatile peroxidase, and laccase) coming from rot fungi (e.g., *Phanerochaete chrysosporium* and *Trametes versicolor*) is gaining increasing importance for bioremediation [76]. These enzymes are involved in the natural degradation pathway of lignin and cellulose, which are structurally and chemically similar to many organic pollutants.

The most popular enzymes used in bioremediation processes are enlisted below:

- Laccases (Lac) belong to a family of copper-containing polyphenol oxidases having a multicopper center in their active sites that catalyze the redox reactions [76].
- Peroxidases are oxidoreductases that utilize hydrogen peroxide to catalyze oxidative reactions [77]. Among the non-fungal enzymes, horseradish peroxidase

(HRP) has been widely studied for bioremediation purposes [80–82]. Versatile peroxidase (VP) oxidizes a wide range of molecules thanks to a surface catalytic tryptophan present in the active site that oxidizes low-redox, and, more significantly, high-redox potential substrates through a long-range electron transfer pathway to the heme complex, where Mn^{2+} is oxidized to Mn^{3+} [77, 83].

- Azoreductases are able to decolorize azo dyes into their corresponding colorless aromatic amines via the hydrolysis of the azo bond [55]. The reaction occurs only in the presence of the reduced form of nicotinamide-based coenzymes [19, 84].
- Other enzymes, like monooxygenase [18] and tyrosinases [85] are also reported in the literature for bioremediation purposes.

As far as the immobilization techniques is regarded, almost all of them have been used with peroxidases and laccases, where a wide variety of supports, going from polymeric beads [79, 86] to different metal oxides [87–90], have been used. A thorough review of the different immobilization techniques used for such enzymes can be found elsewhere [91, 92]. A selection of the most significant works found in the literature for bioremediation of wastewater containing synthetic dyes and phenolic compounds is presented below.

14.3.1.1 Degradation of Synthetic Dyes

Synthetic dyes are important pollutants. Leather, paper, cosmetics, and pharmaceutical industries employ over ten million tons of synthetic dyes per year in their industrial processes [19]. Half of them correspond to azo dyes, which are aromatic compounds with one or more –N=N– groups, many of them having carcinogenic effects [93]. Immobilized azoreductases [84], laccases [79, 91, 94], and horseradish peroxidase [95, 96] have shown great potential for the degradation or decolorization of azo dyes [19]. Some examples are shown below, as well as the immobilization methods and supports used for the preparation of the biocatalysts for the degradation of synthetic dyes.

A CLEA of HRP from *Armoracia rusticana* was used for the degradation of different synthetic dyes (i.e., basic red 9, indigo, methyl orange, rhodamin B, and rhodamine 6G) in a packed bed reactor [97]. The authors observed different decolorization percentages, varying from 73% to 95% according to the dye type and a residual activity of 60% after seven consecutive cycles.

The effective removal of methylene blue and orange II was studied with a laccase-based biocatalyst [94]. Lac was immobilized on a polymethacrylate/carbon nanotubes hybrid material by glutaraldehyde crosslinking. The maximum decolorization yields observed with methylene blue and orange II were 96% and 74%, respectively. The biocatalysts showed a high operational stability, with only 10% of activity loss after 10 successive reaction cycles.

Mesoporous silica particles with two different pore structures: ordered mesoporous silica (SBA-15) and mesocellular foams (MCF) were used to immobilize an

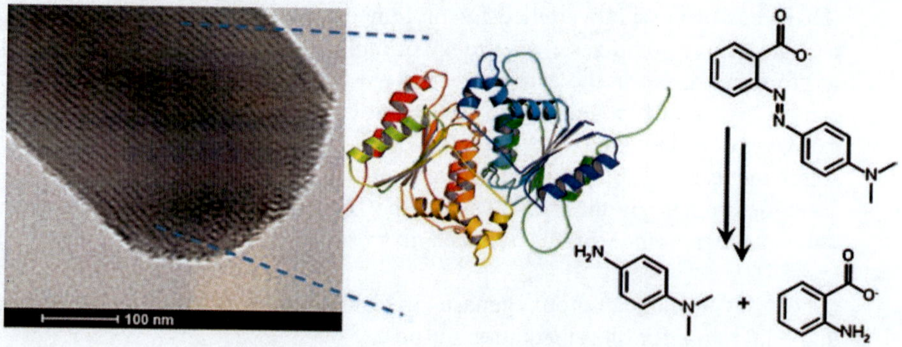

Fig. 14.3 Scheme of the degradation of an azo dye catalyzed by a biocatalyst made of azoreductase from *Rhodococcus opacus* 1CP immobilized on mesoporous silica. (Reprinted from [84])

azoreductase from *Rhodococcus opacus* 1CP [84]. Enzyme immobilization was performed by the functionalization of the support with epoxy and amino groups. Both biocatalysts showed a higher stability in acidic conditions with respect to the free enzyme. The latter was completely inactivated after 35 h of incubation at pH 4, whereas the immobilized enzymes retained 30% of the initial activity after 60 h of incubation. These biocatalysts were studied in the degradation of azo dyes. A scheme of the system is presented in Fig. 14.3.

14.3.1.2 Degradation of Phenolic Compounds

Phenolic compounds are aromatic molecules containing a hydroxyl group attached to the benzene ring structure that are found in several industrial effluents, such as in petrochemical and pharmaceutical plants, pulp mills, mines, and wood preservation plants [82]. The release of these hazardous compounds to the environment may cause serious health effects on the aquatic flora and fauna, and in humans [98]. Because of this, the US Environmental Protection Agency (EPA) has included phenol in the list of priority pollutants [99].

Different immobilized horseradish peroxidases have been studied for the degradation of phenolic compounds [92, 100, 101]. For instance, the removal of 2,4-dichlorophenol was studied with an HRP-based biocatalyst, where the enzyme was immobilized on polyacrylonitrile-based beads, modified with ethanediamine and chitosan, and activated with glutaraldehyde [100]. The results showed a 90% removal of the phenolic compounds by using this enzyme biocatalyst, which was reused up to three cycles with no activity loss, which confirms the robustness of the HRP-based biocatalyst that makes it suitable for large-scale application. A scheme of the biocatalyst preparation method is reported in Fig. 14.4.

Besides, a multienzyme approach was evaluated to study the possible synergistic effect between two or more enzymes. Vishnu et al. [90] studied the co-immobilization of Lac and VP on magnetic silica nanospheres activated with amine and vinyl

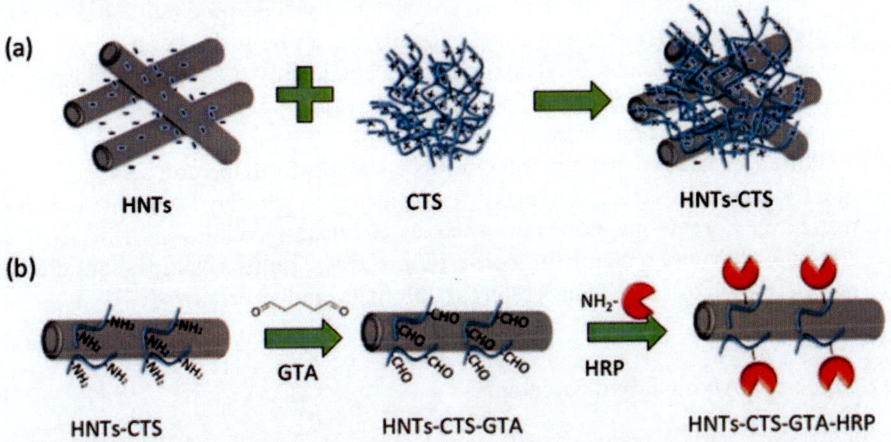

Fig. 14.4 Schematic illustration of preparation and immobilization mechanism. *HNT* halloysite nanotube, *CTS* chitosan, *GTA* glutaraldehyde, *HRP* horseradish peroxidase. (Reprinted from [100])

functional groups, obtaining an immobilization yield of 61% and 76% of retained activity for Lac and VP, respectively. Free enzymes and individually immobilized enzymes showed identical catalytic activity in terms of the degradation of the phenolic compounds in biorefinery wastewater, with 80% of phenols removal in 5 days, while the same reduction was achieved by the co-immobilized biocatalyst after only 1 day.

14.3.2 Biorefinery

An integral waste management is proposed in the context of circular economy that is based on the concept of biorefinery and the approach to reduce, reuse, and recycle waste [75]. Most of the agro-industrial residues are intendent for landfill or are disposed causing environmental damage and economic loss. However, agro-industrial residues are a good source of carbohydrates, proteins, and lipids, thus holding significant potential for enzymatic biotransformation into a variety of high-value compounds.

Food processing wastes rich in carbohydrates can be considered within a biorefinery concept, through their enzymatic transformation into value-added products, such as sweeteners and prebiotics, by the action of hydrolases and isomerases. The upgrading of agro-industrial wastes rich in carbohydrates varies according to their composition; therefore, opportunities for valorization are classified according to the carbohydrate chain length.

Significant amounts of lactose-containing dairy waste streams are generated every year [102, 103]. Whey is the main by-product of the dairy industry, especially in cheese and casein production, and contains many underused nutrients, including 4.5–4.9% lactose [104]. Since a high quantity of lactose is discharged in whey as a waste, the use of immobilized enzymes has been mainly focused on lactose valorization by hydrolysis, transfructosylation, transglycosylation, transgalactosylation, isomerization, and epimerization [105].

Polysaccharides that are present in waste streams from the processing of vegetable, fruit, and crustacean products are considered as attractive substrates for enzymatic transformations. Enzymes such as amylases, cellulases, xylanases, and chitinases have the potential to convert these waste polysaccharides into different products, such as bioplastics, prebiotics, biofuels, and sweeteners [23].

14.3.2.1 Production of Sweeteners

Sweeteners such as fructose and rare sugars may be produced from lactose using different enzymatic biocatalysts, as illustrated in Fig. 14.5.

In the case of the production of fructose syrup from lactose using a bi-enzymatic system, a mixture of fructose, glucose, galactose, and some residual lactose in the final product results in an attractive sweetener for dairy products. The first reaction involved in this process is the hydrolysis of lactose into glucose and galactose by a β-galactosidase (GAL), which is followed by glucose isomerization into fructose by a glucose isomerase (GI) [69, 106, 107–111]. The fructose syrup produced enzymatically from lactose has been applied as a sweetener in ice-cream [107] and yoghurt [112]. Production of fructose syrup from lactose by immobilized forms of GAL and GI has also been reported. Arndt and Wehling [107] immobilized GAL by adsorption in microporous plastic sheets and used the commercial catalyst Maxazyme GI-Immob (Gist-Brocades), where GI is entrapped within gelatin particles and then crosslinked with glutaraldehyde [113]. A trienzymatic system for the

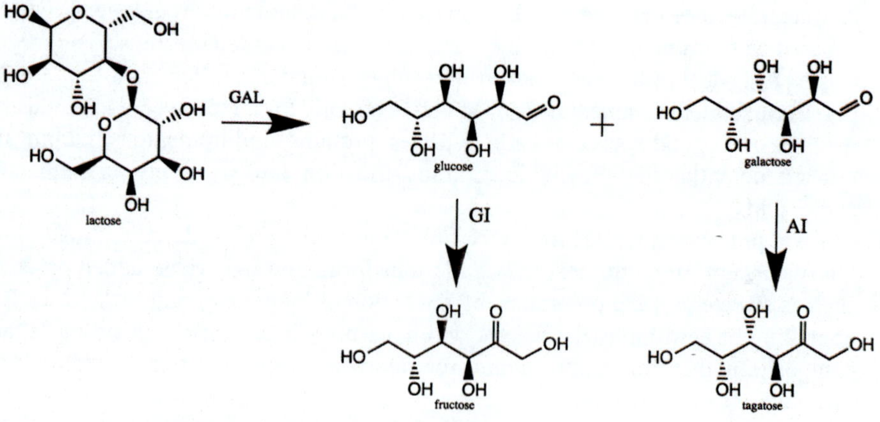

Fig. 14.5 Enzymatic synthesis of fructose and D-tagatose from lactose. *GAL* β-galactosidase, *AI* L-arabinose isomerase, *GI* glucose isomerase

conversion of lactose into fructose and tagatose was reported [111] using GAL, L-arabinose isomerase (AI) and GI immobilized separately on the commercial support Eupergit C. Recently, Araya et al. [69] investigated the co-immobilization of GAL and GI in crosslinked enzyme aggregates (combi-CLEAs) for the production of fructose syrup from lactose. The presence of amino groups on the enzyme surface is a key factor for CLEAs formation, since they are necessary for the crosslinking of the precipitated molecules. Due to the lack of sufficient amine groups on the GI surface, its carboxylic groups were chemically aminated to favor the crosslinking process. Combi-CLEAs preparation was optimized, finding that using a GI-GAL activity ratio of 0.2 and a glutaraldehyde-protein mass ratio of 1.67 resulted in a biocatalyst with good mechanical properties, high expressed activity of both enzymes, and the highest reaction rates of hydrolysis and isomerization among the biocatalysts tested. The selected biocatalyst was utilized in five sequential batch operations obtaining a lactose conversion close to 90% in all batches, with a glucose-fructose conversion close to 45%, which represents approximately 90% of the equilibrium conversion. Results showed that the combi-CLEAs can be used at least for five sequential batches (equivalent to 50 h of operation) with no loss in product quality. Fructose content in the product was around 22% of the total carbohydrates.

Rare sugars synthesized from lactose represent another option to valorize a waste stream. Rare sugars are monosaccharides and their derivatives that rarely exist in nature and are not easily metabolized by the living organisms, though possessing beneficial health effects. Rare sugars such as D-psicose, D-allose, and D-tagatose are quite interesting due to their high relative sweetness (70–92% of sucrose) and low caloric value (0–2 kcal/g) [114, 115]. After a first step of lactose hydrolysis, rare sugars may be produced from glucose or galactose using additional enzymes [114]. In spite of the attractive properties of rare sugars, many of the enzymes that have been utilized for their production are not commercial; therefore, extensive research is still needed [23]. Until now, investigation about the synthesis of rare sugar by immobilized enzymes is limited. In the case of D-tagatose, research has been carried out mainly using permeabilized and immobilized cells [116, 117]. Recently, Torres and Batista-Viera [111] have reported the synthesis of D-tagatose and fructose from lactose using three immobilized enzymes: GAL for the hydrolysis of lactose, AI for the isomerization of galactose to D-tagatose, and GI for the isomerization of glucose into fructose. L-arabinose isomerase was produced from *Enterococcus faecium* and purified by affinity chromatography. Each enzyme was immobilized in Eupergit C and Eupergit C 250 L. Sequential application in separate bio-reactors of immobilized GAL, L-arabinose isomerase, and D-xylose isomerase in the biotransformation of 4.6% lactose in phosphate buffer pH 7.0 resulted in 31% of D-tagatose conversion after 6 h of operation at 50 °C. Under similar operation conditions, lower productivity and conversion were obtained with the soluble enzymes than with the corresponding immobilized biocatalysts.

Starch-rich waste streams are originated from the processing of rice, corn, potato, and sweet potato. This polysaccharide may be hydrolyzed by the use of several enzymes obtaining saccharides of different polymerization degree (Fig. 14.6).

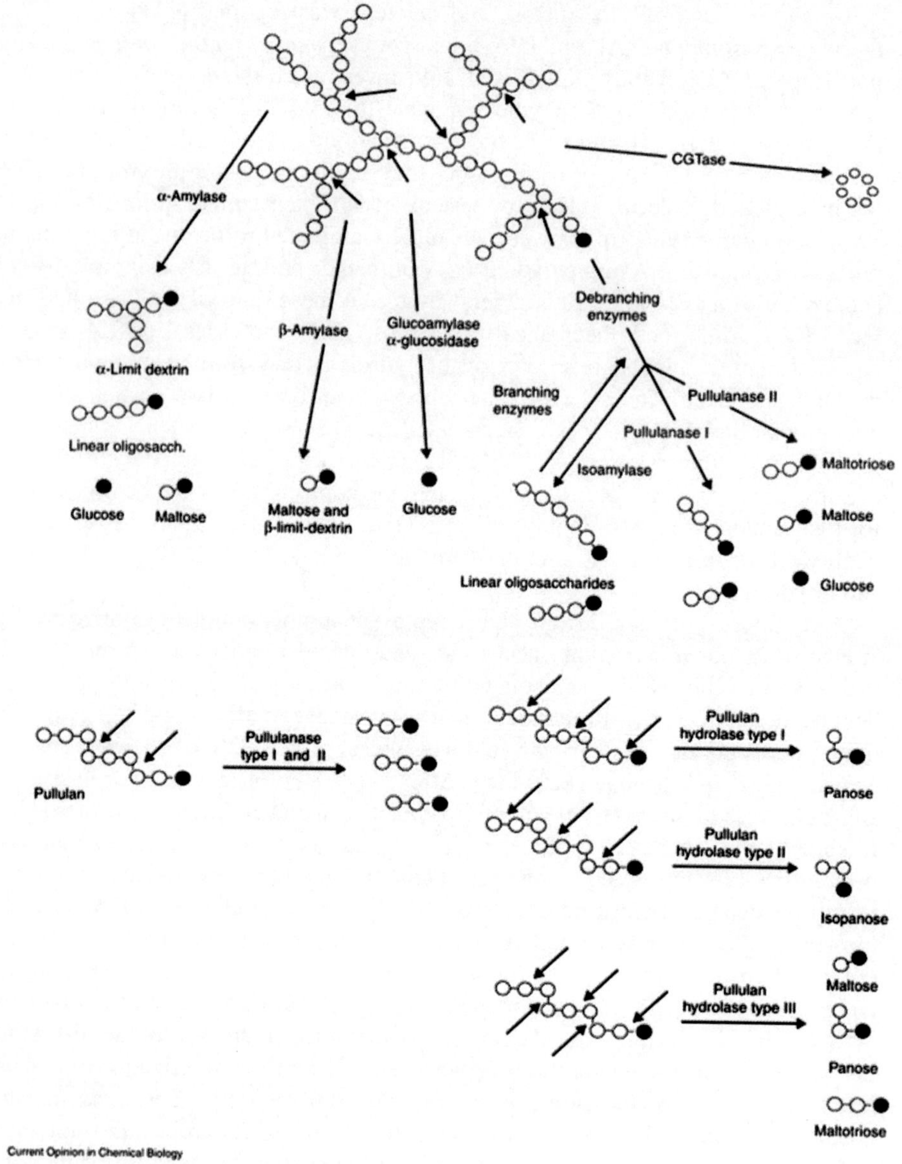

Fig. 14.6 Schematic representation of the action of amylases. Black circles indicate reducing sugars. (Reprinted from [118])

The most common utilization of starch is through its hydrolysis into monosaccharides, as in the case of glucose syrup production. Glucose syrup may be enzymatically produced from starch in a two-step process: α-amylase hydrolyses the α (1–4)glycosidic bonds in starch by a so-called liquefaction process, which is followed by saccharification, in which glucoamylase breaks α (1–6) as well as α (1–4) glycosidic bonds to generate glucose [119]. Both enzymes have been immobilized in order to increase yield and process efficiency. In most of the investigations α-amylase and glucoamylase are immobilized independently using organic or inorganic supports [120–123]. Co-immobilization is another alternative to carry out multi-step cascade reactions but in a single pot. This one-pot strategy has several advantages, such as smaller reactor volumes, fewer unit operations, less solvent usage, shorter reaction time, higher volumetric and space time yields, and less waste generation [124]. Co-immobilization of α-amylase and glucoamylase for their application in starch hydrolysis has been reported using silica gel and DEAE-cellulose entrapped in alginate beads [121] and metal organic frameworks [124]. Edama et al. [125] reported the co-immobilization of α-amylase, glucoamylase, and also cellulase in calcium alginate clay beads for using such biocatalyst in the saccharification of starch. After 7 cycles with 1 h of reaction time each, the biocatalyst still retained 33% of its activity (measured in terms of the release of reducing sugars). In other investigations, α-amylase and glucoamylase were co-immobilized with pullulanase. Pullulanase is a debranching enzyme that has been included since glucoamylase is slower in hydrolyzing α (1–6) bonds. The three enzymes were co-immobilized in the form of combi-CLEAs [71] and in magnetic nanoparticles using glutaraldehyde as crosslinker [126]. In both investigations, the o-immobilizationc of the three enzymes allowed a higher conversion than using the free enzymes in a one-pot reaction.

14.3.2.2 Production of Functional Health-Promoting Oligosaccharides

GAL may also be used as catalyst for the synthesis of galacto-oligosaccharides (GOS) and lactulose (Fig. 14.7), that are recognized as prebiotic [127, 128].

The production of GOS occurs by the transgalactosylation of lactose, which is a kinetically controlled reaction. In the first step of reaction, β-gal forms a galactosyl-enzyme complex after attacking the anomeric center of the galactose residue in lactose, releasing a glucose molecule to the medium [129, 130]. The second step of the reaction depends on the acceptor substrate: if it is water, the galactosyl-enzyme complex undergoes hydrolysis, releasing a galactose molecule; if the acceptor is lactose, transgalactosylation occurs resulting in the production of GOS [130–132]. The predominance of synthesis over hydrolysis depends mainly on the origin of the β-gal [130, 133, 134], the initial sugar concentration [135], and the water thermodynamic activity [136, 137]. GALs of different sources have been immobilized in organic and inorganic carriers for GOS synthesis. Among organic carriers, agarose and chitosan have been reported for the immobilization of GAL from *Aspergillus oryzae* [138–143], *Aspergillus niger* [144], *Bacillus circulans* [145, 146] and *Kluyveromyces lactis* [147, 148]. The enzyme has been covalently bound to both organic carriers activated with aldehyde groups [138–140, 142, 146, 148] or through a two-step process using heterofunctional supports where the enzyme is first adsorbed and then covalently linked [139, 143, 146]. In the case of inorganic sup-

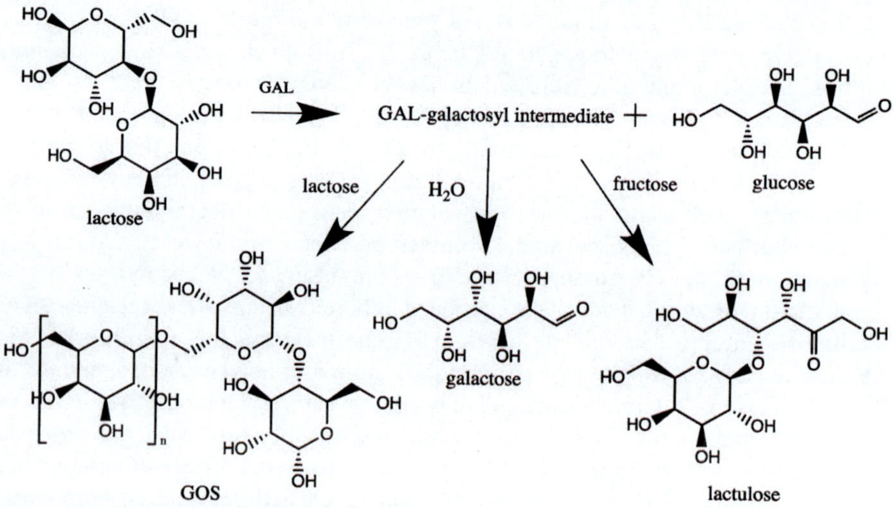

Fig. 14.7 Schematic representation of the enzymatic synthesis of GOS and lactulose from lactose. *GAL* β-galactosidase

ports, synthesis of GOS has been reported for *B. circulans* GAL immobilized in silica supports [149, 150]. It has also been reported for GALs from *A. oryzae* [151], *A. aculeatus* [152] and *K. lactis* [153]. In the case of this siliceous carrier, immobilization is mostly done by adsorption [151, 153] and covalent binding [151, 152]. Mozzafar et al. [149, 150] carried out a two-step process, where the enzyme was first adsorbed and then glutaraldehyde was added to covalently bind the adsorbed enzyme. Among these studies, the investigations of Misson et al. [153] and Banjanac et al. [151] stand out since the utilization of the immobilized enzyme resulted in a production of GOS two or three times higher than obtained with the soluble enzyme at the same conditions, indicating that transgalactosylation was favored over hydrolysis when using the immobilized enzyme. In both cases the best biocatalyst performance was obtained with nanoparticles of silica functionalized with amino groups.

β-galactosidases can be used also in the synthesis of lactulose from lactose through a transgalactosylation reaction, using fructose as galactosyl acceptor [154]. Lactulose is a synthetic ketose disaccharide that is widely used as a drug against constipation and hepatic encephalopathy and also as a prebiotic food additive [155]. Lactulose synthesis using immobilized enzymes has been carried out using mainly organic supports such as agarose [139, 156, 157, 158–160] and chitosan [161]. Immobilization occurred by the covalent binding of GAL, using a one- or two-step process as in GOS synthesis. Only one investigation has been reported for the synthesis of lactulose in silica supports, where GAL from *K. lactis* was covalently immobilized in silica gel functionalized with glutaraldehyde; the immobilized enzyme was reutilized and 52.9% of the initial activity was retained after 10 cycles of use. Continuous synthesis of lactulose was also performed in a packed-bed reactor operated at a flow rate of 0.5 mL/min producing a product stream with 19.1 g/L of lactulose [162].

Pectic substances (polygalacturonic acid, methyl-esterified polygalacturonic acids) extracted from vegetable and fruit wastes can be upgraded to produce fillers, texturizers, thickeners, and glazes. Additionally, the enzymatic hydrolysis of pectin results in the production of pectic oligosaccharides (POS), compounds with potential health benefits. POS are quite variable in structure and degree of polymerization, and include arabinose, xylose, rhamnose and galactose as sugar units [163]. Baldassarre et al. [164] carried out the production of POS from onion skins in order to valorize this agricultural waste. The hydrolysis was carried out using the commercial enzyme preparation Viscozyme L (including carbohydrases such as arabanase, cellulase, β-glucanase, hemicellulase, and xylanase) using a cross-flow continuous membrane enzyme bioreactor. A stable POS production was obtained at a volumetric productivity of 22 g/L/h and 4.5 g/g POS/monosaccharides. Recently, Ramírez-Tapias et al. [165] reported the saccharification of citrus wastes by polygalacturonase immobilized by encapsulation in an alginate matrix. Orange peel represents a large fraction of the by-products generated from citrus processing, and the polysaccharide composition of its albedo is rich in pectin, so it has great potential as raw material for the production of oligogalacturonides. Different bacterial strains of *Streptomyces* were immobilized in alginate gel and the best results in terms of activity and stability were obtained with *Streptomyces halstedii* ATCC 10897 immobilized in the alginate matrix. The hydrolysis of albedo from orange peels with this cell biocatalyst was maximum at 2 h of reaction, generating 1.54 g/L of reducing sugars and decreasing the viscosity of polygalacturonic acid by 98.9%. This immobilized cell biocatalyst with polygalacturonase activity allowed obtaining a product with 9% (w/w) of valuable sugars on a dry basis, which could be used as a nutraceutical food ingredient and as fermentable sugars.

14.3.2.3 Production of Polysaccharide Esters

Starch acylation represents another opportunity for the valorization of starch-containing waste streams. The acylation of starch hydroxyl groups results in different types of polysaccharide esters having a wide range of applications. Acetylated starch with low degree of substitution is used in the food industry to control and adjust the rheological behavior of pastes [166], while succinylated starches reinforce the swelling capacity at lower temperature [167]. Immobilized enzyme catalysts have been utilized to improve enzyme stability in the solvents that are required to solubilize both the starch and the acyl donor substrates. Chakraborty et al. [168] investigated the regioselective modification of starch nanoparticles with *Candida antartica* lipase B in its immobilized (Novozym 435) and free (SP-525) forms. Starch nanoparticles reacted with vinyl stearate, ε-caprolactone, and maleic anhydride using Novozym 435 at 40 °C for 48 h to give starch esters with of 0.8, 0.6, and 0.4 degrees of substitution (DS), respectively. Horchani et al. [169] reported the use of a non-commercial $CaCO_3$-immobilized lipase from *Staphylococcus aureus* (SAL3) to catalyze the esterification reaction between pure oleic acid and starch using microwave heating followed by liquid state esterification. A 76% conversion with a DS of 2.86 was obtained after optimization of the reaction conditions.

14.3.3 Biofuel

The European Commission defines biofuels as liquid or gaseous transport fuels that are made from biomass. Biofuels represent a renewable alternative to fossil fuels in the transport sector and a sound technology to reduce greenhouse gas emissions. If industrial waste is considered as raw material, the production of biofuels by means of enzymatic technology meets all the criteria for environmental sustainability.

14.3.3.1 Production of Bioethanol

Lignocellulose or cellulosic-based waste materials are other raw materials for the production of valuable bioproducts, including rare sugars, surfactants, and biofuels. Lignocelluloses are complex heterogeneous natural composites that comprise three main biopolymers: lignin, cellulose, and hemicellulose. Due to the recalcitrant chemical nature of this material, its valorization requires a multi-enzyme system. Perwez et al. [170] prepared and characterized magnetic combi-CLEAs of pectinases, xylanases and cellulases for the saccharification of wheat straw prior to the fermentation of the resulting sugars. The catalyst was produced adding amino-functionalized magnetic nanoparticles into a mixture of pectinases, xylanases, and cellulases. Using this biocatalyst for the saccharification step, bioethanol concentration was 1.82-fold higher than obtained with free enzymes and could be efficiently reused for 12 cycles, after which pectinase, xylanase, and cellulase retained 86.5%, 90.3%, and 88.6% of activity, respectively. These results show that combi-CLEA methodology can be used for a variety of industrial applications, like food processing, textiles, and bioethanol production. Similarly, Periyasamy et al. [171] reported the immobilization of cellulase, xylanase, and β-1,3-glucanase in silica-amine functionalized iron oxide magnetic nanoparticles and their application in the depolymerization of cellulosic biomass into monomeric sugars. The enzymes were adsorbed and glutaraldehyde was utilized for their crosslinking. The biocatalyst was reused for at least eight consecutive cycles retaining over 70% of its initial activity and the resulting product exhibited approximately 15% increase in carbohydrate digestibility on sugarcane bagasse and eucalyptus pulp with respect to the one obtained with the free enzyme.

14.3.3.2 Biodiesel Production

The replacement of alkaline transesterification for biodiesel production by an enzymatic technology has attracted increasing interest because of its advantages over chemical catalysis. Biodiesel can be produced from fresh and waste oil. The valorization of the latter gives an additional input in the sustainable development direction.

Lipases from different sources and different immobilization methods have been utilized for conducting lipids transesterification, since the cost of the enzyme is a main obstacle for industrial biodiesel production. Badoei-dalfard et al. [22] reported the covalent bonding of Km12 lipase CLEAs in amino-coated magnetite nanoparticles. Covalent linkage of CLEAs to the carrier was conducted by contacting the immobilized enzyme with nanoparticles in the presence of glutaraldehyde. Biodiesel production from waste cooking oils by the immobilized biocatalyst increased about 20% with respect to the free enzyme, and the immobilized biocatalyst remained fully active up to 6 cycles, indicating that crosslinking of lipase and amino-coated magnetite nanoparticles produced operationally stable CLEAs. Another interesting strategy recently reported for lipase immobilization was the use of 5-aminoisophthalic acid as a novel metal-chelating ligand. This acid was successfully grafted onto magnetic nanoparticles (MNP) for the Co^{2+}-chelated affinity immobilization of *Pseudomonas fluorescens* lipase [172]. The MNP-lipase was used for the production of biodiesel from waste cooking oil, and 95% yield was achieved. Biodiesel yield was still 83% after 10 cycles of repeated use, showing a good operational stability. The chelated support could be regenerated and reused after enzyme activity exhaustion, which can reduce the costs associated to the synthesis of the support. This newly designed strategy has great potential in biotechnological applications [172]. A different alternative is the use of a multi-enzyme system for biodiesel production. Babaki et al. [21] studied the production of biodiesel from waste cooking oil using a co-immobilized biocatalyst of lipase from *Rhizomucor miehei* and lipase B from *Candida antarctica* covalently bound onto epoxy-silica. This biocatalyst allowed removing the acyl-migration step, which is the rate-determining step in biodiesel production. The effect of different factors such as enzyme to substrate ratio, *t*-butanol to oil ratio, adsorbed water content, and reaction duration was studied and optimized. A high yield of fatty acid methyl esters (91.5%) was obtained after 10 h of reaction. Zhang et al. [173] analyzed the impact of support characteristics on enzyme performance. The authors compared mesoporous silica supports of varying channel sizes (1.8, 14.0, and 28.0 nm) for lipase (Lipase LVK-S200 from LEVEKING Co. Ltd.) immobilization and reported an optimal 80.1% yield of biodiesel from unrefined waste cooking oil using the enzyme immobilized in the mid-range 14 nm channel size support. The increase of the channel size increased the specific activity of the biocatalyst to a point representing an optimal channel dimension. It was also observed that the need for channel size optimization was conditioned by the nature of the feedstock: the more complex nature of waste cooking oil (with insoluble materials which may block smaller channels) benefited from channel size optimization, while pure olive oil was less sensitive to channel size.

14.4 Conclusions

The impact of heterogeneous enzyme biocatalysts in sustainable development has been highlighted. Different strategies for the attachment of the protein structure of the enzyme to a porous matrix were presented showing the advantages and disadvantages of each one. The procedure of synthesis and the properties of silica, which is one of the most used support materials, was also described. However, the selection of the support and the immobilization strategy will strictly depend on the final application of the biocatalyst. On the one hand, the cost of the biocatalyst is an important issue in environmental applications, which can limit their field implementation. On the other hand, the applications related to the environment and sustainability are quite diverse; thus, they have to comply with different types of regulations and require the use of different enzyme biocatalysts. For bioremediation, oxidoreductases, like laccases and peroxidases, offer good opportunities for the recovery of areas contaminated with different organic pollutants. For biodiesel production, immobilized lipases have been utilized for lipids transesterification. Pectinases, xylanases, and cellulases contribute to bioethanol production by hydrolyzing long-chain carbohydrate polymers into fermentable sugars. β-Galactosidase has a great potential for the revalorization of lactose, as catalyst for the synthesis of prebiotics, such as GOS and lactulose. In addition, when combined with another enzyme, such as L-arabinose isomerase, it can be used to produce rare sugars, like D-tagatose, a sugar that has both sweetening and health-promoting properties.

Despite the difficulties that still have to be overcome, there is no doubt that enzyme biocatalysis will help in achieving a circular economy model and an environmentally sustainable industry.

References

1. K. Buchholz, V. Kasche, U.T. Bornscheuer, *Biocatalysts and Enzyme Technology* (Wiley, New York, 2012)
2. K.M. Koeller, C.-H. Wong, Enzymes for chemical synthesis. Nature **409**(6817), 232–240 (2001)
3. R.D. Ward, Relationship between enzyme heterozygosity and quaternary structure. Biochem. Genet. **15**(1), 123–135 (1977)
4. C. Mateo, J.M. Palomo, G. Fernandez-Lorente, J.M. Guisan, R. Fernandez-Lafuente, Improvement of enzyme activity, stability and selectivity via immobilization techniques. Enzym. Microb. Technol. **40**(6), 1451–1463 (2007)
5. U. Hanefeld, L. Gardossi, E. Magner, Understanding enzyme immobilisation. Chem. Soc. Rev. **38**(2), 453–468 (2009)
6. L. Cao, Immobilised enzymes: science or art? Curr. Opin. Chem. Biol. **9**(2), 217–226 (2005)
7. R. DiCosimo, J. McAuliffe, A.J. Poulose, G. Bohlmann, Industrial use of immobilized enzymes. Chem. Soc. Rev. **42**(15), 6437–6474 (2013)
8. J.R. Cherry, A.L. Fidantsef, Directed evolution of industrial enzymes: an update. Curr. Opin. Biotechnol. **14**(4), 438–443 (2003)

9. A. Kumar, S. Singh, Directed evolution: tailoring biocatalysts for industrial applications. Crit. Rev. Biotechnol. **33**(4), 365–378 (2013)
10. C. Bernal, K. Rodríguez, R. Martínez, Integrating enzyme immobilization and protein engineering: an alternative path for the development of novel and improved industrial biocatalysts. Biotechnol. Adv. **36**(5), 1470–1480 (2018)
11. Y. Huang, J. Ren, X. Qu, Nanozymes: classification, catalytic mechanisms, activity regulation, and applications. Chem. Rev. **119**(6), 4357–4412 (2019)
12. X. Wang, Y. Hu, H. Wei, Nanozymes in bionanotechnology: from sensing to therapeutics and beyond. Inorg. Chem. Front. **3**(1), 41–60 (2016)
13. J.-M. Choi, S.-S. Han, H.-S. Kim, Industrial applications of enzyme biocatalysis: current status and future aspects. Biotechnol. Adv. **33**(7), 1443–1454 (2015)
14. O. Kirk, T.V. Borchert, C.C. Fuglsang, Industrial enzyme applications. Curr. Opin. Biotechnol. **13**(4), 345–351 (2002)
15. M. Alcalde, M. Ferrer, F.J. Plou, A. Ballesteros, Environmental biocatalysis: from remediation with enzymes to novel green processes. Trends Biotechnol. **24**(6), 281–287 (2006)
16. P. Kumar, S. Sharma, Enzymes in green chemistry: the need for environment and sustainability. Int. J. Appl. Res. **2**, 337–341 (2016)
17. C.S. Karigar, S.S. Rao, Role of microbial enzymes in the bioremediation of pollutants: a review. Enzym. Res. **2011**, 805187 (2011)
18. C.H. Okino-Delgado, M.R. Zanutto-Elgui, D.Z. do Prado, M.S. Pereira, L.F. Fleuri, Enzymatic bioremediation: current status, challenges of obtaining process, and applications, in *Microbial Metabolism of Xenobiotic Compounds*, ed. by P. K. Arora, (Springer Singapore, Singapore, 2019), pp. 79–101
19. R.L. Singh, P.K. Singh, R.P. Singh, Enzymatic decolorization and degradation of azo dyes—a review. Int. Biodeterior. Biodegrad. **104**, 21–31 (2015)
20. T. Sutherland, I. Horne, K. Weir, C. Coppin, M. Williams, M. Selleck, R. Russell, J. Oakeshott, Enzymatic bioremediation: from enzyme discovery to applications. Clin. Exp. Pharmacol. Physiol. **31**(11), 817–821 (2004)
21. M. Babaki, M. Yousefi, Z. Habibi, M. Mohammadi, Process optimization for biodiesel production from waste cooking oil using multi-enzyme systems through response surface methodology. Renew. Energy **105**, 465–472 (2017)
22. A. Badoei-Dalfard, S. Malekabadi, Z. Karami, G. Sargazi, Magnetic cross-linked enzyme aggregates of Km12 lipase: a stable nanobiocatalyst for biodiesel synthesis from waste cooking oil. Renew. Energy **141**, 874–882 (2019)
23. M. Bilal, H.M.N. Iqbal, Sustainable bioconversion of food waste into high-value products by immobilized enzymes to meet bio-economy challenges and opportunities—a review. Food Res. Int. **123**, 226–240 (2019)
24. R.A. Sheldon, S. van Pelt, Enzyme immobilisation in biocatalysis: why, what and how. Chem. Soc. Rev. **42**(15), 6223–6235 (2013)
25. J.M. Guisán, Aldehyde-agarose gels as activated supports for immobilization-stabilization of enzymes. Enzym. Microb. Technol. **10**(6), 375–382 (1988)
26. C. Mateo, J.M. Palomo, M. Fuentes, L. Betancor, V. Grazu, F. López-Gallego, B.C.C. Pessela, A. Hidalgo, G. Fernández-Lorente, R. Fernández-Lafuente, J.M. Guisán, Glyoxyl agarose: a fully inert and hydrophilic support for immobilization and high stabilization of proteins. Enzym. Microb. Technol. **39**(2), 274–280 (2006)
27. V.P. Torchilin, E.G. Tischenko, V.N. Smirnov, Effect of electrostatic complex formation prior to immobilization. J. Solid Phase Biochem. **2**(1), 19–29 (1977)
28. N.S. Rios, S. Arana-Peña, C. Mendez-Sanchez, C. Ortiz, L.R.B. Gonçalves, R. Fernandez-Lafuente, Reuse of lipase from *Pseudomonas fluorescens* via its step-by-step coimmobilization on glyoxyl-octyl agarose beads with least stable lipases. Catalysts **9**, 5 (2019)
29. R. Fernandez-Lafuente, P. Armisén, P. Sabuquillo, G. Fernández-Lorente, J.M. Guisán, Immobilization of lipases by selective adsorption on hydrophobic supports. Chem. Phys. Lipids **93**(1–2), 185–197 (1998)

30. J.M. Palomo, G. Fernandez-Lorente, C. Mateo, C. Ortiz, R. Fernandez-Lafuente, J.M. Guisan, Modulation of the enantioselectivity of lipases via controlled immobilization and medium engineering: hydrolytic resolution of mandelic acid esters. Enzym. Microb. Technol. **31**(6), 775–783 (2002)
31. J.M. Palomo, G. Muoz, G. Fernández-Lorente, C. Mateo, R. Fernández-Lafuente, J.M. Guisán, Interfacial adsorption of lipases on very hydrophobic support (octadecyl-Sepabeads): immobilization, hyperactivation and stabilization of the open form of lipases. J. Mol. Catal. B Enzym. **19**(20), 279–286 (2002)
32. A. Care, P.L. Bergquist, A. Sunna, Solid-binding peptides: smart tools for nanobiotechnology. Trends Biotechnol. **33**(5), 259–268 (2015)
33. J. Nilsson, S. Ståhl, J. Lundeberg, M. Uhlén, P.Å. Nygren, Affinity fusion strategies for detection, purification, and immobilization of recombinant proteins. Protein Expr. Purif. **11**(1), 1–16 (1997)
34. S. Lopez, L. Rondot, C. Leprêtre, C. Marchi-Delapierre, S. Ménage, C. Cavazza, Cross-linked artificial enzyme crystals as heterogeneous catalysts for oxidation reactions. J. Am. Chem. Soc. **139**(49), 17994–18002 (2017)
35. L. Cao, F. Van Rantwijk, R.A. Sheldon, Cross-linked enzyme aggregates: a simple and effective method for the immobilization of penicillin acylase. Org. Lett. **2**(10), 1361–1364 (2000)
36. R. Sheldon, *Cross-Linked Enzyme Aggregates (CLEA® s): Stable and Recyclable Biocatalysts* (Portland Press Limited, London, 2007)
37. E.T. Hwang, M.B. Gu, Enzyme stabilization by nano/microsized hybrid materials. Eng. Life Sci. **13**(1), 49–61 (2013)
38. P. Jochems, Y. Satyawali, L. Diels, W. Dejonghe, Enzyme immobilization on/in polymeric membranes: status, challenges and perspectives in biocatalytic membrane reactors (BMRs). Green Chem. **13**(7), 1609–1623 (2011)
39. J. Zdarta, A.S. Meyer, T. Jesionowski, M. Pinelo, A general overview of support materials for enzyme immobilization: characteristics, properties, practical utility. Catalysts **8**, 2 (2018)
40. G. Busca, Catalytic materials based on silica and alumina: structural features and generation of surface acidity. Prog. Mater. Sci. **104**, 215–249 (2019)
41. J. Hou, G. Dong, Y. Ye, V. Chen, Laccase immobilization on titania nanoparticles and titania-functionalized membranes. J. Membr. Sci. **452**, 229–240 (2014)
42. J. Yu, H. Ju, Preparation of porous titania sol–gel matrix for immobilization of horseradish peroxidase by a vapor deposition method. Anal. Chem. **74**(14), 3579–3583 (2002)
43. C. Pizarro, M.A. Fernández-Torroba, C. Benito, J.M. González-Sáiz, Optimization by experimental design of polyacrylamide gel composition as support for enzyme immobilization by entrapment. Biotechnol. Bioeng. **53**(5), 497–506 (1997)
44. W. Jin, J.D. Brennan, Properties and applications of proteins encapsulated within sol-gel derived materials. Anal. Chim. Acta **461**(1), 1–36 (2002)
45. R. Iler, *The Chemistry of Silica: Solubility, Polymerization, Colloid and Surface Properties and Biochemistry of Silica* (John Wiley & Sons, New York, 1979)
46. N. Velikova, Y. Vueva, Y. Ivanova, I. Salvado, M. Fernandes, P. Vassileva, R. Georgieva, A. Detcheva, Synthesis and characterization of sol-gel mesoporous organosilicas functionalized with amine groups. J. Noncrystal. Solids **378**, 89–95 (2013)
47. B. Sun, G. Zhou, H. Zhang, Synthesis, functionalization, and applications of morphology-controllable silica-based nanostructures: a review. Prog. Solid State Chem. **44**(1), 1–19 (2016)
48. A.A. Pisal, A.V. Rao, Comparative studies on the physical properties of TEOS, TMOS and Na_2SiO_3 based silica aerogels by ambient pressure drying method. J. Porous. Mater. **23**(6), 1547–1556 (2016)
49. H. Isobe, S. Utsumi, K. Yamamoto, H. Kanoh, K. Kaneko, Micropore to macropore structure-designed silicas with regulated condensation of silicic acid nanoparticles. Langmuir **21**(17), 8042–8047 (2005)

50. P.S. Nabavi Zadeh, B. Åkerman, Immobilization of enzymes in mesoporous silica particles: protein concentration and rotational mobility in the pores. J. Phys. Chem. B **121**(12), 2575–2583 (2017)
51. N. Zhong, W. Chen, L. Liu, H. Chen, Immobilization of Rhizomucor miehei lipase onto the organic functionalized SBA-15: their enzymatic properties and glycerolysis efficiencies for diacylglycerols production. Food Chem. **271**, 739–746 (2019)
52. C. Bernal, A. Illanes, L. Wilson, Heterofunctional hydrophilic-hydrophobic porous silica as support for multipoint covalent immobilization of lipases: application to lactulose palmitate synthesis. Langmuir **30**(12), 3557–3566 (2014)
53. P. Zucca, E. Sanjust, Inorganic materials as supports for covalent enzyme immobilization: methods and mechanisms. Molecules **19**(9), 14139–14194 (2014)
54. A.C. Pierre, The sol-gel encapsulation of enzymes. Biocatal. Biotransform. **22**(3), 145–170 (2004)
55. S. Pandey, S.B. Mishra, Sol-gel derived organic-inorganic hybrid materials: synthesis, characterizations and applications. J. Sol-Gel Sci. Technol. **59**(1), 73–94 (2011)
56. X. Xiang, H. Suo, C. Xu, Y. Hu, Covalent immobilization of lipase onto chitosan-mesoporous silica hybrid nanomaterials by carboxyl functionalized ionic liquids as the coupling agent. Colloids Surf. B: Biointerf. **165**, 262–269 (2018)
57. H. Dai, S. Ou, Z. Liu, H. Huang, Pineapple peel carboxymethyl cellulose/polyvinyl alcohol/mesoporous silica SBA-15 hydrogel composites for papain immobilization. Carbohydr. Polym. **169**, 504–514 (2017)
58. R. Onbas, O. Yesil-Celiktas, Synthesis of alginate-silica hybrid hydrogel for biocatalytic conversion by β-glucosidase in microreactor. Eng. Life Sci. **19**(1), 37–46 (2019)
59. A. Jędrzak, T. Rębiś, Ł. Klapiszewski, J. Zdarta, G. Milczarek, T. Jesionowski, Carbon paste electrode based on functional GOx/silica-lignin system to prepare an amperometric glucose biosensor. Sens. Actuators B Chem. **256**, 176–185 (2018)
60. J. Luo, A.S. Meyer, R.V. Mateiu, M. Pinelo, Cascade catalysis in membranes with enzyme immobilization for multi-enzymatic conversion of CO2 to methanol. New Biotechnol. **32**(3), 319–327 (2015)
61. J.M. Sperl, V. Sieber, Multienzyme cascade reactions—status and recent advances. ACS Catal. **8**(3), 2385–2396 (2018)
62. C. Schmidt-Dannert, F. Lopez-Gallego, A roadmap for biocatalysis–functional and spatial orchestration of enzyme cascades. Microb. Biotechnol. **9**(5), 601–609 (2016)
63. J. Rocha-Martín, B.L. Rivas, R. Muñoz, J.M. Guisán, F. López-Gallego, Rational co-immobilization of bi-enzyme cascades on porous supports and their applications in bio-redox reactions with insitu recycling of soluble cofactors. ChemCatChem **4**(9), 1279–1288 (2012)
64. J. Rocha-Martin, A. Acosta, J.M. Guisan, F. López-Gallego, Immobilizing systems biocatalysis for the selective oxidation of glycerol coupled to in situ cofactor recycling and hydrogen peroxide elimination. ChemCatChem **7**(13), 1939–1947 (2015)
65. X. Ji, Z. Su, P. Wang, G. Ma, S. Zhang, Tethering of nicotinamide adenine dinucleotide inside hollow nanofibers for high-yield synthesis of methanol from carbon dioxide catalyzed by coencapsulated multienzymes. ACS Nano. **9**(4), 4600–4610 (2015)
66. J. Chung, E.T. Hwang, J.H. Kim, B.C. Kim, M.B. Gu, Modular multi-enzyme cascade process using highly stabilized enzyme microbeads. Green Chem. **16**(3), 1163–1167 (2014)
67. J. Fu, Y.R. Yang, A. Johnson-Buck, M. Liu, Y. Liu, N.G. Walter, N.W. Woodbury, H. Yan, Multi-enzyme complexes on DNA scaffolds capable of substrate channelling with an artificial swinging arm. Nat. Nanotechnol. **9**(7), 531 (2014)
68. R. Xue, J.M. Woodley, Process technology for multi-enzymatic reaction systems. Bioresour. Technol. **115**, 183–195 (2012)
69. E. Araya, P. Urrutia, O. Romero, A. Illanes, L. Wilson, Design of combined crosslinked enzyme aggregates (combi-CLEAs) of β-galactosidase and glucose isomerase for the one-pot production of fructose syrup from lactose. Food Chem. **288**, 102–107 (2019)

70. T.C. Logan, D.S. Clark, T.B. Stachowiak, F. Svec, J.M.J. Fréchet, Photopatterning enzymes on polymer monoliths in microfluidic devices for steady-state kinetic analysis and spatially separated multi-enzyme reactions. Anal. Chem. **79**(17), 6592–6598 (2007)
71. S. Talekar, A. Pandharbale, M. Ladole, S. Nadar, M. Mulla, K. Japhalekar, K. Pattankude, D. Arage, Carrier free co-immobilization of alpha amylase, glucoamylase and pullulanase as combined cross-linked enzyme aggregates (combi-cleas): a tri-enzyme biocatalyst with one pot starch hydrolytic activity. Bioresour. Technol. **147**, 269–275 (2013)
72. J.M. Blamey, F. Fischer, H.-P. Meyer, F. Sarmiento, M. Zinn, Enzymatic biocatalysis in chemical transformations: a promising and emerging field in green chemistry practice, in *Biotechnology of Microbial Enzymes*, ed. by G. Brahmachari, (Elsevier, San Diego, 2017), pp. 347–403
73. M.C. Bryan, P.J. Dunn, D. Entwistle, F. Gallou, S.G. Koenig, J.D. Hayler, M.R. Hickey, S. Hughes, M.E. Kopach, G. Moine, Key green chemistry research areas from a pharmaceutical manufacturers' perspective revisited. Green Chem. **20**(22), 5082–5103 (2018)
74. S. Kobayashi, H. Uyama, J.-I. Kadokawa, *Enzymatic Polymerization Towards Green Polymer Chemistry* (Springer, New York, 2019)
75. T. Fecker, P. Galaz-Davison, F. Engelberger, Y. Narui, M. Sotomayor, L.P. Parra, C.A. Ramírez-Sarmiento, Active site flexibility as a hallmark for efficient PET degradation by I. sakaiensis PETase. Biophys. J. **114**(6), 1302–1312 (2018)
76. D.W. Wong, Structure and action mechanism of ligninolytic enzymes. Appl. Biochem. Biotechnol. **157**(2), 174–209 (2009)
77. J.D.C. Medina, A.L. Woiciechowski, L.R.C. Guimarães, S.G. Karp, C.R. Soccol, 10-Peroxidases, in *Current Developments in Biotechnology and Bioengineering*, ed. by A. Pandey, S. Negi, C. R. Soccol, (Elsevier, New York, 2017), pp. 217–232
78. J. Rocha-Martin, S. Velasco-Lozano, J.M. Guisán, F. López-Gallego, Oxidation of phenolic compounds catalyzed by immobilized multi-enzyme systems with integrated hydrogen peroxide production. Green Chem. **16**(1), 303–311 (2014)
79. M. Bilal, T. Rasheed, F. Nabeel, H.M. Iqbal, Y. Zhao, Hazardous contaminants in the environment and their laccase-assisted degradation—a review. J. Environ. Manag. **234**, 253–264 (2019)
80. W. Chouyyok, J. Panpranot, C. Thanachayanant, S. Prichanont, Effects of pH and pore characters of mesoporous silicas on horseradish peroxidase immobilization. J. Mol. Catal. B Enzym. **56**(4), 246–252 (2009)
81. N.C. Veitch, Horseradish peroxidase: a modern view of a classic enzyme. Phytochemistry **65**(3), 249–259 (2004)
82. L.G.C. Villegas, N. Mashhadi, M. Chen, D. Mukherjee, K.E. Taylor, N. Biswas, A short review of techniques for phenol removal from wastewater. Curr. Pollut. Rep. **2**(3), 157–167 (2016)
83. D. Gonzalez-Perez, M. Alcalde, The making of versatile peroxidase by directed evolution. Biocatal. Biotransform. **36**(1), 1–11 (2018)
84. J. Qi, M.K. Anke, K. Szymańska, D. Tischler, Immobilization of Rhodococcus opacus 1CP azoreductase to obtain azo dye degrading biocatalysts operative at acidic pH. Int. Biodeterior. Biodegrad. **118**, 89–94 (2017)
85. A.T. Biegunski, A. Michota, J. Bukowska, K. Jackowska, Immobilization of tyrosinase on poly(indole-5-carboxylic acid) evidenced by electrochemical and spectroscopic methods. Bioelectrochemistry **69**(1), 41–48 (2006)
86. T. Brugnari, M.G. Pereira, G.A. Bubna, E.N. de Freitas, A.G. Contato, R.C.G. Corrêa, R. Castoldi, C.G.M. de Souza, M.L.T. de Moraes, A. Bracht, A highly reusable MANAE-agarose-immobilized Pleurotus ostreatus laccase for degradation of bisphenol A. Sci. Total Environ. **634**, 1346–1351 (2018)
87. P. Calza, D. Zacchigna, E. Laurenti, Degradation of orange dyes and carbamazepine by soybean peroxidase immobilized on silica monoliths and titanium dioxide. Environ. Sci. Pollut. Res. **23**(23), 23742–23749 (2016)

88. H.-Y. Chen, S.-H. Wu, C.-T. Chen, Y.-P. Chen, F.-P. Chang, F.-C. Chien, C.-Y. Mou, Horseradish peroxidase-ancapsulated hollow silica nanospheres for intracellular sensing of reactive oxygen species. Nanoscale Res. Lett. **13**(1), 123 (2018)
89. J.K. Gill, V. Orsat, S. Kermasha, Screening trials for the encapsulation of laccase enzymatic extract in silica sol-gel. J. Sol-Gel Sci. Technol. **85**(3), 657–663 (2018)
90. D. Vishnu, G. Neeraj, R. Swaroopini, R. Shobana, V.V. Kumar, H. Cabana, Synergetic integration of laccase and versatile peroxidase with magnetic silica microspheres towards remediation of biorefinery wastewater. Environ. Sci. Pollut. Res. **24**(22), 17993–18009 (2017)
91. P. Peralta-Zamora, C.M. Pereira, E.R. Tiburtius, S.G. Moraes, M.A. Rosa, R.C. Minussi, N. Durán, Decolorization of reactive dyes by immobilized laccase. Appl. Catal. B Environ. **42**(2), 131–144 (2003)
92. F. Shakerian, J. Zhao, S.-P. Li, Recent development in the application of immobilized oxidative enzymes for bioremediation of hazardous micropollutants—a review. Chemosphere **239**, 124716 (2019)
93. K. Golka, S. Kopps, Z.W. Myslak, Carcinogenicity of azo colorants: influence of solubility and bioavailability. Toxicol. Lett. **151**(1), 203–210 (2004)
94. Y. Lai, F. Wang, Y. Zhang, P. Ou, P. Wu, Q. Fang, S. Li, Z. Chen, Effective removal of methylene blue and orange II by subsequent immobilized laccase decolorization on crosslinked polymethacrylate/carbon nanotubes. Mater. Res. Exp. **6**(085541), 1–11 (2019)
95. S. Akhtar, A.A. Khan, Q. Husain, Potential of immobilized bitter gourd (Momordica charantia) peroxidases in the decolorization and removal of textile dyes from polluted wastewater and dyeing effluent. Chemosphere **60**(3), 291–301 (2005)
96. S.V. Mohan, K.K. Prasad, N.C. Rao, P. Sarma, Acid azo dye degradation by free and immobilized horseradish peroxidase (HRP) catalyzed process. Chemosphere **58**(8), 1097–1105 (2005)
97. M. Bilal, H.M.N. Iqbal, H. Hu, W. Wang, X. Zhang, Development of horseradish peroxidase-based cross-linked enzyme aggregates and their environmental exploitation for bioremediation purposes. J. Environ. Manag. **188**, 137–143 (2017)
98. M. Ahmaruzzaman, Adsorption of phenolic compounds on low-cost adsorbents: a review. Adv. Colloid Interf. Sci. **143**(1), 48–67 (2008)
99. R. Evans, Revised emergency planning and community right-to-know act (EPCRA), section 313, toxic chemical release reporting for calendar year 1998, Oak Ridge Y-12 Plant, TN (US) (2000)
100. R. Zhai, B. Zhang, Y. Wan, C. Li, J. Wang, J. Liu, Chitosan–halloysite hybrid-nanotubes: horseradish peroxidase immobilization and applications in phenol removal. Chem. Eng. J. **214**, 304–309 (2013)
101. S. Wang, H. Fang, Y. Wen, M. Cai, W. Liu, S. He, X. Xu, Applications of HRP-immobilized catalytic beads to the removal of 2,4-dichlorophenol from wastewater. RSC Adv. **5**(71), 57286–57292 (2015)
102. T. Ahmad, R.M. Aadil, H. Ahmed, U.U. Rahman, B.C.V. Soares, S.L.Q. Souza, T.C. Pimentel, H. Scudino, J.T. Guimarães, E.A. Esmerino, M.Q. Freitas, R.B. Almada, S.M.R. Vendramel, M.C. Silva, A.G. Cruz, Treatment and utilization of dairy industrial waste: a review. Trends Food Sci. Technol. **88**, 361–372 (2019)
103. B.E. Erickson, Acid whey: is the waste product an untapped goldmine? Chem. Eng. News **95**(6), 26–30 (2017)
104. M. Krewinkel, M. Gosch, E. Rentschler, L. Fischer, Epilactose production by 2 cellobiose 2-epimerases in natural milk. J. Dairy Sci. **97**(1), 155–161 (2014)
105. Q. Chen, Y. Xiao, W. Zhang, T. Zhang, B. Jiang, T. Stressler, L. Fischer, W. Mu, Current research on cellobiose 2-epimerase: enzymatic properties, mechanistic insights, and potential applications in the dairy industry. Trends Food Sci. Technol. **82**, 167–176 (2018)
106. J.R. Abril, J.W. Stull, Lactose hydrolysis in acid whey with subsequent glucose isomerisation. J. Sci. Food Agric. **48**(4), 511–514 (1989)

107. E.A. Arndt, R.L. Wehling, Development of hydrolyzed and hydrolyzed-lsomerized syrups from cheese whey ultrafiltration permeate and their utilization in ice cream. J. Food Sci. **54**(4), 880–884 (1989)
108. C.P. Chiu, F.V. Kosikowski, Conversion of glucose in lactase-hydrolyzed whey permeate to fructose with immobilized glucose isomerase. J. Dairy Sci. **69**(4), 959–964 (1986)
109. A. Illanés, Whey upgrading by enzyme biocatalysis. Electron. J. Biotechnol. **14**, 6 (2011)
110. A. Illanes, L. Wilson, L. Raiman, Design of immobilized enzyme reactors for the continuous production of fructose syrup from whey permeate. Bioprocess Eng. **21**(6), 509–515 (1999)
111. P. Torres, F. Batista-Viera, Immobilized trienzymatic system with enhanced stabilization for the biotransformation of lactose. Molecules **22**, 2 (2017)
112. P.C. Lorenzen, J. Breiter, I. Clawin-Rädecker, A. Dau, A novel bi-enzymatic system for lactose conversion. Int. J. Food Sci. Technol. **48**(7), 1396–1403 (2013)
113. J.V. Hupkes, R. van Tilburg, Production and properties of an immobilized glucose isomerase. Starch **28**(10), 356–360 (1976)
114. K. Beerens, T. Desmet, W. Soetaert, Enzymes for the biocatalytic production of rare sugars. J. Ind. Microbiol. Biotechnol. **39**(6), 823–834 (2012)
115. T. Iida, K. Okuma, Properties of three rare sugars D-psicose, D-allose, D-tagatose and their applications. Oleoscience **13**(9), 435–440 (2013)
116. J. Jayamuthunagai, G. Srisowmeya, M. Chakravarthy, P. Gautam, D-Tagatose production by permeabilized and immobilized Lactobacillus plantarum using whey permeate. Bioresour. Technol. **235**, 250–255 (2017)
117. Z. Xu, S. Li, F. Fu, G. Li, X. Feng, H. Xu, P. Ouyang, Production of D-tagatose, a functional sweetener, utilizing alginate immobilized Lactobacillus fermentum CGMCC2921 cells. Appl. Biochem. Biotechnol. **166**(4), 961–973 (2012)
118. C. Bertoldo, G. Antranikian, Starch-hydrolyzing enzymes from thermophilic archaea and bacteria. Curr. Opin. Chem. Biol. **6**(2), 151–160 (2002)
119. A.V. Presecki, Z.F. Blazevic, D. Vasic-Racki, Complete starch hydrolysis by the synergistic action of amylase and glucoamylase: impact of calcium ions. Bioprocess Biosyst. Eng. **36**(11), 1555–1562 (2013)
120. K. Gupta, A.K. Jana, S. Kumar, M. Maiti, Immobilization of α-amylase and amyloglucosidase onto ion-exchange resin beads and hydrolysis of natural starch at high concentration. Bioprocess Biosyst. Eng. **36**(11), 1715–1724 (2013)
121. D. Park, S. Haam, K. Jang, I.S. Ahn, W.S. Kim, Immobilization of starch-converting enzymes on surface-modified carriers using single and co-immobilized systems: properties and application to starch hydrolysis. Process Biochem. **40**(1), 53–61 (2005)
122. I. Roy, M.N. Gupta, Hydrolysis of starch by a mixture of glucoamylase and pullulanase entrapped individually in calcium alginate beads. Enzym. Microb. Technol. **34**(1), 26–32 (2004)
123. M. Soleimani, A. Khani, K. Najafzadeh, α-Amylase immobilization on the silica nanoparticles for cleaning performance towards starch soils in laundry detergents. J. Mol. Catal. B Enzym. **74**(1–2), 1–5 (2012)
124. M. Salgaonkar, S.S. Nadar, V.K. Rathod, Combi-metal organic framework (Combi-MOF) of α-amylase and glucoamylase for one pot starch hydrolysis. Int. J. Biol. Macromol. **113**, 464–475 (2018)
125. N.A. Edama, A. Sulaiman, K.H.K. Hamid, S.N.A. Rahim, A.S. Baharuddin, M.N. Mokhtar, Encapsulation of multi-enzymes on waste clay material: preparation, characterization and application for tapioca starch hydrolysis. Appl. Mech. Mater. **548–549**, 77–82 (2014)
126. S. Talekar, A. Joshi, S. Kambale, S. Jadhav, S. Nadar, M. Ladole, A tri-enzyme magnetic nanobiocatalyst with one pot starch hydrolytic activity. Chem. Eng. J. **325**, 80–90 (2017)
127. M. Roberfroid, Prebiotics: the concept revisited. J. Nutr. **137**(3), 830S–837S (2007)
128. G. Tzortzis, J. Vulevic, Galacto-oligosaccharide prebiotics, in *Prebiotics and Probiotics Science and Technology*, ed. by D. Charalampopoulos, R. A. Rastall, (Springer, New York, 2009), pp. 207–244

129. A. Gosling, G.W. Stevens, A.R. Barber, S.E. Kentish, S.L. Gras, Recent advances refining galactooligosaccharide production from lactose. Food Chem. **121**(2), 307–318 (2010)
130. H. Yin, J.B. Bultema, L. Dijkhuizen, S.S. van Leeuwen, Reaction kinetics and galactooligosaccharide product profiles of the β-galactosidases from Bacillus circulans, Kluyveromyces lactis and Aspergillus oryzae. Food Chem. **225**, 230–238 (2017)
131. A.R. Park, D.K. Oh, Galacto-oligosaccharide production using microbial β-galactosidase: current state and perspectives. Appl. Microbiol. Biotechnol. **85**(5), 1279–1286 (2010)
132. D.P. Torres, M. Gonçalves, J.A. Teixeira, L.R. Rodrigues, Galacto-oligosaccharides: production, properties, applications, and significance as prebiotics. Compr. Rev. Food Sci. Food Saf. **9**(5), 438–454 (2010)
133. M.A. Boon, A.E.M. Janssen, K. Van't Riet, Effect of temperature and enzyme origin on the enzymatic synthesis of oligosaccharides. Enzym. Microb. Technol. **26**(2–4), 271–281 (2000)
134. P. Urrutia, B. Rodriguez-Colinas, L. Fernandez-Arrojo, A.O. Ballesteros, L. Wilson, A. Illanes, F.J. Plou, Detailed analysis of galactooligosaccharides synthesis with β-galactosidase from Aspergillus oryzae. J. Agric. Food Chem. **61**(5), 1081–1087 (2013)
135. R.E. Huber, G. Kurz, K. Wallenfels, A quantitation of the factors which affect the hydrolase and transgalactosylase activities of β-galactosidase (*E. coli*) on lactose. Biochemistry **15**(9), 1994–2001 (1976)
136. C.W. Chen, C.C. Ou-Yang, C.W. Yeh, Synthesis of galactooligosaccharides and transgalactosylation modeling in reverse micelles. Enzym. Microb. Technol. **33**(4), 497–507 (2003)
137. S.X. Chen, D.Z. Wei, Z.H. Hu, Synthesis of galacto-oligosaccharides in AOT/isooctane reverse micelles by β-galactosidase. J. Mol. Catal. B Enzym. **16**(2), 109–114 (2001)
138. R. Gaur, H. Pant, R. Jain, S.K. Khare, Galacto-oligosaccharide synthesis by immobilized Aspergillus oryzae β-galactosidase. Food Chem. **97**(3), 426–430 (2006)
139. C. Guerrero, C. Aburto, S. Suárez, C. Vera, A. Illanes, Effect of the type of immobilization of β-galactosidase on the yield and selectivity of synthesis of transgalactosylated oligosaccharides. Biocatal. Agric. Biotechnol. **16**, 353–363 (2018)
140. L.M. Huerta, C. Vera, C. Guerrero, L. Wilson, A. Illanes, Synthesis of galacto-oligosaccharides at very high lactose concentrations with immobilized β-galactosidases from Aspergillus oryzae. Process Biochem. **46**(1), 245–252 (2011)
141. M.P. Klein, C.R. Hackenhaar, A.S.G. Lorenzoni, R.C. Rodrigues, T.M.H. Costa, J.L. Ninow, P.F. Hertz, Chitosan crosslinked with genipin as support matrix for application in food process: support characterization and β-d-galactosidase immobilization. Carbohydr. Polym. **137**, 184–190 (2016)
142. S. Suárez, C. Guerrero, C. Vera, A. Illanes, Effect of particle size and enzyme load on the simultaneous reactions of lactose hydrolysis and transgalactosylation with glyoxyl-agarose immobilized β-galactosidase from *Aspergillus oryzae*. Process Biochem. **73**, 56–64 (2018)
143. P. Urrutia, C. Bernal, L. Wilson, A. Illanes, Use of chitosan heterofunctionality for enzyme immobilization: β-galactosidase immobilization for galacto-oligosaccharide synthesis. Int. J. Biol. Macromol. **116**, 182–193 (2018)
144. T. Yovcheva, T. Vasileva, A. Viraneva, D. Cholev, I. Bodurov, M. Marudova, V. Bivolarski, I. Iliev, Effect of immobilization conditions on the properties of β-galactosidase immobilized in xanthan/chitosan multilayers. Int. J. Phys. Conf. Ser. (2017). https://doi.org/10.1088/1742-6596/794/1/012032
145. P. Urrutia, C. Bernal, S. Escobar, C. Santa, M. Mesa, L. Wilson, A. Illanes, Influence of chitosan derivatization on its physicochemical characteristics and its use as enzyme support. J. Appl. Polym. Sci. **131**, 8 (2014)
146. P. Urrutia, C. Mateo, J.M. Guisan, L. Wilson, A. Illanes, Immobilization of Bacillus circulans β-galactosidase and its application in the synthesis of galacto-oligosaccharides under repeated-batch operation. Biochem. Eng. J. **77**, 41–48 (2013)
147. C. Giacomini, A. Villarino, L. Franco-Fraguas, F. Batista-Viera, Immobilization of β-galactosidase from Kluyveromyces lactis on silica and agarose: comparison of different methods. J. Mol. Catal. B Enzym. **4**(5–6), 313–327 (1998)

148. M.P. Klein, L.P. Fallavena, J.D.N. Schöffer, M.A.Z. Ayub, R.C. Rodrigues, J.L. Ninow, P.F. Hertz, High stability of immobilized β-D-galactosidase for lactose hydrolysis and galactooligosaccharides synthesis. Carbohydr. Polym. **95**(1), 465–470 (2013)
149. Z. Mozaffar, K. Nakanishi, R. Matsuno, Continuous production of galacto-oligosaccharides from lactose using immobilized β-galactosidase from Bacillus circulans. Appl. Microbiol. Biotechnol. **25**(3), 224–228 (1986)
150. Z. Mozaffar, K. Nakanishi, R. Matsuno, Mechanism for reversible inactivation of immobilized β-galactosidase from Bacillus circulans during continuous production of galacto-oligosaccharides. Appl. Microbiol. Biotechnol. **25**(3), 229–231 (1986)
151. K. Banjanac, M. Carević, M. Ćorović, A. Milivojević, N. Prlainović, A. Marinković, D. Bezbradica, Novel β-galactosidase nanobiocatalyst systems for application in the synthesis of bioactive galactosides. RSC Adv. **6**(99), 97216–97225 (2016)
152. I. González-Delgado, Y. Segura, G. Morales, M.J. López-Muñoz, Production of high galacto-oligosaccharides by Pectinex Ultra SP-L: optimization of reaction conditions and immobilization on glyoxyl-functionalized silica. J. Agric. Food Chem. **65**(8), 1649–1658 (2017)
153. M. Misson, X. Du, B. Jin, H. Zhang, Dendrimer-like nanoparticles based β-galactosidase assembly for enhancing its selectivity toward transgalactosylation. Enzym. Microb. Technol. **84**, 68–77 (2016)
154. C. Guerrero, C. Vera, F. Plou, A. Illanes, Influence of reaction conditions on the selectivity of the synthesis of lactulose with microbial β-galactosidases. J. Mol. Catal. B Enzym. **72**(3–4), 206–212 (2011)
155. P.S. Panesar, S. Kumari, Lactulose: production, purification and potential applications. Biotechnol. Adv. **29**(6), 940–948 (2011)
156. R. Benavente, B.C. Pessela, J.A. Curiel, B. De Las Rivas, R. Muñoz, J.M. Guisán, J.M. Mancheño, A. Cardelle-Cobas, A.I. Ruiz-Matute, N. Corzo, Improving properties of a novel β-galactosidase from Lactobacillus plantarum by covalent immobilization. Molecules **20**(5), 7874–7889 (2015)
157. A. Cardelle-Cobas, A. Olano, G. Irazoqui, C. Giacomini, F. Batista-Viera, N. Corzo, M. Corzo-Martínez, Synthesis of oligosaccharides derived from lactulose (OsLu) using soluble and immobilized Aspergillus oryzae β-galactosidase. Front. Bioeng. Biotechnol. **4**, 21 (2016)
158. C. Guerrero, F. Valdivia, C. Ubilla, N. Ramírez, M. Gómez, C. Aburto, C. Vera, A. Illanes, Continuous enzymatic synthesis of lactulose in packed-bed reactor with immobilized *Aspergillus oryzae* β-galactosidase. Bioresour. Technol. **278**, 296–302 (2019)
159. C. Guerrero, C. Vera, N. Serna, A. Illanes, Immobilization of *Aspergillus oryzae* β-galactosidase in an agarose matrix functionalized by four different methods and application to the synthesis of lactulose. Bioresour. Technol. **232**, 53–63 (2017)
160. C. Guerrero, C. Vera, A. Illanes, Synthesis of lactulose in batch and repeated-batch operation with immobilized β-galactosidase in different agarose functionalized supports. Bioresour. Technol. **230**, 56–66 (2017)
161. V.D. Nguyen, G. Styevkó, L.P. Ta, A.T.M. Tran, E. Bujna, P. Orbán, M.S. Dam, Q.D. Nguyen, Immobilization and some properties of commercial enzyme preparation for production of lactulose-based oligosaccharides. Food Bioprod. Process. **107**, 97–103 (2018)
162. Y.S. Song, Y.J. Suh, C. Park, S.W. Kim, Improvement of lactulose synthesis through optimization of reaction conditions with immobilized β-galactosidase. Korean J. Chem. Eng. **30**(1), 160–165 (2013)
163. O.J. Concha, M.E. Zúñiga Hansen, Enzymatic depolymerization of sugar beet pulp: production and characterization of pectin and pectic-oligosaccharides as a potential source for functional carbohydrates. Chem. Eng. J. **192**, 29–36 (2012)
164. S. Baldassarre, N. Babbar, S. Van Roy, W. Dejonghe, M. Maesen, S. Sforza, K. Elst, Continuous production of pectic oligosaccharides from onion skins with an enzyme membrane reactor. Food Chem. **267**, 101–110 (2018)

165. Y.A. Ramírez-Tapias, A.S. Lapasset Laumann, C.N. Britos, C.W. Rivero, J.A. Trelles, Saccharification of citrus wastes by immobilized polygalacturonase in an improved alginate matrix. 3 Biotech **7**(6), 380 (2017)
166. C.S. Raina, S. Singh, A.S. Bawa, D.C. Saxena, Some characteristics of acetylated, cross-linked and dual modified Indian rice starches. Eur. Food Res. Technol. **223**(4), 561–570 (2006)
167. E. Rudnik, G. Matuschek, N. Milanov, A. Kettrup, Thermal properties of starch succinates. Thermochim. Acta **427**(1–2), 163–166 (2005)
168. S. Chakraborty, B. Sahoo, I. Teraoka, L.M. Miller, R.A. Gross, Enzyme-catalyzed regioselective modification of starch nanoparticles. Macromolecules **38**(1), 61–68 (2005)
169. H. Horchani, M. Chaâbouni, Y. Gargouri, A. Sayari, Solvent-free lipase-catalyzed synthesis of long-chain starch esters using microwave heating: optimization by response surface methodology. Carbohydr. Polym. **79**(2), 466–474 (2010)
170. M. Perwez, M.J. Ahmed, M. Sardar, Preparation and characterization of reusable magnetic combi-CLEA of cellulase and hemicellulase. Enzym. Microb. Technol. **131**, 109389 (2019)
171. K. Periyasamy, L. Santhalembi, G. Mortha, M. Aurousseau, A. Boyer, S. Subramanian, Bioconversion of lignocellulosic biomass to fermentable sugars by immobilized magnetic cellulolytic enzyme cocktails. Langmuir **34**(22), 6546–6555 (2018)
172. J. Wang, K. Li, Y. He, Y. Wang, X. Han, Y. Yan, Enhanced performance of lipase immobilized onto Co2+-chelated magnetic nanoparticles and its application in biodiesel production. Fuel **255**, 115794 (2019)
173. H. Zhang, Y. Zou, Y. Shen, X. Gao, X. Zheng, X. Zhang, Y. Chen, J. Guo, Dominated effect analysis of the channel size of silica support materials on the catalytic performance of immobilized lipase catalysts in the transformation of unrefined waste cooking oil to biodiesel. Bioenergy Res. **7**(4), 1541–1549 (2014)

Correction to: Carbon Nanomaterials for Air and Water Remediation

Sergio Morales-Torres, Luisa M. Pastrana-Martínez, and Francisco José Maldonado-Hódar

Correction to:
Chapter 12 in: M. Piumetti, S. Bensaid (eds.), *Nanostructured Catalysts for Environmental Applications*, https://doi.org/10.1007/978-3-030-58934-9_12

This book was inadvertently published with an incorrect reference [135] in Chapter 12. This has now been updated as follows.

BRILLAS, Enrique; MUR, Eva; CASADO, Juan. Iron (II) Catalysis of the Mineralization of Aniline Using a Carbon-PTFE O 2-Fed Cathode. Journal of the Electrochemical Society, 1996, vol. 143, no 3, p. L49.

The updated online version of this chapter can be found at https://doi.org/10.1007/978-3-030-58934-9_12

M. Piumetti, S. Bensaid (eds.), *Nanostructured Catalysts for Environmental Applications*, https://doi.org/10.1007/978-3-030-58934-9_15

Index

M. Piumetti, S. Bensaid (eds.), *Nanostructured Catalysts for Environmental Applications*, https://doi.org/10.1007/978-3-030-58934-9

C

D

E

Q

R

Z